Lecture Notes in Networks and Systems 655

The series "Lecture Notes in Networks and Systems" publishes the latest developments in Networks and Systems—quickly, informally and with high quality. Original research reported in proceedings and post-proceedings represents the core of LNNS.

Volumes published in LNNS embrace all aspects and subfields of, as well as new challenges in, Networks and Systems.

The series contains proceedings and edited volumes in systems and networks, spanning the areas of Cyber-Physical Systems, Autonomous Systems, Sensor Networks, Control Systems, Energy Systems, Automotive Systems, Biological Systems, Vehicular Networking and Connected Vehicles, Aerospace Systems, Automation, Manufacturing, Smart Grids, Nonlinear Systems, Power Systems, Robotics, Social Systems, Economic Systems and other. Of particular value to both the contributors and the readership are the short publication timeframe and the world-wide distribution and exposure which enable both a wide and rapid dissemination of research output.

The series covers the theory, applications, and perspectives on the state of the art and future developments relevant to systems and networks, decision making, control, complex processes and related areas, as embedded in the fields of interdisciplinary and applied sciences, engineering, computer science, physics, economics, social, and life sciences, as well as the paradigms and methodologies behind them.

Indexed by SCOPUS, INSPEC, WTI Frankfurt eG, zbMATH, SCImago.

All books published in the series are submitted for consideration in Web of Science.

For proposals from Asia please contact Aninda Bose (aninda.bose@springer.com).

Leonard Barolli

Editor

Advanced Information Networking and Applications

Proceedings of the 37th International Conference on Advanced Information Networking and Applications (AINA-2023), Volume 3

 Springer

Editor
Leonard Barolli
Department of Information and Communication
Engineering, Faculty of Information Engineering
Fukuoka Institute of Technology
Fukuoka, Japan

ISSN 2367-3370   ISSN 2367-3389 (electronic)
Lecture Notes in Networks and Systems
ISBN 978-3-031-28693-3   ISBN 978-3-031-28694-0 (eBook)
https://doi.org/10.1007/978-3-031-28694-0

This Springer imprint is published by the registered company Springer Nature Switzerland AG
The registered company address is: Gewerbestrasse 11, 6330 Cham, Switzerland

Welcome Message from AINA-2023 Organizers

Welcome to the 37th International Conference on Advanced Information Networking and Applications (AINA-2023). On behalf of AINA-2023 Organizing Committee, we would like to express to all participants our cordial welcome and high respect.

AINA is an International Forum, where scientists and researchers from academia and industry working in various scientific and technical areas of networking and distributed computing systems can demonstrate new ideas and solutions in distributed computing systems. AINA was born in Asia, but it is now an international conference with high quality thanks to the great help and cooperation of many international-friendly volunteers. AINA is a very open society and is always welcoming international volunteers from any country and any area in the world.

AINA International Conference is a forum for sharing ideas and research work in the emerging areas of information networking and their applications. The area of advanced networking has grown very rapidly, and the applications have experienced an explosive growth especially in the areas of pervasive and mobile applications, wireless sensor and ad hoc networks, vehicular networks, multimedia computing, social networking, semantic collaborative systems, as well as IoT, Big Data and Cloud Computing. This advanced networking revolution is transforming the way people live, work, and interact with each other and is impacting the way business, education, entertainment, and health care are operating. The papers included in the proceedings cover theory, design, and application of computer networks, distributed computing, and information systems.

Each year AINA receives a lot of paper submissions from all around the world. It has maintained high-quality accepted papers and is aspiring to be one of the main international conferences on the information networking in the world.

We are very proud and honored to have two distinguished keynote talks by Dr. Leonardo Mostarda, Camerino University, Italy, and Prof. Flávio de Oliveira Silva, Federal University of Uberlândia, Brazil, who will present their recent work and will give new insights and ideas to the conference participants.

An international conference of this size requires the support and help of many people. A lot of people have helped and worked hard to produce a successful AINA-2023 technical program and conference proceedings. First, we would like to thank all the authors for submitting their papers, the session chairs, and the distinguished keynote speakers. We are indebted to Program Track Co-Chairs, Program Committee Members, and Reviewers, who carried out the most difficult work of carefully evaluating the submitted papers.

We would like to thank AINA-2023 General Co-Chairs, PC Co-Chairs, and Workshops Co-chairs for their great efforts to make AINA-2023 a very successful event. We have special thanks to Finance Chair and Web Administrator Co-Chairs.

We do hope that you will enjoy the conference proceedings and readings.

Organization

AINA-2023 Organizing Committee

Honorary Chair

Makoto Takizawa — Hosei University, Japan

General Co-chairs

Mario A.R. Dantas — Federal University of Juiz de Fora, Brazil
Tomoya Enokido — Rissho University, Japan
Isaac Woungang — Toronto Metropolitan University, Canada

Program Committee Co-chairs

Victor Ströele — Federal University of Juiz de Fora, Brazil
Flora Amato — University of Naples "Federico II", Italy
Marek Ogiela — AGH University of Science and Technology, Poland

International Journals Special Issues Co-chairs

Fatos Xhafa — Technical University of Catalonia, Spain
David Taniar — Monash University, Australia
Farookh Hussain — University of Technology Sydney, Australia

Award Co-chairs

Arjan Durresi — Indiana University Purdue University in Indianapolis (IUPUI), USA
Fang-Yie Leu — Tunghai University, Taiwan
Kin Fun Li — University of Victoria, Canada

Publicity Co-chairs

Markus Aleksy — ABB Corporate Research Center, Germany
Omar Hussain — University of New South Wales, Australia

Lidia Ogiela   AGH University of Science and Technology,
                                    Poland
Hsing-Chung Chen   Asia University, Taiwan

International Liaison Co-chairs

Nadeem Javaid   COMSATS University Islamabad, Pakistan
Wenny Rahayu   La Trobe University, Australia
Beniamino Di Martino   University of Campania "Luigi Vanvitelli", Italy

Local Arrangement Co-chairs

Regina Vilela   Federal University of Juiz de Fora, Brazil
José Maria N. David   Federal University of Juiz de Fora, Brazil

Finance Chair

Makoto Ikeda   Fukuoka Institute of Technology, Japan

Web Co-chairs

Phudit Ampririt   Fukuoka Institute of Technology, Japan
Kevin Bylykbashi   Fukuoka Institute of Technology, Japan
Ermioni Qafzezi   Fukuoka Institute of Technology, Japan

Steering Committee Chair

Leonard Barolli   Fukuoka Institute of Technology, Japan

Tracks and Program Committee Members

1. Network Protocols and Applications

Track Co-chairs
Makoto Ikeda   Fukuoka Institute of Technology, Japan
Sanjay Kumar Dhurandher   Netaji Subhas University of Technology,
                                    New Delhi, India
Bhed Bahadur Bista   Iwate Prefectural University, Japan

TPC Members

Admir Barolli	Aleksander Moisiu University of Durres, Albania
Elis Kulla	Fukuoka Institute of Technology, Japan
Keita Matsuo	Fukuoka Institute of Technology, Japan
Shinji Sakamoto	Kanazawa Institute of Technology, Japan
Akio Koyama	Yamagata University, Japan
Evjola Spaho	Polytechnic University of Tirana, Albania
Jiahong Wang	Iwate Prefectural University, Japan
Shigetomo Kimura	University of Tsukuba, Japan
Chotipat Pornavalai	King Mongkut's Institute of Technology Ladkrabang, Thailand
Danda B. Rawat	Howard University, USA
Amita Malik	Deenbandhu Chhotu Ram University of Science and Technology, India
R. K. Pateriya	Maulana Azad National Institute of Technology, India
Vinesh Kumar	University of Delhi, India
Petros Nicopolitidis	Aristotle University of Thessaloniki, Greece
Satya Jyoti Borah	North Eastern Regional Institute of Science and Technology, India

2. Next-Generation Wireless Networks

Track Co-chairs

Christos J Bouras	University of Patras, Greece
Tales Heimfarth	Universidade Federal de Lavras, Brazil
Leonardo Mostarda	University of Camerino, Italy

TPC Members

Fadi Al-Turjman	Near East University, Cyprus
Alfredo Navarra	University of Perugia, Italy
Purav Shah	Middlesex University London, UK
Enver Ever	Middle East Technical University, Northern Cyprus Campus, Cyprus
Rosario Culmone	University of Camerino, Italy
Antonio Alfredo F. Loureiro	Federal University of Minas Gerais, Brazil
Holger Karl	University of Paderborn, Germany
Daniel Ludovico Guidoni	Federal University of São João Del-Rei, Brazil
João Paulo Carvalho Lustosa da Costa	Hamm-Lippstadt University of Applied Sciences, Germany
Jorge Sá Silva	University of Coimbra, Portugal

Apostolos Gkamas	University Ecclesiastical Academy of Vella, Greece
Zoubir Mammeri	University Paul Sabatier, France
Eirini Eleni Tsiropoulou	University of New Mexico, USA
Raouf Hamzaoui	De Montfort University, UK
Miroslav Voznak	University of Ostrava, Czech Republic
Kevin Bylykbashi	Fukuoka Institute of Technology, Japan

3. Multimedia Systems and Applications

Track Co-chairs

Markus Aleksy	ABB Corporate Research Center, Germany
Francesco Orciuoli	University of Salerno, Italy
Tomoyuki Ishida	Fukuoka Institute of Technology, Japan

TPC Members

Tetsuro Ogi	Keio University, Japan
Yasuo Ebara	Osaka Electro-Communication University, Japan
Hideo Miyachi	Tokyo City University, Japan
Kaoru Sugita	Fukuoka Institute of Technology, Japan
Akio Doi	Iwate Prefectural University, Japan
Hadil Abukwaik	ABB Corporate Research Center, Germany
Monique Duengen	Robert Bosch GmbH, Germany
Thomas Preuss	Brandenburg University of Applied Sciences, Germany
Peter M. Rost	NOKIA Bell Labs, Germany
Lukasz Wisniewski	inIT, Germany
Angelo Gaeta	University of Salerno, Italy
Graziano Fuccio	University of Salerno, Italy
Giuseppe Fenza	University of Salerno, Italy
Maria Cristina	University of Salerno, Italy
Alberto Volpe	University of Salerno, Italy

4. Pervasive and Ubiquitous Computing

Track Co-chairs

Chih-Lin Hu	National Central University, Taiwan
Vamsi Paruchuri	University of Central Arkansas, USA
Winston Seah	Victoria University of Wellington, New Zealand

TPC Members

Hong Va Leong	Hong Kong Polytechnic University, Hong Kong
Ling-Jyh Chen	Academia Sinica, Taiwan
Jiun-Yu Tu	Southern Taiwan University of Science and Technology, Taiwan
Jiun-Long Huang	National Chiao Tung University, Taiwan
Thitinan Tantidham	Mahidol University, Thailand
Tanapat Anusas-amornkul	King Mongkut's University of Technology North Bangkok, Thailand
Xin-Mao Huang	Aletheia University, Taiwan
Hui Lin	Tamkang University, Taiwan
Eugen Dedu	Universite de Franche-Comte, France
Peng Huang	Sichuan Agricultural University, China
Wuyungerile Li	Inner Mongolia University, China
Adrian Pekar	Budapest University of Technology and Economics, Hungary
Jyoti Sahni	Victoria University of Technology, New Zealand
Normalia Samian	Universiti Putra Malaysia, Malaysia
Sriram Chellappan	University of South Florida, USA
Yu Sun	University of Central Arkansas, USA
Qiang Duan	Penn State University, USA
Han-Chieh Wei	Dallas Baptist University, USA

5. Web-Based and E-Learning Systems

Track Co-chairs

Santi Caballe	Open University of Catalonia, Spain
Kin Fun Li	University of Victoria, Canada
Nobuo Funabiki	Okayama University, Japan

TPC Members

Jordi Conesa	Open University of Catalonia, Spain
Joan Casas	Open University of Catalonia, Spain
David Gañán	Open University of Catalonia, Spain
Nicola Capuano	University of Basilicata, Italy
Antonio Sarasa	Complutense University of Madrid, Spain
Chih-Peng Fan	National Chung Hsing University, Taiwan
Nobuya Ishihara	Okayama University, Japan
Sho Yamamoto	Kindai University, Japan
Khin Khin Zaw	Yangon Technical University, Myanmar
Kaoru Fujioka	Fukuoka Women's University, Japan

Kosuke Takano	Kanagawa Institute of Technology, Japan
Shengrui Wang	University of Sherbrooke, Canada
Darshika Perera	University of Colorado at Colorado Spring, USA
Carson Leung	University of Manitoba, Canada

6. Distributed and Parallel Computing

Track Co-chairs

Naohiro Hayashibara	Kyoto Sangyo University, Japan
Minoru Uehara	Toyo University, Japan
Tomoya Enokido	Rissho University, Japan

TPC Members

Eric Pardede	La Trobe University, Australia
Lidia Ogiela	AGH University of Science and Technology, Poland
Evjola Spaho	Polytechnic University of Tirana, Albania
Akio Koyama	Yamagata University, Japan
Omar Hussain	University of New South Wales, Australia
Hideharu Amano	Keio University, Japan
Ryuji Shioya	Toyo University, Japan
Ji Zhang	The University of Southern Queensland, Australia
Lucian Prodan	Universitatea Politehnica Timisoara, Romania
Ragib Hasan	The University of Alabama at Birmingham, USA
Young-Hoon Park	Sookmyung Women's University, South Korea
Dilawaer Duolikun	Cognizant Technology Solutions, Hungary
Shigenari Nakamura	Tokyo Metropolitan Industrial Technology Research Institute, Japan

7. Data Mining, Big Data Analytics, and Social Networks

Track Co-chairs

Omid Ameri Sianaki	Victoria University, Australia
Alex Thomo	University of Victoria, Canada
Flora Amato	University of Naples "Frederico II", Italy

TPC Members

Eric Pardede	La Trobe University, Australia
Alireza Amrollahi	Macquarie University, Australia
Javad Rezazadeh	University of Technology Sydney, Australia

Farshid Hajati	Victoria University, Australia
Mehregan Mahdavi	Sydney International School of Technology and Commerce, Australia
Ji Zhang	University of Southern Queensland, Australia
Salimur Choudhury	Lakehead University, Canada
Xiaofeng Ding	Huazhong University of Science and Technology, China
Ronaldo dos Santos Mello	Universidade Federal de Santa Catarina, Brazil
Irena Holubova	Charles University, Czech Republic
Lucian Prodan	Universitatea Politehnica Timisoara, Romania
Alex Tomy	La Trobe University, Australia
Dhomas Hatta Fudholi	Universitas Islam Indonesia, Indonesia
Saqib Ali	Sultan Qaboos University, Oman
Ahmad Alqarni	Al Baha University, Saudi Arabia
Alessandra Amato	University of Naples "Frederico II", Italy
Luigi Coppolino	Parthenope University, Italy
Giovanni Cozzolino	University of Naples "Frederico II", Italy
Giovanni Mazzeo	Parthenope University, Italy
Francesco Mercaldo	Italian National Research Council, Italy
Francesco Moscato	University of Salerno, Italy
Vincenzo Moscato	University of Naples "Frederico II", Italy
Francesco Piccialli	University of Naples "Frederico II", Italy

8. Internet of Things and Cyber-Physical Systems

Track Co-chairs

Euripides G. M. Petrakis	Technical University of Crete (TUC), Greece
Tomoki Yoshihisa	Osaka University, Japan
Mario Dantas	Federal University of Juiz de Fora (UFJF), Brazil

TPC Members

Akihiro Fujimoto	Wakayama University, Japan
Akimitsu Kanzaki	Shimane University, Japan
Kawakami Tomoya	University of Fukui, Japan
Lei Shu	University of Lincoln, UK
Naoyuki Morimoto	Mie University, Japan
Yusuke Gotoh	Okayama University, Japan
Vasilis Samolada	Technical University of Crete (TUC), Greece
Konstantinos Tsakos	Technical University of Crete (TUC), Greece
Aimilios Tzavaras	Technical University of Crete (TUC), Greece

Spanakis Manolis	Foundation for Research and Technology Hellas (FORTH), Greece
Katerina Doka	National Technical University of Athens (NTUA), Greece
Giorgos Vasiliadis	Foundation for Research and Technology Hellas (FORTH), Greece
Stefan Covaci	Technicak University of Berlin (TUB), Germany
Stelios Sotiriadis	University of London, UK
Stefano Chessa	University of Pisa, Italy
Jean-Francois Méhaut	Université Grenoble Alpes, France
Michael Bauer	University of Western Ontario, Canada

9. Intelligent Computing and Machine Learning

Track Co-chairs

Takahiro Uchiya	Nagoya Institute of Technology, Japan
Omar Hussain	UNSW, Australia
Nadeem Javaid	COMSATS University Islamabad, Pakistan

TPC Members

Morteza Saberi	University of Technology Sydney, Australia
Abderrahmane Leshob	University of Quebec in Montreal, Canada
Adil Hammadi	Curtin University, Australia
Naeem Janjua	Edith Cowan University, Australia
Sazia Parvin	Melbourne Polytechnic, Australia
Kazuto Sasai	Ibaraki University, Japan
Shigeru Fujita	Chiba Institute of Technology, Japan
Yuki Kaeri	Mejiro University, Japan
Zahoor Ali Khan	HCT, UAE
Muhammad Imran	King Saud University, Saudi Arabia
Ashfaq Ahmad	The University of Newcastle, Australia
Syed Hassan Ahmad	JMA Wireless, USA
Safdar Hussain Bouk	Daegu Gyeongbuk Institute of Science and Technology, South Korea
Jolanta Mizera-Pietraszko	Military University of Land Forces, Poland
Shahzad Ashraf	NFC Institute of Engineering and Technology, Pakistan

10. Cloud and Services Computing

Track Co-chairs

Asm Kayes	La Trobe University, Australia
Salvatore Venticinque	University of Campania "Luigi Vamvitelli", Italy
Baojiang Cui	Beijing University of Posts and Telecommunications, China

TPC Members

Shahriar Badsha	University of Nevada, USA
Abdur Rahman Bin Shahid	Concord University, USA
Iqbal H. Sarker	Chittagong University of Engineering and Technology, Bangladesh
Jabed Morshed Chowdhury	La Trobe University, Australia
Alex Ng	La Trobe University, Australia
Indika Kumara	Jheronimus Academy of Data Science, The Netherlands
Tarique Anwar	Macquarie University and CSIRO's Data61, Australia
Giancarlo Fortino	University of Calabria, Italy
Massimiliano Rak	University of Campania "Luigi Vanvitelli", Italy
Jason J. Jung	Chung-Ang University, South Korea
Dimosthenis Kyriazis	University of Piraeus, Greece
Geir Horn	University of Oslo, Norway
Gang Wang	Nankai University, China
Shaozhang Niu	Beijing University of Posts and Telecommunications, China
Jianxin Wang	Beijing Forestry University, China
Jie Cheng	Shandong University, China
Shaoyin Cheng	University of Science and Technology of China, China

11. Security, Privacy, and Trust Computing

Track Co-chairs

Hiroaki Kikuchi	Meiji University, Japan
Xu An Wang	Engineering University of PAP, P.R. China
Lidia Ogiela	AGH University of Science and Technology, Poland

TPC Members

Takamichi Saito	Meiji University, Japan
Kouichi Sakurai	Kyushu University, Japan
Kazumasa Omote	University of Tsukuba, Japan
Shou-Hsuan Stephen Huang	University of Houston, USA
Masakatsu Nishigaki	Shizuoka University, Japan
Mingwu Zhang	Hubei University of Technology, China
Caiquan Xiong	Hubei University of Technology, China
Wei Ren	China University of Geosciences, China
Peng Li	Nanjing University of Posts and Telecommunications, China
Guangquan Xu	Tianjing University, China
Urszula Ogiela	AGH University of Science and Technology, Poland
Hoon Ko	Chosun University, Republic of Korea
Goreti Marreiros	Institute of Engineering of Polytechnic of Porto, Portugal
Chang Choi	Gachon University, Republic of Korea
Libor Měsíček	J. E. Purkyně University, Czech Republic

12. Software-Defined Networking and Network Virtualization

Track Co-chairs

Flavio de Oliveira Silva	Federal University of Uberlândia, Brazil
Ashutosh Bhatia	Birla Institute of Technology and Science, Pilani, India
Alaa Allakany	Kyushu University, Japan

TPC Members

Rui Luís Andrade Aguiar	Universidade de Aveiro (UA), Portugal
Ivan Vidal	Universidad Carlos III de Madrid, Spain
Eduardo Coelho Cerqueira	Federal University of Pará (UFPA), Brazil
Christos Tranoris	University of Patras (UoP), Greece
Juliano Araújo Wickboldt	Federal University of Rio Grande do Sul (UFRGS), Brazil
Yaokai Feng	Kyushu University, Japan
Chengming Li	Chinese Academy of Science (CAS), China
Othman Othman	An-Najah National University (ANNU), Palestine
Nor-masri Bin-sahri	University Technology of MARA, Malaysia
Sanouphab Phomkeona	National University of Laos, Laos
Haribabu K.	BITS Pilani, India

Shekhavat, Virendra	BITS Pilani, India
Makoto Ikeda	Fukuoka Institute of Technology, Japan
Farookh Hussain	University of Technology Sydney, Australia
Keita Matsuo	Fukuoka Institute of Technology, Japan

AINA-2023 Reviewers

Admir Barolli	Evjola Spaho	Kin Fun Li
Ahmed Bahlali	Fabian Kurtz	Kiplimo Yego
Aimilios Tzavaras	Farookh Hussain	Kiyotaka Fujisaki
Akihiro Fujihara	Fatos Xhafa	Konstantinos Tsakos
Akimitsu Kanzaki	Feilong Tang	Kouichi Sakurai
Alaa Allakany	Feroz Zahid	Lei Shu
Alba Amato	Flavio Corradini	Leonard Barolli
Alberto Volpe	Flavio Silva	Leonardo Mostarda
Alex Ng	Flora Amato	Libor Mesicek
Alex Thomo	Francesco Orciuoli	Lidia Ogiela
Alfredo Navarra	Gang Wang	Lucian Prodan
Anne Kayem	Goreti Marreiros	Luciana Oliveira
Antonio Esposito	Hadil Abukwaik	Makoto Ikeda
Arcangelo Castiglione	Hiroaki Kikuchi	Makoto Takizawa
Arjan Durresi	Hiroshi Maeda	Marek Ogiela
Ashutosh Bhatia	Hiroyoshi Miwa	Marenglen Biba
Asm Kayes	Hiroyuki Fujioka	Mario Dantas
Bala Killi	Hsing-Chung Chen	Markus Aleksy
Baojiang Cui	Hyunhee Park	Masakatsu Nishigaki
Beniamino Di Martino	Indika Kumara	Masaki Kohana
Bhed Bista	Isaac Woungang	Masaru Kamada
Bruno Zarpelão	Jabed Chowdhury	Mingwu Zhang
Carson Leung	Jana Nowaková	Minoru Uehara
Chang Choi	Ji Zhang	Miroslav Voznak
Changyu Dong	Jiahong Wang	Mohammad Faiz Iqbal Faiz
Chih-Peng Fan	Jianfei Zhang	Nadeem Javaid
Christos Bouras	Jolanta Mizera-Pietraszko	Naohiro Hayashibara
Christos Tranoris	Jörg Domaschka	Neder Karmous
Chung-Ming Huang	Jorge Sá Silva	Nobuo Funabiki
Darshika Perera	Juliano Wickboldt	Omar Hussain
David Taniar	Julio Costella Vicenzi	Omid Ameri Sianaki
Dilawaer Duolikun	Jun Iio	Paresh Saxena
Donald Elmazi	K. Haribabu	Pavel Kromer
Elis Kulla	Kazunori Uchida	Petros Nicopolitidis
Eric Pardede	Keita Matsuo	Philip Moore Fatos Xhafa
Euripides Petrakis	Kensuke Baba	Purav Shah

Rajdeep Niyogi
Rodrigo Miani
Ronald Petrlic
Ronaldo Mello
Rui Aguiar
Ryuji Shioya
Salimur Choudhury
Salvatore Venticinque
Sanjay Dhurandher
Santi Caballé
Satya Borah
Shahriar Badsha
Shengrui Wang

Shigenari Nakamura
Shigetomo Kimura
Somnath Mazumdar
Sriram Chellappan
Stelios Sotiriadis
Takahiro Uchiya
Takamichi Saito
Takayuki Kushida
Tetsuya Oda
Tetsuya Shigeyasu
Thomas Dreibholz
Tomoki Yoshihisa
Tomoya Enokido

Tomoyuki Ishida
Vamsi Paruchuri
Wang Xu An
Wei Lu
Wenny Rahayu
Winston Seah
Yoshihiro Okada
Yoshitaka Shibata
Yusuke Gotoh
Zahoor Khan
Zia Ullah

AINA-2023 Keynote Talks

Blockchain and IoT Integration: Challenges and Future Directions

Leonardo Mostarda

Camerino University, Camerino, Italy

Abstract. Massive overhead costs, concerns about centralized data control, and single point of vulnerabilities are significantly reduced by moving IoT from a centralized data server architecture to a trustless, distributed peer-to-peer network. Blockchain is one of the most promising and effective technologies for enabling a trusted, secure, and distributed IoT ecosystem. Blockchain technology can allow the implementation of decentralized applications that not only perform payments but also allow the execution of smart contracts. This talk will investigate the state of the art and open challenges that are related to IoT and blockchain integration. We review current approaches and future directions.

Toward Sustainable, Intelligent, Secure, Fully Programmable, and Multisensory (SENSUOUS) Networks

Flávio de Oliveira Silva

Federal University of Uberlândia, Uberlândia, Brazil

Abstract. In this talk, we will discuss and present the evolution of current networks toward sustainable, intelligent, secure, fully programmable, and multisensory (SENSUOUS) networks. The evolution of networks happens through these critical attributes that will drive the next-generation networks. Here networks consider data networks capable of transmitting audio and video in computer or telecommunication systems. While there is an established process for the evolution of telecommunication networks, regarding computer networks, this area is still open and has several challenges and opportunities. So far, networks can transmit audio and video data, which sensitize only part of our senses. Still, new senses must be considered in the evolution of networks, expanding the multisensory experience. SENSUOUS networks will shape and contribute to scaling our society's sustainable, smart, and secure digital transformation.

Contents

Next Generation Mobile Sensors: Review Regarding the Significance
of Deep Learning and Privacy Techniques for Data-Driven Soft Sensors 1
   Razvan Bocu and Dorin Bocu

Simulation Modeling of Human Aortic Valve Blood Flow 12
   Ilya Kudrenok, Maxim Davidov, and Manuel Mazzara

An Integrated System for Vibration Suppression Using Fuzzy Control
and 2D-LiDAR 28
   Masahiro Niihara, Yuma Yamashita, Chihiro Yukawa,
   Kyouhei Toyosima, Yuki Nagai, Tetsuya Oda, and Leonard Barolli

Prediction in Smart Environments and Administration: Systematic
Literature Review 36
   Mohamed Krichene, Nesrine Khabou, and Ismael Bouassida Rodriguez

Protect Trajectory Privacy in Food Delivery with Differential Privacy
and Multi-agent Reinforcement Learning 48
   Suleiman Abahussein, Tianqing Zhu, Dayong Ye, Zishuo Cheng,
   and Wanlei Zhou

Enhanced Machine Learning-Based SDN Controller Framework
for Securing IoT Networks 60
   Neder Karmous, Mohamed Ould-Elhassen Aoueileyine,
   Manel Abdelkader, and Neji Youssef

Sustainment of Military Operations by 5G and Cloud/Edge Technologies 70
   Souradip Saha, Warren Low, and Beniamino Di Martino

Federated Learning of Predictive Models from Real Data on Diabetic
Patients 80
   Gennaro Junior Pezzullo, Antonio Esposito, and Beniamino di Martino

Design of a Process and a Container-Based Cloud Architecture
for the Automatic Generation of Storyline Visualizations 90
   Emilio Di Giacomo, Beniamino Di Martino, Walter Didimo,
   Antonio Esposito, Giuseppe Liotta, and Fabrizio Montecchiani

Cycle Detection and Clustering for Cyber Physical Systems 100
   Gabriel Iuhasz, Silviu Panica, and Alecsandru Duma

Cloud Computing and Critical Infrastructure Resilience 115
  Oronzo Mazzeo, Antonella Longo, and Marco Zappatore

Towards a Parallel Graph Approach to Drug Discovery 127
  Dario Branco, Beniamino Di Martino, Sandro Cosconati,
  Dieter Kranzlmueller, and Salvatore D'Angelo

Experiences in Architectural Design and Deployment of eHealth
and Environmental Applications for Cloud-Edge Continuum 136
  Atakan Aral, Antonio Esposito, Andrey Nagiyev, Siegfried Benkner,
  Beniamino Di Martino, and Mario A. Bochicchio

Programming Paradigms for the Cloud Continuum 146
  Geir Horn, Beniamino Di Martino, Salvatore D'Angelo,
  and Antonio Esposito

Worker-to-Task Skill-Based Assignment 157
  Vlad Rochian, Cosmin Bonchis, and Ionut Tepeneu

Prototype for Controlled Use of Social Media to Reduce Depression 169
  Furqan Haider, Hamna Aslam, Rabab Marouf, and Manuel Mazzara

Why Zero Trust Framework Adoption has Emerged During and After
Covid-19 Pandemic 181
  Abeer Z. Alalmaie, Priyadarsi Nanda, Xiangjian He,
  and Mohrah Saad Alayan

An Interoperable Microservices Architecture for Healthcare Data Exchange ... 193
  Allender V. de Alencar, Marcus M. Bezerra, Dalton C. G. Valadares,
  Danilo F. S. Santos, and Angelo Perkusich

D-insta: A Decentralized Image Sharing Platform 206
  Yadagiri Shiva Sai Sashank, Ankit Agrawal, Ritika Bhatia,
  Ashutosh Bhatia, and Kamlesh Tiwari

Ramification of Sentiments on Robot-Based Smart Agriculture:
An Analysis Using Real-Time Tweets 218
  Tajinder Singh, Amar Nath, and Rajdeep Niyogi

The Digital Humanities Trend in Chinese Film History: A Case Study
of Filmmaker Lvban 228
  Zitong Zhu

A Tool for Creation of Virtual Exhibits Presented as IIIF Collections
by Intelligent Agents 241
  Dario Branco, Rocco Aversa, and Salvatore Venticinque

Recommender Systems in the Museum Sector: An Overview 251
  Alba Amato

Towards the Enrichment of IIIF Framework with Semantically Annotated
and Geo-Located images 261
  Alba Amato and Giuseppe Cirillo

Comparison of ML Solutions for HRIR Individualization Design
in Binaural Audio 271
  Simone Angelucci, Claudia Rinaldi, Fabio Franchi, and Fabio Graziosi

Performance Analysis of a BESU Permissioned Blockchain 279
  Leonardo Mostarda, Andrea Pinna, Davide Sestili, and Roberto Tonelli

AI-Powered Drone to Address Smart City Security Issues 292
  Ramiz Salama, Fadi Al-Turjman, and Rosario Culmone

Range Proofs with Constant Size and Trustless Setup 301
  Emanuele Scala and Leonardo Mostarda

Sensorless Predictive Maintenance: An Example on a 'Not 4.0' Coffee
Machine Production Process 311
  Diletta Cacciagrano, Flavio Corradini, and Marco Piangerelli

Attendance System via Internet of Things, Blockchain and Artificial
Intelligence Technology: Literature Review 321
  Sarumi Usman Abidemi, Auwalu Saleh Mubarak, Olukayode Akanni,
  Zubaida Said Ameen, Diletta Cacciagrano, and Fadi Al-turjman

An Overview and Current Status of Blockchains Performance 331
  Hamza Salem, Manuel Mazzara, and Siham Hattab

AgriBIoT: A Blockchain-Based IoT Architecture for Crop Insurance 340
  Oumayma Jouini and Kaouthar Sethom

Distribution of the Training Data Over the Shortest Path Between
the Servers 351
  Ibrahim Dahaoui, Mohamed Mosbah, and Akka Zemmari

A Decentralized Architecture for Electric Vehicle Charging Platform 357
  Marlon Rodrigues Martin and Fabiano Hessel

Services and Operations of Electric Vehicle System by Virtual Power
Plant in Rural Area 370
  Yoshitaka Shibata, Masahiro Ueda, and Akiko Ueda

A Triangulation Based Water Level Measuring System for a Water
Reservoir Tank 376
  Yuki Nagai, Tetsuya Oda, Kyohei Toyoshima, Chihiro Yukawa,
  Kei Tabuchi, Tomoaki Matsui, and Leonard Barolli

A System Architecture for Heterogeneous Time- Sensitive Networking
Based on SDN 384
  Hongrui Nie

A Parking System Based on Priority Scheme 396
  Walter Balzano, Antonio Lanuto, Erasmo Prosciutto,
  Biagio Scotto di Covella, and Silvia Stranieri

DTAG: A Dynamic Threshold-Based Anti-packet Generation Method
for Vehicular DTN 406
  Shota Uchimura, Masaya Azuma, Makoto Ikeda, and Leonard Barolli

Optimal and Suboptimal Routing Protocols for WSN 415
  Rahil Bensaid, Adel Ben Mnaouer, and Hatem Boujemaa

CL-DECCM-SA: A Cluster-based Delaunay Edge and Simulated
Annealing Approach for Optimization of Mesh Routers Placement
in WMNs 427
  Aoto Hirata, Yuki Nagai, Kyohei Toyoshima, Chihiro Yukawa,
  Tetsuya Oda, and Leonard Barolli

An Analytical Queuing Model Based on SDN for IoT Traffic in 5G 435
  Aliyu Lawal Aliyu and Jim Diockou

An Expert Survey for the Evaluation of 5G Adoption in Bangladesh 446
  Md. Zahirul Islam, Md. Abdur Rahim, Md. Salahuddin,
  Syed Md. Galib, and Rahamatullah Khondoker

Optical Advanced Hybrid Phase Shift Approach for RF Beamforming
and 5G Wideband Radar 458
  Yosra Bouchoucha, Dorsaf Omri, and Taoufik Aguili

Quo Vadis, Web Authentication? – An Empirical Analysis of Login
Methods on the Internet 471
  Andreas Grüner, Alexander Mühle, Nils Rümmler, Adnan Kadric,
  and Christoph Meinel

Device Tracking Threats in 5G Network 480
   Maksim Iavich, Giorgi Akhalaia, and Razvan Bocu

Trusted and only Trusted. That is the Access! Improving Access Control
Allowing only Trusted Execution Environment Applications 490
   Dalton C. G. Valadares, Álvaro Sobrinho, Newton C. Will,
   Kyller C. Gorgônio, and Angelo Perkusich

A Survey of Intrusion Detection-Based Trust Management Approaches
in IoT Networks 504
   Meriem Soula, Bacem Mbarek, Aref Meddeb, and Tomáš Pitner

Context-Aware Security in the Internet of Things: A Review 518
   Everton de Matos, Eduardo Viegas, Ramão Tiburski, and Fabiano Hessel

Cybersecurity Attacks and Vulnerabilities During COVID-19 532
   Sharmin Akter Mim, Roksana Rahman, Md. Rashid Al Asif,
   Khondokar Fida Hasan, and Rahamatullah Khondoker

Identifying Fake News in the Russian-Ukrainian Conflict Using Machine
Learning 546
   Omar Darwish, Yahya Tashtoush, Majdi Maabreh, Rana Al-essa,
   Ruba Aln'uman, Ammar Alqublan, Munther Abualkibash,
   and Mahmoud Elkhodr

Challenges of Managing an IoT-Based Biophilic Services in Green Cities 558
   Farhad Daneshgar, Rahim Foroughi, Nava Tavakoli-Mehr,
   and Atefa Youhangi

Control and Diagnosis of Brain Tumors Using Deep Neural Networks 565
   Alireza Izadi, Farshid Hajati, Roohollah Barzamini, Negar Janpors,
   Babak Farjad, and Sahar Barzamini

Co-evolution Genetic Algorithm Approximation Technique for ROM-Less
Digital Synthesizers 573
   Soheila Gheisari, Alireza Rezaee, and Farshid Hajati

Application of Generalized Deduplication Techniques in Edge Computing
Environments 585
   Ryu Watanabe, Ayumu Kubota, and Jun Kurihara

On the Realization of Cloud-RAN on Mobile Edge Computing 597
   Andres F. Ocampo and Haakon Bryhni

TEATOM: A True Zero Touch Intent Based Multi-cloud Framework   609
   B. Ramesh Ramanathan and P. Preethika

Utility Function Creator for Cloud Application Optimization   619
   Marta Różańska, Kyriakos Kritikos, Jan Marchel, Damian Folga,
   and Geir Horn

A Review of Monitoring Probes for Cloud Computing Continuum   631
   Yiannis Verginadis

Multi Languages Pattern Matching-Based Scraping of News and Articles
Websites   644
   Hamza Salem and Manuel Mazzara

Decoding COVID-19 Vaccine Hesitancy Using Multiple Regression
Analysis with Socioeconomic Values   649
   Wei Lu, Ling Xue, and Bria Shorten

Video Indexing for Live Nature Camera on Digital Earth   660
   Hiroki Mimura, Masaya Tahara, Kosuke Takano, Nobuya Watanabe,
   and Kin Fun Li

Sports Data Mining for Cricket Match Prediction   668
   Antony Anuraj, Gurtej S. Boparai, Carson K. Leung,
   Evan W. R. Madill, Darshan A. Pandhi, Ayush Dilipkumar Patel,
   and Ronak K. Vyas

Author Index   681

Next Generation Mobile Sensors: Review Regarding the Significance of Deep Learning and Privacy Techniques for Data-Driven Soft Sensors

Razvan Bocu[1,2(✉)] and Dorin Bocu[1]

[1] Department of Mathematics and Computer Science,
Transilvania University of Brasov,
500036 Brașov, Romania
razvan.bocu@unitbv.ro, dorin@bocu.ro
[2] Department of Research and Technology, Siemens Industry Software,
500203 Brașov, Romania

Abstract. The increasing usage of mobile devices amounts to around 6.8 billion by 2022. This implies a substantial increase in the quantity of personal data that are managed. The paper surveys the most relevant contributions that pertain to human activity, behavioural patterns detection, demographics, health and body parameters. Moreover, significant aspects regarding data privacy are also analyzed. The paper also defines relevant research questions and challenges.

1 Introduction

Mobile devices, such as numerous wearables or smartphones, provide certain sensors, which gather a significant quantity of personal data [1, 2]. Moreover, the enhancement of the respective processing capacities mediates their application to numerous real-world situations [3, 4]. Thus, it can be noticed that the readily available mobile devices are prone to illegitimate access attempts, which are related to personal information and assets [5].

The General Data Protection Regulation (GDPR), which was turned into legislation by the European Union (EU), regards personal information as any unique item, which is related to a certain natural person [6]. Moreover, the GDPR refers to private data as a subset of personal items, which also pertains to the following categories: personal data concerning race or ethnicity, political opinions, and religious or philosophical beliefs; trade-union membership; genetic data and biometric data, which is processed solely for the identification of a human being; personal health information, and also data related to aspects of sex life [6]. The automatic management of these relevant data items, which is also refered to as user profiling [6], can define relevant data fields, which are populated following the end users' interaction with their mobile devices. Additionally, it is important to note that this is also generated by the inadequate consolidation of

L. Barolli (Ed.): AINA 2023, LNNS 655, pp. 1–11, 2023.
https://doi.org/10.1007/978-3-031-28694-0_1

collected personal data items [7,8]. The improper accesses of personal data items are considered by certain scientific and technical projects, such as PriMa [9], and TReSPAsS [10].

The rest of this paper considers the following structure. First, relevant details concerning contributions that aim to propose full privacy preserving models are discussed. Following, the most relevant real-world use case scenarios are presented, and the problematic of personal user data processing is also approached. Furthermore, the methods that are useful in order to collect and process the data are analyzed. Moreover, the general data privacy methods are discussed. The last section concludes the paper and discusses about certain open research problems that have been determined during the comprehensive research effort, which was allocated to this survey.

2   Proper Management of Sensitive Private Data

The automatic processing of personal data collected through mobile sensors and devices implies the existence of numerous personal data items. Some mobile sensors, such as GPS, microphones, and cameras, are particularly difficult to access, without necessary permissions, in order to access the related personal data. Unfortunately, other mobile sensors or assets, such as the touchscreen, accelerometer, and the logged network data, may be accessed in an easier fashion. These may allow for unwanted backdoors to be created, which may consequently be used for personal data reidentification.

The intrinsic nature of personal information implies that proper data privacy management models are designed and implemented. This is particularly important considering that these data items uniquely identify private persons. Thus, biometric data are particularly significant. Furthermore, the type and time reliability of this data constitute an important research aspect [11]. The next subsections analyze relevant kinds of personal data items, which can be generated by the sensors of mobile devices.

2.1   Demographic Data

Generally, the scope of demographics determines the most populated category of personal data entities, through attributes like ethnicity, age, or gender.

2.2   Sensors that Detect Movement

Article [12] relates to the detection of the users' age categories, which uses the accelerometer data. The experimental setup considers a preset series of taps on the touchscreen, in connection to certain contact spots. The used algorithm is k-Nearest Neighbors (k-NN), and the accuracy is 85.3%. Furthermore, paper [13] propose an algorithmic model, which distinguishes an adult from a child, based on certain behavioural aspects, which are analyzed using mobile motion sensors. The assumption is that children's hands are smaller, and move in a shakier pattern. The generated accuracy is 96% using Random Forest (RF). Moreover, article [14] regards the gender of the end users. Thus, their walking routines data is used, which are acquired by mobile motion sensors. The accuracy is 76.8%, which is generated using Support Vector Machines (SVMs), and also bagging algorithms. Article [15] presents a model that relates to the determination of gender data through the study of gait (walking) data, as they are provided by the respective mobile sensors. The accuracy is 96.3% using a bagged tree classifier.

Paper [16] describes a gender detection model that considers gyroscope and accelerometer data. The obtained accuracy is 80% through the usage of a Principal Component Analysis (PCA) model. Furthermore, article [17] describes the detection of gender and age using Hidden Markov Models (HMMs). More precisely, a competition is configured, which uses the data that are gathered by accelerometer and gyroscope. The determined error is 24.23% concerning the gender, and also 5.39% regarding the age. Deep learning also contributes to the analyzed field through enhanced results, which are reported. Paper [18] discusses about an accuracy of 94.11% relative to the processing of gait (walking) data that is used in the context of gender classification. The paper considers Long Short-Term Memory (LSTM) Recurrent Neural Networks (RNNs). These are particularly suitable in order to determine the temporal dependencies, which are detected in the analyzed dataset.

2.2.1 Touchscreen Data

Article [19] classifies the end users in two categories, adults and children, relative to the mechanics of their tap and swipe movements. The algorithm is Active User Detection (AUD), and the accuracy is 97%. Additionally, article [2] refers to a database that stores mobile interaction data generated by children. The respective touch interaction data assigns the children to three categories, which pertain to an aggregated age range between 18 months and 8 years. The approach is based on a Support Vector Machine (SVM) model, and the accuracy is 90.45%. Moreover, article [13] reports a Random Forest (RF) approach, which considers tap gestures in order to classify between adults and children. The obtained accuracy is 99%. The determination of the gender using touchscreen data represents the object of several surveyed papers. Thus, article [20] discusses about the generation of soft biometrics data related to swipe gestures. The obtained accuracy is 78%, and the algorithmic core is based on a decision voting scheme with four classifiers: Decision Tree (DT), Naive Bayes (NB), Support Vector Machine (SVM), and Logistic Regression (LR). Moreover, article [21] relates to the collection of behavioral data through the usage of accelerometers, gyroscopes, and orientation sensors. The reported approach considers a k-NN classifier, and the obtained accuracy is 93.65%.

2.2.2   Sensors Data Related to Mobile Applications, Location and Network

Existing papers demonstrate the presence of a logical link between geolocation data, and the end users' demographics and usage patterns. Thus, article [22] discusses about the significance of data produced by mobile devices relative to demographic modeling and measurement of related data. This may circumvent the need to organize classical sociological researches, and traditional censuses, which may enhance the efficiency and timeliness of respective political decisions. Moreover, article [23] relates to radius, eccentricity, and entropy, in connection to travel behavior. More precisely, the authors study the correlation between the usage of mobile devices, and the personal travel behavioural patterns. Thus, among other variables, the authors assess the link that exists between the frequency of the phone calls, and particular demographic parameters, such as age, gender, and the determining features of the environment.

Article [24] relates to an unsupervised, data-driven model, which generates user categories based on high-resolution mobility data. The relevant data are gathered using mobile navigation applications. Moreover, paper [25] proposes an approach concerning the determination of demographic data based on photos, which are uploaded on social networks. They also contain geographical tagging data. Thus, this proves the determination of ethnical features using the mentioned data obtained from two metropolitan zones. The approach discriminates between three ethnic groups, and the accuracy is 72% relative to a Logistic Regression (LR) model.

Article [26] analyzes the appropriateness of geolocation data for the inference of data concerning marital status and residence. The reported research is based on the determination of spatial and temporal features relative to human mobility patterns, and to the geographical context. This approach provides relevant data regarding the places that are visited by a certain person, such as private homes, leisure facilities, and hospitals. The generated accuracy is 80% using an eXtreme Gradient Boosting (XGBoost) algorithmic model [27]. Moreover, article [28] initially describes an analysis of behavioral data related to gender, which are generated by mobile applications. The paper reports on the possible prediction of the end users' gender, and the obtained accuracy is 91.8%. The algorithmic model considers Random Forest (RF), and multinomial Naive Bayes (NB). The data are extracted from the networking logs. The events are sorted based on the occurrence of the calculated frequencies. Furthermore, an assessment of 1,000 selected events is conducted relative to the respective time patterns. This may prove useful in various real-world use cases, such as the customization of the advertisements, and the home screens personalization.

3   Study of Human Behaviour

The reviewed literature suggests that the patterns of the users' daily activities, and behavioural traits can be determined using the data that are acquired by mobile sensors [29]. This immediately suggests issues concerning the privacy of the respective personal data.

3.1   Motion Sensors

Article [30] proposes a software architecture, which is proper for the assessment of the end users' spatial mobility status. More precisely, it assesses whether the person is stationary, walking, running, riding a bicycle, climbing stairs, going downstairs, or driving, based on the data collected using the accelerometer. Their algorithmic approach considers a Support Vector Machine (SVM) model, and the obtained accuracy attains a maximum of 93.2%. Moreover, article [31] considers data that are generated by the gyroscope and accelerometer. Furthermore, an application is developed, which tracks data regarding the users' daily routines. They used a Decision Tree (DT) classifier. The average Area Under the Receiver Operating Characteristic (AUROC) curve is in excess of 99.0%.

Paper [32] analyzes the users' mobility while eating. The data are collected using the accelerometer. Moreover, article [33] defines a taxonomy of human drinking behavior, which also uses accelerometer data that are generated by young people during their nightlife activities. The determined accuracy is 76.1%, and the algorithmic model considers a Density-based Spatial Clustering of Applications (DBSCAN) model. The approach is also proper for the assessment of the consumed alcohol.

The evaluation of the end users' mood and physical state, which may be sober, tipsy, or drunk, is performed according to paper [34], which uses accelerometer data, and also the users' self-reported behavioural data. Essentially, this defines an auxiliary feature, which cannot be considered an objective data source. The approach uses a Random Forest (RF) model, and the accuracy is 70%. Moreover, motion sensors are used in order to collect data, which pertain to sleep, in oder to evaluate sleep habits and postures. Moreover, article [35] is based on accelerometer, gyroscope, and orientation data, which creates a taxonomy of sleep postures with the following categories: supine, left lateral, right lateral, and prone. The obtained accuracy is in excess of 95%, and the model uses Euclidean distances. The approach is also proper to evaluate the position of the users' hand, relative to the following three states: placed on the abdomen, chest, or head. The approach is based on a k-NN algorithm, and the accuracy is in excess of 88%.

3.1.1   Mobile Applications, Location and Network

Article [36] uses GPS data to assess if the user stands, walks, or uses other means of transportation. The used fuzzy classifier computes the speed and angle of the person relative to the ground, and the accuracy is 96%. Radio receivers and transmitters may also be used to intercept personal behavioural data. Thus, article [37] assesses the Received Signal Strength Indicator (RSSI), as a parameter to detect the users' activity types from the following possible values: lying down, falling, walking, running, sitting down, and standing up. The used Convolutional Neural Network (CNN) produces an accuracy of 97.7%. Article [38] pertains to three neural networks concerning the Channel State Information (CSI) of the Wi-Fi module, and the accuracy rate is 83%.

4   Body Features and Health Parameters

4.1   Motion Sensors

The Body Mass Index (BMI) is a mathematical ratio that links the body mass to the height of a person. Thus, article [39] describes a hybrid model that uses a Convolutional Neural Network and Long Short-Term Memory (CNN-LSTM) architecture. It estimate the BMI using accelerometer and gyroscope data, and the accuracy attains a maximum of 94.8%. Articles [40,41] determine other relevant health parameters, which are logically connected to the BMI. Article [42] refers to an accuracy of 71% using the mentioned technical approach, and an algorithmic model based on the Naive Bayes model.

4.2   Remarks Concerning the Touchscreen

Mobile sensors data may be used in order to evaluate certain medical conditions. It is feasible to assess Parkinson's disease based on the users' keystroke writing pattern, independent from the semantics of the text. Article [43] uses an SVM algorithm, and the value of the AUROC is 0.88. Moreover, article [44] evaluates various classes of features for particular handwriting patterns. Article [45], proves and postulates that people with longer thumbs need less time for their swipe movements.

4.3   Mobile Applications, Location and Network

Article [46] presents an approach for the detection of times of psychological depression based on geolocation data. The target group of individuals suffer from bipolar disorder (BD). The linear regression algorithm, and the quadratic discriminant analysis algorithm determines an accuracy of 85%. Also, GPS data may help detect sleep disorders, like sleep-wake stages, and sleep-disordered breathing disorders (SRBD) from the class of Obstructive Sleep Apnea (OSA). The model uses SVM algorithms, and the accuracy is, at most, 92.3% [47,48].

4.3.1   Detection of Psychological Mood and Emotions

The psychological mood directly influences the human behavioural performance.

Article [50] discusses about the influence of psychological mood regarding the mobile biometric systems' detection accuracy. The authors describe an RF classifier, and they suggest that the users that determine an accuracy rate of less than 70% manifest the least psychological mood variations. Moreover, article [51] evaluates the state of the psychological mood through an RF model, and the obtained value of the AUROC is 81%. Additionally, interesting related contributions are reported in articles [52] and [53], which generate accuracy values of 90.31% and 86%, respectively.

The finger strokes patterns represent the object of scientific research projects, which are connected to gameplays. Article [54] defines four psychological states:

excited, relaxed, frustrated, and bored. They specify an SVM-based algorithm, and the obtained accuracy is 69%. Consequently, article [55] quantitatively evaluates the available finger strokes patterns, which indicate the psychological state of the end user. The taxonomy is determined by three values: positive, negative, or neutral. The obtained accuracy is 90.47%, and the algorithm is based on a linear regression model.

5 Conclusions and Open Research Avenues

This review suggests that even apparently legitimate collection of personal data may offer the opportunity to conduct the illegitimate access of personal data, which must be protected relative to specific regulations. The conducted state-of- the-art review effort relates to some classic, and mostly up-to-date scientific contributions, which also present highly relevant machine learning and artificial intelligence models. Some open research issues include the analysis of the statistical and logical link between sensitive attributes, efficient data change algorithms for the protection of data privacy, and also the research on the unified quantitative assessment metrics, which pertain to data anonymization methods, and the study of relevant ethical implications.

References

1. Rajkumar, N., Kannan, E.: Attribute-based collusion resistance in group-based cloud data sharing using LKH model. J. Circ. Syst. Comput. **29**(02), 2030001 (2020)
2. Tolosana, R., et al.: Child-computer interaction: recent works, new dataset, and age detection. arXiv (2021)
3. Abuhamad, M., Abusnaina, A., Nyang, D., Mohaisen, D.: Sensor-based continuous authentication of smartphones' users using behavioral biometrics: a contemporary survey. arXiv. (2020)
4. Hussain, A., et al.: Security framework for IOT based real-time health applications. Electronics **10**(6), 719 (2021)
5. Ellavarason, E., Guest, R., Deravi, F., Sanchez-Riello, R., Corsetti, B.: Touch-dynamics based behavioural biometrics on mobile devices-a review from a usability and performance perspective. ACM Comput. Surv. (CSUR) **53**(6), 1–36 (2020)
6. General Data Protection Regulation (2022). https://gdprinfo.eu/ro
7. Aljeraisy, A., Barati, M., Rana, O., Perera, C.: Privacy laws and privacy by design schemes for the Internet of Things: a developer's perspective. ACM Comput. Surv. **54**(5), 1–38 (2021). Article 102
8. Barth, S., de Jong, M.D.T., Junger, M., Hartel, P.H., Roppelt, J.C.: Putting the privacy paradox to the test: online privacy and security behaviors among users with technical knowledge, privacy awareness, and financial resources. Telematics Inform. **41**, 55–69 (2019)
9. European Commission: PriMa: Privacy Matters, H2020-MSCA-ITN-2019-860315 (2022). https://www.prima-itn.eu/
10. European Commission: TReSPAsS-ETN: TRaining in Secure and PrivAcy-preserving biometricS, H2020-MSCAITN-2019-860813 (2022). https://www.trespass-etn.eu/

11. Labati, R.D., Piuri, V., Scotti, F.: Biometric privacy protection: guidelines and technologies. In: Obaidat, M.S., Sevillano, J.L., Filipe, J. (eds.) ICETE 2011. CCIS, vol. 314, pp. 3–19. Springer, Heidelberg (2012). https://doi.org/10.1007/978-3-642-35755-8_1

12. Davarci, E., Soysal, B., Erguler, I., Aydin, S.O., Dincer, O., Anarim, E.: Age group detection using smartphone motion sensors. In: Proceedings of the European Signal Processing Conference (2017)

13. Nguyen, T., Roy, A., Memon, N.: Kid on the phone! Toward automatic detection of children on mobile devices. Comput. Secur. **84**, 334–348 (2019)

14. Jain, A., Kanhangad, V.: Investigating gender recognition in smartphones using accelerometer and gyroscope sensor readings. In: Proceedings of the International Conference on Computational Techniques in Information and Communication Technologies (2016)

15. Meena, T., Sarawadekar, K.: Gender recognition using in-built inertial sensors of smartphone. In: Proceedings of the IEEE Region 10 Conference, pp. 462–467 (2020)

16. Singh, S., Shila, D.M., Kaiser, G.: Side channel attack on smartphone sensors to infer gender of the user: poster abstract. In: Proceedings of the Conference on Embedded Networked Sensor Systems, pp. 436–437 (2019)

17. Ngo, T.T., et al.: OU-ISIR wearable sensor-based gait challenge: age and gender. In: Proceedings of the International Conference on Biometrics (2019)

18. Sabir, A., Maghdid, H., Asaad, S., Ahmed, M., Asaad, A.: Gait-based gender classification using smartphone accelerometer sensor. In: Proceedings of the International Conference on Frontiers of Signal Processing, pp. 12–20 (2019)

19. Acien, A., Morales, A., Fierrez, J., Vera-Rodriguez, R., Hernandez-Ortega, J.: Active detection of age groups based on touch interaction. IET Biom. **8**(1), 101–108 (2019)

20. Miguel-Hurtado, O., Stevenage, S., Bevan, C., Guest, R.: Predicting sex as a soft-biometrics from device interaction swipe gestures. Pattern Recogn. Lett. **79**, 44–51 (2016)

21. Jain, A., Kanhangad, V.: Gender recognition in smartphones using touchscreen gestures. Pattern Recogn. Lett. **125**, 604–611 (2019)

22. Almaatouq, A., Prieto-Castrillo, F., Pentland, A.: Mobile communication signatures of unemployment. In: Spiro, E., Ahn, Y.-Y. (eds.) SocInfo 2016. LNCS, vol. 10046, pp. 407–418. Springer, Cham (2016). https://doi.org/10.1007/978-3-319-47880-7_25

23. Yuan, Y., Raubal, M., Liu, Y.: Correlating mobile phone usage and travel behavior-a case study of Harbin, China. Comput. Environ. Urban Syst. **36**(2), 118–130 (2012)

24. Scherrer, L., Tomko, M., Ranacher, P., Weibel, R.: Travelers or locals? Identifying meaningful sub-populations from human movement data in the absence of ground truth. EPJ Data Sci. **7**(1), 1–21 (2018). https://doi.org/10.1140/epjds/s13688-018-0147-7

25. Riederer, C., Zimmeck, S., Phanord, C., Chaintreau, A., Bellovin, S.: I don't have a photograph, but you can have my footprints. Revealing the demographics of location data. In: Proceedings of the ACM on Conference on Online Social Networks, pp. 185–195 (2015)

26. Wu, L., et al.: Inferring demographics from human trajectories and geographical context. Comput. Environ. Urban Syst. **77**(2019), 101368 (2019)

27. The eXtreme Gradient Boosting library (2022). https://xgboost.ai/about

28. Neal, T., Woodard, D.: A gender-specific behavioral analysis of mobile device usage data. In: Proceedings of the International Conference on Identity, Security, and Behavior Analysis, pp. 1–8 (2018)

29. Chen, K., Zhang, D., Yao, L., Guo, B., Yu, Z., Liu, Y.: Deep learning for sensor-based human activity recognition: overview, challenges, and opportunities. ACM Comput. Surv. **54**(4), 1–40 (2021). https://doi.org/10.1145/3447744. Article 77

30. Sun, L., Zhang, D., Li, B., Guo, B., Li, S.: Activity recognition on an accelerometer embedded mobile phone with varying positions and orientations. In: Yu, Z., Liscano, R., Chen, G., Zhang, D., Zhou, X. (eds.) UIC 2010. LNCS, vol. 6406, pp. 548–562. Springer, Heidelberg (2010). https://doi.org/10.1007/978-3-642-16355-5_42

31. Anjum, A., Ilyas, M.: Activity recognition using smartphone sensors. In: Proceedings of the IEEE Consumer Communications and Networking Conference, pp. 914–919 (2013)

32. Thomaz, E., Essa, I., Abowd, G.D.: A practical approach for recognizing eating moments with wrist-mounted inertial sensing. In: Proceedings of the ACM International Joint Conference on Pervasive and Ubiquitous Computing, pp. 1029–1040 (2015)

33. Santani, D., Do, T., Labhart, F., Landolt, S., Kuntsche, E., Gatica-Perez, D.: DrinkSense: characterizing youth drinking behavior using smartphones. IEEE Trans. Mob. Comput. **17**(10), 2279–2292 (2018)

34. Arnold, Z., Larose, D., Agu, E.: Smartphone inference of alcohol consumption levels from gait. In: 2015 International Conference on Healthcare Informatics, pp. 417–426 (2015)

35. Chang, L., et al.: SleepGuard: capturing rich sleep information using smartwatch sensing data. Proc. ACM Interact. Mob. Wearable Ubiquit. Technol. **2**(3), 1–34 (2018)

36. Wan, N., Lin, G.: Classifying human activity patterns from smartphone collected GPS data: a fuzzy classification and aggregation approach. Trans. GIS **20**(6), 869–886 (2016)

37. Chen, Z., Zhang, L., Jiang, C., Cao, Z., Cui, W.: WiFi CSI based passive human activity recognition using attention based BLSTM. IEEE Trans. Mob. Comput. **18**(11), 2714–2724 (2018)

38. Ma, Y., et al.: Location-and person-independent activity recognition with WiFi, deep neural networks, and reinforcement learning. ACM Trans. Internet Things **2**(1), 1–25 (2021)

39. Yao, Y., Song, L., Ye, J.: Motion-To-BMI: using motion sensors to predict the body mass index of smartphone users. Sensors **20**(4), 1134 (2020)

40. Albanese, E., et al.: Body mass index in midlife and dementia: systematic review and meta-regression analysis of 589,649 men and women followed in longitudinal studies. Alzheimer's Dement. Diagn. Assess. Dis. Monit. **8**, 165–178 (2017)

41. Dobner, J., Kaser, S.: Body mass index and the risk of infection-from underweight to obesity. Clin. Microbiol. Infect. **24**(1), 24–28 (2018)

42. Garcia-Ceja, E., Riegler, M., Nordgreen, T., Jakobsen, P., Oedegaard, K.J., Torresen, J.: Mental health monitoring with multimodal sensing and machine learning: a survey. Pervasive Mob. Comput. **51**, 1–26 (2018)

43. Arroyo-Gallego, T., et al.: Detection of motor impairment in Parkinson's disease via mobile touchscreen typing. IEEE Trans. Biomed. Eng. **64**(9), 1994–2002 (2017)

44. Castrillon, R., et al.: Characterization of the handwriting skills as a biomarker for Parkinson disease. In: IEEE International Conference on Automatic Face and Gesture Recognition (FG 2019) - Human Health Monitoring Based on Computer Vision (2019)
45. Bevan, C., Fraser, D.: Different strokes for different folks? Revealing the physical characteristics of smartphone users from their swipe gestures. Int. J. Hum Comput Stud. **88**, 51–61 (2016)
46. Palmius, N., et al.: Detecting bipolar depression from geographic location data. IEEE Trans. Biomed. Eng. **64**(8), 1761–1771 (2016)
47. Tal, A., Shinar, Z., Shaki, D., Codish, S., Goldbart, A.: Validation of contact-free sleep monitoring device with comparison to polysomnography. J. Clin. Sleep Med. **13**(3), 517–522 (2017)
48. Behar, J., et al.: SleepAp: an automated obstructive sleep apnoea screening application for smartphones. IEEE J. Biomed. Health Inform. **19**(1), 325–331 (2014)
49. Kostopoulos, P., Nunes, T., Salvi, K., Togneri, M., Deriaz, M.: StayActive: an application for detecting stress. In: Proceedings of the International Conference on Communications, Computation, Networks and Technologies (2015)
50. Neal, T., Canavan, S.: Mood versus identity: Studying the influence of affective states on mobile biometrics. In: Proceedings of the IEEE International Conference on Automatic Face and Gesture (2020)
51. Quiroz, J.C., Geangu, E., Yong, M.H.: Emotion recognition using smart watch sensor data: mixed-design study. JMIR Ment. Health **5**(3), e10153 (2018)
52. Cao, B., et al.: DeepMood: modeling mobile phone typing dynamics for mood detection. In: Proceedings of the ACM SIGKDD International Conference on Knowledge Discovery and Data Mining (2017)
53. Hung, G., Yang, P., Chang, C., Chiang, J., Chen, Y.: Predicting negative emotions based on mobile phone usage patterns: an exploratory study. JMIR Res. Protoc. **5**(3), e160 (2016)
54. Gao, Y., Bianchi-Berthouze, N., Meng, H.: What does touch tell us about emotions in touchscreen-based gameplay? ACM Trans. Comput.-Hum. Interact. **19**(4), 1–30 (2012)
55. Shah, S., Teja, J., Bhattacharya, S.: Towards affective touch interaction: predicting mobile user emotion from finger strokes. J. Interact. Sci. **3**(1), 1–15 (2015). https://doi.org/10.1186/s40166-015-0013-z
56. Zhang, X., Li, W., Chen, X., Lu, S.: MoodExplorer: towards compound emotion detection via smartphone sensing. Proc. ACM Interact. Mob. Wearable Ubiquit. Technol. **1**(4), 1–30 (2018)
57. Nguyen, K.A., Akram, R.N., Markantonakis, K., Luo, Z., Watkins, C.: Location tracking using smartphone accelerometer and magnetometer traces. In: Proceedings of the International Conference on Availability, Reliability and Security (2019)
58. Hua, J., Shen, Z., Zhong, S.: We can track you if you take the metro: tracking metro riders using accelerometers on smartphones. IEEE Trans. Inf. Forensics Secur. **12**(2), 286–297 (2017)
59. Han, J., Owusu, E., Nguyen, L.T., Perrig, A., Zhang, J.: ACComplice: location inference using accelerometers on smartphones. In: Proceedings of the 4th International Conference on Communication Systems and Networks (2012)
60. Singh, V., Aggarwal, G., Ujwal, B.V.S.: Ensemble based real-time indoor localization using stray WiFi signal. In: Proceedings of the IEEE International Conference on Consumer Electronics (ICCE 2018), pp. 1–5 (2018)
61. Cai, L., Chen, H.: TouchLogger: inferring keystrokes on touch screen from smartphone motion. HotSec **11**, 9 (2011)

62. Owusu, E., Han, J., Das, S., Perrig, A., Zhang, J.: ACCessory: password inference using accelerometers on smartphones. In: Proceedings of the Workshop on Mobile Computing Systems and Applications (2012)
63. Aviv, A.J., Sapp, B., Blaze, M., Smith, J.M.: Practicality of accelerometer side channels on smartphones. In: Proceedings of the Annual Computer Security Applications Conference (2012)

Simulation Modeling of Human Aortic Valve Blood Flow

Ilya Kudrenok[1]([✉]), Maxim Davidov[1], and Manuel Mazzara[2]

[1] Belarusian State University of Informatics and Radioelectronics, Petrusya Brovki Street, Minsk, Republic of Belarus
ilyakudrenok999@gmail.com, davydov-mv@bsuir.by
[2] Innopolis University, University Street, Innopolis, Republic of Tatarstan, Russia
m.mazzara@innopolis.ru

Abstract. In this work we performed simulation modeling of the blood flow of the human aortic valve under pathologies. Such results as: velocity, pressure were visualized. Based on the results, it was found that when the blood flow in the area of the aortic valve, there is a change in velocity and pressure. From what the conclusions about the necessity of the valve replacement were made.

1 Introduction

The treatment of vascular and cardiovascular diseases is a topical process and the study of blood flow processes in the heart and the behavior of its components, in particular valves, requires a comprehensive study. Creation of analytical models of biological tissue behavior due to the complexity of geometry requires numerical solutions in CAE systems.

The aortic valve is located at the border between the left ventricle of the heart and the aorta, the largest artery in the body. The aortic valve consists of three tightly fitting, triangular flaps of tissue called flaps. These flaps are attached to the aorta through what is called a ring. The heart valves only open in one direction. The flaps of the aortic valve can only open into the left ventricle and blood is released into the aorta. When the blood has passed through the valve and the left ventricle has relaxed, the flaps will close so that the blood that has just passed into the aorta does not go into the left ventricle [1].

As a result of successful modeling, it becomes possible to create a technology of artificial valve design that takes into account individual features of a particular organism. Such three-dimensional hydrodynamic calculations before the appearance of modern medical scanners and most powerful computing supercomputer complexes were difficult even to imagine.

Aortic valve insufficiency is a type of acquired heart defect. At this time, valve defects account for 25% of all heart disease. In insufficiency, the valve is incompletely opened or closed, resulting in backflow of blood from the aorta into the left ventricle.

Aortic insufficiency is common in patients of all ages. There are cases where children are born with a 2-branch aortic valve, making them ill from birth, often the condition is detected when a formidable complication - infective.

endocarditis or aortic dissection – joins [2] (Fig. 1).

L. Barolli (Ed.): AINA 2023, LNNS 655, pp. 12–27, 2023.
https://doi.org/10.1007/978-3-031-28694-0_2

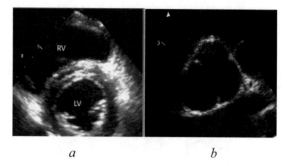

a b

Fig. 1. Aortic valve: a - open; b - closed

Degree of defect according to the amount of blood regurgitated:

- Grade I - the amount of blood regurgitated does not exceed 15%;
- Grade II - the amount of blood ranges from 15% to 30%;
- Grade III - blood volume is up to 50% of cardiac output;
- IV degree - more than half of the blood is returned to the ventricle [3].

2   Modelling Methodology

The aorta with aortic valve pathology has been modelled using SolidWorks software for a better understanding.

The aorta is divided into 3 component parts: ascending part, aortic arch and descending part.

To determine the degree of pathology we will be guided by the data from the Table 1 [4].

Table 1. Classification by severity.

Stenosis degree	Hole area, cm^2	Average transvalvular pressure gradient
Easy (I)	$\geq 1,5$	$0-20$
Moderate (II)	$1-1,5$	$20-40$
Heavy (III)	≤ 1	$40-50$
Critical (IV)	$<0,7$	>50

2.1   Construction of a Closed Valve in Pathologies

We will consider the construction of the valve using a model of a closed valve with stenosis. Since the aortic valve is under the aortic root, which has a diameter of 30 cm.

Fig. 2. Extended sketch

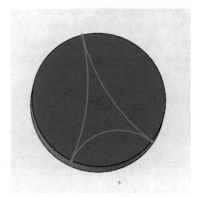

Fig. 3. Sketch of the pathology

We first sketch a circle with a radius of 15 cm and use the "Extended boss/base" operation to draw the sketch (Fig. 2).

On one of the edges, create a sketch of the pathology (Fig. 3).

Next we cut out our sketch (Fig. 4).

To determine the degree of pathology, we first calculate the area of the entire circle with SolidWorks software which equals 706,86 mm^2. Next, we calculate the area not included in the hole (Fig. 5).

Then we determine the degree of pathology.

$$S = S_{\text{all}} - S_{\text{without hole}} = 706, 86 - 596, 2 = 110, 66 \, \text{mm}^2 \tag{1}$$

Based on the value, this model corresponds to II degrees of stenosis.

By analogy, we determine the remaining 3 degrees of stenosis (Fig. 6).

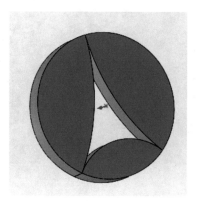

Fig. 4. Closed aortic valve model with stenosis

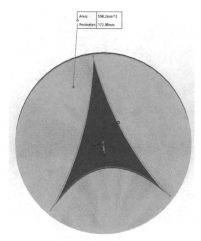

Fig. 5. Area without hole

2.2 Construction of a Opened Valve in Pathologies

Since the aortic valve is under the aortic root, which has a diameter of 30 cm. We first sketch a circle with a radius of 15 cm and use the "Extended boss/base" operation to draw the sketch (Fig. 7).

On one of the edges, create a sketch of the pathology (Fig. 8).

Next we cut out our sketch (Fig. 9).

To determine the degree of pathology, we first calculate the area of the entire circle with SolidWorks software which equals 706,86 mm^2. Next, we calculate the area not included in the hole (Fig. 10).

Then we determine the degree of pathology.

$$S = S_{all} - S_{without\ hole} = 706,86 - 538,98 = 167,88\ mm^2 \tag{2}$$

Based on the value, this model corresponds to I degrees of stenosis.

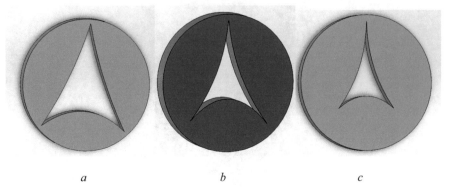

a b c

Fig. 6. Closed valve model in pathologies: a - I degree of stenosis; b- III degree of stenosis; c-IV degree of stenosis

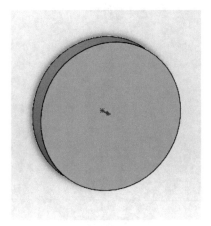

Fig. 7. Extended sketch

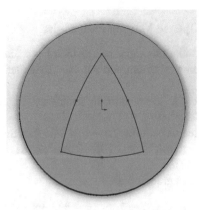

Fig. 8. Sketch of the pathology

Fig. 9. Closed aortic valve model with stenosis

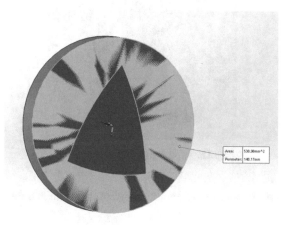

Fig. 10. Area without hole

By analogy, we determine the remaining 3 degrees of stenosis (Fig. 11).

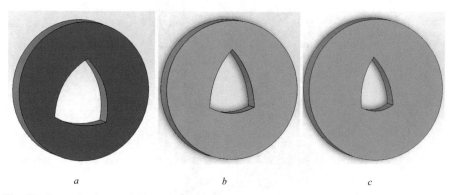

Fig. 11. Opened valve model in pathologies: a- II degree of stenosis; b- III degree of stenosis; c-IV degree of stenosis

2.3   Simulation of Blood Flow

To simulate blood flow, a model of the aorta with the left ventricle must be built (Fig. 12).

Fig. 12. Model aorta with left ventricle

Next, you need to create an assembly model of the aorta and the pathology model (Fig. 13).

Fig. 13. Model of aorta with open stenosis valve of the 1st degree

Initial conditions at FlowSimulation setup: inlet mass flow rate 10 kg/s, static pressure at start-up 101325 Pa.

To start the blood flow simulation visualisation, select the fluid type and set its parameters (Figs. 14, 15, 16, 17, 18, 19, 20, 21, 22, 23, 24, 25, 26, 27, 28, 29 and 30).

Items Item Properties Tables and Curves	
Property	Value
Name	Blood
Comments	
Density	1050 kg/m^3
Specific heat	3600 J/(kg*K)
Thermal conductivity	0.53 W/(m*K)
Viscosity	Herschel-Bulkley model
Consistency coefficient	0.003 Pa*s
Yield stress	0 Pa
Power-law index	0

Fig. 14. Basic blood parameters

Fig. 15. Aortic blood flow pressure with opened valve with I degree of stenosis

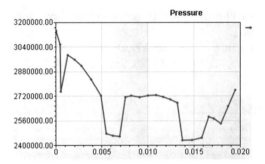

Fig. 16. Graph of changes in aortic blood flow pressure with opened valve with I degree of stenosis

Fig. 17. Aortic blood flow pressure with opened valve with II degree of stenosis

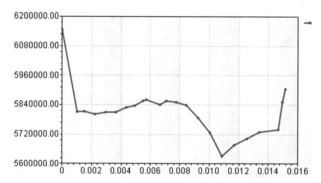

Fig. 18. Graph of changes in aortic blood flow pressure with opened valve with II degree of stenosis

Fig. 19. Aortic blood flow pressure with opened valve with III degree of stenosis

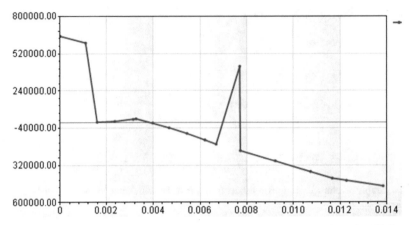

Fig. 20. Graph of changes in aortic blood flow pressure with opened valve with III degree of stenosis

Fig. 21. Aortic blood flow pressure with opened valve with IV degree of stenosis

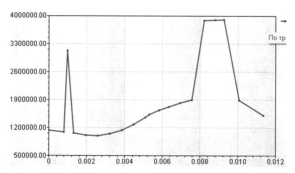

Fig. 22. Graph of changes in aortic blood flow pressure with opened valve with IV degree of stenosis

Fig. 23. Aortic blood flow pressure with closed valve with I degree of stenosis

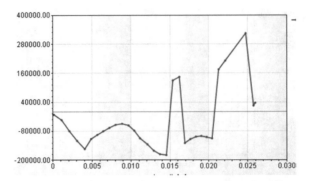

Fig. 24. Graph of changes in aortic blood flow pressure with closed valve with I degree of stenosis

Fig. 25. Aortic blood flow pressure with closed valve with II degree of stenosis

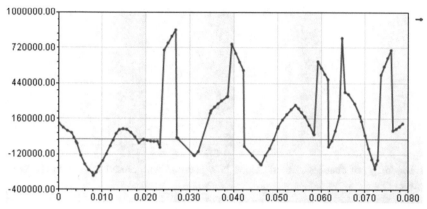

Fig. 26. Graph of changes in aortic blood flow pressure with closed valve with II degree of stenosis

Fig. 27. Aortic blood flow pressure with closed valve with III degree of stenosis

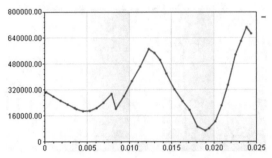

Fig. 28. Graph of changes in aortic blood flow pressure with closed valve with III degree of stenosis

Fig. 29. Aortic blood flow pressure with closed valve with IV degree of stenosis

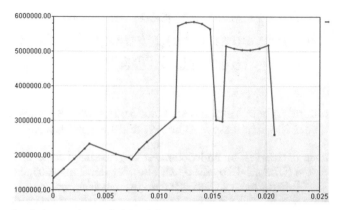

Fig. 30. Graph of changes in aortic blood flow pressure with closed valve with IV degree of stenosis

3 Conclusion

The proposed simulation model allows the appropriate estimation of pressure and velocity in the aortic circulation in normal and various valve pathologies. The resulting model can be personalised on the basis of cardiac ultrasound data. The obtained data can be used when deciding on the necessity of heart valve prosthesis in aortic insufficiency in order to avoid regurgitation and to ensure stable human life activity. A tricuspid mechanical prosthesis is recommended as a valve prosthesis.

Thanks to the visual data, it is clearer why aortic valve replacement is necessary and how much blood is regurgitated.

References

1. The aorta, parts of the aorta. Arteries and veins of the heart [Electronic resource]. Access mode. www.bsmu.by/downloads/kafedri/k_anatomia/stud/2018-1/zan-9.pdf
2. The aortic valve [Electronic resource]. Access mode. www.cardiokurort.ru/encyclopedia/448-poroki-klapanov-serdtsa-prichiny-simptomy-diagnostika
3. Aortic heart failure [Electronic resource]. Access mode. www.medpractic.com/rus/153/13093/ Современные%20представления%20о%20норме%20и%20пато логических%20отклоненях/article.more.html
4. Aortic valve stenosis [Electronic resource]. Access mode. https://cardiograf.com/bolezni/pat ologii/aortalnyj-stenoz.html?ysclid=lary3fm9os77056060

An Integrated System for Vibration Suppression Using Fuzzy Control and 2D-LiDAR

Masahiro Niihara[1], Yuma Yamashita[1], Chihiro Yukawa[2], Kyouhei Toyosima[2], Yuki Nagai[2], Tetsuya Oda[1(✉)], and Leonard Barolli[3]

[1] Department of Information and Computer Engineering, Okayama University of Science (OUS), 1-1 Ridaicho, Kita-ku, Okayama 700-0005, Japan
{t19j061nm,t20j091yy}@ous.jp, oda@ous.ac.jp
[2] Graduate School of Engineering, Okayama University of Science (OUS), 1-1 Ridaicho, Kita-ku, Okayama 700-0005, Japan
{t22jmm19st,t22jm24jd,t22jm23rv}@ous.jp
[3] Department of Information and Communication Engineering, Fukuoka Institute of Technology, 3-30-1 Wajiro-higashi, Higashi-ku, Fukuoka 811-0295, Japan
barolli@fit.ac.jp

Abstract. Different robots such as Automatic Guided Vehicles (AGVs) and Autonomous Mobile Robots (AMRs) need to know their position in a building in order to operate automatically and autonomously. They can receive the environmental data by Light Detection And Ranging (LiDAR) based on Simultaneous Localization and Mapping (SLAM) to perform autonomous control. Recently, 3D-LiDAR capable for three-dimensional measurement has been manufactured, but it is very expensive. In this paper, we propose an integrated system consisting of a 2D-LiDAR, a servo motor and Fuzzy Control module. Three-dimensional measurement is performed by fixing and rotating 2D-LiDAR by 1 Degree of Freedom (DOF) rotation mechanism. Fuzzy control is applied to suppress the vibration of the servo motor. We carried out an experiment to evaluate the proposed system. The experimental results show that when applying fuzzy control, the acceleration is decreased compared with the case of no fuzzy control.

1 Introduction

Swarm robots such as Automatic Guided Vehicles (AGVs)/Autonomous Mobile Robots (AMRs) [1–3] are utilized in factories, warehouses and buildings o support employees, prevent human error and reduce labor costs. Operating swarm robots based on county intelligence can improve work efficiency by simultaneously observing the ambient at multiple locations and sharing logistics information, something that has been difficult with stand-alone robots. For swarm robots to act autonomously, each robot needs to avoid persons and obstacles by considering the surrounding environment and estimating its own position where

L. Barolli (Ed.): AINA 2023, LNNS 655, pp. 28–35, 2023.
https://doi.org/10.1007/978-3-031-28694-0_3

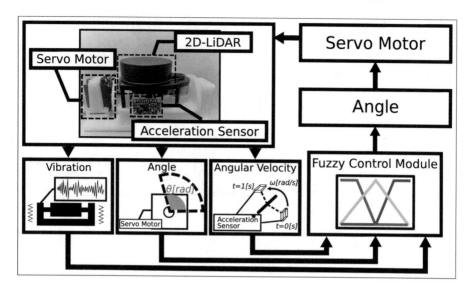

Fig. 1. Proposed system.

Global Navigation Satellite System (GNSS) cannot be used. Therefore, Simultaneous Localization and Mapping (SLAM) [4–6], a self-positioning estimation method and Light Detection And Ranging (LiDAR) [7–9], which can immediately acquire environmental data around robots, have attracted attention and they can be used by swarm robots to perform autonomous behavior.

The 3D-LiDAR is capable of three-dimensional (3D) measurement, but it very expensive. Therefore, we consider to develop a low cost system for 3D measurement. In this paper, we propose an integrated system consisting of a 2D-LiDAR, a servo motor and Fuzzy Control module [10–14]. Three-dimensional measurement is performed by fixing and rotating 2D-LiDAR by 1 Degree of Freedom (DOF) rotation mechanism. Fuzzy control is applied to suppress the vibration of the servo motor. We carried out an experiment to evaluate the proposed system. The experimental results show that when applying fuzzy control, the acceleration is decreased compared with the case of no fuzzy control.

The structure of the paper is as follows. In Sect. 2, we describe the proposed method. In Sect. 3, we present the experimental results. Finally, conclusions and future work are given in Sect. 4.

2   Proposed System

2.1   Proposed System Structure and 3D Measurement

Figure 1 shows the proposed system structure. The proposed system consists of a 2D-LiDAR, a servo motor and the Fuzzy Control module. The 3D measurement is performed by fixing and rotating 2D-LiDAR by 1 Degree of Freedom (DOF) rotation mechanism.

Figure 2 shows the image of three-dimensional measurement using 2D-LiDAR. The 3D coordinates are determined from the rotation angle of the 2D-LiDAR, the measured distance and the angle of servo motor. Using the distance $D \in \mathbb{N}$ and angle $\theta \in \{0 < \mathbb{N} < 360\}$, the z coordinate is calculated by Eq. (1).

$$Z = D \times \cos \theta \tag{1}$$

By using the angle of servo motor $\varphi \in \{0 \leq \mathbb{N} \leq 180\}$, the lx and ly coordinates of LiDAR can be calculated by Eq. (2) and Eq. (3). Where d is the distance from the servo motor rotation axis to the 2D-LiDAR measurement position and φ is the angle of servo motor.

$$lx = d \times \cos \varphi \tag{2}$$

$$ly = d \times \sin \varphi \tag{3}$$

The x and y coordinates of the obstacle are calculated by Eq. (4) and Eq. (5).

$$x = D \times \sin \theta \times \cos \varphi + lx \tag{4}$$

$$y = D \times \sin \theta \times \cos \varphi + ly \tag{5}$$

Figure 3 shows 2D-LiDAR with different servo motor angles: 0, 45, 90, 135 and 180 [deg.]. The 2D-LiDAR is fixed to a rotation mechanism with 1-DOF by the servo motor, which is rotated in the range of $\varphi \in \{0 \leq \mathbb{N} \leq 180\}$. Thus, it is possible to measure the distance to surrounding obstacles by using rotation angle of 2D-LiDAR.

The 3D coordinates are calculated from the measured distance, 2D-LiDAR rotation angle and the servo motor angle by Eq. (1), Eq. (4) and Eq. (5). The rotating mechanism of the proposed system is developed by using a 3D printer.

2.2   Fuzzy Control for Vibration Suppression

The proposed system performs vibration suppression of the servo motor by fuzzy control. Figure 4 shows the fuzzy membership functions. Figure 4(a), Fig. 4(b) and Fig. 4(c) show the input membership functions, while Fig. 4(d) shows the output membership functions.

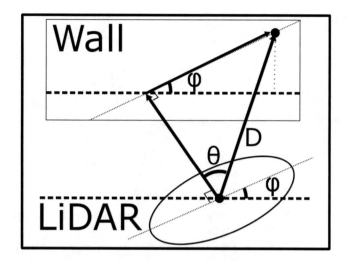

Fig. 2. Image of 3D surveying measurement by 2D-LiDAR.

Table 1. Fuzzy rule-base.

SA	ASM	AVA	AVS	SA	ASM	AVA	AVS	SA	ASM	AVA	AVS
High	High	High	High	Middle	High	High	High	Low	High	High	Middle
High	High	Middle	High	Middle	High	Middle	Low	Low	High	Middle	Low
High	High	Low	High	Middle	High	Low	Low	Low	High	Low	Middle
High	Middle	High	High	Middle	Middle	High	Low	Low	Middle	High	Low
High	Middle	Middle	Middle	Middle	Middle	Middle	Middle	Low	Middle	Middle	Low
High	Middle	Low	Middle	Middle	Middle	Low	low	Low	Middle	Low	Low
High	Low	High	Midole	Middle	Low	High	Middle	Low	Low	High	Low
High	Low	Middle	High	Middle	Low	Middle	Low	Low	Low	Middle	Low
High	Low	Low	Middle	Middle	Low	Low	Low	Low	Low	Low	Low

Table 1 shows the fuzzy rule-base. The proposed system defines the acceleration for each angle as a vibration. Therefore, the rule-base for fuzzy control is built to decrease the acceleration. The input parameters for fuzzy control consist of Swing of Acceleration (SA), Angle of Servo Motor (ASM) and Angular Velocity at the Axis rotation by the servo motor (AVA). The output parameter is the Angle for Vibration Suppression (AVS).

3   Experimental Results

The experimental environment is shown in Fig. 5, considering a real environment. In the experiment, we measure the change in acceleration for two cases: with fuzzy control and without fuzzy control. The 2D-LiDAR is rotated in the range of 0 to 180 [$deg.$]. In the proposed system, the angle of the servo motor is updated based on the output of fuzzy control.

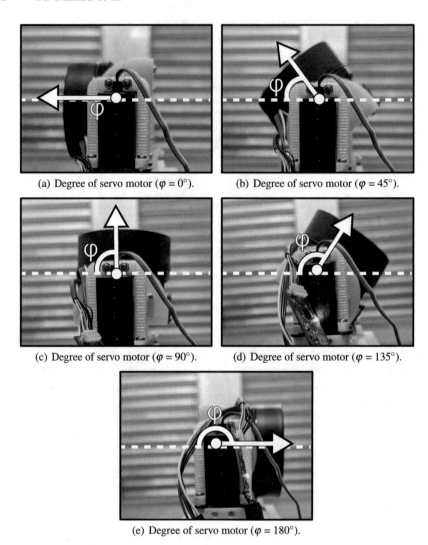

(a) Degree of servo motor ($\varphi = 0°$).   (b) Degree of servo motor ($\varphi = 45°$).

(c) Degree of servo motor ($\varphi = 90°$).   (d) Degree of servo motor ($\varphi = 135°$).

(e) Degree of servo motor ($\varphi = 180°$).

Fig. 3. Position of 2D-LiDAR when the angle of the servo motor is changed.

Figure 6 shows the acceleration generated for each angle. We consider the average results of 10 measurements. When applying fuzzy control, the acceleration is decreased compared with the case of no Fuzzy control.

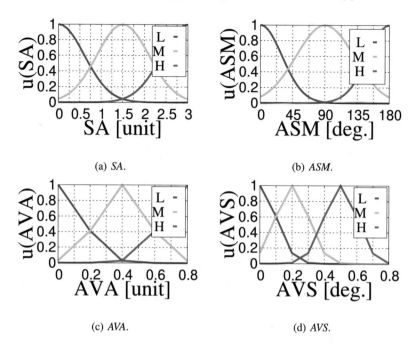

(a) *SA.*   (b) *ASM.*

(c) *AVA.*   (d) *AVS.*

Fig. 4. Membership functions.

Fig. 5. Experimental environment.

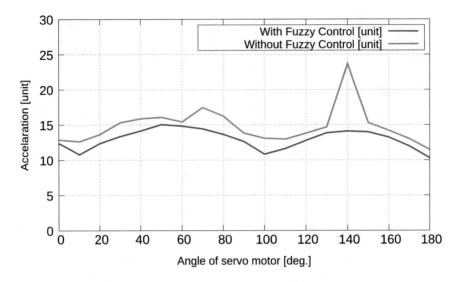

Fig. 6. Experimental result of acceleration measurement.

4   Conclusions

In this paper, we proposed an integrated system consisting of a 2D-LiDAR, a servo motor and Fuzzy Control module. The 3D measurement is performed by fixing and rotating 2D-LiDAR by 1 DOF rotation mechanism. Fuzzy control is applied to suppress the vibration of the servo motor. We carried out an experiment to evaluate the proposed system. The experimental results show that when applying fuzzy control, the acceleration is decreased compared with the case of no fuzzy control.

In the future work, we would like to consider different scenarios and mount the proposed system in a drone.

References

1. Saito, N., et al.: A LiDAR based mobile area decision method for TLS-DQN: improving control for AAV mobility. In: Advances on P2P, Parallel, Grid, Cloud and Internet Computing, pp. 30–42 (2022)
2. Saito, N., et al.: A movement adjustment method for LiDAR based mobile area decision: improving control for AAV mobility. In: Barolli, L. (ed.) Innovative Mobile and Internet Services in Ubiquitous Computing, pp. 41–53. Springer, Cham (2022). https://doi.org/10.1007/978-3-031-08819-3_5
3. Liu, H., et al.: Enhancing LIDAR performance metrics using continuous-wave photon-pair sources. Optica **6**(10), 1349–1355 (2019)
4. Xu, W., et al.: FAST-LIO2: fast direct lidar-inertial odometry. IEEE Trans. Robot. **38**(4), 2053–2073 (2022)
5. Ismail, H., et al.: Exploration-based SLAM (e-SLAM) for the indoor mobile robot using lidar. Sensors **22**(4), 1689 (2022)

6. Li, Y., et al.: Deepfusion: lidar-camera deep fusion for multi-modal 3D object detection. In: Proceedings of the IEEE/CVF Conference on Computer Vision and Pattern Recognition, pp. 17182–17191 (2022)

7. Xu, X., et al.: A review of multi-sensor fusion slam systems based on 3D LIDAR. Remote Sens. **14**(12), 2835 (2022)

8. Li, Y., Ibanez-Guzman, J.: Lidar for autonomous driving: the principles, challenges, and trends for automotive lidar and perception systems. IEEE Signal Process. Mag. **37**(4), 50–61 (2020)

9. Karimi, M., Oelsch, M., Stengel, O., Babaians, E., Steinbach, E.: LoLa-SLAM: low-latency LiDAR SLAM using continuous scan slicing. IEEE Robot. Autom. Lett. **6**(2), 2248–2255 (2021)

10. Mehdi, S., et al.: Paradoxes of gender, technology, and the pandemic in the Iranian music industry. Pop. Music Soc. **44**(1), 1–13 (2021)

11. Yukawa, C., et al.: Design of a fuzzy inference based robot vision for CNN training image acquisition. In: Proceedings of the IEEE 10th Global Conference on Consumer Electronics, pp. 806–807 (2021)

12. Yukawa, C., et al.: Evaluation of a fuzzy-based robotic vision system for recognizing micro-roughness on arbitrary surfaces: a comparison study for vibration reduction of robot arm. In: Barolli, L., Miwa, H., Enokido, T. (eds.) Advances in Network-Based Information Systems. LNCS, vol. 526, pp. 230–237. Springer, Cham (2022). https://doi.org/10.1007/978-3-031-14314-4_23

13. Hayashi, K., et al.: A fuzzy control based cluster-head selection method for CNN distributed processing: improving QoS of limited computing resources. In: Proceedings of the IEEE 11th Global Conference on Consumer Electronics (2022)

14. Hayashi, K., et al.: A fuzzy control based cluster-head selection and CNN distributed processing system for improving performance of computers with limited resources. In: Advances on P2P, Parallel, Grid, Cloud and Internet Computing, pp. 232–239 (2022)

Prediction in Smart Environments and Administration: Systematic Literature Review

Mohamed Krichene$^{(\boxtimes)}$, Nesrine Khabou, and Ismael Bouassida Rodriguez

ReDCAD, Sfax, Tunisia
krichenemed@proton.me, {nesrine.khabou,bouassida}@redcad.org

Abstract. We broad an expanding connected and computerized community. Smart environments embody this tendency by associating computers and other devices with everyday settings and prevalent tasks. Predictive analysis in smart environment (PASE) helps planning and making a higher cognitive process associated with controlling smart devices using different analysis models. Administration element is associated with context-aware application which is used in a variety of industries, and can be relevant and applied in lots of sectors including Retails, Pharmaceuticals, Healthcare, Government and Public sectors. The main contribution of this paper is presenting results of a Systematic Literature Review (SLR) that feature the various models applied in analyzing data collected from smart devices, sensor-enabled and networked devices, and also the significant role of integrating an administration element to administrate the use of the applications.

Keywords: Systematic Literature Review (SLR) · Predictive analysis in smart environment (PASE) · Context-aware systems

1 Introduction

Prediction is a technique that uses historical data as inputs to make informed estimates that are predictive in determining the direction of future trends. The term predictive analytics refers to the use of standard statistical techniques, machine learning, deep learning and other types of artificial intelligence technologies to predict future outcomes based on current and past data. It builds on descriptive analytics, which describes what happened, and is the precursor to prescriptive analytics, analyzes why something happened or to make predictions about future outcomes and performance and eventually what should be done. For instance, predictive analytics can help identify conditions where environmental pollution risks are high, relate agriculture actions to environmental footprint, and reduce detrimental environmental decisions. We intend to discuss predictive analytics models and exactly in time series analysis. Therefore, the need for development of an ability to analyze and predict upcoming events within smart environment (inc. Smart Grid, intelligent buildings, smart water network,

© The Author(s), under exclusive license to Springer Nature Switzerland AG 2023
L. Barolli (Ed.): AINA 2023, LNNS 655, pp. 36–47, 2023.
https://doi.org/10.1007/978-3-031-28694-0_4

and intelligent transportation network) is evident. Furthermore, for establishing a better control of the collected results it should be more convenient to include more administration elements (which are for a fact the solutions implemented by researchers) as means of controls to facilitate the process of using models created by those who made research in this matter.

This study helps discovering the different administration elements applied in the studied research papers. Additionally it shows the different used prediction models that contribute to making context-aware systems based on used algorithms. This paper is organized as follows: In Sect. 2, we represent the Systematic Literature Review planning. Results are presented in Sect. 3. Finally a conclusion with general highlights and perspectives for further work in Sect. 4.

2   Systematic Literature Review Planning

2.1   Research Questions

Aiming to find all relevant primary studies related to the different types of models used to describe the phenomenon of prediction in a smart environment, the following research questions (RQ) were established:

RQ1: Did prediction solutions integrate an administration element?
RQ2: What are the prediction methods applied in the smart environment and administration in a data-series case?

Subsequently, we determined the initial research in the databases. About the keywords, we have formed three groups:

- Group 1: ("prediction" OR "predict")
- Group 2: ("smart city" OR "smart cities" OR "smart environment" OR "smart environments" OR "smart spaces" OR "smart space")
- Group 3: ("time series" OR "date series")

Table 1. Search results by resources

Resources	Number of paper(s)
Springer	1
IEEE Xplore Digital library	29
ACM	67
Science Direct	387
HyperArticle en Ligne (HAL)	8
Total	492

2.2 Search Strategy

The search strategy combines the key concepts of our search question to retrieve accurate results. It is an organized structure of keywords, which are "smart environments", "prediction" and "data-series", used to search a database. Then, we added synonyms, variations, and related terms for each keyword. A Boolean operator ("AND" and "OR") allow us to try different combinations of search terms. The final search string is [("prediction" OR "predict") AND ("smart city" OR "smart cities" OR "smart environment" OR "smart environments" OR "smart spaces" OR "smart space") AND ("time series" OR "data series")].

NOTE: we did not include all the synonyms of the word "prediction" (such as Forecasting, prognosis. prognosticate..) because the list of research studies will become too much to handle.

2.3 Selection Criteria

After obtaining the search results from the different sources, a set of exclusion/inclusion criteria was applied to help in the identification of relevant primary studies. Therefore, Inclusion Criteria (IC) are used to select primary studies which indicate related data analysis techniques, purpose, or the prediction in a smart environment. The Exclusion Criteria (EC) are used to remove those primary studies that do not address the main topics searched in this SLR, are not available, or are directly related to an included primary study of the same author.

Inclusion Criteria are as follow: (IC1) Publications that match one of the search items, (IC2) Publications that have best practices version, (IC3) Publications that are related to prediction in smart environment, (IC4) Publications that are related to context-aware computing and (IC5) Publications that are related to the research questions.

Exclusion Criteria are as follow: (EC1) Publications that do not match one of the search items, (EC2) Publications that do not have best practices version, (EC3) Publications that are published before or on the 31.12.2009 and (EC4) Publications that are not related to the research questions.

2.4 Data Collection

The number of papers resulting in the search is summarized in Table 1. After filtering irrelevant, duplicate and incomplete papers, a total of 60 papers were selected for the reviewing process (Table 2).

Table 2. Filtered search results

Filtering method	Number of paper(s)
Irrelevant and duplicates	12
Incomplete and not related to RQ, excluded by reading title and abstract	375
File not found	4
Total for Introduction reading	101
Not related to RQ, excluded by reading Introduction	41
Total for reading	60

3   Results and Discussion

3.1   Introduction

Predictive analysis (PA) covers many fields and domains and in general, PA aims to cover all aspects of each domain. Thus, after carefully reading all the relevant research papers and aiming for an optimal balance between generality and concreteness, we have tried to keep the number of domains to a minimum while keeping them self-explanatory so that the scope of each domain's is always evident. As a result, we have divided predictive analysis according to the domains listed in Table 3:

NOTE: Total of 60 papers. A paper may belong to one or more domain.

Table 3. Number of papers with Prediction solution by domain

Domain	Number of paper(s)	References(s)
Smart economy	2	[29, 44]
Smart transport and mobility	17	[2–4, 8, 13, 16, 23, 35, 36, 43, 46–48, 53–55, 59]
Smart healthcare	6	[5, 14, 21, 26, 52, 58]
Smart buildings and living	15	[1, 6, 15, 17–19, 27, 29, 31, 32, 39, 42, 55–57]
Smart energy	11	[7, 9, 11, 12, 17, 22, 24, 28, 38, 39, 41]
Smart agriculture	3	[30, 34, 49]
Smart city	10	[10, 20, 27, 33, 37, 40, 45, 50, 51, 60]

In our opinion, these seven domains are diverse enough to cover all the PA application that can meet the needs of our research. In addition, we wanted to divide the papers according to their domains.

3.2   RQ1: Did Prediction Solutions Integrate an Administration Element?

After filtering admissible articles and journals we have obtained a total number of 6 papers that include administration element. The obtained results are divided according to the application's type. The results can be categorized as follows:

- **Web application:** Salotti et al. [4] created a web application for short term traffic state forecasting based on K-NN and a new framework (Multi-horizon framework for congestion forecasting) that they have developed for traffic regulation and it was integrated in Lyon Metroplis for an Interactive Training System.
- **Desktop application:** In the domain of smart energy A. Maté et al. [41] developed "weka explorer" a desktop administration interface to predict energy consumption in form of graph. On the same subject Iftikhar et al. [25] have designed and have created "smart meter tool" to control and administrate the energy consumption to evaluate the efficiency of their system.
- **Mobile application:** In the domain of healthcare Zhang et al. [58] have invented a new IoT device using Raspberry Pi and Arduino mega and GPS to monitor the quality of air, this device collects and sends data to mobile application called "WeAIR" to check the air quality immediately.
- **Other types of application:** In the healthcare domain, Munish et al. [5] have invented a "UbD (Urine-based Diabetes) monitoring system" using raspberry pi device with liquid crystal display to visualize the data collected from diabetic patients to help in the prediction process. However in the domain of smart transport and mobility, Fernandez-Ares et al. [16] have created a new monitoring and tracking system called "MOBYWIT" using Raspberry Pi board with RFID, wifi and BT sensors to scan the radio-electric space searching for people devices to ameliorate traffic flow in terms of trip-time and security.

3.3   RQ2: What Are the Prediction Methods Applied in the Smart Environment and Administration in a Data-Series Case?

In this section, we discuss the different proposed models and the algorithms used in creating these models.

In the domain of **smart economy**: Kadar et al. [29] have presented their model "imbalance-aware hyper-ensemble" which was based on K-NN algorithm. In the same domain H. Al-Nuaimy et al. [44] created model named "MSM30 index" using ARIMA and ANN algorithms to detect anomaly.

Within the field of **smart transport and mobility**: H. Erroussoa et al. [13] have made a "parking occupancy prediction" model using decision tree and K-NN algorithms. In the work of Attila M. Nagy et al. [43] they have created a Model named "Congestion-based Traffic Prediction Model (CTPM)" using CPMA and XGboost algorithms. In Eric Lin and Jinhyung D. Park's paper [36] they have developed Bayesian model using Apriori algorithm, it is called "traffic outflow model". P. Shu et al. [47] have combined CNN and LSTM algorithms to create LSTM-CNN (model). R. Chawuthai et al. [8] they have created a model called "KM-Stone" based on linear regression. Unlike the work of A.J. Fernandez-Ares et al. [16] they have used ARIMA and ETS algorithms to create their model named "MOBYWIT". Bahman Askari et al. [3] the authors have created "lstm-deep sequence" model based on deep learning and LSTM. K. Tekouabou et al. [48] had based on decision tree, KNN and regression to create thier model. In article [35] Zutai Li and Xiaoping Liu made "Dynamic Emergency Navigation (DENP)" model Based on Prediction via Wireless Sensor Networks. Michel Chammas et al. [53] created "att-DCRNN" model based on RNN. Xiaojie Yu and Lijun Sun [54] produced a model called TFPM-STC using CNN. In paper [23] P. Huang et al. model called "CL-IncNet" using LSTM.

On the subject of **smart healthcare**: In the work of M. Bhatia et al. [5] they used RNN algorithm to create UbD "Monitoring and Prediction System" as well as its model for diabetic people with the same name as the system. In another matter D. Zhang et al. [58] have additionally created the Mobile interface "WeAIR" they created a model with same name as the device based on SVM and RFR. In the work of Rui Xu et al. [52] they created a "SARIMA-LSTM" model as the name suggest it is based on SARIMA and LSTM algorithms. R. Ali et al. [21] have used POI and MDA and HR methods to create "Matter Prediction model". In the work of Lei Ji et al. [26] they have used ARIMA, CNN and LSTM to "ARIMA-CNN-LSTM" model.

In the area of **smart energy**: H. Fangwan et al. [22] created "Clock-Work RNN (CW-RNN)" model using RNN and LSTM. In another point I. Nadeem et al. [25] have made a "PAR-based" model and KNN. Into J.C Kabugo et al. paper [28] we can see that they have developed the "NN-NARX" model using neural network. L. Xiufeng et al. [38] have created a "periodic auto-regression with exogenous variables (PARX)" model. In the work of C. Emanuele et al. [12] have developed a 'context-aware power forecasting' module. Investigating X.J. Luo et al. paper [39] we can see that they have used K-means clustering and artificial neural network (ANN) to create model called sub-ANN predictive. Y. Zhan and H. Hamed paper [7] have used RNN to create "MLP" model. In the paper [9] Run-Qing Chen et al. created a model called "Predictor and Anomaly Detection (PAD)".

Regarding the domain of **smart living and buildings**: In the work of Z. Tianzi et al. [55] they have invented a "Double-Encoder" model using ST-Encoder, FR-Encoder and RNN. Into C. Charlie et al.'s work [6] they have used spatial analysis and auto-regressive models to create new model called "spatio-temporal crime forecasting" using SARIMA and DBSCAN to detect anomaly. In M. Bryan et al.'s work [42] they have used regression tree and KNN to create "AF algorithm model". U. Fattore et al. [15] have used RNN to develop "autoMEC" model. As for K. Nesrine et al.'s work [32] they have designed and have created a model named "ANALOG" using WMA, MA, and EWMA. In D. Ivan et al.'s paper [1] they have produced a "Proactive autonomic manager" as a model for self-adapted sensor. As for E. Raquel et al. [56] they have used RNN to develop two variants from LSTM called "1lstm" and "2lstm".

Regarding to **smart city domain**: L. Chengxi et al. [37] have invented a better version of "LHCnn" model using LSTM and Convolutional neural network (CNN). In a similar matter V. John et aL. [50] have used ANN and LSTM to create LSTM ANN model. Z. Jingbo et al. [60] have used KNN, GPs and SVMs algorithms to produce a model called "SMiLer". As for C. Yuming et al. in their work [10] they have created a GRU-ES model using KNN, LR, ARIMA and LSTM. In R. Rege et al. paper [45] they have created a "ContextPerf automation tool" using transfer learning and LSTM. W. et al. [51] have introduced two variants from the same model "ctGAN-S2S" and "ctGAN".

On the domain of **smart agriculture**: S.Bangaru et al. [30] have produced a ANN-based model for weather forecasting. In the work of A. Kociana et al. [34], they have created a model called "Dynamic Bayesian network model (DBNM)" based on naive Bayes algorithm.

3.4   Learned Lesson

This section articulate findings which are not included in the research questions. We found out that not many papers included an administration element which brings out the importance of this section. Also, we wanted to give an overview (Table 4) of predictive analysis models used in this research to respond to "what are the most used algorithms in predictive analysis models ?" question.

Finally, we concluded that to improve PA outcomes, we recommend the use of LSTM and GRU (GRU is a variant of LSTM) algorithms improves accuracy and increases the confidence level allowing for very reliable and stable long-term results. It is worth mentioning that LSTM was used in 17 scientific paper in these research [3, 9, 10, 14, 15, 18, 22, 23, 26, 37, 45, 47, 50, 52, 56, 57, 59] and nearly every time including algorithms with LSTM perform significantly better than other algorithms.

Table 4. Classification of analysis models based on its type

Type of predictive analysis models		Used model	Number of time(s) it was used
Classification models		K-NN	10
		Decision Tree	6
		SVR	4
		Random Forest	4
		Naive Bayes	4
Clustering models		K-means	4
		DBSCAN	1
Forecasting models	Time Series	Moving average	2
		WMA	2
	Econometric methods	ARIMA	11
		SARIMA	3

4 Conclusion

Despite the efforts made to create a better environment, there is still a shortage in the number of administration solutions related to the smart environment. In conclusion, studying the models and what they consist of, and looking for integrated administration solutions within these models, our research methodology consists of creating an SLR to answer the above question, and we have found that although there is a different solution, there is still a lack of understanding administration elements with developed models. So we believe that there will be an increase in number of solutions in the near future. Finally in the learned lesson section we presented an overview of predictive analysis models used in this research and showed the importance of integration LSTM in models.

Acknowledgements. This work was partially supported by LABEX-TA project MeFoGL: Méthodes Formelles pour le Génie Logiciel.

References

1. Anaya, I.D.P., Simko, V., Bourcier, J., Plouzeau, N., Jézéquel, J.-M.: A prediction-driven adaptation approach for self-adaptive sensor networks. In: Proceedings of the 9th International Symposium on Software Engineering for Adaptive and Self-Managing Systems, pp. 145–154 (2014)
2. Arabghalizi, T., Labrinidis, A.: Data-driven bus crowding prediction models using context-specific features. ACM Trans. Data Sci. **1**(3), 1–33 (2020)
3. Askari, B., Le Quy, T., Ntoutsi, E.: Taxi demand prediction using an LSTM-based deep sequence model and points of interest. In: 2020 IEEE 44th Annual Computers, Software, and Applications Conference (COMPSAC), pp. 1719–1724. IEEE (2020)
4. Attoui, S.-E., Meddeb, M.: A generic framework for forecasting short-term traffic conditions on urban highways. In: 2021 IEEE 8th International Conference on Data Science and Advanced Analytics (DSAA), pp. 1–10. IEEE (2021)
5. Bhatia, M., Kaur, S., Sood, S.K., Behal, V.: Internet of Things-inspired healthcare system for urine-based diabetes prediction. Artif. Intell. Med. **107**, 101913 (2020)
6. Catlett, C., Cesario, E., Talia, D., Vinci, A.: Spatio-temporal crime predictions in smart cities: a data-driven approach and experiments. Pervasive Mob. Comput. **53**, 62–74 (2019)
7. Chammas, M., Makhoul, A., Demerjian, J.: An efficient data model for energy prediction using wireless sensors. Comput. Electr. Eng. **76**, 249–257 (2019)
8. Chawuthai, R., Pruekwangkhao, K., Threepak, T.: Spatial-temporal traffic speed prediction on Thailand roads. In: 2021 7th International Conference on Engineering, Applied Sciences and Technology (ICEAST), pp. 58–62. IEEE (2021)
9. Chen, R.-Q., Shi, G.-H., Zhao, W.-L., Liang, C.-H.: A joint model for IT operation series prediction and anomaly detection. Neurocomputing **448**, 130–139 (2021)
10. Cheng, Y., Wang, C., Yu, H., Hu, Y., Zhou, X.: GRU-ES: resource usage prediction of cloud workloads using a novel hybrid method. In: 2019 IEEE 21st International Conference on High Performance Computing and Communications; IEEE 17th International Conference on Smart City; IEEE 5th International Conference on Data Science and Systems, pp. 1249–1256. IEEE (2019)

11. Chrysopoulos, A., Diou, C., Symeonidis, A.L., Mitkas, P.A.: Bottom-up modeling of small-scale energy consumers for effective demand response applications. Eng. Appl. Artif. Intell. **35**, 299–315 (2014)
12. Cuncu, E., Manca, M.M., Pes, B., Riboni, D.: Towards context-aware power forecasting in smart-homes. Procedia Comput. Sci. **198**, 243–248 (2022)
13. Errousso, H., Malhene, N., Benhadou, S., Medromi, H.: Predicting car park availability for a better delivery bay management. Procedia Comput. Sci. **170**, 203–210 (2020)
14. Espinosa, R., Palma, J., Jiménez, F., Kamińska, J., Sciavicco, G., Lucena-Sánchez, E.: A time series forecasting based multi-criteria methodology for air quality prediction. Appl. Soft Comput. **113**, 107850 (2021)
15. Fattore, U., Liebsch, M., Brik, B., Ksentini, A.: AutoMEC: LSTM-based user mobility prediction for service management in distributed MEC resources. In: Proceedings of the 23rd International ACM Conference on Modeling, Analysis and Simulation of Wireless and Mobile Systems, pp. 155–159 (2020)
16. Fernández-Ares, A., et al.: Studying real traffic and mobility scenarios for a smart city using a new monitoring and tracking system. Future Gener. Comput. Syst. **76**, 163–179 (2017)
17. González-Vidal, A., Moreno-Cano, V., Terroso-Sáenz, F., Skarmeta, A.F.: Towards energy efficiency smart buildings models based on intelligent data analytics. Procedia Comput. Sci. **83**, 994–999 (2016)
18. Hadri, S., Naitmalek, Y., Najib, M., Bakhouya, M., Fakhri, Y., Elaroussi, M.: A comparative study of predictive approaches for load forecasting in smart buildings. Procedia Comput. Sci. **160**, 173–180 (2019)
19. Hajela, G., Chawla, M., Rasool, A.: A clustering based hotspot identification approach for crime prediction. Procedia Comput. Sci. **167**, 1462–1470 (2020)
20. Hauser, M., Flath, C.M., Thiesse, F.: Catch me if you scan: data-driven prescriptive modeling for smart store environments. Eur. J. Oper. Res. **294**(3), 860–873 (2021)
21. Honarvar, A.R., Sami, A.: Towards sustainable smart city by particulate matter prediction using urban big data, excluding expensive air pollution infrastructures. Big Data Res. **17**, 56–65 (2019). ISSN: 2214-5796
22. Huang, F., Zhuang, S., Yu, Z.: Power load prediction based on an improved clockwork RNN. In: 2019 IEEE SmartWorld, Ubiquitous Intelligence & Computing, Advanced & Trusted Computing, Scalable Computing & Communications, Cloud & Big Data Computing, Internet of People and Smart City Innovation, pp. 596–601. IEEE (2019)
23. Huang, P., Huang, B., Zhao, F., Zhang, Y., Chen, M.: Deep ConvLSTM-inception network for traffic prediction in smart cities. In: 2020 IEEE 22nd International Conference on High Performance Computing and Communications; IEEE 18th International Conference on Smart City; IEEE 6th International Conference on Data Science and Systems (HPCC/SmartCity/DSS), pp. 1211–1218. IEEE (2020)
24. Huotari, M., Arora, S., Malhi, A., Främling, K.: Comparing seven methods for state-of-health time series prediction for the lithium-ion battery packs of forklifts. Appl. Soft Comput. **111**, 107670 (2021)
25. Iftikhar, N., Liu, X., Nordbjerg, F.E., Danalachi, S.: A prediction-based smart meter data generator. In: 2016 19th International Conference on Network-Based Information Systems (NBiS), pp. 173–180. IEEE (2016)
26. Ji, L., Zou, Y., He, K., Zhu, B.: Carbon futures price forecasting based with ARIMA-CNN-LSTM model. Procedia Comput. Sci. **162**, 33–38 (2019)

27. Johanna, G.-R., et al.: Predictive model for the identification of activities of daily living (ADL) in indoor environments using classification techniques based on machine learning. Procedia Comput. Sci. **191**, 361–366 (2021)
28. Kabugo, J.C., Jämsä-Jounela, S.-L., Schiemann, R., Binder, C.: Industry 4.0 based process data analytics platform: a waste-to-energy plant case study. Int. J. Electr. Power Energy Syst. **115**, 105508 (2020)
29. Kadar, C., Maculan, R., Feuerriegel, S.: Public decision support for low population density areas: an imbalance-aware hyper-ensemble for spatio-temporal crime prediction. Decis. Support Syst. **119**, 107–117 (2019)
30. Kamatchi, S.B., Parvathi, R.: Improvement of crop production using recommender system by weather forecasts. Procedia Comput. Sci. **165**, 724–732 (2019)
31. Karunaratne, P., Moshtaghi, M., Karunasekera, S., Harwood, A., Cohn, T.: Multi-step prediction with missing smart sensor data using multi-task Gaussian processes. In: 2017 IEEE International Conference on Big Data (Big Data), pp. 1183–1192. IEEE, 2017
32. Khabou, N., Rodriguez, I.B., Jmaiel, M.: A novel analysis approach for the design and the development of context-aware applications. J. Syst. Softw. **133**, 113–125 (2017)
33. Khabou, N., Rodriguez, I.B., Jmaiel, M.: An overview of a novel analysis approach for enhancing context awareness in smart environments. Inf. Softw. Technol. **111**, 131–143 (2019)
34. Kocian, A., et al.: Dynamic Bayesian network for crop growth prediction in greenhouses. Comput. Electron. Agric. **169**, 105167 (2020)
35. Li, Z., Liu, X., Wu, S.: Dynamic emergency navigation based on prediction via wireless sensor networks. In: The 8th Conference on Information Technology: IoT and Smart City, pp. 210–215 (2020)
36. Lin, E., Park, J.D., Züfle, A.: Real-time Bayesian micro-analysis for metro traffic prediction. In: Proceedings of the 3rd ACM SIGSPATIAL Workshop on Smart Cities and Urban Analytics, pp. 1–4 (2017)
37. Liu, C., Li, K., Liu, J., Chen, C.: LHCnn: a novel efficient multivariate time series prediction framework utilizing convolutional neural networks. In: 2019 IEEE 21st International Conference on High Performance Computing and Communications; IEEE 17th International Conference on Smart City; IEEE 5th International Conference on Data Science and Systems (HPCC/SmartCity/DSS), pp. 2324–2332. IEEE (2019)
38. Xiufeng Liu and Per Sieverts Nielsen: Scalable prediction-based online anomaly detection for smart meter data. Inf. Syst. **77**, 34–47 (2018)
39. Luo, X.J., et al.: Development of an IoT-based big data platform for day-ahead prediction of building heating and cooling demands. Adv. Eng. Inform. **41**, 100926 (2019)
40. Macaš, M.: Variable selection for prediction of time series from smart city. In: 2015 Smart Cities Symposium Prague, pp. 1–5. IEEE (2015)
41. Mate, A., Peral, J., Ferrandez, A., Gil, D., Trujillo, J.: A hybrid integrated architecture for energy consumption prediction. Future Gener. Comput. Syst. **63**, 131–147 (2016)
42. Minor, B., Cook, D.J.: Forecasting occurrences of activities. Pervasive Mob. Comput. **38**, 77–91 (2017)
43. Nagy, A.M., Simon, V.: Improving traffic prediction using congestion propagation patterns in smart cities. Adv. Eng. Inform. **50**, 101343 (2021)

44. Al-Nuaimy, L.A.H.: Muscat securities market index (MSM30) prediction using single layer linear counterpropagation (SLLIC) neural network. In: 2016 3rd MEC International Conference on Big Data and Smart City (ICBDSC), pp. 1–5. IEEE (2016)

45. Rege, M.R., Handziski, V., Wolisz, A.: Generation of realistic cloud access times for mobile application testing using transfer learning. Comput. Commun. **172**, 196–215 (2021)

46. Salotti, J., Fenet, S., Billot, R., El Faouzi, N.-E., Solnon, C.: Comparison of traffic forecasting methods in urban and suburban context. In: 2018 IEEE 30th International Conference on Tools with Artificial Intelligence (ICTAI), pp. 846–853. IEEE (2018)

47. Shu, P., Sun, Y., Zhao, Y., Xu, G.: Spatial-temporal taxi demand prediction using LSTM-CNN. In: 2020 IEEE 16th International Conference on Automation Science and Engineering (CASE), pp. 1226–1230. IEEE (2020)

48. Tekouabou, S.C.K., Cherif, W., Silkan, H., et al.: Improving parking availability prediction in smart cities with IoT and ensemble-based model. J. King Saud Univ.-Comput. Inf. Sci. **34**(3), 687–697 (2020)

49. Vijai, P., Sivakumar, P.B.: Performance comparison of techniques for water demand forecasting. Procedia Comput. Sci. **143**, 258–266 (2018)

50. Violos, J., Tsanakas, S., Androutsopoulou, M., Palaiokrassas, G., Varvarigou, T.: Next position prediction using LSTM neural networks. In: 11th Hellenic Conference on Artificial Intelligence, pp. 232–240 (2020)

51. Wang, Z., Jia, H., Min, G., Zhao, Z., Wang, J.: Data-augmentation-based cellular traffic prediction in edge-computing-enabled smart city. IEEE Trans. Ind. Inf. **17**(6), 4179–4187 (2020)

52. Xu, R., Xiong, Q., Yi, H., Wu, C., Ye, J.: Research on water quality prediction based on SARIMA-LSTM: a case study of Beilun Estuary. In: 2019 IEEE 21st International Conference on High Performance Computing and Communications; IEEE 17th International Conference on Smart City; IEEE 5th International Conference on Data Science and Systems, pp. 2183–2188. IEEE (2019)

53. Yin, S., Wang, J., Cui, Z., Wang, Y.: Attention-enabled network-level traffic speed prediction. In: 2020 IEEE International Smart Cities Conference, pp. 1–8. IEEE (2020)

54. Xiaojie, Yu., Sun, L., Yan, Y., Liu, G.: A short-term traffic flow prediction method based on spatial-temporal correlation using edge computing. Comput. Electr. Eng. **93**, 107219 (2021)

55. Zang, T., Zhu, Y., Yanan, X., Jiadi, Yu.: Jointly modeling spatio-temporal dependencies and daily flow correlations for crowd flow prediction. ACM Trans. Knowl. Discov. Data (TKDD) **15**(4), 1–20 (2021)

56. Zhan, Y., Haddadi, H.: Activity prediction for mapping contextual-temporal dynamics. In: Proceedings of the 2019 ACM International Joint Conference on Pervasive and Ubiquitous Computing and of the 2019 ACM International Symposium on Wearable Computers, pp. 246–249 (2019)

57. Zhan, Y., Haddadi, H.: Towards automating smart homes: contextual and temporal dynamics of activity prediction. In: Adjunct Proceedings of the 2019 ACM International Joint Conference on Pervasive and Ubiquitous Computing and Proceedings of the 2019 ACM International Symposium on Wearable Computers, pp. 413–417 (2019)

58. Zhang, D., Woo, S.S.: Real time localized air quality monitoring and prediction through mobile and fixed IoT sensing network. IEEE Access **8**, 89584–89594 (2020)

59. Zheng, G., Chai, W.K., Katos, V., Walton, M.: A joint temporal-spatial ensemble model for short-term traffic prediction. Neurocomputing **457**, 26–39 (2021)

60. Zhou, J., Tung, A.K.H.: SMiLer: a semi-lazy time series prediction system for sensors. In: Proceedings of the 2015 ACM SIGMOD International Conference on Management of Data, pp. 1871–1886 (2015)

Protect Trajectory Privacy in Food Delivery with Differential Privacy and Multi-agent Reinforcement Learning

Suleiman Abahussein[(✉)], Tianqing Zhu, Dayong Ye, Zishuo Cheng, and Wanlei Zhou

School of Computer Science, University of Technology, Sydney, Australia
Suleiman.Abahussein@student.uts.edu.au,
{tianqing.zhu,dayong.ye,zishuo.cheng,Wanlei.Zhou}@uts.edu.au

Abstract. Today, multiple food delivery companies work globally in different regions, and this expansion could expose users' data to danger. This data could be stored by a third party and could be used in further analysis. The stored data needs to be stored in a proper way to prevent any other from identifying the real data if this data is disclosed. This work considers this issue to maintain the data privacy of stored customer data by leveraging differential privacy and multi-agent reinforcement learning. In the beginning, the agent delivers the food to the customer. Then the agent constructs N of obfuscated trajectories with different privacy budgets. The multi-agent reinforcement learning then chooses one trajectory from the constructed trajectories. The trajectory is then evaluated by considering three factors: the similarity between the selected trajectory and the original trajectory, the sensitivity of destination location and the frequency of the number of orders by the customer. We implemented our experiment on meal delivery data sets in Iowa City, USA.

Keywords: Multi-agent reinforcement learning · Privacy · Differential privacy · Trajectory

1 Introduction

Many companies today operate globally, resulting in several headquarters with numerous employees to operate their business in each region. Different access authorities are granted access to various data, which gives the possibility of illegal access to customer data. Moreover, this data could be stored and analysed by third parties, or the IT department could be operated by a third party as well.

A malicious entity could obtain this access authority, and the data could be used illegally. Some of the data could be related to important individuals or linked with sensitive places such as some government offices, military or security.

© The Author(s), under exclusive license to Springer Nature Switzerland AG 2023
L. Barolli (Ed.): AINA 2023, LNNS 655, pp. 48–59, 2023.
https://doi.org/10.1007/978-3-031-28694-0_5

Disclosing this information could reveal private and sensitive information such as user location, pattern or job location.

In this work, the method of Protect trajectory in food delivery (PTFD) has been proposed in order to protect privacy in such of this data. The Differential privacy Laplace mechanism and multi-agent reinforcement learning leveraged in this method to obfuscate the customer location and trajectory. In the beginning, the courier receives and then delivers the food order. After that, the agent constructs N obfuscated trajectories by injecting Laplace noise to a selected point of the original trajectory and destination point, the multi-agent reinforcement learning then selects one of the constructed trajectories. The selected trajectory is then evaluated based on three factors: the similarity between the selected trajectory and the original trajectory, the sensitivity of destination location and the frequency of the number of orders by the customer. The contribution of this work is as follows:

(a) This work is the first one which leverages the multi-agent reinforcement learning in trajectory privacy protection and we used QMIX method to train decentralised policies in a centralised end-to-end fashion.
(b) To have a strong obfuscated trajectory, we used the differential privacy Laplace mechanism to construct N number of obfuscated trajectories.
(c) To gain highly private results, we used the similarity of the selected trajectory with the original trajectory, the sensitivity of the destination location and the frequency of the customer orders as factors used in the reward function.

The rest of this article is organized as follows. The reviews of the related works presented in Sect. 2. The Background and Problem statement presented in Sect. 3 and Sect. 4. The methodology presented in Sect. 5. Sections 6, 7 and 8 show the experiment design, results and conclusion.

2 Related Work

The proposed method of protecting trajectory can be divided into two categories: protecting location privacy using anonymity or pseudonyms, and protecting location privacy using Differential privacy [2]. The following presents some of the research in this field.

There are many privacy location algorithms, based on anonymity and pseudonyms, which have been proposed in the past ten years. Khacheba et al. [7] proposed a context-based location privacy scheme (CLPS). The CLPS provides a strategy of pseudonym changing that makes a vehicle based on its context, change pseudonyms. In the user trajectory to prevent a malicious LBS, reconstructing Hwang et al. [6] proposed a trajectory privacy technique to cloak location information by combining ambient conditions and using user privacy profiles. In the beginning to blur the actual trajectory of a service user they preprocess a set of similar trajectories R by r-anonymity mechanism. They then combine s road segments with k-anonymity to protect the user's privacy. The query issue time sequence for a service user is broken by the technique of time-obfuscated to confuse the LBS.

Moreover, many privacy location algorithms based on differential privacy have been proposed. Min et al. [10] proposed a sensitive semantic location privacy protection scheme based on reinforcement learning. This scheme selects the perturbation policy based on the sensitivity of the semantic location and the attack history and uses the idea of differential privacy to randomize the released vehicle locations. A deep deterministic policy gradient based semantic location perturbation scheme (DSLP) is developed to solve the location protection problem with high-dimensional and continuous-valued perturbation policy variables. Chen et al. [2] consider vehicular ad hoc network vehicle trajectory protection. They proposed an optimized privacy differential privacy scheme with reinforcement learning in vehicular ad hoc network. To achieve a better balance between geolocation obfuscation and semantic security, the privacy budget allocation dynamically optimizes by the proposed schema for each location on the vehicle trajectory. The experiment results show that the proposed scheme can reduce the risk of geographical and semantic location leakage and ensure the balance between utility and privacy.

3   Background

3.1   Multi-agent Reinforcement Learning

Reinforcement learning is a popular solution for the problem of sequential decision-making. We move from problems concerning a single agent, to problems concerning multiple agents in multi-agent reinforcement learning (MARL). Multi-agent reinforcement learning is about the intersection of two concepts where the agents interact with other agents, and reinforcement learning is used to achieve their tasks [3]. We have multiple agents in this setting interacting in their environment, the main goal is to find a policy that maximizes its own reward [14].

3.2   Trajectory Protection

let $\mathcal{L} = \{L_1, L_2, \cdots, L_{|\mathcal{L}|}\}$ is universe of locations and $|\mathcal{L}|$ is the size of the universe, we consider locations as discrete spatial areas in a map. The trajectory modeled as an ordered list of locations drawn from the universe.

$Definition$ 3.1 (Trajectory): A trajectory T of length $|T|$ is an ordered list of locations $T = t_1 \rightarrow t_2 \rightarrow \cdots \rightarrow t_{|T|}, where \; \forall_1 \leqslant i \leqslant |T|, \; t_i \in \mathcal{L}$

A location may occur consecutively in T, and may occur multiple times in T. Thus, $\mathcal{L} = \{L_1, L_2, L_3, L_4\}$, T = L1 $\rightarrow$ L2 $\rightarrow$ L3 is a valid trajectory. The trajectory in some cases may include timestamps. The database of the trajectory is formed of a multiset of trajectories. Every trajectory shows the owner's record of movement history. A formal definition is as follows:

Definition 3.2 (Trajectory Database): A trajectory database D of size $|D|$ is a multiset of trajectories $D = \{D1, D2, \cdots, D_{|D|}\}$.

The user privacy may be obtained by the attacker through analysis of trajectory data. In this work we provide defense methods for some attack models existing in trajectory analysis based on differential private Laplace mechanism.

3.3 Differential Privacy

The data privacy threats recently increased, particularly with data mining and aggregation growth. Differential privacy is generally a mathematical model that guarantees a statistical dataset privacy [9]. A robust standard has been provided by Dwork et al. [4,5] to preserve privacy in data analysis. Without knowing anything about the trader's background knowledge, differential privacy is a common privacy model that provides privacy guarantees [15]. The following are Differential privacy, Sensitivity and Laplace Noise definitions.

Definition 1. A randomized mechanism $\mathcal{M} : \mathcal{D} \rightarrow \mu$ satisfies differential privacy (ϵ, δ) if for any two neighboring inputs d *and* d' and any subset of outputs $\mathcal{Z} \subseteq \mu$ it holds that

$$\mathbb{P}(\mathcal{M}(d) \in \mathcal{Z}) \leq exp(\epsilon)\mathbb{P}(\mathcal{M}(d') \in \mathcal{Z}) + \delta \qquad (1)$$

Definition 2 (Sensitivity) for neighboring inputs $(d, d' \in \mathcal{D})$, the mechanism $\mathcal{M}$ sensitivity f defines as:

$$\Delta f = \max_{D,D'} \|f(D) - f(D')\| \qquad (2)$$

Laplace Noise: The Laplace noise is drawn from the probability density function of the Laplace distribution. The Laplace noise is said to be $\epsilon-$ differential private if Theorem 1 is satisfied.

Theorem 1. For a given function $f : D \rightarrow R$ over a dataset $\mathcal{D}$, the mechanism $\mathcal{M}$ in Eq. (3) provides ϵ-differential privacy [1].

$$\mathcal{M}(D) = f(D) + Lap(\frac{\Delta f}{\epsilon}) \qquad (3)$$

4 Problem Statement

This section explains in detail the problem statement, overview of this model, its component, and how it works. Also, it explains the formulation Markov decision process in detail.

4.1 Model Overview

Today, multiple food delivery companies work globally in different countries, with different headquarters and numerous employees to operate their business in each region, and different access authorities are granted to access various data. This creates the opportunity for illegal access to customer data in different regions. Moreover, this data could be stored and analysed by third parties, or the IT department could be operated by a third party as well. Disclosing this information could reveal private or sensitive information such as user location, pattern or job location. Moreover, this data could be related to important individuals or linked with sensitive places such as some government offices, military or security sites. This work focuses on this issue and proposes a solution to maintain customer data privacy and not disclose their private information. We leverage differential privacy and multi-agent reinforcement learning in this work to maintain the customer's privacy.

We consider each courier as an independent agent. The courier receives and then delivers the food order to the customer. To obfuscate the customer location and the trajectory, our method creates N number of obfuscated trajectories by adding Laplace noise into a selected point of the trajectory using a different amount of privacy budget in each trajectory. The reinforcement learning then picks up one of the created trajectories and uses it instead of the original trajectory. The reinforcement learning then evaluates the selected trajectory in the reward function based on three criteria: the similarity between the original trajectory and the selected trajectory, the sensitivity of destination location, and the frequency of the number of orders by the customer.

4.2 DRL Formulation

Our method is formulated as the Markov decision process. The following presents more details about the Markov decision process formulation and the major components of the method.

Agent: In our problem, we used multi-agent reinforcement learning, and we considered each courier as an independent agent. We formed more than two agents $\{A_1, A_2, \cdots A_n\}$, to assign the learning policy of the agent, respectively.

The agent aims to save the obfuscated trajectory instead of the original one. The agent constructs N different obfuscated trajectories by using differential privacy and injecting Laplace noise to selected points into the trajectory. The agent then picks up one of the constructed trajectories.

State: The state in our dynamic model is defined based on three components: the selected trajectory, the frequency of the customer order and the sensitivity of destination location. The following are more details for each of them.

(a) The selected trajectory: Each time the courier receives a new delivery request and then delivers it to the customer, the agent at this point constructs N number of obfuscated trajectories. Then the agent selects one of the obfuscated trajectories. In our state, we use the chosen trajectory as a component of the state.

(b) The frequency of the customer order: An important factor that has been considered is the frequency of orders by the same customer. This factor shows the number of repeat orders by a particular customer. A high number of repeated orders could increase the probability of revealing the customer's information as the attacker can link the records and infer the customer's information.

(c) The sensitivity of destination location: The food delivery order could be delivered to a sensitive location such as some government offices, police or security offices. Thus, this factor considers to provide more data protection.

Action: The agent, in each time step t has N actions. The agent action is to select one of the constructed obfuscated trajectories.

Reward Function: After taking action, all agents receive an immediate reward from the environment to evaluate the collective actions. In each time step t an instant reward is received by the agent. Only if the agent is able to achieve the desired privacy preservation will the agent receive a positive reward, otherwise negative reward is received. The reward function is defined based on the following:

$$(SimTr \times 0.5) + (SenLoc \times 0.25) + (FreOr \times 0.25) \tag{4}$$

where $SimTr$ represents the similarity between the selected trajectory and the original trajectory. $SenLoc$ represents the sensitivity of destination location, and $FreOr$ represents the frequency of the number of orders by the customer.

The Similarity of the Selected Trajectory: We compared the similarity between the original trajectory and the chosen obfuscated trajectory using the Hausdorff Distance(HD). Hausdorff distance is broadly used to measure the similarity of two data sets of points of D with respect to data set $\tilde{D}$ [8]:

$$similarity(D) = (max\{h(D, \tilde{D}), h(\tilde{D}, D)\} \times 100) \tag{5}$$

The Sensitivity of Destination Location: Some online food orders could be delivered to sensitive locations such as government offices or security sites like police or national security centres. Disclosing such information could reveal some private information, like the customer's job. Thus, we consider this factor to provide more protection for customers' information. Haversine formula was used to measure $100\,m$ from the entered coordinate of each sensitive location entered. Each time, our method checks if the start point or the endpoint of the courier is sensitive or not.

Frequency of the Number of Orders by the Customer: An important factor that has been considered is the frequency of orders by the same customer. The high number of repeat orders by a particular customer could increase the chance of inferring the customer's information by linking the record. The frequency of orders by the customer can be calculated based on the following.

$$y = ((x - min(x))/(max(x) - min(x))) \tag{6}$$

where x is the number of frequent orders, $min(x)$ is the minimum frequent order in the data sets and $max(x)$ is the maximum frequent order in the data sets.

5   Methodology

5.1   QMIX Algorithm

Each agent in a multi-agent framework chooses an action, forming a collective action a_t, and shares a global immediate reward r_t that evaluates the collective action taken previously. There is collective agent-value function $Q_{tot}(s_t, a_t) = \mathbb{E}_{s_t+1: \infty, a_t+1: \infty}[R_t|s_t, a_t]$ for the collective action, as R_t is the discounted return at time t. Evaluating the contribution of each agent separately and accurately, to obtain individual value function from the collective action-value function is the main challenge in the MARL framework. To represent the individual value functions of agents A_i, it used $Q_i(o_t, a_t)$. QMIX is a novel value-based technique for training decentralized policies in a centralized end-to-end way [11].

5.2   Protect Trajectory Privacy in Food Delivery with Multi-agent Reinforcement Learning

This section explains how multi-ageing reinforcement learning and protect trajectory in food delivery (PTFD) work. Each time, the agent receives an order and then delivers it to the customer. The agent then performs the obfuscated method using Algorithm 2. The following shows more details on how both multi-ageing reinforcement learning and protect trajectory in food delivery (PTFD) algorithms work.

QMIX Multi-agent Reinforcement Learning: In the training process, we minimize the loss function by using gradient descent with respect to the parameter θ at iteration i. The neural network for this purpose is used as an approximator in reinforcement learning. Algorithm 1 shows the overview of the QMIX algorithm for Multi-agent reinforcement learning. Initially, the reply buffer $\mathcal{D}$ with size $\mathcal{N}$ is initialized. In line 2 to 4, initialize action value function with random weight θ and target action value function with $\bar{\theta} \leftarrow \theta$ and mixing network of the QMIX. The environment then is reset and while not in termination the agents observe the current observation and the action selected based on probability ϵ, or otherwise select the highest value from $argmax_{a_t}(o_t, a_t : \theta)$.

Algorithm 1. Overview of QMIX Multi agent reinforcement learning

---

1: **Initialize**: reply buffer $\mathcal{D}$ with size $\mathcal{N}$
2: **Initialize**: action value function with random weight θ
3: **Initialize**: target action value function with $\bar{\theta} \leftarrow \theta$
4: **Initialize**: mixing network ϕ
5: **while** Training episodes not complete **do**
6:     Reset environment
7:     **while** not in terminal state **do**
8:         **for** agent in Agents **do**
9:             Agent observe state o_i^t
10:             With probability ε select a random action a_t
11:             Otherwise select $argmax_{a_t}(o_t, a_t : \theta)$
12:         Execute joint action a_t and observe next state s_{t+1}, receive reward r_t and termination info
13:         Add (o_t, a_t, s_{t+1}, r_t) to reply buffer $\mathcal{D}$
14:         Sample random minibatch $\mathcal{K}$ from $\mathcal{D}$
15:         $Q_{tot} = mixing((Q_1(o_t^1, a_t^1) + Q_2(o_t^2, a_t^2) + ...Q_n(o_t^n, a_t^n))$
16:         Calculate loss function $\mathcal{L} = \left(r_t + \gamma max_{a_t} Q_{tot}(s_t, a_t | \bar{\theta}) - Q_{tot}(s_t, a_t | \theta)\right)^2$
17:         Perform a gradient descent step on $\mathcal{L}$ to the network parameter θ
18:         Update target network parameter $\bar{\theta}$
19:         Update exploration rate ε

Execute the action, receive corresponding rewards, and observe the next state. Add current observations, action, state and reward to reply buffer $\mathcal{D}$. Extra sample from reply buffer $\mathcal{D}$, calculate loss function $\mathcal{L}$. Perform a gradient descent step with network parameter θ and update target network parameter and then update exploration rate ε.

Protect Trajectory in Food Delivery (PTFD): Preserving privacy is crucial, this algorithm aims to maintain privacy in food delivery services. To improve our result, some vital factors are considered, such as the similarity between the selected trajectory and the original trajectory, the sensitivity of destination location, and the frequency of the number of orders by the customer. In Algorithm 2, after the agent receives the food delivery request, customer location coordinates are sent to Google maps API, which then fetches the trajectory as a dataset of coordinates. The courier then delivers the order to the destination. The agent uses the Laplace mechanism to construct N number of obfuscated trajectories with different privacy budgets by injecting noise into a selected point of the trajectory. Multi-agent reinforcement learning selects one of the constructed obfuscated trajectories. The multi-agent reinforcement learning reward function evaluates the selected obfuscated trajectory based on the similarity between the selected trajectory and the original trajectory, the sensitivity of destination location, and the frequency of the number of orders by the customer.

Algorithm 2. Protect trajectory in food delivery (PTFD)
___
1: Courier receives food delivery order
2: Courier receives trajectory dataset
3: Courier delivers the order
4: Construct N obfuscated trajectory by injecting Laplace noise into a selected point of the original trajectory
5: Check if the order destination is a sensitive location
6: Use multi-agent reinforcement learning to select trajectory
7: Used the selected obfuscated trajectory

---

6   Experiment Design

6.1   Protect Trajectory in Food Delivery (PTFD)

We ran the experiment for 1500 run times, with a list of the expected sensitive locations entered manually. Our experiment was implemented on Iowa City, USA datasets. The result was evaluated by computing the similarity between the original and selected trajectory. We used two similarity measure methods: Hausdorff distance and FastDTW.

Dynamic time warping (DTW) is a technique that finds an optimal alignment between two time series in which one time series may be warped. This alignment can be used to find corresponding regions or to determine the similarity between the two time series. FastDTW algorithm is used to find an accurate approximation of the optimal warp (DTW) path between two time series [12].

In our experiment, we also analyzed the distribution of privacy parameters resulting from choosing a trajectory each time. We used in this experiment the data from Iowa City datasets, USA.

6.2   Data Set Description

In this work, we used the meal delivery services dataset provided by Ulmer et al. [13] from Iowa City, USA. The dataset contains 1,200,391 records and 111 restaurants. The used locations in this data set are real. The orders are randomly generated with equal request probability for every location and every point of time.

7   Experiment Results

7.1   Trajectory Similarity and Data Utility

This section shows the similarity result between the selected and the original trajectory by using protect trajectory in food delivery method (PTFD). The result was conducted by using two measuring method: Hausdorff distance and FastDTW. Figure 1 shows the results for 100 samples after training the model for 1,500 run times. The result in Fig. 1a, shows the values of Hausdorff distance for

100 samples in blue and the average in orange colour. The values are approximate fluctuations between 0.01 and 0.1. When the value of Hausdorff distance is small, this means the similarity between the original trajectory and the obfuscated trajectory is high, and if the value is zero, this means both trajectories are identical. Figure 1b shows the result of using FastDTW in blue and the average in orange colour. It shows the values of FastDTW are approximate fluctuations between 0.02 to 0.2, and the smaller value means more similarity.

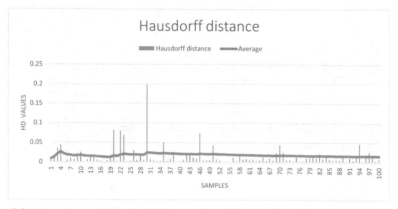

(a) The values of Hausdorff distance for 100 samples with average value

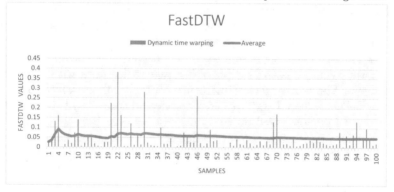

(b) The FastDTW values for 100 samples with average value

Fig. 1. The result for trajectory similarity between the original trajectory and selected trajectory using two measuring method: Hausdorff distance and FastDTW. (Color figure online)

Analyze Privacy Parameter. Each time the PTFD method selects one obfuscated trajectory, and each trajectory has a different privacy parameter. Figure 2 demonstrates the frequency of the used privacy parameter in 100 samples where the privacy parameter between 0.5 to 1.5 was used 42 times, the privacy parameter between 1.5 to 2.5 was used 20 times, and the privacy parameter between 2.5 to 3.5 was used 17 times, privacy parameter between 3.5 to 4.5 was used 21 times.

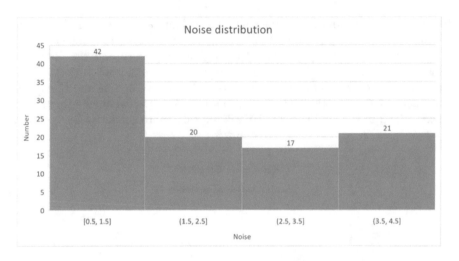

Fig. 2. Distribution of the used privacy parameter in 100 samples of the selected trajectories by (PTFD) method.

8   Conclusion

We consider the issue of maintaining the data privacy of saved customer information in food delivery services. We propose protect trajectory in food delivery (PTFD) method that uses differential privacy Laplace mechanism and multi-agent reinforcement learning to protect customer information privacy in food delivery services. In the beginning, the agent delivers the food to the customer, and then the agent constructs N of obfuscated trajectories with different privacy budgets. The multi-agent reinforcement learning then chooses one trajectory from the constructed trajectories. The selected trajectory is then evaluated in the reward function based on the similarity between the selected trajectory and the original trajectory, the sensitivity of the destination location, and the frequency of the number of orders by the customer.

References

1. Assam, R., Hassani, M., Seidl, T.: Differential private trajectory protection of moving objects. In: Proceedings of the 3rd ACM SIGSPATIAL International Workshop on GeoStreaming, pp. 68–77 (2012)
2. Chen, X., Zhang, T., Shen, S., Zhu, T., Xiong, P.: An optimized differential privacy scheme with reinforcement learning in VANET. Comput. Secur. **110**, 102446 (2021)
3. Cheng, Z., Ye, D., Zhu, T., Zhou, W., Yu, P.S., Zhu, C.: Multi-agent reinforcement learning via knowledge transfer with differentially private noise. Int. J. Intell. Syst. **37**(1), 799–828 (2022)
4. Dwork, C., Kenthapadi, K., McSherry, F., Mironov, I., Naor, M.: Our data, ourselves: privacy via distributed noise generation. In: Vaudenay, S. (ed.) EUROCRYPT 2006. LNCS, vol. 4004, pp. 486–503. Springer, Heidelberg (2006). https://doi.org/10.1007/11761679_29

5. Dwork, C., McSherry, F., Nissim, K., Smith, A.: Calibrating noise to sensitivity in private data analysis. In: Halevi, S., Rabin, T. (eds.) TCC 2006. LNCS, vol. 3876, pp. 265–284. Springer, Heidelberg (2006). https://doi.org/10.1007/11681878_14
6. Hwang, R.H., Hsueh, Y.L., Chung, H.W.: A novel time-obfuscated algorithm for trajectory privacy protection. IEEE Trans. Serv. Comput. **7**(2), 126–139 (2013)
7. Khacheba, I., Yagoubi, M.B., Lagraa, N., Lakas, A.: CLPS: context-based location privacy scheme for VANETs. Int. J. Ad Hoc Ubiquitous Comput. **29**(1–2), 141–159 (2018)
8. Li, M., Zhu, L., Zhang, Z., Xu, R.: Achieving differential privacy of trajectory data publishing in participatory sensing. Inf. Sci. **400**, 1–13 (2017)
9. Ma, P., Wang, Z., Zhang, L., Wang, R., Zou, X., Yang, T.: Differentially private reinforcement learning. In: Zhou, J., Luo, X., Shen, Q., Xu, Z. (eds.) ICICS 2019. LNCS, vol. 11999, pp. 668–683. Springer, Cham (2020). https://doi.org/10.1007/978-3-030-41579-2_39
10. Min, M., Wang, W., Xiao, L., Xiao, Y., Han, Z.: Reinforcement learning-based sensitive semantic location privacy protection for VANETs. China Commun. **18**(6), 244–260 (2021)
11. Rashid, T., Samvelyan, M., De Witt, C.S., Farquhar, G., Foerster, J., Whiteson, S.: Monotonic value function factorisation for deep multi-agent reinforcement learning. J. Mach. Learn. Res. **21**(1), 7234–7284 (2020)
12. Salvador, S., Chan, P.: Toward accurate dynamic time warping in linear time and space. Intell. Data Anal. **11**(5), 561–580 (2007)
13. Ulmer, M.W., Thomas, B.W., Campbell, A.M., Woyak, N.: The restaurant meal delivery problem: dynamic pickup and delivery with deadlines and random ready times. Transp. Sci. **55**(1), 75–100 (2021)
14. Vamvoudakis, K.G., Wan, Y., Lewis, F.L., Cansever, D.: Handbook of Reinforcement Learning and Control. Springer, Cham (2021). https://doi.org/10.1007/978-3-030-60990-0
15. Zhu, T., Li, G., Zhou, W., Philip, S.Y.: Differentially private data publishing and analysis: a survey. IEEE Trans. Knowl. Data Eng. **29**(8), 1619–1638 (2017)

Enhanced Machine Learning-Based SDN Controller Framework for Securing IoT Networks

Neder Karmous[(⊠)], Mohamed Ould-Elhassen Aoueileyine[(⊠)], Manel Abdelkader[(⊠)], and Neji Youssef[(⊠)]

InnovCOM Laboratory, SUPCOM, Carthage University, Kenosha, Tunisia
`{nader.karmous,mohamed.ouldelhassen,neji.youssef}@supcom.tn,`
`manel.abdelkader@gmail.com`

Abstract. The Internet of Things (IoT) ecosystem consists of interconnected devices that work together. It facilitates communication between devices, real-time data exchange and cloud computing. The occurrence risk of cyber attacks grows exponentially with these interconnected systems. Devices, networks and data could easily come under attack, wherefore they become vulnerable and could be compromised by hackers. To address this problem, we propose an enhanced Framework used on Software Defined Network (SDN) environment-based Intrusion Detection System (IDS) for securing IoT devices from malicious activity. We implement the machine learning (ML) method as part of the SDN controller's Network Intrusion Detection System (NIDS). Our enhanced ML-based SDN Controller Framework (Improved ML-SDN) classify the data traffic and makes a real-time prediction. It is based on K-Nearest Neighbor (kNN) supervised learning algorithm with others improving model Accuracy. It has produced an accuracy of 99.7%, 0.02304 s of building model time, 0.2997 s of detection time and a false alarm rate (FAR) of 0.34%.

Keywords: Internet of Things · Machine learning · Cybersecurity · Intrusion detection system · Artificial intelligence · And software defined networks

1 Introduction

In recent years, SDN [1] has broken the Traditional Network, especially in managing and maintaining the networks and become one of the most important technologies for handling networks such as IoT Networks. SDN is software based and depends on a physical separation between the control plane and the data plane [2], which can enhance the network's visibility, adaptability, and other local security operations. Moreover, it allows the customization of networks, a short start-up time and the deployment of the network with the right quality of service. ML is the most effective way to detect intrusion in IoT devices [3–5], especially when combined with software-based. Thanks to these combinations could create a network access control, configure, manage, monitor traffic on the SDN network, and analyze traffic to detect attacks. An ML-based SDN Controller [6]

L. Barolli (Ed.): AINA 2023, LNNS 655, pp. 60–69, 2023.
https://doi.org/10.1007/978-3-031-28694-0_6

is a Self-driving network that would learn to drive itself and decision-making automatically. In this context, this paper proposed an improved ML-SDN framework, a network
intrusion detection system (NIDS) [7] used for securing IoT Networks Using SDN and
ML. Our improved ML-SDN framework is based on the kNN methods with reduced
features using both cross-validation and gridsearch hyperparameter optimization. The
proposed work was trained on NSL-KDD [8], a data set created by researchers at the
University of New Brunswick Canadian Institute for Cyber- security. It solves some
of the inherent problems of the KDD'99 data [9] by removing redundant records from
the original KDD'99 dataset to obtain a more concise and efficient intrusion detection
dataset that gets more frequent records and eliminates biased classification results. NSL-
KDD includes a total of 39 attacks, each of which is classified into one of the following
four categories: Denial of Service (DoS) [10] is designed to make a network resource
inaccessible to its intended users. Probe [11] is an attack that tries to steal important
information from a network. User to Root(U2R) [12], in which an attacker attempts
to get unauthorized access to the targeted system. Remote to Local (R2L) [13] which
an attacker sends packets to a machine over a network that exploits their vulnerability
to gain access to an unknown user network. These datasets contain the internet traffic
records seen by a simple intrusion detection network. We used the training and testing
datasets, namely, KDDTrain + and KDDTest +. The number of records in the training
set is 125,973, and the testing set is 22,544. It generates 42 features per record, with 41
of the features referring to the traffic input itself, and the last one is labels (whether it is
normal traffic or an attack).

This paper is consisted of four sections. The first section reviews the work related
to our improved ML-SDN framework. The second section discusses the hardware and
software used to create our SDN topology in IoT networks, and we define the improved
ML-SDN framework method. Finally, we analyzed the results we obtained based on
Evaluation Metrics and compared these results with the optimal existing ML algorithms
based on the IDS and other related works.

2 Related Work

The author in [14] proposed NIDS architecture based on SDN. The proposed System
Architecture of SDN consists of three main layers: the application layer, which contains
the Rule management system and NIDS; the control layer contains the SDN controller;
and the infrastructure layer includes the switches, servers and attackers that may exist
in one of those switches. The author suggested a process to perform the attack detection
named XGBoost algorithm [15], which is a decision-tree (DT) [16] based ensemble ML
algorithm. The proposed process started with the data source, the NSL-KDD dataset, followed by data analysis and some pre-processing techniques for data cleaning. Then, feature normalization and feature selection were performed to implement the later selected
tree-based ML algorithm to classify whether there is an attack or not and the types of
attack. The presented XGBoost algorithm with the selected five features achieved an
accuracy score of 95.55%. This work's shortcomings need work with other classification algorithms, other methods and approaches to get more accuracy, lower consumption
of time and resources, and focus on the probability of FAR, which is essential to identify
real alarms.

The author [17] compared ML algorithms over SDN to detect intrusion attacks. In his work, he selected the NSL-KDD dataset from training and testing with 41 features and 125000 samples and 10 Kfold cross-validations. Although he used different feature classification approaches, he selected only seven features for testing data. He tested the different ML classification algorithms, which are Random Forest RF [18], J48 [19], Bayesian Networks (BN) [20], DT, Naive Bayes (NB) [21] and Radial Basis Function Networks (RBFN) [22]. The author found that Random Forest (RF) is the optimal algorithm compared to other ML algorithms based on its high accuracy and low FAR, with a very low FAR of 0.28% and an accuracy of 81.94%. The limitation of this work is the need for working on other classification ML algorithms, using other feature selection methods that reduce features without decreasing accuracy.

The author in [23] recommended a Network Intrusion Detection System based on Deep Learning methods (NIDS-DL) to detect a specific type of attack, such as (DOS) attacks. Also, he applied this approach inside the SDN environment to an NSL-KDD dataset, and his suggested method combines Network Intrusion Detection Systems (NIDS) with many types of deep learning algorithms. He applied feature selection methods to train data on high correlation features; he selected 12 features extracted from 41 features. He employed classifiers Convolutional Neural Network (CNN) [24], Deep Neural Networks DNN [25], Recurrent Neural Network RNN [26], Long Short-Term Memory (LSTM) [27], and Gated Recurrent Unit (GRU) [28]). CNN produced accuracy results of 98.63%. The author made pre-processing dataset to obtain the best results. This work needs to decrease training time by using classification ML algorithms which give less training time and interest in FAR, which is a principal key limiting factor to measure the effectiveness of the NIDS.

3   Methodology

3.1   Software and Hardware Environment

Software. We have used python 3.8 and Jupyter notebooks for software, and we installed mininet 2.3.0 [29] in Ubuntu Virtual Machine (VM) to create our topology. The topology includes one OpenFlow kernel switch connected with hosts and the OpenFlow reference controller, the Ryu controller [30]. The last one is also installed in Ubuntu VM on version 1.2. It is a Python-based Open Flow controller with a robust API that enables developers to create their applications to manage the network with Python language. It can be configured as a traffic monitor, a firewall, or a switch. Therefore, we can write our topology and our improved ML-SDN algorithms directly.

Hardware. For hardware we used a machine that includes an Intel i3-115G4 processor (3.4 GHz) with 32 GB RAM, a 64- bit Ubuntu 14.04 operating system, and an Nvidia GeForce GTX 1080 graphics card.

Installation of Ryu controller and mininet [31] in two VM Ubuntu Servers help us to create our topology.

3.2 SDN Architecture

As Fig. 1 shows, our architecture comprises two layers: infrastructure and control. The infrastructure layer is composed of the physical switch in the IoT network, which forwards traffic based on the instructions from the control layer. We have one switch (S1), IoT hosts, a Ryu controller and an ML-based SDN Ryu Controller. All hosts are connected to the switch (S1). The switch and Ryu controller communicate by OpenFlow protocol [32]. We run the Ryu controller using the switching application. In live traffic, the switch sends the packet information to the ML classifier installed in the Ryu controller. It detects whether this is an attack using our improved ML-SDN model. If it is legal traffic, the controller looks at the destination Mac of the packet and decides on the output port, adding a new rule to the infrastructure layer to allow the traffic. If it is malicious traffic, it sends the rule to the infrastructure layer to block packets (send a flow entry to drop a packet).

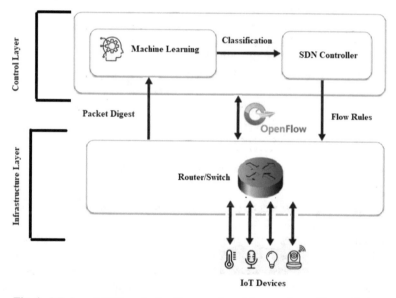

Fig. 1. ML based SDN controller Framework architecture for traffic prediction.

3.3 Proposed ML-SDN Framework

Our proposed ML-SDN Framework is shown in Fig. 2:

Our improved ML-SDN is composed of six steps. The description of the steps is shown below:

Step 1: NSL-KDD Dataset Select. We Used KDDTrain + datasets for training and KDDTest + datasets for testing.

Step 2: Categorical Encoding and Scaling. Before fitting and evaluating the model, ML models require all input and output variables to be numeric. Therefore, we used the

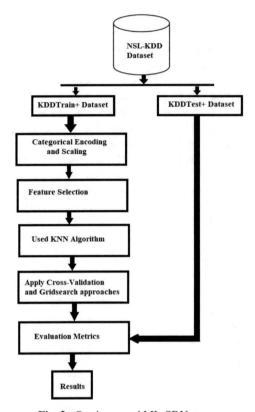

Fig. 2. Our improved ML-SDN steps.

OneHotEncoder function to encode categorical columns and the StandardScaler function on numerical columns.

Step 3: Feature Selection. Based on the Principal Component Analysis (PCA) method [33], we dropped the features with a more than 95% correlation. The features were reduced from 42 inputs to 38.The dropped features are: num_root,srv_serror_rate,srv_rerror_rate,dst_host_serror_rate,dst_host_srv_serror_rate and dst_host_srv_rerror_rate.

Step 4: Used K-Nearest Neighbour. In our work, we choose kNN [34], which is considered one of the most powerful ML algorithms to detect IDS [35–37]. It is a supervised learning algorithm used to solve classification problems, and their equation is written by:

$$D(x, y) = \sum_{i=1}^{n} \sqrt{(x_i - y_i)^2} \tag{1}$$

where is the ith featured element of the instance x, is the ith featured element of the instance y, n is the total number of features.and D(x,y) is The Euclidean distance between the two instances x and y.

Step 5: Apply Cross-Validation and Gridsearch Approaches. For hyperparameters, we used Stratified K-fold cross-validation [38] and Gridsearch [39]. The Stratified 5-fold cross-validation helps us avoid overfitting and is used to select the best model. The Gridsearch approach facilitates searching the best parameter and selects it from the list of parameters by a grid of parameters; in our work, we define k between 1–30 kNN model, and the best score given by grid-search is $k = 1$.

Step 6: Evaluation Metrics and results. This step is explained below in the fourth section of this paper.

4 Simulation Results

4.1 Evaluation Fundamental

Evaluation metrics were used to evaluate the effectiveness of ML methods on our improved ML-SDN model. It is listed and defined below:

Fit time. It denotes the time of creating a model based on the training dataset.

Predict time. It denotes the number of seconds to predict a single sample of data.

Confusion matrix. As shown in Table 1, the Confusion matrix is an N x N matrix used for evaluating the performance of a classification model, where:

N: It is the number of target classes
TP: Rule matched, and an attack is present
FP: Rule matched, and no attack present
TN: No rule matched, and no attack present
FN: No rule matched, and an attack is present.

Table 1. Confusion matrix.

	Predicted +	Predicted-
Actual +	True positive (TP)	False Negative (FN)
Actual -	False positive (FP)	True Negative (TN)

Accuracy. It is the ratio of correctly classified instances to the total number of instances. The accuracy equation is defined as:

$$\text{Accuracy} = \frac{TP + TN}{TP + FN + FP + TN} \tag{2}$$

Precision. It is the fraction of correctly predicted Attacks to all the samples predicted as Attacks. The precision equation is written by:

$$\text{Precision} = \frac{TP}{TP + FP} \tag{3}$$

Recall. It is a fraction of all samples correctly classified as Attacks to all the samples that are actual attacks. Recall equation is written by:

$$Recall = \frac{TP}{TP + FN} \tag{4}$$

F1 Score. The F1 score is the harmonic mean of precision and recall. The equation is written by:

$$F1\ Score = \frac{2 * precision * recall}{precision + recall} \tag{5}$$

AUC-ROC. The (Area Under the Receiver Operating Characteristics) is one of the most crucial evaluation metrics for checking any classification model's performance.

FAR. It is also called the false positive rate (FPR), which is the ratio of the total number of false positives divided by the sum of false positives and true negatives. The equation is written by:

$$FAR = \frac{FP}{FP + TN} \tag{6}$$

4.2 Results

We compare our improved ML-SDN method to most of the popular ML algorithms used for intrusion detection [40–42], which are RF, kNN and DT with their default parameter.

Table 2. Shows that all algorithms have almost the same prediction time. RF gives the best accuracy result compared to other algorithms and a good rate of FAR. However, it does not have a good result in terms of the fit time. DT has an adequate fit time and good accuracy result, especially in the recall, but it has not a good result in the FAR. On the contrary, kNN has acceptable results in precision, the best fit time and FAR. Finally, our Improved ML-SDN gives the best result compared to all other algorithms, especially in precision with 0.997 and has a fit time of 0.002304 s, prediction time of 0.29976 and FAR of 0.0034.

Table 2. Performance comparison of our improved ML-SDN model and others ML algorithms.

Model Name	Fit time	Predict time	Accuracy	Precision	Recall	F1-Score	ROC-AUC	FAR
RF	4.116297	0.299535	0.9931	0.9907	0.9945	0.9926	0.9932	0.012307
DT	0.313964	0.300483	0.9926	0.9898	0.9942	0.9920	0.9927	0.804040
kNN	0.050673	0.302168	0.9846	0.9927	0.9739	0.9832	0.9838	0.011953
Improved ML-SDN	**0.02304**	**0.29976**	**0.9968**	**0.9970**	**0.9970**	**0.9970**	**0.9966**	**0.0034**

Table 3 shows that in [23], the author used a deep learning method with 12 features extracted. His work gives 98.63% of accuracy, and it is considered the best result compared to [14] and [17]. The author in [17] used feature selection. He selected only seven features from the dataset and used a ten kfold cross-validation hyperparameter. He did not reach a good result on accuracy compared to [14] and [23]. However, he got a good result on FAR compared to our work with 0.28%. The author in [14] used only five feature selections. He gets 95% accuracy, which is a good result compared to [17]. Our improved ML-SDN used three methods: feature selection, cross-validation hyperparameter and grid search approach. It gives the best accuracy result compared to all other work and has a good FAR that is very close to [17].

Table 3. Comparison between related works and our improved ML-SDN model.

Model Name	Accuracy	FAR	Used methods
Improved ML-SDN	99.70%	0.34%	– Feature selection - 5-fold cross-validation hyperparameter – Gridsearch approach
[14]	95.95%	–	– Feature selection
[17]	81.94%	0.28%	– Feature selection -10- fold cross-validation
[23]	98.63%	–	– Feature extraction - Deep Learning approach

5 Conclusion and Future Works

This paper presents an improved ML-SDN framework to protect IoT devices from attacks over SDN. The novelty of our new approach (improved ML-SDN) is the mix-up of the feature selection with gridsearch hyperparameters to pick the best results. The feature selection reduces data, helps us train our model faster, and gets a good fit time. Cross-validation helps us to evade overfitting and to select the most appropriate model. The gridsearch helps us to find the optimal k parameter in order to get an improved accuracy. The simulation results show that our model could detect the malicious with an accuracy of 99.7%, a 0.34% FAR, taking 0.02304 s to fit the model and 0.29976 s to make a prediction. The future work is to create our datasets for IoT devices based on our topology, working on multiclass classification methods to detect the types of attacks and trying other methods to improve latency time and accuracy and minimize the FAR.

References

1. Liatifis, A., Sarigiannidis, P., Argyriou, V., Lagkas, T.: Advancing SDN: from OpenFlow to P4, a survey. ACM Computing Surveys (CSUR) (2022)
2. Manguri, K.H., Omer, S.M.: SDN for IoT environment: a survey and research challenges. In: ITM Web of Conferences, vol. 42, p. 01005. EDP Sciences (2022)

3. Karmous, N., Aoueileyine, M.-E., Abdelkader, M., Youssef, N.: A proposed intrusion detection method based on machine learning used for internet of things systems. In: Barolli, L., Hussain, F., Enokido, T. (eds.) Advanced Information Networking and Applications. LNNS, vol. 451, pp. 33–45. Springer, Cham (2022). https://doi.org/10.1007/978-3-030-99619-2_4
4. Gad, A.R., Nashat, A.A., Barkat, T.M.: Intrusion detection system using machine learning for vehicular ad hoc networks based on ToN-IoT dataset. IEEE Access **9**, 142206–142217 (2021)
5. Malhotra, P., et al.: Internet of things: evolution,concerns and security challenges. Sensors. **21**(5), 1809 (2021)
6. Dietz, K., Gray, N., Seufert, M., Hossfeld, T.: ML-based performance prediction of SDN using simulated data from real and syntheticnetworks. In: NOMS 2022–2022 IEEE/IFIP Network Operations and Management Symposium, pp. 1–7. IEEE, April 2022
7. Kim, T., Pak, W.: Robust network intrusion detection system based on machine-learning with early classification. IEEE Access **10**, 10754–10767 (2022)
8. NSL-KDD | Datasets | Research | Canadian Institute for Cybersecurity | UNB, 2017. http://www.unb.ca/cic/datasets/nsl.html
9. Tavallaee, M., Bagheri, E., Lu, W., Ghorbani, A.: A detailed analysis of the KDD CUP 99 data set. In: Second IEEE Symposium on Computational Intelligence for Security and Defense Applications (CISDA) (2009)
10. What is a DDoS Attack? - DDoS Meaning. usa.kaspersky.com. 2021–01–13. Accessed 05 Sep 2021
11. Ambedkar, C., Kishore Babu, V.: Detection of probe attacks using machine learning techniques. Int. J. Res. Stud. Comput. Sci. Eng. (IJRSCSE) **2**(3), 25–29 (2015)
12. Revathi, S., Malathi, A.: Detecting user-to-root (U2R) attacks based on various machine learning techniques. Int. J. Adv. Res. Comput. Commun. Eng . **3**(4), 6322–6324 (2014)
13. Ahmad, I.,Abdullah, A.B., Alghamdi, A.S.: Remote to local attack detection using supervised neural network. In: 2010 International Conference for Internet Technology and Secured Transactions. IEEE (2010)
14. Alzahrani, A.O., Alenazi, M.J.F.: Designing a network intrusion detection system based on machine learning for software defined networks. Future Internet. **13**(5), 111 (2021). https://doi.org/10.3390/fi13050111
15. Mitchell, R., Frank, E.: Accelerating the XGBoost algorithm using GPU computing. PeerJ. Comput. Sci. **3**, e127 (2017)
16. Charbuty, B., Abdulazeez, A.: Classification based on decision tree algorithm for machine learning. J. Appl. Sci. Technol. Trends **2**(01), 20–28 (2021)
17. Batra, R., Mahajan, M., Goel, A.: Implementation of SDN-Based Feature Selection Approaches on NSL-KDD Dataset for Anomaly Detection
18. Denisko, D., Hoffman, M.M.: Classification and interaction in random forests. Proc. Natl. Acad. Sci. USA. **115**(8), 1690–1692 (2018). https://doi.org/10.1073/pnas.180025 6115. PMC 5828645. PMID 29440440
19. Bhargava, N., et al.: Decision tree analysis on j48 algorithm for data mining. In: Proceedings of International Journal of Advanced Research in Computer Science and Software Engineering, vol. 3.6 (2013)
20. Heckerman, D.: A Tutorial on Learning with Bayesian Networks. Innovations Bayesian Network, pp. 33–82(2008)
21. Wickramasinghe, I., Kalutarage, H.: Naive Bayes: applications, variations and vulnerabilities: a review of literature with code snippets for implementation. Soft. Comput. **25**(3), 2277–2293 (2020). https://doi.org/10.1007/s00500-020-05297-6
22. Chenou, J., Hsieh, G., Fields, T.: Radial basis function network: its robustness and ability to mitigate adversarial examples." In: 2019 International Conference on Computational Science and Computational Intelligence (CSCI). IEEE (2019)

23. Hadi, M.R., Mohammed, A.S.: A novel approach to network intrusion detection system using deep learning for SDN: Futuristic approach." arXiv preprint arXiv:2208.02094 (2022)

24. Khan, A., Chase, C.: Detecting attacks on IoT devices using featureless 1D-CNN. In: 2021 IEEE International Conference on Cyber Security and Resilience(CSR). IEEE (2021)

25. Swarnalatha, G.: Detect and classify the unpredictable cyber-attacks by using DNN model. Turkish J. Comput. Math. Educ. (TURCOMAT) **12**(6), 74–81 (2021)

26. Park, S.H., Hyun, J.P., Young-June, C.: RNN-based prediction for network intrusion detection. In: 2020 International Conference on Artificial Intelligence in Information and Communication (ICAIIC). IEEE (2020)

27. Hochreiter, S., Schmidhuber, J.: 'Long short-term memory.' Neural Comput. **9**(8), 1735–1780 (1997)

28. Kasongo, S.M., Sun, Y.: A deep gated recurrent unit based model for wireless intrusion detection system." ICT Express. **7**(1), 81–87 (2021)

29. Sood, M.: SDN and mininet: some basic concepts. Int. J. Adv. Netw. Appl. **7**(2), 2690 (2015)

30. Asadollahi, S., Goswami, B., Sameer, M.: Ryu controller's scalability experiment on software-defined networks. In: 2018 IEEE International Conference on Current Trends in Advanced Computing (ICCTAC). IEEE (2018)

31. Gupta, N., Maashi, M.S., Tanwar, S., Badotra, S., Aljebreen, M., Bharany, S.: A comparative study of software defined networking controllers using Mininet. Electronics **11**, 2715 (2022). https://doi.org/10.3390/electronics11172715

32. Shang, Z.: Performance Evaluation of the Control Plane in OpenFlow Networks.Freie Universitaet Berlin, Germany (2019)

33. Kurita, T.: Principal component analysis (PCA). In: Computer Vision: A Reference Guide, pp. 1–4 (2019)

34. Alfarshouti, A.M., Almutairi, S.M.: An intrusion detection system in IoT environment using KNN and SVM classifiers. Webology **19**(1), 3500–3517 (2022). https://doi.org/10.14704/WEB/V19I1/WEB19231

35. Salih, A.A., Abdulazeez, A.M.: Evaluation of classification algorithms for intrusion detection system: a review. J. Soft Comput. Data Min. **02**(01), 31–40 (2021). https://doi.org/10.30880/jscdm.2021.02.01.004

36. Syamsuddin, I., Barukab, O.M.: SUKRY: Suricata IDS with enhanced kNN algorithm on raspberry Pi for classifying IoT botnet attacks. Electron. **11**(5), 737 (2022)

37. Alhammadi, M., Ali, N.: Comparative study between (SVM) and (KNN) classifiers by using (PCA) to improve intrusion detection system. Iraqi J. Intell. Comput. Inform. (IJICI). **1**(1), 22–33 (2022)

38. Rhohim, A., Vera, S., Muhammad Arief, N.: Denial of service traffic validation using K-fold cross-validation on software-defined network. eProc. Eng. **8**(5), 1–10 (2021)

39. Godalle, E.: How to find optimal parameters using GridSearchCV in ML in python, 1 January 2023. https://www.projectpro.io/recipes/find-optimal-parameters-using-gridsearchcv

40. Kilincer, I.F., Ertam, F., Sengur, A.: Machine learning methods for cyber security intrusion detection: datasets and comparative study. Comput. Netw. **188**, 107840 (2021). https://doi.org/10.1016/j.comnet.2021.107840

41. Sarker, I.H.: CyberLearning: effectiveness analysis of machine learning security modelling to detect cyber-anomalies and multi-attacks. Internet of Things. **14**, 100393 (2021). https://doi.org/10.1016/j.iot.2021.100393

42. Karmous, N., Aoueileyine, M.O.-E., Abdelkader, M., Youssef, N.: IoT real-time attacks classification framework using machine learning. In: 2022 IEEE Ninth International Conference on Communications and Networking (ComNet), pp. 1–5 (2022).https://doi.org/10.1109/ComNet55492.2022.9998441

Sustainment of Military Operations by 5G and Cloud/Edge Technologies

Souradip Saha[1]([✉]), Warren Low[2], and Beniamino Di Martino[3]

[1] Communication Systems, Fraunhofer FKIE, Wachtberg, Germany
`souradip.saha@fkie.fraunhofer.de`
[2] Capabilities Division, Allied Command Transformation, Norfolk, USA
`warren.low@act.nato.int`
[3] Engineering Department, University of Campania "Luigi Vanvitelli", Aversa, Italy
`beniamino.dimartino@unicampania.it`

Abstract. We present military use cases that have the potential of supporting Sustainment functions through the use of 5G enabling technologies, especially edge computing, that have been developed to support civilian use cases. Specifically personnel training enhanced by augmented reality/virtual reality (AR/VR), healthcare, and logistics use cases, referred together as Sustainment functions, have analogous military utilities especially in permissive electromagnetic (EM) threat environments. The unique nature of military applications presents challenges similar in some respects, but additive in others for areas such as privacy, security, network architecture, and quality of services. The military need for potential operations together with forces from other nations in case of allied joint operations will necessitate enhanced interoperability. In this paper, we aim to address these aspects from the perspective of 5G and trends in future technologies.

Keywords: Sustainment · 5G · AR/VR · Logistics · Healthcare · MEC · Interoperability · Security

1 Introduction

5G is a congregation of a wide range of technologies that have mainly/initially been developed around verticals, for e.g., vehicular communications, entertainment, Industry 4.0, healthcare etc. for which a shift towards higher level of automation/digitization is envisioned, with improvement of wireless network design towards maximizing the overall throughput and coverage while maintaining low latency. Within the foreseeable future, 5G technologies would be a key mode of wireless connectivity for electronic systems in almost all aspects of life.

In this paper, we describe the viability of 5G and edge computing along with other contemporary technologies which can improve military communications-information systems (CIS) to support Sustainment function that includes comprehensive training, logistics, medical, and military engineering. To limit the scope of this vast research area, we focus just on a subset of Sustainment military

L. Barolli (Ed.): AINA 2023, LNNS 655, pp. 70–79, 2023.
https://doi.org/10.1007/978-3-031-28694-0_7

use cases. The rest of the paper is organized as follows. Sections 2, 3 and 4 provide use case descriptions of personnel training enhanced by AR/VR, healthcare and logistics respectively. In Sect. 5, how some of the key 5G enabling technologies can support the aforementioned use cases are discussed. The corresponding challenges associated are provided in Sect. 6 along with interoperability issues for joint force operations in Sect. 7. Section 8 briefly details the future trends and finally, Sect. 9 provides the concluding remarks.

2 Use Case Scenario 1: Personnel Training Enhanced by AR/VR

Military Training programs ensure cohesion, effectiveness and readiness of forces. Nowadays, military training is provided in traditional in-person classrooms and on-line platforms. The COVID-19 Pandemic have negatively impacted organizations worldwide to engage in face to face activities. Organizations have mitigated the impact of the Pandemic on business continuity through the enhancing online presence through the use of AR/VR technologies. The application of VR to replace academic classroom interaction and lab based courses achieved positive outcomes and lessons about proper preparation of the technology and courseware before going live with courses [1–3]. AR was described to support a number of Industry 4.0 applications within a shipyard in the areas of maintenance and control tasks [4]. Today, the U.S. Marine Corps uses the Indoor Simulated Marksmanship Training for virtual marksmanship training. More novel command and control enhancements have been proposed through AR/VR solutions to provide resilient multi-domain C2 from the perspective of, for e.g., the Air Commander in [5]. As the commercial rollout of 5G services and specifically Business-to-Business (B2B) services accelerate, 5G together with AR/VR have the potential greatly enhance the effectiveness and resilience of military education, training, operations and maintenance.

A deployable HQ or mechanized unit may typically employ a platoon size unit to operate and maintain a variety of complicated equipment. These maintenance personnel normally receive training in the classical classroom setting with reading materials and physical devices. Once deployed in their units, the military technicians will be responsible to maintain sophisticated military hardware, operating systems, and applications software supporting operational personnel. The sophisticated equipment will go through periodic technology upgrades requiring additional training. Military personnel technical training delivered by AR/VR and 5G technologies would provide military benefit, especially in garrison, as follows and illustrated in Fig. 1,

1. *Virtual Instruction.* VR courseware could be used as complementary or replacement instruction to military personnel on complex equipment installation, maintenance or update at their duty location. Courseware could be updated at the same time as equipment updates across all locations where equipment resides. The courseware would be integrated with learning management system to serve as a post instruction reference, track student progress and identify areas of improvement.

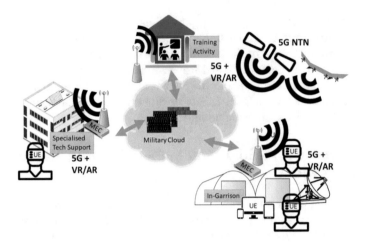

Fig. 1. 5G and VR enabled training and remote support.

2. *Collaborative Maintenance.* Often times, the installation and troubleshooting of networks involve addressing actions at multiple dispersed locations. The use of AR head mounted displays, tablet devices or even mobile phones by maintenance personnel at multiple locations would allow for sharing views and collaboratively troubleshooting in real time reducing operational down time.

3. *Hazardous Environments.* Military operating environment are inherently dangerous. 5G enabled devices, using mobility features, together with autonomous vehicles could be sent into hazardous areas to provide assessment, guide and even implement remediation actions.

4. *Expert Assistance.* Now a days, there are basic level of maintenance that can be performed by military personnel outside of depot locations. AR could serve as the over the shoulder "eyes" of remote expert technicians whom can provide guidance through virtual overlays on the physical equipment. Garrison military personnel would follow the instructions and can greatly reduce equipment downtime not waiting for equipment to be sent to depot, repaired and returned. Also, expert personnel will not require travel to the garrison locations to effect repairs as they can conduct virtual troubleshooting and provide detailed instructions to the local technicians.

3   Use Case Scenario 2: Healthcare

Any military operation, either battlefield operations or humanitarian ones, requires an extensive level of medical support. This includes a wide range of personnel, resources and equipment. With the advent of 5G technologies and standards enabling more efficient CIS, a lot of healthcare facilities previously deemed impractical can be enabled. Tele-mentoring/consultation/surgery can be added to existing healthcare infrastructure due to 5G features like high data rates, low latency, multi-connectivity, mobility and ubiquitous network access. 5G can also enhance the opportunity for remote training with visual and tactile communication as described in Sect. 2. This is especially beneficial for tactical units where specialists may not be physically present. Some of the key tele-medicine sub-scenarios are as follows:

1. *Rehabilitation and Tele-consultation.* AR/VR is critical to provide virtual training for physical rehabilitation.Psychological analysis and corresponding help can be provided to a unit who has just undergone trauma from an active conflict operation. Availability of such consultation in a timely manner can help units cope with ailments such as, Post-Traumatic Stress Disorder (PTSD).
2. *Surgery.* With the lack of medical/surgical experts close to conflict zones or Joint Operations Area (JOA), the units can be provided instructions and training from remote specialists to perform complex medical procedures required on an urgent basis. Remote operation of electro-mechanical arms capable of performing such procedures, can be safely and accurately executed. Ultra-responsive network of information transfer between the surgeon and patient forms a communication loop for remote surgery [7].

A doctor-centric diagnosis and treatment model is not optimal for military personnel especially when involved in JOAs. Whereas, the self-determination medical model can significantly enhance the patient's autonomy subject participation throughout the entire treatment system because the diagnosis and treatment plan are timely, dynamic, and interactive, allowing for individual status feedback with regards to lifestyle elements, behavioral factors, and treatment effect, to help patients become more autonomous from clinical services [8]. 5G is seen to immensely enhance data processing services for patient monitoring, promotion of the rational allocation of quality medical resources, and efficient management of massive wearable and monitoring devices. A routine data-driven approach allows assessment of a large data set of medical records to for improvement through reflections on past and evidence-based healthcare decisions [6].

The outbreak of a pandemic/epidemic amongst troops deployed in remote or uninhabited geographic areas, where the existence of highly infectious strains of viruses or bacteria is not a far-fetched possibility and the lessons we have learnt from the COVID-19 pandemic could be applied to minimize non-combat casualties. Limited access to conventional medical infrastructure makes monitoring of troops' health conditions all the more necessary.

Recent advances in microelectronics and nano-fabrication expanded the scope of sensors from just wearables to implants, ingestibles or epidermal devices, which transmit health-related information from inside the body. Such sensors have enhanced capabilities to capture vital health-related signs along with critically accelerated detection of failing implants, thereby minimizing healthcare hazard [9]. Such health monitoring services can also assess psychological health by analysing behavioral patterns to measure stress and trauma caused by involvement in conflicts. This could enable a timely rotation of personnel to avoid physical/psychological exhaustion amongst soldiers.

4   Use Case Scenario 3: Logistics

The Fourth Industrial Revolution or 'Industry 4.0' is the terminology used to describe the advancement in manufacturing industry with aid of automation for efficient and quick production and supply of goods. Many of the novel 5G technologies can enable agile and flexible supply chains, with reduced management, energy and storage costs [10]. This aids supply chains to pivot from a dispersed series of independently managed location networks to an increasingly connected ecosystem of devices that share knowledge across the network in (pseudo) real-time [11].

In retails logistics, manufacturing and retail logistics systems are relatively independent units, arranged sequentially, where the supply chain is open ended. However, with the advent of Industry 4.0, these supply chain units can be more integrated and inter-dependent, adaptable to fast changing parameters to mitigate some of the supply chain issues. This would transition retail logistics to Smart Logistics. The difference in their visualizations is provided in Fig. 2.

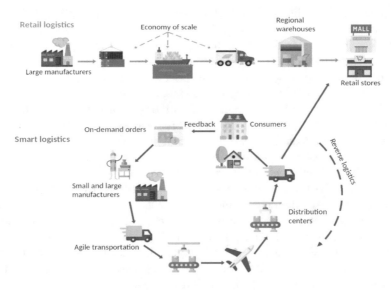

Fig. 2. Layout of Retail and Smart Logistics [10].

Depending on the military operation, the mobilization of an enormous amount of resources might need to be undertaken. Manufacturing military equipment, transporting and storing them at military warehouses or distribution centers and requesting re-supply in due time, all form a logistical supply chain which is the backbone for sustainment of any military operation. An infrastructure of Smart Logistics would be highly beneficial and crucial for optimal resource management. Along with the flow of products, in Smart Logistics, there is a flow of data about operations in logistics hubs, supply of products to operational units and corresponding feedback for re-supply. This results in constant generation and processing of a massive amount of data from multiple entities within the supply chain that located very far apart from each other.

Some key features of Smart Logistics, include,

1. Asset tracking. SmartTags based on barcodes (e.g. QR codes) or electronic tags (e.g. RFID tags), depending on cost efficiency and requirement, can be used to track equipment at logistics hubs as well as during transport. Tracking the status of devices, while in use, can indicate the need of repairs/replacements/updates, thus effectively optimizing the device lifecycle.
2. Automated Guided Vehicles (AGV) and drones play a major role in handling objects within a manufacturing or distribution center, particularly those deemed hazardous for human handling, e.g. munitions. They not only aid in minimizing casualties but also in faster accomplishment of the tasks.
3. Remote assistance for employees through AR/VR support aids in the execution of procedural tasks in design, maintenance and repair. Additionally, it can support in training inexperienced military personnel to handle heavy machinery without the risk of accidents.
4. Video surveillance of various supply chain units to maintain necessary synchronization and avoid bottlenecks.
5. Ambient sensors and data processing to monitor the conditions of the infrastructure like manufacturing/distribution centres, containers, etc. to predict functional problems before failure and order repair/replacement, thus minimizing the financial and temporal impact of non-functional organ in the supply chain ecosystem.

During the lifetime of ongoing military operations which the corresponding logistics hubs are supporting, the network parameters may change depending on the ongoing military operations. Manufacturing and supply priorities are prone to drastic changes within a short period of time. Thus, the infrastructure must be able to adapt to such dynamic circumstances and tune the logistics supply chain accordingly. The traffic profile of the equipment involved in the logistics supply chain is indicative of such changes.

5   Use Cases and Their Enabling Technologies

The enabling technologies were introduced in general terms previously. In this section, the specific use cases supported by the technologies will be summarised as follows,

5.1 Personnel Training Enhanced by AR/VR

MEC is considered to be an enabler to overcome limitations of computational overhead, by building on high bandwidth and ultra-low latency provided by 5G to dynamically offload computational tasks from remote cloud servers to small edge-servers near to the players' devices and training fields, and thus to perform tracking, annotation and rendering near to the video data sources with remote data (e.g. annotations) dynamically offloaded together with computational components with ultra-low latency. Cloud-Edge computing paradigm together with ML/AI algorithms will support the instantaneous and accurate identification of objects in the visual field of view and accurately overlay virtual content [12]. As a 5G physical network is expected to support a number of different user communities requiring different quality of service, the use of network slicing will be useful in providing guaranteed computing, radio access network, storage and communications resources.

5.2 Healthcare

Using MEC, a significant portion of computational overhead from the data received by a large number of devices distributed over a wide geographic area, can be offloaded to the network edge. This would aid in network provisioning, maintenance and timely response to anomalies, elements that are necessary for deployability of a medical ecosystem monitoring a tactical military unit. Additionally, edge computing allows processing on the edge of the network data such as patient data resulting in privacy-preserving benefits

5.3 Logistics

Monitoring the status of hundreds of thousands of IoT and mobile assets associated within the logistics use case is a primary criterion. MEC can aid with computational offloading, to track all active devices within an operational area, without the requirement of reachback to a remote cloud servers. Additionally, any anomalies in the manufacturing and supply chain aspects can be determined quickly, based on-going military operations, especially within an ecosystem of a mesh of many tactical units. Timely feedback on such changes would support the adaptation of logistics ecosystem such that scarcity of military equipment can be minimized and punctual production and supply of military equipment can be achieved. Further as logistics is an important to sustain combat operations, the use of MEC can also provide continuity of operations in the event of disconnection from central cloud servers or communications.

6 Challenges

The integrity of AR/VR environments during run time needs to be developed to exercise access control and monitoring and mitigation of aberrant applications behaviour within virtual environments. In addition, with the use of high fidelity avatars, privacy concerns on digital representation of individuals could arise. Confidentiality and reliability of the data in medical archives are important for accurate diagnosis and treatment. On top of vulnerable exposure of medical information of military units, manipulation of which can result in wrong diagnosis and treatment, cyberattacks can also render the network incapable of providing services like tele-medicine/consultation/surgery, by mechanisms like denial-of-service (DoS) or distributed denial-of-service (DDoS). Information on hundreds to thousands of devices interconnected within a common wireless network ecosystem, if acquired at the manufacturing end, in between the supply chain (at distribution centres or transportation units) or at the operational area itself, can provide a layout/prediction of the future operations/activities to adversarial units by methods like man-in-the-middle attack for production data theft, or critical system access to third-party contractors [11]. Additionally, DoS and DDoS attacks on one or more entities can cause major disruption to the entire supply chain.

When large number of devices with varying types of data constitute within a network infrastructure, constant data storage and analysis, identification of anomalies, as well as maintaining security and integrity of such confidential data becomes necessary. If the network is centralized, then the system may suffer from single-point failure (data breach) that will affect the availability of the whole system. Data stored on a centralized server can be easily deleted or manipulated and changed, by malicious system administrators [13]. A distributed network architecture or use of MEC are necessary to maintain the network integrity, but, this kind of cloud/fog based network architecture may also increases the scope of vulnerability of the network to external cybersecurity threats.

7 Interoperability

The realization of these scenarios are dependent upon the continued enhancement of existing technologies or entirely novel features that 5G networks are expected to provision. A military unit on occasion may operate with other military forces from other nations and/or also civilian organization, so the procurement of capabilities that conform to mutually recognized and utilized standards would support the ability of the military forces to inter-operate. However, owing to the potential for optional and conditional 5G specifications, additional effort would be required to enable 5G technologies from different national military networks to interoperate and provide 5G capabilities. Within NATO such a Framework called the Federated Mission Networking supports interoperability between nations and partner processes and CIS, including new technologies, to enable deployment of a mission network to support joint operations.

An additional challenge is represented by the potential lock-in of commercial technology providers, with the accompanying cyber-threats represented by providers' equipment which may contain "back-doors" introduced through intentional or unintentional action. These challenges may be avoided, or at least mitigated, by the definition and adoption of standards (de-iure or at least de-facto) enabling seamless interoperability, composition, integration and replacement of diverse HW and SW components and services, coming from diverse providers. Two promising initiatives in this respect are Open Network Automation Platform (ONAP) [14] and Open Random Access Network (O-RAN) [15].

The ONAP initiative aims at defining a holistic and open platform (fully virtualized, based on containers and micro-services technology) and accompanying use case scenarios for orchestration, management, and automation of network and edge computing services to be adopted by network operators, cloud providers, and enterprises, including real-time, policy-driven orchestration and automation of physical and virtual network functions; thus enabling rapid automation of new services and complete lifecycle management for 5G and next-generation networks. The O-RAN Alliance is defining and open architecture, specifications and use cases, for building virtualized RANs on open HW and Cloud.

8 Outlook

Cloud-edge computing is clearly a key enabler to mitigate certain network challenges within a military operational area. The aforementioned use cases involve either the need for high data rate, low-latency, resilience or large number of devices to be supported, both of which can benefit significantly from computational offloading at a local level. Nevertheless, the deployment-ready status of an entirely digitized military ecosystem is yet far from reality, including the maturity of 5G edge computing and network slicing capabilities that provide military vertical with the required quality of service. A number of initiatives in nations are underway to further study and prototype unique 5G applications for military and civilian use case. Prototyping interoperability of 5G technologies and then develop military system of systems is a prudent path to ensuring a positive outlook for the application of revolutionary 5G technologies to improve military effectiveness.

9 Conclusion

5G and enabling technologies will deliver advanced wireless CIS that support Military sustainment functions and ultimately operationally capable forces. Upgrading military communication to high amount of digitization and automation is an enormous task with a high cost implementation. However, in the long-term, this level of automation would make it easier for deploying the right resources to support operations safely and efficiently. It is clear that further research should be undertaken into the designing 5G-based networks in operational environments to reveal the reliability, latency, jamming mitigation methods and security future military wireless systems. Evaluating the heterogeneous interoperability of 5G and existing communication protocols currently deployed in military sustainment environments is needed.

References

1. Attallah, B.: Post COVID-19 higher education empowered by virtual worlds and applications. In: Seventh International Conference on Information Technology Trends (ITT) 2020, pp. 161–164 (2020). https://doi.org/10.1109/ITT51279.2020.9320772
2. Abichandani, P., Mcintyre, W., Fligor, W., Lobo, D.: Solar energy education through a cloud-based desktop virtual reality system. IEEE Access **7**, 147081–147093 (2019). https://doi.org/10.1109/ACCESS.2019.2945700
3. Borst, C.W., Ritter, K.A., Chambers, T.L.: Virtual energy center for teaching alternative energy technologies. In: Proceedings of IEEE Virtual Reality (VR), pp. 157–158, March 2016
4. Fraga-Lamas, P., FernáNdez-CaraméS, T.M., Blanco-Novoa, Ó., Vilar-Montesinos, M.A.: A review on industrial augmented reality systems for the industry 4.0 shipyard. IEEE Access **6**, 13358–13375 (2018). https://doi.org/10.1109/ACCESS.2018.2808326
5. Fullingim, D.D., II, GS-15, DAFC: Resilient Multi-Domain Command and Control: Enabling Solutions for 2025 with Virtual Reality. Air University, April 2017. https://apps.dtic.mil/sti/pdfs/AD1041954.pdf. Accessed 10 Aug 2021
6. Latif, S., Qadir, J., Farooq, S., Imran, M.A.: How 5G wireless (and concomitant technologies) will revolutionize healthcare? MDPI J. Future Internet **9**, 93 (2017). https://doi.org/10.3390/fi9040093
7. Chen, M., Yang, J., Hao, Y., Mao, S., Hwang, K.: A 5G cognitive system for healthcare. MDPI J. Big Data Cogn. Comput. **1** (2017). https://doi.org/10.3390/bdcc1010002
8. Li, D.: 5G and intelligence medicine-how the next generation of wireless technology will reconstruct healthcare? Precis. Clin. Med. **2**(4), 205–208 (2019). https://doi.org/10.1093/pcmedi/pbz020
9. Andreu-Perez, J., Leff, D.R., Ip, H.M.D., Yang, G.: From wearable sensors to smart implants-toward pervasive and personalized healthcare. IEEE Trans. Biomed. Eng. **62**(12), 2750–2762 (2015). https://doi.org/10.1109/TBME.2015.2422751
10. Khatib, E., Barco, R.: Optimization of 5G networks for smart logistics. MDPI J. Energ. **14**, 1758 (2021). https://doi.org/10.3390/en14061758
11. O'Connell, E., Moore, D., Newe, T.: Challenges associated with implementing 5G in manufacturing. MDPI J. Telecom **1**, 48–67 (2020). https://doi.org/10.3390/telecom1010005
12. Siriwardhana, Y., Porambage, P., Liyanage, M., Ylianttila, M.: A survey on mobile augmented reality with 5G mobile edge computing: architectures, applications, and technical aspects. IEEE Commun. Surv. Tutor. **23**(2), 1160–1192 (2021). https://doi.org/10.1109/COMST.2021.3061981
13. Zhang, C., Xu, C., Sharif, K., Zhu, L.: Privacy-preserving contact tracing in 5G-integrated and blockchain-based medical applications. Comput. Stand. Interfaces **77**, 103520 (2021). https://doi.org/10.1016/j.csi.2021.103520
14. ONAP, Open Network Automation Platform. https://www.onap.org/. Accessed 04 Aug 2021
15. O-RAN Alliance. https://www.o-ran.org/. Accessed 04 Aug 2021

Federated Learning of Predictive Models from Real Data on Diabetic Patients

Gennaro Junior Pezzullo[1], Antonio Esposito[1(✉)],
and Beniamino di Martino[1,2,3]

[1] Department of Engineering, University of Campania Luigi Vanvitelli, Caserta, Italy
gennaro.pezzullo@unicampus.it,
{antonio.esposito,beniamino.dimartino}@unicampania.it
[2] Department of Computer Science and Information Engineering, Asia University,
Taichung City, Taiwan
[3] Department of Computer Science, University of Vienna, Vienna, Austria

Abstract. Today Federated Learning is gaining more and more notoriety in the medical field, especially since it can guarantee the privacy of sensitive data, unlike Machine Learning. Furthermore, training a model on a single patient does not always guarantee acceptable results as we are different from each other. Starting from this assumption, the idea would therefore be to create a system that allows diabetic patients to share their data on the cloud (also using the Cloud/Edge Continuum technology and clearly going to respect all the constraints related to privacy by differentiating the use of the public cloud from the private one), collected via smartwatch, and, through Federated Learning, to train a centralized model that allows them to go to predict when one of the patients may need to inject insulin into his/her body. In particular, this document represents a part of a larger work in which various Time Series Analysis and Deep Learning algorithms were tested on real data from a diabetic patient, making predictions to understand, among those analyzed, which was the best algorithm on a single patient and then generalize by applying a multi-client approach.

1 Introduction

As already mentioned in the abstract, this work is a piece of a larger project. To create this tool, it is advisable to divide the work into the following 4 phases:

Algorithm Study: In this first step, a preliminary study was made of all the algorithms which can be used to predict time series.

FL Implementation: In this phase, the work will be about trying to understand how to bring this algorithm into Federated Learning in a virtual environment [1].

Porting to the Cloud: Once everything is working, the idea would be to move everything to the cloud in order to have greater reliability in terms of security, precision and performance.

L. Barolli (Ed.): AINA 2023, LNNS 655, pp. 80–89, 2023.
https://doi.org/10.1007/978-3-031-28694-0_8

User and Administrator Interface Creation: Finally, after creating the backend, the last point would be the development of an interface to always notify users of any predictions via smartwatch.

In particular, in this document, we will deal with showing all the work that has been done regarding phase 1, which is the study of the algorithms. In this regard, the project focuses mainly on Time Series Prediction and Deep Learning algorithms, in particular, the following have been used: AR, MA, ARMA, ARIMA, SAR, SARMA, SARIMA, VAR, VARMA, VARIMA and LSTM. These algorithms were first applied to the data and subsequently evaluated to verify their accuracy. To do this, the first step was to find the data. A quantity of about 30,000 samples collected was generated through a 4-month observation of a diabetic patient. Clearly, after doing this, the raw data could not be used in a raw way, so some modifications were made to be able to best adapt them. In fact, after having tested and calibrated the algorithms, they were evaluated through appropriate metrics.

2   Data Analysis and Preprocessing

All data captured by the sensors could be exported via a CSV file. However, the reading of this proved to be complex because of the null values and inconsistencies in writing. In particular, citing the most relevant:

> on all float values the transcription took place with the comma instead of the dot;
> the days, months and years were incorrectly formatted;
> when a sample was not taken the cell was generated but it was null;
> different data taken in the same moments of time were placed on different tables, giving rise to numerous duplicates;

So, the first step was to format all the data in order to be able to display them correctly on the algorithm. Most of the correction operations were done either through proprietary Excel functions or through the libraries **NumPy** and **Pandas** in Python. Then, the following values were isolated:

Bolus: the amount of insulin that is introduced into a diabetic subject.

SensorGlucose: the amount of glucose in the blood.

ISIG: Varying electrical signals in relation to glucose.

3   Performance Analysis Methods

As regards the measurement of the performance of the algorithms, reference was made to different methods, in order to have a vision as objective as possible of the results obtained through the performance. In particular, the following were calculated for each prediction:

Mean Error (ME): it is the arithmetic mean of the forecast errors committed in n intervals (all of the equal duration) up to the period t to which the forecast was made.

Mean Absolute Percentage Error (MAPE): this is the arithmetic mean of the ratios between the absolute value of the forecast errors and the actual demand occurring in n intervals (all of the equal duration) up to the period t to which the prediction was made.

Root Mean Squared Error (RMSE): this is the square root of the mean squared error and is, therefore, more sensitive to outliers than other accuracy metrics. A lower value indicates a more accurate model.

All these calculations have always been performed in a python environment with the help of the NumPy library.

4   Application of the ARIMA Model

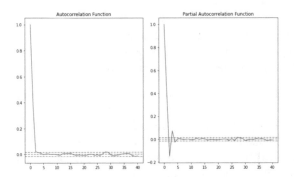

Fig. 1. ACF and PACF of complete Bolus data

The first data that have been used and manipulated are those concerning the bolus. In this first experience, all the data were taken into consideration. In particular, an initial analysis was carried out on the type of data. It was necessary to ensure the **stationarity** of the data which is a necessary condition. This was made possible through the Dickey-Fuller test by verifying the p-value. If the p-value of the test is less than a significance level ($x = 0.05$), then we can reject the null hypothesis and conclude that the time series is stationary. A script able to perform this test was therefore created using the **statsmodel library** and the result was a value tending to 0 with an order less than 10e-8. This leads us to two conclusions: the former is that the series is stationary and the latter is that we can set the parameter D of the ARIMA model [2] equal to 0 since the data must not be integrated. What remains to be done is, therefore, to calculate the remaining parameters of the ARIMA model (P and Q). To do this we rely on two graphs which are respectively **ACF** and **PACF** [3]. Also for this, a script was created using the statsmodels library and the results are shown in Fig. 1

From the figure, we can deduce that the autocorrelation graph and the partial autocorrelation graph both decay to a value of approximately 2. Thus, it is an ARIMA(2,0,2) or ARMA(2,2) model. It remains to create a script that implements this model, analyze the first results of this experience and in turn draw the first conclusions. Furthermore, the same tests and operations were also performed on the other datasets (SensorGlucose and ISIG). For the test set, there were used the latest 800 data of about the 15000 available. In Figs. 2, 3, 4 we can see the results of these predictions:

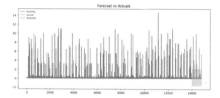

Fig. 2. Bolus prediction   **Fig. 3.** SensorGlucose prediction

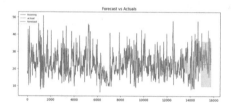

Fig. 4. ISIG prediction

The results will also be reported below, both in tabular form and all the measurements made for completeness. However, already from the observation of the graphs it can be deduced that the results are not very encouraging as the forecasts are constant and give life to a linear graph that does not vary over time. The results of the Forecasts are shown in Tables 1, 2, 3.

Table 1. Bolus   **Table 2.** Sensor Glucose   **Table 3.** ISIG

Index	Value	Forecast
14000	0.100	0.203
14001	0.100	0.255
14002	0.100	0.258
...	...	...
14797	0.100	0.262
14798	0.000	0.265
14799	5.500	0.262

Index	Value	Forecast
14901	120.0	120.251
14902	123.0	121.527
14903	122.0	122.660
...	...	...
15749	137.0	136.021
15750	137.0	136.021
15751	135.0	136.021

Index	Value	Forecast
14901	22.72	22.555
14902	23.08	22.593
14903	22.99	22.629
...	...	...
15749	26.376	22.866
15750	26.12	22.866
15751	25.62	22.866

Subsequently, starting from the data, the various metrics were calculated. The results are shown in Table 4.

Table 4. Forecast Metrics

Type	SG	ISIG	Bolus
ME	3.31	−3.18	0.048
MAPE	0.22	0.16	inf
RSME	32.86	6.01	0.84

From here on the idea was to analyze the data and try to make predictions in a less dilated way from a temporal point of view to see how the ARIMA algorithm behaves with a smaller amount of data. Then, we moved on to carry out a resample on the data, passing from samples every 5 min to every day fields, passing from a quantity of 14,000 data to a quantity of 115 data. For this type of analysis, being it experimental, only the SensorGlucose and ISIG data were analyzed; this is because, in a multivariable perspective, the fact that both were taken in the same time interval made it easier to adapt the data. Once all 2 graphs have been plotted, the results can be seen in Figs. 5, 6

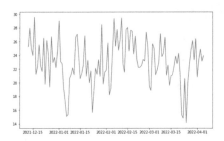

Fig. 5. ISIG after resample   **Fig. 6.** SensorGlucose after resample

Once again the Dickey-Fuller test was carried out, the plots of the ACF and PCF Graphs were analyzed and, after having correctly calibrated the parameters of the ARIMA algorithm, the whole dataset was divided into training set, with a number of samples equal to 100, and a test set, with a number of samples equal to 14. After the computation it is clear that we are still far from an accurate forecast; however, from a visual point of view, the forecast graphs already seem to somehow follow the trend of the graph more faithfully. Below the predictions are shown in Figs. 7, 8

Moving on to the analysis in tabular form, the results are shown in Tables 5, 6, 7.

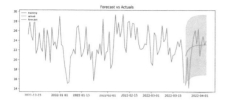

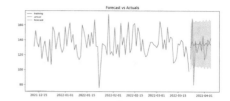

Fig. 7. ISIG resampled Forecast   **Fig. 8.** SensorGlucose resampled forecast

Table 5. ISIG after resample

Index	Value	Forecast
1	14.80	20.91
2	20.61	21.85
3	14.14	22.21
...	...	...
12	24.87	23.14
13	23.11	23.15
14	23.92	23.16

Table 6. Sensor Glucose after resample

Index	Value	Forecast
1	114.66	129.89
2	168.63	137.99
3	79.05	132.43
...	...	...
12	140.45	132.90
13	136.28	136.11
14	142.23	133.31

Table 7. Metrics after resample

Type	SG	ISIG
ME	4.59	0.65
MAPE	0.12	0.12
RSME	19.02	3.20

We have therefore had notable improvements not only from the point of view of the pursuit of the function of the forecasts to that of the test set, but also from the point of view of the data; in fact, there has been a significant improvement both on the MAPE and on the RSME.

5   Application of SARIMA and VARIMA Models

Now, we wanted to try out the use of 2 variants of the ARIMA algorithm which were respectively SARIMA and VARMA. The SARIMA algorithm introduces 4 other parameters into the variable which is seasonality [4]. In particular, it was decided to use 5 seasons being 5 the months on which the data were analyzed. The test was performed on SensorGlucose data. Also in this case the algorithm was made available through the Statsmodel library (in a Python environment) and the results obtained from this processing were: ME = 1.89, MAPE = 0.11, RSME = 19.82. The results were therefore very similar to those we obtained with ARIMA, but the negative aspect is the amount of time necessary to carry out this processing. So, for this reason, at least for the moment, it has not been taken into consideration. Then, we moved on to the VARIMA algorithm [5] in which both SensorGlucose and ISIG were taken as data inputs. Before using the algorithm, a feasibility check was carried out by analyzing the **Granger causality test** provided once again by the statsmodel library. The test gave a positive result and then the algorithm was used. In particular, in this specific case, a VAR model was used as, after a series of tests and monitoring on the AIC values, this proved to be the one that had the best response. After executing the algorithm the result was analyzed Fig. 9. Table 8 and Fig. 9 show the forecasts in a timely manner plus the measurement of errors.

Table 8. Metrics recalculated with SARIMA and VARIMA

Type	SG	ISIG
ME	1.89	−1.21
MAPE	0.11	0.11
RSME	18.18	3.83

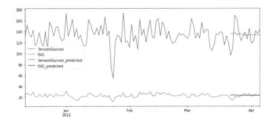

Fig. 9. VARMA algorithm predictions

6   Deep Learning with LSTM

A final technique employed was to use Deep Learning, in particular referring to the LSTM algorithm [6] and a walk-forward approach. Specifically, first of all, the problem has been transformed from a time series problem to a supervised Learning problem. To do this, the series to be predicted has been copied and shifted by one in order to have for each pair an input that represents the previous value and an output that represents the next value. After doing this we have passed to convert each value into a value of the domain of the tangent $[-1, 1]$. Then, a fit of the algorithm was carried out on the data with a period of epochs equal to 40 and on 2 neurons (which, after a series of tests, were the parameters that we performed best in terms of reliability and performance on the algorithm). Everything about the algorithm processing has been done through the Keras library. After having carried out all the operations previously mentioned, the results are shown in tabular and graphical form in Fig. 10. Finally, for completeness, the same steps have been carried out to see the forecasts made with the insertion of ISIG input data. The results are shown in Tables 9, 10, 11 and Fig. 11.

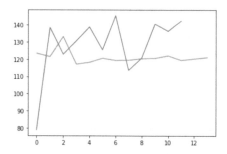

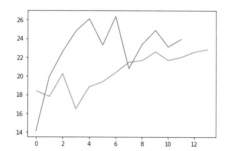

Fig. 10. SensorGlucose Resampled forecasts with LSTM algorithm

Fig. 11. ISIG Resampled forecasts with LSTM algorithm

Table 9. Sensor Glucose with LSTM

Value	Forecast
114.66	114.66 123.38
168.63	121.46
79.05	133.10
...	...
140.45	119.42
136.28	120.23
142.23	121.05

Table 10. ISIG with LSTM

Value	Forecast
14.80	18.40
20.61	17.79
15.14	20.26
...	...
24.87	22.00
23.11	22.56
23.92	22.83

Table 11. Metrics recalculated with LSTM

Type	SG	ISIG
ME	−1.03	1.60
MAPE	0.11	0.18
RMSE	20.26	4.38

7 Sliding Window Approach

From a less theoretical and more practical point of view, in an environment where we need to predict the progress of insulin, it could easily be up to 30 min from now. The idea therefore would be to understand how an ARIMA-type algorithm could perform if developed not in a long time frame but in a short slide window, [7]which is shifted gradually with the time it takes. We then perform the ARIMA test in 2 different ways. We do not carry out any resample but at first, we take the first 200 data and we try with these to predict the next 50 data; on the other hand, we try to make 50 forecasts, but 5 at a time always shifting the inputs by 5. The reason why we chose 5 as initial parameter is given by the fact that the forecasts are taken every 5 min, so we will have a time span of 25 min ahead in progressive time. Everything has been applied to the SensorGlucose data, so we are going to analyze the two results. In Fig. 12 we find the forecasts with a train set of 200 data and a forecast of 50 of them in Fig. 13 the sum of the forecasts of 5 subsequent data with a trainset of 200 data.

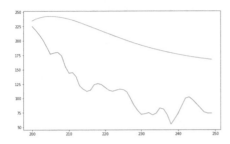
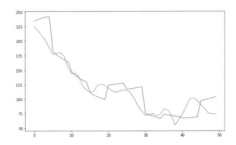

Fig. 12. ARIMA algorithm on the first 200 real data with forecasts of the next 50

Fig. 13. ARIMA algorithm on the first 200 real data with forecasts of the next 50 through a 5-min shifted time window

Since there are 50 forecasts, we re-report only the results of the measurements for easier reading in Table 12.

Table 12. Measurement results with Sliding Windows

Type	Normal	Windows
ME	89.21	4.62
MAPE	0.92	0.13
RMSE	92.68	19.21

The forecasts clearly show a trend much more faithful to the real value, this is due above all to the dynamism of the algorithm which, taking the data as input each time, uses them to predict the next ones.

8   Conclusion

After this analysis, there may be many future developments. The main idea would be to use either the LSTM or ARIMA algorithm, which are the two that have given the best results in terms of reliability and speed of execution. Obviously, it will be advisable to implement the time window method to have more precise results in any case. All this will be brought into a Federated Learning type environment [8] to ensure that the measurements take place more precisely and with greater privacy. Subsequently, everything will be brought to a cloud environment using the cloud edge to make all measurements more efficient [9]. Finally, a front-end interface for the devices worn by patients will be created in order to be able to notify when a patient may need an insulin dose.

Acknowledgements. Gennaro Junior Pezzullo is a PhD student enrolled in the National PhD in Artificial Intelligence, XXXVII cycle, course on Health and life sciences, organized by "Università Campus Bio-Medico di Roma".

References

1. Bonawitz, K., et al.: Towards federated learning at scale: system design. Proc. Mach. Learn. Syst. **1**, 374–388 (2019)
2. Ho, S.L., Xie, M.: The use of ARIMA models for reliability forecasting and analysis. Comput. Ind. Eng. **35**(1–2), 213–216 (1998)
3. Jain, G., Mallick, B.: A study of time series models ARIMA and ETS. Available at SSRN 2898968 (2017)
4. Zhang, N., Zhang, Y., Lu, H.: Seasonal autoregressive integrated moving average and support vector machine models: prediction of short-term traffic flow on freeways. Transp. Res. Rec. **2215**(1), 85–92 (2011)
5. Barcelo-Vidal, C., Aguilar, L., Martín-Fernández, J.A.: Compositional VARIMA time series. Compositional Data Analysis: Theory and Applications. Wiley, Hoboken (2011)
6. Bruneo, D., De Vita, F.: On the use of LSTM networks for predictive maintenance in smart industries. In: 2019 IEEE International Conference on Smart Computing (SMARTCOMP), pp. 241–248. IEEE (2019)
7. Dong, H., Guo, X., Reichgelt, H., Hu, R.: Predictive power of ARIMA models in forecasting equity returns: a sliding window method. J. Asset Manag. **21**(6), 549–566 (2020)
8. Li, T., Sahu, A.K., Talwalkar, A., Smith, V.: Federated learning: challenges, methods, and future directions. IEEE Signal Process. Mag. **37**(3), 50–60 (2020)
9. Sun, L., Jiang, X., Ren, H., Guo, Y.: Edge-cloud computing and artificial intelligence in internet of medical things: architecture, technology and application. IEEE Access **8**, 101079–101092 (2020)

Design of a Process and a Container-Based Cloud Architecture for the Automatic Generation of Storyline Visualizations

Emilio Di Giacomo[1]([✉]), Beniamino Di Martino[2,3,4], Walter Didimo[1], Antonio Esposito[2], Giuseppe Liotta[1], and Fabrizio Montecchiani[1]

[1] Department of Engineering, University of Perugia, Perugia, Italy
{emilio.digiacomo,walter.didimo,giuseppe.liotta,
fabrizio.montecchiani}@unipg.it
[2] Department of Engineering, University of Campania "Luigi Vanvitelli",
Caserta, Italy
{beniamino.dimartino,antonio.esposito}@unicampania.it
[3] Department of Computer Science and Information Engineering, Asia University,
Taichung City, Taiwan
[4] Department of Computer Science, University of Vienna, Vienna, Austria

Abstract. This paper addresses the goal of automated creation of intuitive graphical representations of textual descriptions, in particular utilizing the popular storyline visualization paradigm - a diagram that describes a temporal sequence of interactions among several actors. We propose a container-based architecture that integrates natural language processing and information visualization components, which is flexible enough to be deployed on cloud platforms, including Hybrid and Multi-Cloud ones. The methodology and the architecture that we propose are driven by a motivating example, focusing on the automatic creation of visualizations for textual descriptions of movie synopses. Anyway, the approach devised would be relevant in several domains, such as Cultural Heritage and Smart Tourism, e.g. to create an effective and easy to understand visualization of historical events.

1 Introduction

A *storyline visualization* is a popular paradigm introduced to describe a temporal sequence of events that involve several actors. This type of visualization was originally introduced to describe the narrative of a movie, where characters appear together and interact in different scenes [7]. Subsequent research works successfully adopted storyline visualizations to represent the temporal dynamics of the interactions between actors (individuals or organizations) in a social network or in a working environment [6,8–13]. In a classical storyline visualization, the narrative unfolds from left to right, each actor is represented as a line,

L. Barolli (Ed.): AINA 2023, LNNS 655, pp. 90–99, 2023.
https://doi.org/10.1007/978-3-031-28694-0_9

and two lines may converge or diverge at a time instant, based on whether the two corresponding actors interact or not at that instant. See Fig. 2 for an illustrative example.

Since there is a common consensus about the usefulness of storyline visualizations to convey the temporal development of a narrative in a friendly and intuitive fashion, a natural and fascinating problem is how to automatically generate such a visualization from an unstructured text. In terms of applications, this type of technology would be relevant in several domains. For example, other than illustrating the dynamics of an artistic opera, such as a movie or a literary work, it could be used in the context of smart tourism to create compact pictorial representations of a sequence of historical events related to the cultural heritage of a certain geographic area.

This paper addresses the aforementioned problem, namely how to develop automatic tools that create a storyline visualization starting from a textual description of a narrative in natural language. We highlight two main challenges behind this problem: (*i*) How to automatically extract from the text the most relevant actors and their interactions over time; (*ii*) How to produce an effective illustration based on the storyline visualization paradigm, which can be interactively explored through an intuitive user interface.

About point (*i*), several tools have been proposed in the literature, which can represent a basis for our solution. For example, in the works presented in [3,4] the authors apply Natural Language Processing technique to unstructured and heterogeneous sources, such a textual documents and Tweet, to recognize the main actors involved in interactions in two different contexts, that is Road Accidents and news regarding Energy Communities. The tools used by the authors represent the current state of the art, and consist in the application of the Natural Language Toolkit (NLTK) platform [1], of the SciPy Python library [14] and of the Bert algorithm for the training of neural networks [5].

About point (*ii*), as already mentioned above, there is an array of papers that propose algorithms and systems for computing and interactively exploring storyline visualizations. Recently, the classical storyline visualization paradigm has also been extended to support the representation of actors that participate in multiple events at the same time instant [2]. However, all these works assume that a structured description of the actors and of their interactions over time is part of the input and not automatically extracted from a text. As a consequence, the graphical interfaces designed to interactively explore storyline visualizations do not offer the possibility of enhancing the diagram with information directly taken from the textual narrative.

Contribution. The main contribution of this paper is twofold:

- We describe a process for the automatic creation of storyline visualizations from textual descriptions. It consists in a linear workflow, where three main logical components are involved.

- We propose a container-based architecture that integrates natural language processing and information visualization tools to implement our workflow. Our architecture supports the design of an interactive exploration of storyline visualizations through multiple coordinated views that allow users to enrich the diagrammatic part with information coming from the textual description, and vice versa. The devised architecture is flexible enough to be deployed on different Cloud Architectural Models, including Hybrid and Multiple Cloud ones.

For the sake of concreteness, the design of our architecture is driven by a motivating example, focusing on the prominent domain for which storyline visualizations have been originally conceived, namely the illustration of movie plots.

2   A Motivating Example

We envision the following scenario. A system allows the user to enter a textual description of a movie synopsis. In response, the system extracts from the text the main characters of the movie together with their interactions over time, and generates a storyline visualization that the user can interactively explore. As an example, Fig. 1 shows a portion of the synopsis of the popular 1997 American movie "Titanic", directed by James Cameron, taken from the Web site IMDb (https://www.imdb.com/). In particular, the figure shows the results of the manual annotation of a small portion of the textual description of a scene, extracted from the aforementioned movie. The main characters (Actors) have all been annotated with a generic **Actor** label, while a **Scene** tag has been used to annotate the reference scenarios. A relation **is_in_scene** connects one of the Actors to the specific scene. In this example, the annotations have been done manually, through the Doccano text annotator, which also allows us to visualize the annotations and to import/export them in a JSON like format. Our future plan is to apply NLP techniques to automatically identify Actors, Scenes and their relations, in order to build a consistent storytelling directly from the text analysis.

An example of visualization that we expect from the system, according to the storyline paradigm, is depicted in Fig. 2. Each character of the movie is represented by a line with a distinct color and each group, corresponding to a scene of the movie in which several characters occur together, is shown as a gray box. The graphical interface of the system should allow users to interact with the representation in several ways. For example:

- The system could provide the user with an interface having two coordinated views, the storyline visualization on one side and the text on the other side. Clicking on an element (character or scene) of one of the two views, would highlight the corresponding element in the other view.

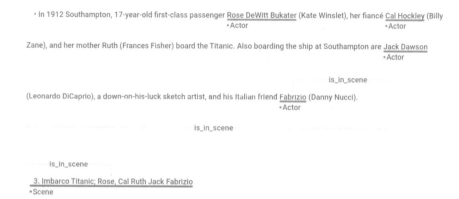

Fig. 1. Annotation of a textual description of a movie synopsis.

- By selecting a specific character, the interface should highlight the corresponding line in the visualization and all occurrences of the character in the text view. When a specific group is selected in the storyline visualization, the interface should provide additional information about the corresponding scene such as, for instance, movie snapshots or biographic data about whom is acting in the scene.
- When the number of lines (characters) in the visualization becomes too large, the user should be able to compact the layout horizontally or vertically, to make it fit the size of the drawing window.

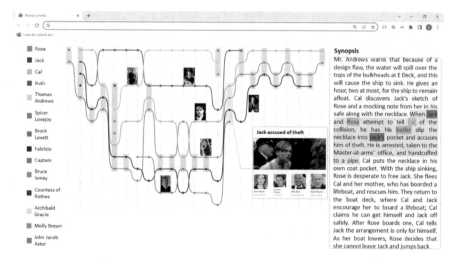

Fig. 2. Example of interactive visualization with coordinated views.

Fig. 3. Components of the envisioned framework.

3   Design of a Container-Based Architecture

In order to create an end-to-end framework that can analyze textual descriptions and visualize a storytelling extracted from it, some basic components are needed. Figure 3 provides a description of the components of the envisioned framework, that are linearly connected to one another.

- The **Text Annotator** is the entry point component, used to annotate the story text with relevant entities and relationships. Actors, groups, scenes and their relationships can be annotated here, so that all the relevant elements for the story are accurately identified.
- A **Parser** that takes as input a specific file text annotation format, exported from the annotator directly, and transforms it into a pure different format, compatible with the Storytelling Visualizer. Therefore, the Parser acts as an **Adapter** between the text annotator and the storytelling visualizer.
- The **Storytelling Visualizer** is the end point component, in charge of computing an interactive diagram in the storyline visualization paradigm.

Workflow and Related Technologies. Figure 4 focuses on the workflow of the envisioned framework, stressing the linearity of the approach, which is fundamental for the complete independence of the adopted components, but also addressing the fact that the text annotator can be used both as a mere visualizer of the annotation and as a manual validator of automatically inferred tags. The three components that have been briefly presented have the characteristic to be completely independent from one another, so that they can run on different containers, exposing their mutual interfaces to interact with each other, and can be seamlessly integrated or even substituted if needed, as long as the interfaces of the new components stay the same. This modularity is fundamental for a micro-service oriented implementation, where each component just shows a simplified and easily accessible interface, which is completely independent on the actual implementation of the service.

A set of possible technologies have already been identified to implement each of the agnostic components reported in Fig. 3. In particular, we envision to use:

- The Doccano Text Annotator, which is able to both support the visualization of existing tag on text and to produce new annotations. Doccano allows the export of the annotated text in the JSONL machine readable format, which is completely JSON compatible. In the future, annotations will be

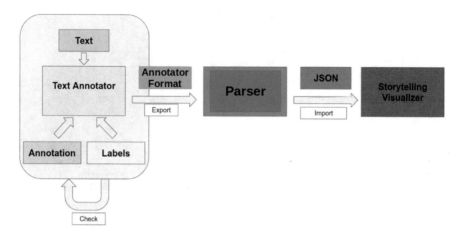

Fig. 4. Workflow of the operations within the envisioned Framework

automatically identified and only visualized through Doccano to validate and integrate them.

- A JSONL/JSON Parser, in charge of translating the JSONL format exported by Doccano, to the JSON format used by the Storytelling Visualizer. Such a parser needs to be created, as no solutions exist to automatically adapt Doccano's formats to a generic JSON.
- A Storyline Visualizer interface partially based on the algorithmic framework presented in [2].

Container-Based Architecture. Figure 5 shows a container-based architecture that can be used as a reference to deploy the framework. The central hub of the architecture is the UI Composition Service, which is in charge of building an interactive Web application by suitably merging and integrating the three services already described. Indeed, the UI Composition Service, which is still to be developed, will orchestrate the workflow of Fig. 4. Thanks to this architecture, the Text Annotator service and the Storyline service can focus on specific views of the data, while the coordination among these views and the potential integration of additional services can be managed in the UI Composition Service. In terms of technological solutions, for the Text Annotator service Doccano is already available as a containerized solution, but it does not provide REST interfaces to upload texts, annotations, and labels, as it uses a more complex GUI for this. For the other services the implementing technologies can be easily containerized and used to expose REST interfaces for communications.

In the following we give some details about the different logical components of the architecture and their related technologies.

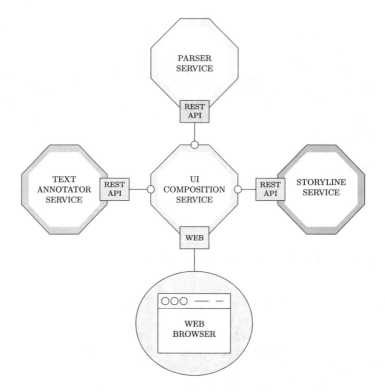

Fig. 5. Schematic illustration of the container-based architecture.

3.1 The Text Annotator Service

As previously mentioned, the Doccano Text Annotator can be used for this service. It produces a standard representation of the annotations that have been associated with any analysed text. In particular, Doccano allows the export of the annotated text and of all the associated annotation in the JSONL format. JSONL reports the text that has been annotated, the labels that have been used for annotation, and the position of the text that has been annotated with a specific label. A JSONL file consists of two different sections, as also shown graphically in Fig. 6:

- A **Text** section that reports the text without any annotations.
- A **Labels** section that reports the tags used within the annotated texts.

The Label section is structured differently, depending on the fact that the labels only refer to entities or can also address relationships. Since we need relationships to address the connections existing among actors and scenes in a story, we will refer to a structure in which the Labels section is divided into two subparts:

- An **Entities** part that reports the tags associated to a single element in the text.

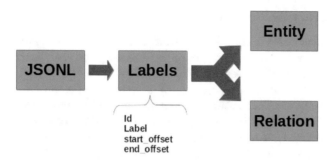

Fig. 6. High level description of a JSONL file

- A **Relationships** section that reports the tags used to connect different elements. A Relationship has a domain and a range, consisting in specific entities tagged within the text.

Both Entities and Relationships expose indexes and offsets, used by Doccano to determine the exact position of the labeled text and to visualize it accordingly.

3.2 The Parser Service

The availability of the JSONL format allows the Doccano Text Editor to clearly visualise the annotations, but can also be exploited to export the very same annotations to different tools. For our purposes, we want to use the JSON based representation of a storyline described in [2]. Since the two representations were born with different objectives, there is the need to map the concepts. To provide the actual mapping, some rules can be defined for the annotation phase. In particular, all occurrences of the same character in a story will be annotated with the very same label. Scenes are labelled in the same way, with a different and unique label for each scene, and actors that are present in a scene will be connected to it by a **is_in_scene** relations. Such a relation is intended to have Actors as a Domain, and Scenes as a Range. By proceeding in this way, it is possible to define the actual mapping.

- The Entities elements in a JSONL file can be interpreted in multiple ways: they can refer to an actor or to a scene. To discriminate between the two possibilities, it is necessary to look at the "relationship" element of the JSONL: If a Label is tied to another one by a "is_in_scene" relation and such a Label represents the Domain of the relation, than it can be considered as an Actor; if it represents the Range of the relation, then it is automatically considered as a Scene.
- Scenes are automatically translated into Groups, that represent a concurrent interaction among different Actors.
- The "is_in_scene" relation is not translated into a specific element of the Storyteller's JSON, but it is exploited to distinguish between actors and scenes, and to build the groups.

3.3   The Storyline Service

The storyline service is in charge of creating an interactive pictorial representation of the story described by the text, based on the storyline visualization paradigm. To this aim, it can offer a service method (based on a REST API) that takes a JSON string as input, describing the list of actors and a temporal sequence of groups (scenes), and that returns as output an HTML resource containing an interactive visualization. Each group in the JSON string consists of the subset of actors that interplay at the time instant associated with the group. The resource returned by the service is then integrated in the user interface by the composition service, which, in turns, returns the final user interface to the client.

The diagrammatic representation of the storyline can be created by means of several existing algorithms, including the recent algorithm described in [2]. The diagrammatic view should be suitably coupled and synchronized with a textual view, in such a way that the user can interactively enhance the visualization with information coming from the text, and vice versa.

4   Conclusions and Future Work

We described the design of a container-based architecture for the automatic generation of storyline visualizations starting from textual descriptions in natural language. Our design has been driven by a scenario in which this kind of technology is used to graphically illustrate the dynamics of a movie. Nonetheless, we believe that similar solutions can be effectively extended to illustrate other types of operas of higher complexity, such as for example a literary work. Furthermore, this type of technology could be adopted in the context of smart tourism to create compact pictorial representations of a sequence of historical events related to the cultural heritage of a certain geographic area. In this respect, our work can be regarded as a building block for the goals of a wide project for the promotion of the Italian cultural heritage[1].

As a future work, we plan to implement a system according to the container-based architecture, and to asses its effectiveness on real data sources.

Acknowledgments. This work is partially supported by the research project RASTA: Realtà Aumentata e Story-Telling Automatizzato per la valorizzazione di Beni Culturali ed Itinerari; Italian MUR PON Proj. ARS01_00540.

References

1. Bird, S.: NLTK: the natural language toolkit. In: Proceedings of the COLING/ACL 2006 Interactive Presentation Sessions, pp. 69–72 (2006)
2. Di Giacomo, E., Didimo, W., Liotta, G., Montecchiani, F., Tappini, A.: Storyline visualizations with ubiquitous actors. In: GD 2020. LNCS, vol. 12590, pp. 324–332. Springer, Cham (2020). https://doi.org/10.1007/978-3-030-68766-3_25

---

[1] RASTA: Realtà Aumentata e Story-Telling Automatizzato per la valorizzazione di Beni Culturali ed Itinerari; Italian MUR PON Proj. ARS01_00540.

3. Di Martino, B., et al.: Machine learning, big data analytics and natural language processing techniques with application to social media analysis for energy communities. In: Barolli, L. (ed.) Complex, Intelligent and Software Intensive Systems. LNNS, vol. 497, pp. 425–434. Springer, Cham (2022). https://doi.org/10.1007/978-3-031-08812-4_41

4. Di Martino, B., et al.: Semantic based knowledge management in e-government document workflows: a case study for judiciary domain in road accident trials. In: Barolli, L. (ed.) Complex, Intelligent and Software Intensive Systems. LNNS, vol. 497, pp. 435–445. Springer, Cham (2022). https://doi.org/10.1007/978-3-031-08812-4_42

5. Horev, R.: Bert explained: state of the art language model for NLP. Towards Data Sci. **10** (2018)

6. Liu, S., Wu, Y., Wei, E., Liu, M., Liu, Y.: Storyflow: tracking the evolution of stories. IEEE Trans. Vis. Comput. Graph. **19**(12), 2436–2445 (2013). https://doi.org/10.1109/TVCG.2013.196

7. Munroe, R.: Movie narrative charts, vol. 1 (2009). Diagram. https://xkcd.com/657

8. Ogawa, M., Ma, K.: Software evolution storylines. In: Telea, A., Görg, C., Reiss, S.P. (eds.) Proceedings of the ACM 2010 Symposium on Software Visualization, pp. 35–42. ACM (2010). https://doi.org/10.1145/1879211.1879219

9. Padia, K., Bandara, K.H., Healey, C.G.: A system for generating storyline visualizations using hierarchical task network planning. Comput. Graph. **78**, 64–75 (2019)

10. Qiang, L., Chai, B.: Storycake: a hierarchical plot visualization method for storytelling in polar coordinates. In: CW, pp. 211–218. IEEE Computer Society (2016)

11. Tanahashi, Y., Hsueh, C., Ma, K.: An efficient framework for generating storyline visualizations from streaming data. IEEE Trans. Vis. Comput. Graph. **21**(6), 730–742 (2015). https://doi.org/10.1109/TVCG.2015.2392771

12. Tanahashi, Y., Ma, K.: Design considerations for optimizing storyline visualizations. IEEE Trans. Visual Comput. Graphics **18**(12), 2679–2688 (2012)

13. Tong, C., et al.: Storytelling and visualization: an extended survey. Information **9**(3), 65 (2018). https://doi.org/10.3390/info9030065

14. Virtanen, P., et al.: SciPy 1.0: fundamental algorithms for scientific computing in python. Nat. Methods **17**(3), 261–272 (2020)

Cycle Detection and Clustering for Cyber Physical Systems

Gabriel Iuhasz[1]([✉]), Silviu Panica[1], and Alecsandru Duma[2]

[1] Institute eAustria Timisoara/West University of Timisoara,
Vasile Parvan Nr 4, Timisoara, Romania
{iuhasz.gabriel,silviu.panica}@e-uvt.ro
[2] ETA2U Computers, Gh. Dima Nr 1, Timisoara, Romania
aduma@eta2u.ro

Abstract. In this paper we present our work on cycle detection and clustering using unsupervised Machine Learning methods on manufacturing data. First we discuss the overall architecture of our cyber-physical system specially designed to gather large quantities of heterogeneous industrial data. Next, we detail several analysis steps, focusing on core tasks such as cycle detection and identification. Finally we show that even relatively simple data can be successfully used for predictive maintenance and fault detection.

1 Introduction

In recent years a wide range of technologies have led to heavy digitisation of industrial practices, resulting in the collection of various manufacturing data. In the industrial domain the term Industry 4.0 has been used to describe the integration between physical and digital systems of a production environment [3]. Access to this kind of data enable the improvement of several key areas ranging from business decisions to production planning by Cyber-Physical Systems (CPS) designed for this purpose.

One of the most important issue in Industry 4.0 is that of predictive maintenance (PdM) and early anomaly detection (AD). Maintenance in itself can make up around 16–20% of all operational costs in manufacturing [23]. PdM aims to improve the normal maintenance process as these operations can pe planed well in advance based around an estimation of a systems remaining useful life. This in turn leads to increased equipment uptime and availability. One prerequisite for PdM is the ability to detect any patterns or behaviours which do not conform to expectations. In short we require some form of AD which in turn can be used for the identification of faulty pieces.

However, there is an issue when applying any Machine Learning (ML) based AD methods, as these are knowledge-driven and can require significant effort both for data understanding and predictive model training. Furthermore, these methods are often used on multivariate time-series data further increasing processing/analysis pipeline complexity.

L. Barolli (Ed.): AINA 2023, LNNS 655, pp. 100–114, 2023.
https://doi.org/10.1007/978-3-031-28694-0_10

In this paper we aim to present a microservice based CPS which we use to break down both PdM and AD into basic tasks. Automatic production cycle detection and identification lie at the core of both PdM and AD in the context of Industry 4.0. We aim to show how these tasks can be accomplished using a relatively limited dataset. Furthermore, we show how AD techniques coupled with Explainable AI methods can be used for faulty cycle identification and causal analysis.

The paper is organized as follows: Sect. 2 details the state of the art for ML based detection techniques for AD and PdM in manufacturing as well as CPS designed for manufacturing. Section 3 describes our proposed CPS for data gathering and analysis. In Sect. 4 we describe experiments conducted for production cycle detection and identification/clustering as well as ML based AD methods. Finally, in Sect. 5 we present a conclusive summary and identification of future research directions.

2   State of the Art

Data streams from Industry 4.0 scenarios are large scale and heterogeneous. Typically, these are highly correlated and full of noisy data. This means that selecting analysis and detection machine learning (ML) methods is a difficult task [2]. Both supervised and unsupervised methods are fairly common for various analysis and detection tasks [5]. The task of detecting and classifying production cycles and motifs are in many ways the cornerstones of industrial data analysis [4,5]. CPS's must be capable to reliably accomplish these tasks.

AD methods have been widely used in cyber physical systems. Historically, methods such as k-Means, kNN [19], DBSCAN [20]. Explainable AI based methods were also used in conjunction with AD methods. In [10] a variety of methods including DBSCAN, HDBSCAN, One class SVM where used in conjunction with Shapley value calculations with the ultimate goal of reducing false alarms in heavy industry. Particularly IF [11] has been used for a variety of CFS designed for PdM and fault detection. Usages range from milling force and faulty bearing detection [11] to plasma etching [18].

ML method require substantial computational resources, especially during the training and validation steps. Due to these computational constraints training is offloaded to a dedicated compute environment, typically a cloud based solution. Inference requires less computational resources thus it is possible, even desired to deploy this task as close as possible to the origins of the data, in a Edge/Fog type scenario [8,22]. One of the key benefits of Edge/Fog deployments is that processing latency is significantly reduced and some privacy related issues usually associated with Cloud solutions are diminished as data is not sent to a centralized location for inference [17].

There are a number of CPS reference architectures who's goal is to connect industrial environments to analysis and data processing service deployed on a cloud infrastructure. These include but are not limited to; Azure IIoT[1], IAAS [9], WSO2 [7] etc. All of these reference architectures vary significantly in implementation however, their overall design is similar. These have several layers in common particularly the ones dealing with monitoring data, communication, connectivity and data analysis layer. However, these are difficult to deploy and in some cases result in vendor lock-in.

3   Architecture

Typically, in CPS low complexity operations are performed on-premises, near the data source but the complex data analysis has to be performed away from the data source, where the computational and storage resources are sufficient to handle this complexity. AD is a perfect example of this because it involves complex ML algorithms and techniques that requires both computational power and massive storage resources to keep the monitoring data acquired from the bespoken specialized sensors.

The proposed solution follows a microservice architecture where the core functionality is divided into stateless components, connected through a distributed communication bus. All these components are part of complex workflows, linked together to fulfil a common goal. These components are hosted on a distributed platform that is able to orchestrate complex scenarios and react dynamically based on the computational load generated by the executed workflow. Figure 1 depicts a high-level overview for data ingestion.

The monitoring data is ingested into the data analytics infrastructure through a gateway microservice which further publishes it into a stream processor. From the stream processor the data is routed, using a publish/subscribe model, to specific services using one or more dedicated stream topics.

There are a few components which are minimally required in order to create an on-premises data warehouse: Data Gateway which collects various payloads and pushes them into the stream processor; Stream processor which is a distributed and scalable publish/subscribe platform tasked with aggregation and

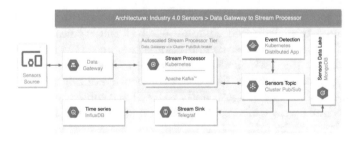

Fig. 1. High-level architectural details.

---

[1] https://learn.microsoft.com/en-us/azure/architecture/reference-architectures/iot.

stream data routing; Time series database used to store sensor data; Data lake platform is a persistent storage of sensor data required by analytical and ML model training; Event Detection Engine (EDE), is the data analytics engine used for analysis. These components can be seen in Fig. 1.

3.1 Platform Implementation Details

The data analytics platform load is unpredictable as it is influenced by two factors: (i) the amount of data to be collected and (ii) the complexity of the data analysis. In both cases processing and storage resources are involved dynamically. Thus resources have to be dynamically scaled. In this direction two technologies have been tested, Docker Swarm[2] and Kubernetes[3]. Both enables the containers orchestration process using a different level of dynamic resource allocation granularity and security hardening. We have found that Kubernetes offer far more advanced features to handle processing and storage resources with proven production automation tools that simplifies the deployment of the data analytics platform.

The stream processor is an important component of the proposed platform and in our tests we evaluated two different technologies, Apache Kafka and Pulsar. Both technologies are offering similar performance for the presented use case however, Apache Kafka proved to have better support for Kubernetes deployment. For the time series database we are using InfluxDB[4] technology and a distribute cluster of MongoDB NoSQL to store the historical monitoring data.

All the bespoken technologies are using on top of Kubernetes, using configurable deployment templates that can be replicated on any Kubernetes installation even with multi-tier support for the data lake platform and the stream processing (multi-tier support means than we can have several on-premises deployments that are linked together).

EDE represents a data analysis engine for identifying anomalous events in large datasets by applying supervised and unsupervised AD methods. EDE consists of several core components depicted in Fig. 2. The main components of EDE are: Training specialized in ML model training and validation with advanced techniques like model ensembling or model enhancement using hyper-parameter optimization methods; Prediction used to detect anomalous event in a pool of data or a live stream of monitoring data by selecting the proper detection model and to report detected events using the data bus.

Both Training and Prediction components are compute resources consumers and are set to run on top of a distributed framework (Dask[5]) which autoscales using the Kubernetes mechanisms based on the load and type of resources requested. The components are consuming monitoring data from the timeseries

---

[2] https://docs.docker.com/engine/swarm/.

[3] https://kubernetes.io/.

[4] https://www.influxdata.com/.

[5] https://www.dask.org/.

database and use an internal distributed object storage for ML related data payloads store. In case of training phase historical data is needed so it will have to
have access to the data lake repository.

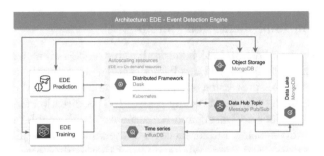

Fig. 2. Event Detection Engine Workflow

4   Experiments

In Sect. 3 we described the overall architecture of the proposed CPS. This section
will focus on the experiments done in order to solve two problems common
in cyberphysical systems. Namely, Cycle detection and Cycle identification in
manufacturing.

4.1   Data

The data available for our experiments consists of several time-series. In many
CPS each device is instrumented with several types of sensors ranging from
torque sensors to microphones. In our case we only have one type of sensor for
power consumption. This at first glance may seem a hindrance however, over-
instrumentation of production infrastructure can be costly and time consuming.
Minimal instrumentation is much more cost effective. The experiments detailed
in the following paragraphs will show that some problems can still be effectively
handled with limited monitoring data-sources.

We use the data lake to fetch data for further analysis (Fig. 1). In our experiments we have identified 3 devices from a production workflow for which we
have not only known cycles but also a clear overview of what pieces have been
manufactured in a given time-frame.

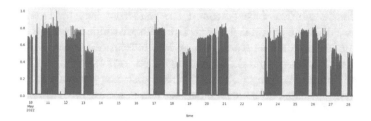

Fig. 3. Example time series of power consumption for Device 355

Figure 3 shows an example time series for a monitored device. The time-frame covers 10th to 28th of May 2022. During this time 6 unique types of pieces where manufactured on device with id 355. We should note that piece types are manufactured on a per need basis. In the given time-frame some pieces are manufactured multiple times while others are manufactured on a limited scale.

4.2 Cycle Detection

While there are many problems which occur in CPS's one of the more fundamental ones is production cycle detection. It represents a prerequisite for other problems such as PdM, machine and piece fault detection. By production cycle detection we understand the process of identifying when a pieces has been manufactured using a particular device in a production workflow.

The first step in detecting these types of cycles is the definition of a pattern. The goal is to count how many times this pattern is repeated in a given time series and mark the beginning and end of each pattern/cycle occurrence.

Figure 4 shows examples of the user defined pattern used for cycle detection in case of Device 355. The length of each cycle is important later in order to fix detected cycle alignment. In our case cycles length for device 355 is 88. We should note that we present the experiments done on Device 355 only, results for the other 2 devices (Devices 359 and 369) are similar thus for the sake of brevity these where not included.

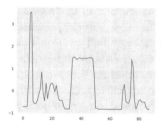

Fig. 4. Cycle Pattern as defined for Device 355

We define a window based around the pattern's length. This windows is
then compared to the original pattern. Different distance measures can be used
for this comparison such as Euclidean or Manhattan distances. However, these
distance measures have several disadvantages. For our experiments we have chosen Dynamic Time Warping (DTW) [21] as a distance measure as it allows the
comparison of two time-series which can vary in speed or have different lengths.

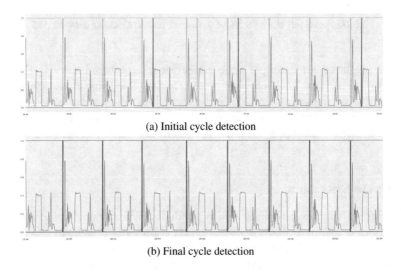

(a) Initial cycle detection

(b) Final cycle detection

Fig. 5. Production cycle detection example

We utilized a Numba[6] implementation of DTW as this allowed for considerable speed-up. Before we computed DTW Z-score normalization was applied as
follows:

$$x_z = (x - \mu)/\sigma \tag{1}$$

where x_z is the normalized value, x is the original, μ represents the mean of
the data and σ the standard deviation. This results in the z-score having a zero
mean and standard deviation of 1. This is desirable with ML based methods
because if the input data is close to zero models tend to converge faster.

Figure 5a shows the initial results of cycle detection for device 355. We can
see that DTW did a good job of identifying production cycles, in some instances
there is an overlap between identified cycles. This is caused by DTW itself.
There are several ways we can go about fixing this issue. First we can define a
threshold to the detected matches as the overlapping cycles invariably have a
lower similarity score. Applying this threshold runs the risk of also eliminating
unusual cycles which can be used for other analysis tasks such as PdM and

---

[6] https://numba.pydata.org/.

fault detection. Seeing that it is impossible for two cycles happening at the same time we define a heuristic which takes into account the distance between the last cycles starting point and the length of the newly detected cycle. Thus we defined the max distance between the start of one cycle and another as:

$$\Delta_{max} = \delta_p - (\mu - 4 * \sigma) \tag{2}$$

where Δ_{max} represents the max distance, δ_p represents pattern length while μ and σ are the same as for Eq. 1. The results can be seen in Fig. 5b. Overlapping cycles have been completely eliminated, while at the same time still being able to detect cycles with higher DTW distance measures, preserving potentially anomalous cycles.

4.3 Cycle Clustering

Once we have successfully detected production cycles we wanted to see if we can detect how many types of cycles we have in our dataset. Here each production cycle represents a type of piece produced. Applying unsupervised clustering ML methods on these cycles has several advantages. Primarily, it doesn't require pre-labelled training data. Arguably this makes unsupervised methods easier to use in CPS.

When trying to effectively cluster the detected production cycles there are several considerations we need to take into account. First, because we used DTW we should not use k-means as it may not converge. Mean is a least-square estimator it minimizes variance not distance. Instead we should compute a distance matrix using DTW and apply this to hierarchical clustering methods.

Second, clustering methods which are also capable of detecting the number of distinct clusters without needing to specify them a priori and the capability to deal with noise or anomalous data instances is also an important consideration.

Clustering algorithms such as DBSCAN [6] and Optics [1] showed initial promise. Optics is capable of dealing with large differences in densities while DBSCAN can not. At the same time Optics tends to mark over 50% of cycles as noise. A more appropriate clustering method called HDBSCAN [16] was found to produce the best results. It is in fact a hierarchical version of DBSCAN which no longer uses border points, only core points to define a cluster.

In case of HDBSCAN we used DTW to pre-compute a distance matrix. Next we set the minimum cluster size to 30, other then this we obtained the best results with default parameters. The resulting clusterer detected 6 unique cycle types beside the noise cluster.

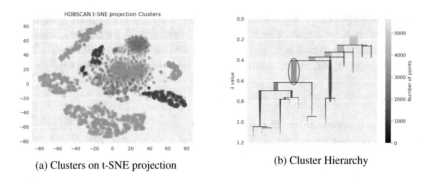

(a) Clusters on t-SNE projection

(b) Cluster Hierarchy

Fig. 6. HDBSCAN Clusters

Figure 6a shows a 2D re-projection of the detected clusters. We used t-SNE [15] which is a non-linear dimensionality reduction technique which tries to preserve the local structure of the original data. The labels shown in Fig. 6a are computed using the original data structure not the re-projected one, we only use the re-projection for visualization purposes.

Figure 6b show a dendrogram representing the cluster hierarchy for HDB-SCAN where each branch represents the number of points in a cluster. We can actually see that some clusters are considerably larger than others. This behaviour is consistent across all datasets described in Sects. 4.1 and 4.2.

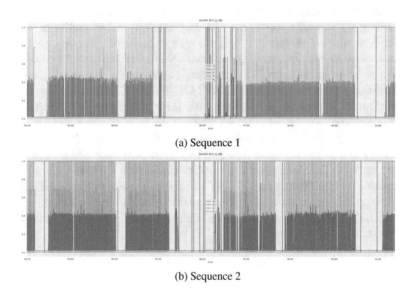

(a) Sequence 1

(b) Sequence 2

Fig. 7. Production cycle detection example sequences

From the 6 clusters detected one contains more datapoints than the others combined. Furthermore, noise is the second most represented cluster. There are several contributing factors to this behaviour. We can see from Fig. 3 that there are long periods where no energy consumption has been measured. These roughly correspond to weekend and free days. Also, some pieces are manufactured for several days while others only for a couple of hours. Some pieces are fairly similar and can be only small variations to other pieces. All of this leads to very disproportionate distribution of clusters.

Anomalies

Figures 7a and 7b show two sequences where we know a particular pieces was manufactured. We where able to successfully identify a contiguous piece type for these sequences. However, some anomalous cycles where also detected. These usually occur at the beginning of long sequence of normal cycles or, infrequently at the end.

We decided to have a more focused look at anomalous instances. For this we chose a specialized anomaly detection algorithm. Isolation Forest (IF) is an outlier detection algorithm which is constructed from multiple isolation trees [13]. It explores random subspaces from the data, each tree explores different splits thus exploring random local subspaces. Scoring is done by qualifying how easy it is to find a local subspace of low dimensionality for a given isolated event [12].

We trained an IF model on all of the detected cycles. We selected a contamination factor of 0.05 and the number of estimators to 100. We detected a total of 597 anomalies. The anomalous instances are fairly isolated from more densely packed regions, similar to what Fig. 6a showed for HDBSCAN. Our goal with IF is to identify those cycles which are true anomalies. We hope to gain insight into why these particular cycles are marked as anomalies.

These anomalous instances where in large part aligned with the noise detected by HDBSCAN. Next we wanted to further analyse these anomalous cycles using Shapely values so that we can see what part of a cycle is most influential in marking it as anomalous. Figure 8 shows a comparison of normal Fig. 8a and anomalous cycle Fig. 8c together with a waterfall plot depicting the Shapley [14] value derived ranking of each point in the cycle, Fig. 8b and 8c respectively. The latter two figures show if a feature pushes a cycle toward being a normal instance (positive values marked in red) or an anomalous cycle (negative values marked with blue). From these figures we can see that in case of the anomalous cycles points 86, 74, 85, 78 are the most influential ones. We can infer from this that the end of the cycle is significantly different than a normal one.

Figure 9 show the optimum alignment path between two cycles the line from the cost matrix represents the minimum path from the first to the last cycle point. Figure 9a represents a normal cycle and Fig. 9b an anomalous cycle. The analysis steps shown in Figs. 8 and 9 can be used for root cause analysis. We can say with a high degree of confidence why a particular cycle is marked as an anomaly and which part of the cycle is the most influential in this decision.

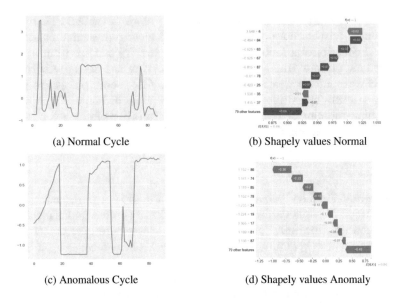

(a) Normal Cycle

(b) Shapely values Normal

(c) Anomalous Cycle

(d) Shapely values Anomaly

Fig. 8. Comparison of normal versus anomalous cycles

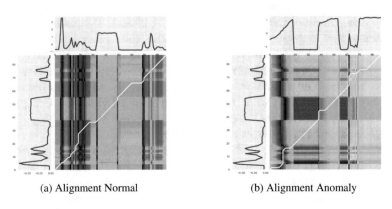

(a) Alignment Normal

(b) Alignment Anomaly

Fig. 9. Alignment Paths relative to original pattern

Discussion

The experiments presented in this paper have yielded some promising results. First we selected a distance measure namely DTW which is suitable for production cycle detection. Although it is more complex to compute then Euclidean distance we managed to reduce computation time by using a Numba based implementation. If we wish to move computation closer to the data source in an Edge/Fog type scenario we can do that by using CUDA enabled embedded systems such as NVIDIAs Jetson NANO [7]. We also managed to preserve

---

[7] https://developer.nvidia.com/embedded/jetson-nano-developer-kit.

anomalous cycles by specifying a permissive threshold while computing DTW. As mentioned in Sect. 4.2 we defined a heuristic which eliminated overlapping false cycles. This allows for a better detection and analysis of faulty pieces and PdM. The only required user input is the definition of an initial cycle pattern.

Experiments dealing with unsupervised pieces identification also show promising results. During our experiments we could successfully detect the correct number of manufactured piece types. We have 6 identified sequences which have a known piece type being manufactured. Table 1 shows the scores obtained for cycle identification compared to a hand labeled ground truth. We can see that there is some issue regarding false positives. This is most likely caused by anomalous cycles which, as stated before are grouped at the beginning or end of long cycle sequences (see Fig. 7). The current dataset has some limitations in that it doesn't provide information regarding manufactured pieces quality which is necessary to fully quantify cycle type identification performance. From the total of 5646 identified cycles 7% are identified as anomalies or noise. These aspects requires further research.

Table 1. Cycle detection scores for each sequence

Seq	Precision	Recall	Specificity	f1	Geometric
S1	0.94	1.00	1.00	0.97	1.00
S2	0.92	1.00	1.00	0.96	1.00
S3	0.64	1.00	1.00	0.78	1.00
S4	0.84	1.00	1.00	0.91	1.00
S5	0.91	1.00	1.00	0.95	1.00
S6	0.87	1.00	1.00	0.93	1.00

IF was tested as an AD method on the detected cycles. Although HDB-SCAN has the concept of noise, a specialised AD method was required as these are easier to analyse. We compared the anomalies detected by IF to the noise detected by HDBSCAN an found that we obtained an F1 score of 0.92 indicating substantial consensus between the two models. Additional experimentation is required with other AD methods particularly Deep Learning based models such as Variational-Auto Encoders. These have been shown to perform well on a large array of problem domains. Their generalization capability being superior to that of IF which is susceptible to poor out of sample performance, requiring extensive hyper-parameter optimization.

In order to use the AD results for PdM and/or fault detection we used Shapely values to compute feature importance. In our case features are represented by the datapoints which make up a cycle. This way we can specify which part of a cycle is the most anomalous and which is closer to a normal cycle. We can do this because Shapely values are symmetrical, equal contribution of a feature results in equal Shapely values while non-contributing features have a

Shapely value of 0. This interpretation lends itself to both PdM and manufacturing fault detection. A dataset containing data on manufactured piece quality and device status is required and will be the focus of future research.

5   Conclusions

In this paper we have presented our work with CPS and industrial manufacturing data. We presented our proposed architecture based around micro-services. Making this solution deployable both on-premises or on Cloud. The main goal was to enable the collection and analysis of data from various devices used for manufacturing.

We described several cycle detection, clustering and AD methods. The experiments where designed to maximize the performance of these methods on low dimensionality data while at the same time limit as much as possible the necessity for users to configure method parameter. We where able to obtain results with minimal user interaction. Further research will tackle open research question identified in this paper, such as; complete dataset for use in PdM, expanding the number of sensors types, investigation of Deep Learning methods for AD.

Acknowledgements. This work was supported by the POC SCAMP-ML Project SMIS 2014+120725, and partially supported by the H2020 research project SERRANO (101017168) and by the Chist-Era DIPET (62652/15.11.2019) project funded via PN 124/2020.

References

1. Ankerst, M., Breunig, M.M., Kriegel, H.P., Sander, J.: Optics: ordering points to identify the clustering structure. In: Proceedings of the 1999 ACM SIGMOD International Conference on Management of Data, SIGMOD 1999, pp. 49–60. Association for Computing Machinery, New York (1999). https://doi.org/10.1145/304182.304187

2. Carvalho, T.P., Soares, F.A.A.M.N., Vita, R., Francisco, R.D.P., Basto, J.P., Alcalá, S.G.S.: A systematic literature review of machine learning methods applied to predictive maintenance. Comput. Ind. Eng. **137**, 106024 (2019). https://doi.org/10.1016/j.cie.2019.106024. https://www.sciencedirect.com/science/article/pii/S0360835219304838

3. Cinar, Z.M., Zeeshan, Q., Solyali, D., Korhan, O.: Simulation of factory 4.0: a review. In: Calisir, F., Korhan, O. (eds.) GJCIE 2019. LNMIE, pp. 204–216. Springer, Cham (2020). https://doi.org/10.1007/978-3-030-42416-9_19

4. Dalzochio, J., et al.: Machine learning and reasoning for predictive maintenance in industry 4.0: current status and challenges. Comput. Ind. **123**, 103298 (2020). https://doi.org/10.1016/j.compind.2020.103298. https://www.sciencedirect.com/science/article/pii/S0166361520305327

5. Erhan, L., et al.: Smart anomaly detection in sensor systems: a multi-perspective review. Inf. Fusion **67**, 64–79 (2021). https://doi.org/10.1016/j.inffus.2020.10.001. https://www.sciencedirect.com/science/article/pii/S1566253520303717

6. Ester, M., Kriegel, H.P., Sander, J., Xu, X.: A density-based algorithm for discovering clusters in large spatial databases with noise. In: Proceedings of the Second International Conference on Knowledge Discovery and Data Mining, KDD 1996, pp. 226–231. AAAI Press (1996)
7. Fremantle, P.: A reference architecture for the internet of things. Whitepaper (2015)
8. Greco, L., Percannella, G., Ritrovato, P., Tortorella, F., Vento, M.: Trends in IoT based solutions for health care: moving AI to the edge. Pattern Recognit. Lett. **135**, 346–353 (2020). https://doi.org/10.1016/j.patrec.2020.05.016. https://www.sciencedirect.com/science/article/pii/S0167865520301884
9. Guth, J., et al.: A detailed analysis of iot platform architectures: concepts, similarities, and differences. In: Di Martino, B., Li, K.-C., Yang, L.T., Esposito, A. (eds.) Internet of Everything. IT, pp. 81–101. Springer, Singapore (2018). https://doi.org/10.1007/978-981-10-5861-5_4
10. Hermansa M., K.M.: Sensor-based predictive maintenance with reduction of false alarms-a case study in heavy industry. Sensors **22**(226) (2022). https://doi.org/10.3390/s22010226
11. Li, C., Guo, L., Gao, H., Li, Y.: Similarity-measured isolation forest: anomaly detection method for machine monitoring data. IEEE Trans. Instrum. Meas. **70**, 1–12 (2021). https://doi.org/10.1109/TIM.2021.3062684
12. Liu, F.T., Ting, K.M., Zhou, Z.H.: Isolation forest. In: 2008 Eighth IEEE International Conference on Data Mining, pp. 413–422 (2008). https://doi.org/10.1109/ICDM.2008.17
13. Liu, F.T., Ting, K.M., Zhou, Z.H.: Isolation-based anomaly detection. ACM Trans. Knowl. Discov. Data **6**(1) (2012). https://doi.org/10.1145/2133360.2133363
14. Lundberg, S.M., et al.: From local explanations to global understanding with explainable AI for trees. Nat. Mach. Intell. **2**(1), 2522–5839 (2020)
15. van der Maaten, L.: Accelerating t-SNE using tree-based algorithms. J. Mach. Learn. Res. **15**(93), 3221–3245 (2014). http://jmlr.org/papers/v15/vandermaaten14a.html
16. McInnes, L., Healy, J.: Accelerated hierarchical density based clustering. In: 2017 IEEE International Conference on Data Mining Workshops (ICDMW), pp. 33–42. IEEE (2017)
17. Sanchez-Iborra, R., Skarmeta, A.F.: TinyML-enabled frugal smart objects: challenges and opportunities. IEEE Circuits Syst. Mag. **20**(3), 4–18 (2020). https://doi.org/10.1109/MCAS.2020.3005467
18. Susto, G.A., Beghi, A., McLoone, S.: Anomaly detection through on-line isolation forest: an application to plasma etching. In: 2017 28th Annual SEMI Advanced Semiconductor Manufacturing Conference (ASMC), pp. 89–94 (2017). https://doi.org/10.1109/ASMC.2017.7969205
19. Verdier, G., Ferreira, A.: Adaptive mahalanobis distance and k -nearest neighbor rule for fault detection in semiconductor manufacturing. IEEE Trans. Semicond. Manuf. **24**(1), 59–68 (2011). https://doi.org/10.1109/TSM.2010.2065531
20. Yoon, H.S., Han, S.S.: Clustering parameter optimization of predictive maintenance algorithm for semiconductor equipment using one-way factorial design. In: 2019 19th International Conference on Control, Automation and Systems (ICCAS), pp. 1219–1221 (2019). https://doi.org/10.23919/ICCAS47443.2019.8971633
21. Zhang, H., Dong, Y., Li, J., Xu, D.: Dynamic time warping under product quantization, with applications to time-series data similarity search. IEEE Internet Things J. **9**(14), 11814–11826 (2022). https://doi.org/10.1109/JIOT.2021.3132017

22. Zhang, W., Yang, D., Wang, H.: Data-driven methods for predictive maintenance of industrial equipment: a survey. IEEE Syst. J. **13**(3), 2213–2227 (2019). https://doi.org/10.1109/JSYST.2019.2905565
23. Zonta, T., da Costa, C.A., da Rosa Righi, R., de Lima, M.J., da Trindade, E.S., Li, G.P.: Predictive maintenance in the industry 4.0: a systematic literature review. Comput. Ind. Eng. **150**, 106889 (2020). https://doi.org/10.1016/j.cie.2020.106889. https://www.sciencedirect.com/science/article/pii/S0360835220305787

Cloud Computing and Critical Infrastructure Resilience

Oronzo Mazzeo, Antonella Longo, and Marco Zappatore[✉]

Department of Engineering for Innovation, University of Salento, Via Monteroni Sn,
73100 Lecce, Italy
{oronzo.mazzeo,antonella.longo,
marcosalvatore.zappatore}@unisalento.it

Abstract. The protection of critical infrastructures and entities from threats and hazards has become increasingly critical in modern society, which is more and more dependent on supplied services. The significance of the topic has been proved by the growing interest of the European Union in developing a common policy to address critical entities protection. This paper aims to introduce the reader to the resilience of critical entities, a key concept in critical entities protection, and to highlight the relevance of cloud computing as a resource for implementing solutions aimed to enhance resilient capabilities. In the paper, not only resilience conceptual models and corresponding analysis dimensions will be discussed, but also resilience indicators and assessment frameworks will be addressed.

1 Introduction

Modern Societies heavily depend on the so-called Critical Infrastructures (CI), namely physical resources, services or structures whose malfunctioning or destruction would have a serious effect on the availability and deliverability of essential services, whose interruption would affect strategic fields (economy, health, security, etc.), which, in turn, would have implication for citizens and societies' wellness. Energy production plants and distribution network, communication systems and networks, security systems, industrial plants, health and emergency facilities are some examples of critical infrastructures. All these infrastructures are exposed to potential threats, whose origin might be either natural (floods, landslides, earthquakes, etc.) or man-made (terrorist attacks, cyberattacks, etc.). Threats can interrupt or limit the availability of services of critical infrastructures, with catastrophic consequences for the delivery of essential services and the well-being of people and society.

In recent years, the relationship between critical infrastructure protection and the well-being of citizens has gained considerable importance. As early as 2004, the Council of Europe commissioned experts to formulate general strategies for critical infrastructure protection. It also underlined the importance of making critical infrastructures able to tolerate ad eventually fix the damage produced by their critical service interruption.

The classical approach to improve critical infrastructures security against disruptive event consists in employing preventive and protective programs focused on minimizing

L. Barolli (Ed.): AINA 2023, LNNS 655, pp. 115–126, 2023.
https://doi.org/10.1007/978-3-031-28694-0_11

the possible disruptive events' probability and consequences. However, this risk management strategy solutions – like proactive data-driven risk prevention employing historical data, analytics and expert systems able to identify behaviours and patterns that might resulting systems' damage – has been proved ineffective in protecting systems against rare events with major consequences, which happened in recent years. We refer to events like, in example, big electric power outages or blackouts like the one that affected 15 million European people in 2006, the one which lasted for three months in Tanzania in 2009; the severe floods in UK in 2007, that brought lack of water and electricity, transport network's failure and caused emergency facilities to stop operation, the Tohoku earthquake and the following tsunami in Japan in 2011, which resulted in accidents' chains (i.e., water e power outages, and transport network failure), the hurricane Sandy in US (2012) that had outcomes like losses in terms of electricity and water supplies, the recent COVID-19 pandemic that had serious consequences, impacts and damages in the health, social and economic fields all over the world. These kinds of rare events highlighted it is impossible to anticipate and prevent all kinds of disruptive events (and hazard) and consequences, at least not in all cases [1, 2].

The previous observations necessarily lead to the conclusion that is important to develop an approach for critical infrastructures' security based on both risk-management and resilience concepts: critical infrastructures designed in this way would be best equipped to guarantee service continuity even in the case of threats due to rare events with major consequences, like those listed in the quote above.

The first section depicts the critical infrastructures/entities' scenario, illustrates the reasons that make them worthy of protection, gives some clues to traditional protection approaches and related limitations. The second section presents a conceptual model of resilience along with its analysis dimensions, while the third section illustrates the resilience indicators and the related assessment framework. A section focusing on the role cloud computing may provide to the critical infrastructure protection and resilience follows. In section five, further considerations about how critical infrastructure protection is currently addressed in the scientific literature are presented. The last section, preceding the conclusions, consists of a brief excursus on European Union policies on critical infrastructure protection.

2   Resilience

2.1   The Concept

The concept of resilience has run in several definition in the past decades. In engineering domain, the resilience concept is based on the ability of the system to maintain or return to a dynamically stable state, which allows it to continue operating after a major accident and/or in presence of continuous stress [3].

Wied and colleagues [4], in their paper *Conceptualizing resilience in engineering systems: An analysis of the literature*, developed a conceptual framework for analysing the resilience concept by looking for answers to the question: *"Resilience of what to what, and how?"*. Figure 1 shows their conceptual model.

Fig. 1. A conceptual model for understanding system resilience (source: [4]).

In this model, resilience (R) is the mediator between the effect on uncertain conditions (C) – the possible threat to the system – and the system performance (P) – let's say the system's functioning or service output. In this view, "the resilience of a system is determined by its ability to mediate between performance and uncertain conditions" [4]. In other words, from the engineering point of view, a resilient system is characterized by the ability to cope with threats and uncertainty in order to continue its operations and deliver its services.

Among the several models about the systems resilience, the multi-phase resilience trapezoid of infrastructure resilience in power systems, presented by Panteli and colleagues [5], can be easily generalized to other infrastructures. It shows the effect of the resilience over the time on a system that undergoes to a critical event.

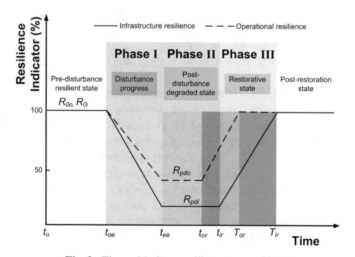

Fig. 2. The multi-phase resilience trapezoid [5].

The three-phase model is depicted in Fig. 2 and it distinguishes resilience in *operational* and *infrastructure* resilience. The first refers to the characteristics that would secure operational capacity to the system (i.e., online load, online generation capacity and online transmission lines in a power system), the latter refers to the capacity of the system to limit the portion of the system that is damaged, collapsed or in general becomes non-functional.

The figure depicts all the phases, and transitions between the associated states, that a critical infrastructure may reside at the happening of a critical event. Looking at dynamics of the resilience, the three-phase model shows that, a full operational infrastructure can undergo to critical event at time t_{oe}. As the disturbance persists, the system's resilience percentage drops (t_{oe}-t_{ee}) (*Phase I*), characterized by a fast reduction in system's ability to continue operations. This dropping in resilience percentage and service availability tends to stabilize during the so-called post-disturbance degraded state (t_{ee}-t_{ir}) (*Phase II*), where a limited, if any, operational capacity can be available. The restorative state (t_{ir}-T_{ir}) (*Phase III*) follows, when resilience and operations ability increase again until they reach their pre-disturbance levels (after time T_{ir}).

2.2   The Dimensions of Resilience

Research identified five dimensions featuring the concept of resilience: robustness, rapidity, redundancy, resourcefulness, and protectiveness. Robustness is defined as the strength of the system (or its elements) to withstand external stress or demand without degradation of functioning; rapidity is the speed with which disruption can be overcome and services restored; redundancy is the extent to which the elements of the system can be substituted; resourcefulness is the capacity to identify problems, establish priorities, and mobilize resources in the case of crisis; and, finally, protectiveness is the capacity of external works or equipment to protect the system from threats ([6, 7]).

Another approach to the definition of resilience dimensions in critical infrastructures sheds light on the aspect of the management process, the components and involved domains [7].

As presented in Fig. 3, the first dimension, named management phases, distinguishes the phases starting from the perturbative event to the time in which the system regained its operations capabilities and resilience and it is characterized by the definition of specific strategy to manage and/or prevent the critical event. Therefore, the process can be split in planning/preparation (ex-ante phase), absorption (during the event phase), and recovery and adaption (ex post phases). The management components (second dimension) involve anticipation (i.e., event's occurrence prediction), monitoring/detection (identification and interpretation of precursory signs), control (using the defined indicators to implement actions focused on system's recovery or adaptation), collection of feedback from experience (useful for the anticipation, monitoring and detection of future events). Finally, the field dimension of resilience refers to the different domains impacting resilience: technical, organizational, human, and economic. These dimensions, with relative examples, are depicted in the following figure.

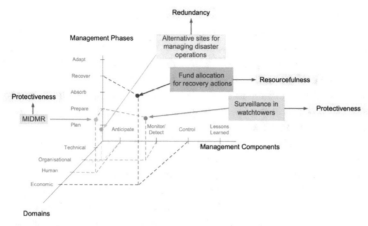

Fig. 3. Different dimensions of resilience (illustration by examples [7]).

3 Critical Infrastructures' Resilience Indicators and Assessment Framework

In order to define the level of resilience of a critical infrastructure, the system must be assessed. A resilience assessment framework for critical infrastructures [1] is presented in Fig. 4. It is based on four dimensions: technical, organizational, social, and economic.

The *technical dimension* refers to physical system's capacity to maintain an acceptable level of performance when it is affected by a disruptive event. Thus, this dimension focuses on the vulnerability and recovery of the entire system, its components and the related interconnections and interaction. This dimension includes the following indicators: *robustness* (capacity of the system to withstand critical events without

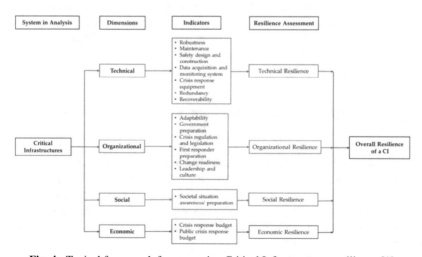

Fig. 4. Typical framework for measuring Critical Infrastructures resilience [1].

compromising its performance or functionality), *maintenance* (preventive and correc-
tive), *safety design and construction* (system design characteristics supporting a high
level of resilience), *data acquisition and monitoring systems* (monitoring system uses
data collected by the *data acquisition system* to check whether system critical parts are
functioning in the correct value range), *redundancy* (availability to alternative resources
able to substitute the part of the infrastructure damaged by the disruptive event in order
to continue operations), *recoverability* (capacity of a system or component to restore its
original functioning and performance).

The *organizational dimension* of resilience is related to those organizations in charge
of responding to disasters or critical events. In instance, it refers to the capacity of those
organization to decide and take actions to prevent (or prepare for) and respond to a
disruptive event involving critical infrastructures. Some indicators of the organizational
dimensions are the following: *adaptability* (capacity of the critical infrastructure organization to dynamically adapt to undesirable circumstances and/or uncertain environment
by undergoing some change), *government preparation* (government's preparedness to
anticipate critical events and the capacity to act quickly when they occur), *crisis regulation and legislation* (levels of maturity and compliance of laws and regulation), *first
responder preparation* (level of first responders preparation, training, commitment, crisis
and situational awareness), *change readiness* (capacity of the organization to change in
response to changes in, and to perturbations of, the environment), *leadership and culture*
(capacity of an organization to promote a transparent organizational commitment to a
resilient culture, vision, and values).

The social dimension of resilience refers to groups or community's ability to cope
with external pressures and disturbances [8] and to the societal capability to reduce the
impact of the disrupting event by helping first responders or acting as volunteers [9].
Societal situation awareness/preparation, namely the public awareness level of the risks
and vulnerability they may face in unfavourable situation, is its unique indicator.

The economic resilience's dimension concerns the capacity of minimize direct and
indirect losses consequent to a crisis" [1]. The two indicators are *crisis response budget*,
namely the size of the critical infrastructure's funds destined to absorb the impact of the
disruptive event and repair/replace facilities in order to restore and acceptable state as
soon as possible, and *public crisis response budget*, namely the size of public funds set
aside as crisis response budget.

4   Cloud Computing as Resource for Critical Infrastructure Resilience

The assessment of critical infrastructure resilience is challenge in term of volume of
collected data requirement and computing power. Often data must be collected from
several different, process which generates high volume data stream, and computing
power must support real-time processing. Typically, critical infrastructures rarely have
the resources (bandwidth, computing power, storage, etc.) needed to accomplish these
challenging tasks by themselves.

A cloud-computing solution, like the monitoring and threat-detection system presented by Chen and colleagues in 2015 [10], can secure critical infrastructure systems without overwhelming their data management and computing power resources. Moreover, a cloud computing solution lets integrate simulated environment, by assembling complex and distributed experimental critical infrastructures scenarios, to help in evaluating the vulnerability of each interacting critical systems [11]. Lastly, critical infrastructures (and/or some of their services) can be hosted in the cloud [12].

It seems that the cloud computing can be considered a useful resource for assessing and monitoring critical infrastructures' health and security and provide computation power to their services. But this makes the cloud-computing platform a critical infrastructure itself, which in turn, arises questions about their design, security and protection.

A study about cloud as critical infrastructure [13] list few key considerations that must be taken into account when considering deploying critical infrastructures' components and/or services on the cloud. First of all, when we talk about cloud, the term refers to different type of providers, services, deployment models, functions and infrastructures that make difficult to identify where one element of the cloud ends and the other begins. Second, continuous evolving technologies and business practices make difficult government to keep up with evolving technology when choosing requirements for critical infrastructures designation, so the used criteria risk to become obsolete very fast. Third, critical infrastructure designation can address security, robustness and resilience concerns but may neglect to address other cloud governance issues (consumer and enterprise protection, sustainability, human and civil right, etc.). Lastly, critical infrastructure designation may bring increased potential government intervention which in turn may affect service functionality, consumer trust, and privacy.

A paper published by Al-Garibi and colleagues (2020) [14], focused on the risks involved in moving a critical infrastructure (or some part of and/or services it provides) in the cloud, highlight the point that a cloud-hosted critical infrastructure is subject to cybersecurity risks (i.e., confidentiality, availability and integrity if data and information) e these risks are increased because of their services are accessible by any device connected to the Internet. Moreover, being critical infrastructure services, as well as the cloud-computing itself, accessible through the internet, they heavily depend on the stability and quality of the network connection. There is also a cascading effect risk due to the interconnected system's nature of the cloud-hosted critical infrastructure and/or service; in instance, any success of an attacker to destruct or disrupt the cloud services brings the possibilities of negative cascading consequences affecting the different sector of the critical infrastructures.

In the end, cloud computing is a great resource while deploying critical infrastructures (and/or their services), in protecting them by assessing threat's vulnerability and systems' resilience, but at the condition to consider the cloud computing platform itself a critical infrastructure (or entity, as defined in the next section), thus something which must be in turn protected. The inclusion of the cloud computing in the European proposal directive for critical entities resilience (topic of the next section) acknowledge the critical infrastructure status to the cloud computing and, in general, its important role in critical infrastructures' protection and resilience promotion.

5   Further Considerations on Critical Infrastructure Protection

We think that is worth to mention that published papers in the critical infrastructure protection fields are mostly related to theoretical research aspects, as they are aimed at illustrating new paradigms or modelling techniques to assess or evaluate critical infrastructure resilience. Similarly, a considerable amount of research works compare critical infrastructure protection techniques based on risk management solutions with resilience-based solutions in order to promote either the latter typology or a mixed technique implementation by evaluating the respective pros and cons. For instance, just to cite few of them, we can recall a scenario-independent vulnerability method to assess the vulnerability of water distribution networks (WDNs) based on a network entropy model [15]. Another interesting work presents a new method for risk and resilience analysis of gas transmission systems focused on the mitigation of gas transmission disturbances [16]. To the same publication stream belongs the work presented in [17], where a data-driven and digital-twin-based framework for resilience management in human-cyber-physical systems is discussed.

Noteworthy, a large extent of the current scientific literature on these topics revolves around some typical technological solutions and enablers, like as cloud computing [18], Internet of Things [18, 19], digital twins [17–19], artificial intelligence in general [19], and machine learning in most of its flavours, ranging from transfer learning to online learning [17, 18, 20].

This scenario surely opens the doors to interesting and promising further researches, especially those aimed at implementing and applying models, technologies and techniques to real use cases.

6   Critical Infrastructure Protection in the European Union

As introduced in the first section, starting from 2004 the importance of critical infrastructures came to awareness in the European Union. The first framework for critical infrastructure protection was developed in the years 2004–2006 with the initial focus on protecting these infrastructures from terrorism[1], then extending its protection target on all possible threats, with the *European Programme on Critical Infrastructure Protection* and the *Directive on European Critical Infrastructures*[2], including network and information security (NIS Directive) hazards[3] [21].

---

[1] Refer to Commission of the European Communities, "Communication from the Commission to the Council and the European Parliament: Critical Infrastructure Protection in the fight against terrorism" COM(2004)702. Brussels, 20 October 2004.

[2] See Commission of the European Communities, "Communication from the Commission on a European Programme for Critical Infrastructure Protection" COM(2006)702 and Directive of the European Council and the Parliament 2008/114/EC on "the identification and designation of European Critical infrastructures and the assessment of the need to improve their protection".

[3] Refer also to Directive (EU) 2016/1148 of the European Parliament and of the Council of 6 July 2016 concerning measures for a high common level of security of network and information systems across the Union.

The *European Programme on Critical Infrastructure Protection* and the *Directive on European Critical Infrastructures*[4] created a list of the critical infrastructure classified by sectors as follows: Energy, including electricity (generation and transmission infrastructures), oil (production, refining, treatment, storage, transmission) and gas (production, refining, treatment, storage, transmission), and transport, comprising road, rail, air, inland waterways transports, ocean and short shipping and ports.

In 2012, the European Commission published the "Seveso Directive"[5] on the control of major-accident hazards. This directive can be considered a milestone in the previous European protection policies because it extends their field to health, safety, and environment.

A major step and change of direction in the area of security, resilience and cooperation took place on December 16, 2020, with the publication of two proposals for new directives by the Commission. These proposals aimed to promote security and resilience improvement in both physical and cyber domain and in essential services. In detail, the first proposal's aim was toward improving the network information systems protection by repealing the old NIS directive and proposing an updated version (NIS 2.0)[6]. The second proposal, extended the need of protection to a wider class of "objects" called "critical entities". A synthesis of the critical entity's characteristics defined by the European Commission, in their *Proposal for a Directive of the European Parliament and of the Council on the resilience of critical entities*[7], would make a definition like the following:

A *critical entity* is a public or private entity which has been identified as such by an EU Member State taking into account the outcomes of the risk assessment and applying the following criteria: (a) the entity provides one or more essential services; (b) the provision of that service depends on infrastructures located in the Member State; and (c) an incident would have significant disruptive effect on the provision of the service or of other essential services in the sectors that depend on the service.

The NIS 2.0 and the critical entity resilience directives are expected to be promulgated in late 2022 – early 2023 [21]. With those directives' promulgation, European Member States can refer to a complete and inclusive framework useful to face the challenge in the years to come. For an extensive analysis of the normative evolution towards the regulations of critical entities resilience in the EU, see [22].

---

[4] Refer also to Commission of the European Communities, "Communication from the Commission on a European Programme for Critical Infrastructure Protection" COM(2006)702 and Directive of the European Council and the Parliament 2008/114/EC on "the identification and designation of European Critical infrastructures and the assessment of the need to improve their protection".

[5] See Directive 2012/18/EU of the European Parliament and of the Council of 4 July 2012 on the control of major-accident hazards involving dangerous substances, amending a subsequently repeating Council Directive 96/82/EC.

[6] European Commission, Proposal for a Directive of the European Parliament and the Council on measures for a high common level of cybersecurity across the Union, repeating Directive (EU) 2016/1148 (Brussels: European Commission, 2020).

[7] European Commission, Proposal for a Directive of the European Parliament and of the Council on the resilience of critical entities (Brussels: European Commission, 2020).

7 Conclusion

The heavy dependence of modern Societies, and the wellness of their citizens, on services (material and immaterial) and goods provided by the so-called Critical Infrastructures and, more in general, by Critical Entities is well acknowledged. Their vulnerability to many kinds of hazards and threats, whose origin might be either man-made (i.e., terrorist attacks, cyberattacks) or natural (floods, landslides, earthquakes, etc.) is also so well acknowledged that, in the past decades, several risk management techniques have been employed to preserve critical infrastructure's service continuity.

Risk management techniques, however, proved not to be able to anticipate rare events with major consequences (i.e., earthquakes, tsunamis, and, recently, pandemics and wars). To overcome these limits the concept of resilience – namely the capacity of an entity to mediate between performance and uncertain conditions (i.e., critical and disrupting events, major accidents or continuous stress) so to maintain or regain a dynamically stable state that allows it to continue operations – was explored. Several models have been identified to support resilience management to protect critical entities. Typically, critical infrastructures rarely have the resources (bandwidth, computing power, storage, etc.) required by these models. Cloud storage and cloud computing are useful resources to overcome these limits, and they can also host critical infrastructures (or their parts) in the cloud. This makes these cloud entities critical infrastructures as well, which in turn bring them protection needs. This is also acknowledged by the European Commission by including cloud computing services in European critical entity sector listing.

It seems that National approaches to critical entity protection are not anymore sufficient because of the involved entities and the complexity of the threats. Moreover, having different protection policies and approaches in different European Nations became cumbersome to manage especially when considering complex infrastructures interdependences crossing national boundaries. These are some of the considerations that lead to the need of building a coherent and cooperative approach to critical entities' security and protection shared and shareable within the EU State Members. This has driven the European Commission to discuss a critical entity resilience directive which is expected to be promulgated in late 2022 – early 2023.

Acknowledgement. This research activity is partially funded by the Italian research programmes "Brindisi Smart City Port", in the framework of the PAC - PON "Infrastrutture e reti" 2014–2020, and "PON Ricerca e Innovazione 2014–2020 (DM n.1062, 10 August 2021), in the framework of "The Italian Data Lake for Energy (ItaDL4E)" project.

References

1. Guo, D., Shan, M., Owusu, E.: Resilience assessment frameworks of critical infrastructures: state-of-the-art review. Buildings **11**(10), 464 (2021). https://doi.org/10.3390/buildings111 00464
2. Mottahedi, A., Sereshki, F., Ataei, M., Qarahasanlou, A.N., Barabadi, A.: The resilience of critical infrastructure systems: a systematic literature review. Energies **14**(6), 1571 (2021). https://doi.org/10.3390/en14061571

3. Wears, R.L.: Resilience engineering: concepts and precepts. Qual. Saf. Health Care **15**(6), 447–448 (2006). https://doi.org/10.1136/qshc.2006.018390
4. Wied, M., Oehmen, J., Welo, T.: Conceptualizing resilience in engineering systems: An analysis of the literature. Syst. Eng. **23**(1), 3–13 (2020). https://doi.org/10.1002/sys.21491
5. Panteli, M., Mancarella, P., Trakas, D.N., Kyriakides, E., Hatziargyriou, N.D.: Metrics and quantification of operational and infrastructure resilience in power systems. IEEE Trans. Power Syst. **32**(6), 4732–4742 (2017). https://doi.org/10.1109/TPWRS.2017.2664141
6. Bruneau, M., et al.: A framework to quantitatively assess and enhance the seismic resilience of communities. Earthq. Spectra **19**(4), 733–752 (2003). https://doi.org/10.1193/1.1623497
7. Curt, C., Tacnet, J.M.: Resilience of critical infrastructures: review and analysis of current approaches. Risk Anal. **38**(11), 2441–2458 (2018). https://doi.org/10.1111/risa.13166
8. Adger, W.N.: Social and ecological resilience: are they related? Prog. Hum. Geogr. **24**(3), 347–364 (2000). https://doi.org/10.1191/030913200701540465
9. Labaka, L., Hernantes, J., Sarriegi, J.M.: A holistic framework for building critical infrastructure resilience. Technol. Forecast. Soc. Change **103**, 21–33 (2016). https://doi.org/10.1016/j.techfore.2015.11.005
10. Chen, Z., et al.: A cloud computing based network monitoring and threat detection system for critical infrastructures. Big Data Res. **3**, 10–23 (2016). https://doi.org/10.1016/j.bdr.2015.11.002
11. Ficco, M., Choraś, M., Kozik, R.: Simulation platform for cyber-security and vulnerability analysis of critical infrastructures. J. Comput. Sci. **22**, 179–186 (2017). https://doi.org/10.1016/j.jocs.2017.03.025
12. MacDermott, Á., Shi, Q., Merabti, M., Kifayat, K.: Hosting critical infrastructure services in the cloud environment considerations. Int. J. Crit. Infrastruct. **11**(4), 365 (2015). https://doi.org/10.1504/IJCIS.2015.073843
13. Cloud as Critical Infrastructure Key Considerations. https://cloud.carnegieendowment.org/macro-issues/critical-infrastructure/. Accessed 28 Nov 2022
14. AlGharibi, M., Warren, M., Yeoh, W.: Risks of critical infrastructure adoption of cloud computing by government. Int. J. Cyber Warfare Terror. **10**(3), 47–58 (2020). https://doi.org/10.4018/IJCWT.2020070104
15. Wang, F., Zheng, X., Li, N., Shen, X.: Systemic vulnerability assessment of urban water distribution networks considering failure scenario uncertainty. Int. J. Crit. Infrastruct. Prot. **26**, 100299 (2019). https://doi.org/10.1016/j.ijcip.2019.05.002
16. Zalitis, I., et al.: Mitigation of the impact of disturbances in gas transmission systems. Int. J. Critic. Infrastruct. Protect. **39**, 100569 (2022). https://doi.org/10.1016/j.ijcip.2022.100569
17. Bellini, E., et al.: Resilience learning through self adaptation in digital twins of human-cyber-physical systems. In: 2021 IEEE International Conference on Cyber Security and Resilience (CSR), July 2021, pp. 168–173 (2021). https://doi.org/10.1109/CSR51186.2021.9527913
18. Azari, M.S., Flammini, F., Santini, S.: Improving resilience in cyber-physical systems based on transfer learning. In: 2022 IEEE International Conference on Cyber Security and Resilience (CSR), July 2022, pp. 203–208 (2022). https://doi.org/10.1109/CSR54599.2022.9850282
19. Argyroudis, S.A., et al.: Digital technologies can enhance climate resilience of critical infrastructure. Clim. Risk Manag. **35**, 100387 (2022). https://doi.org/10.1016/j.crm.2021.100387
20. Derras, B., Makhoul, N.: An overview of the infrastructure seismic resilience assessment using artificial intelligence and machine-learning algorithms. In: Proceedings of the International Conference on Natural Hazards and Infrastructure (2022)
21. Castiglioni, M., Lazari, A.: The normative landscape in security and resilience: the future of critical infrastructures and essential services in the EU. In: Martino, L., Gamal, N., (Eds.) European Cybersecurity in Context A Policy-Oriented Comparative Analysis, European Liberal Forum, pp. 37–42 (2022)

22. Pursiainen, C., Kytömaa, E.: From European critical infrastructure protection to the resilience of European critical entities: what does it mean? Sustain Resilient Infrastruct. **08**, 1–17 (2022). https://doi.org/10.1080/23789689.2022.2128562

Towards a Parallel Graph Approach to Drug Discovery

Dario Branco[3(✉)], Beniamino Di Martino[1,2,3], Sandro Cosconati[4], Dieter Kranzlmueller[5], and Salvatore D'Angelo[3]

[1] Department of Engineering, University of Campania "Luigi Vanvitelli", Caserta, Italy
beniamino.dimartino@unicampania.it
[2] Department of Computer Science and Information Engineering, Asia University, Taichung, Taiwan
[3] Department of Engineering, University of Campania, via Roma 29, 81031 Aversa, Italy
dario.branco@unicampania.it
[4] DiSTABiF, University of Campania Luigi Vanvitelli, Via Vivaldi 43, 81100 Caserta, Italy
sandro.cosconat@unicampania.it
[5] Leibniz Supercomputing Centre (LRZ) of the Bavarian Academy of Sciences and Humanities, Boltzmannstrasse 1, 85748 Garching bei München, Germany
dieter.kranzlmueller@lrz.de

Abstract. This work deals with the problem of recognising chemical entities that are able to act as polypharmacological compounds by binding and modulating two different proteins that have been demonstrated to be critical for cancer metastatic potential and aggressivity. In particular, the aim is to develop a method that automatically indicates which of the given compounds are likely to be 'active' with respect to the two proteins in question. In medicinal chemistry, an active compound is defined as a ligand that is capable of modulating the e-mail: dieter.kranzlmueller@lrz.de. In this work, on the other hand, we aim to develop a parallel algorithm that is able to provide an exact solution to the problem using classical techniques of isomorphism and similarity between graphs.

1 Introduction

This work deals with the problem of recognising chemical entities that are able to act as polypharmacological compounds by binding and modulating two different proteins that have been demonstrated to be critical for cancer metastatic potential and aggressivity. In particular, the aim is to develop a method that automatically indicates which of the given compounds are likely to be 'active' with respect to the two proteins in question. In medicinal chemistry, an active compound is defined as a ligand that is capable of modulating the activity a given biological target (i.e. proteins and nucleic acids). In the field of university

© The Author(s), under exclusive license to Springer Nature Switzerland AG 2023
L. Barolli (Ed.): AINA 2023, LNNS 655, pp. 127–135, 2023.
https://doi.org/10.1007/978-3-031-28694-0_12

research, therefore, a tool is needed that, upstream of the experimental verification process, returns with an acceptable degree of certainty a list of compounds that are likely to be active. It is therefore necessary to draw up algorithms, based on a criterion provided by a domain expert, to define a probability percentage for which a compound is active. In the literature, this problem is often addressed by machine learning techniques and in particular by graph-based convolutional neural network. In this work, on the other hand, we aim to develop a parallel algorithm that is able to provide an exact solution to the problem using classical techniques of isomorphism and similarity between graphs.

2   Problem Overview

The drug discovery pipeline requires the selection of a specific disease to cure, the identification of a biological target to modulate, and the search for potential chemical entities that, by modulating the biological target, can ameliorate or cure the disease The structure and physicochemical properties of molecules that are already described as being active against the target are key information for the rational design of new and more potent compounds. To use this information, it is critical to translate it into significant data for further estimation and analysis. Chemoinformatics is the field that allows to link computational and chemical data and allows for predicting the chemical and pharmacological properties in the drug discovery frame.

To date, there are almost 92 million compounds and billions of chemical entities that can be potentially synthesized. Conversely, almost 10000 drugs have been discovered so far that bind and modulate 60% of the human proteome to treat human diseases. In this context, computational methods allow to in silico predict the biological as well as physical properties of a chemical entity. To date, these methods permeate all the aspects of modern drug discovery spanning from target identification and validation, ligand design and optimization as well as prediction of pharmacokinetic parameters. Very recently, there has been an increasing interest in deploying of new computational methods to overcome the intrinsic limitations of classical computational methods. One of the most striking examples is the employment of artificial intelligence algorithms (machine learning and deep learning) to complement classical methods in the prediction of the biological and physical properties of a new chemical entity. Despite encouraging results in this direction, the application of AI in silico drug discovery is still in its infancy and much is still needed to substantiate the actual utility of these new methods in this field.

Thus, the development of new alternative methods is still in great demand, and, in this respect, this contribution aims at providing a new computational method that, by making use of classical techniques of isomorphism and similarity between graphs can be implemented in the sub-field of virtual screening (VS).

The main idea of this ligand discovery approach is to use a computational method to predict if a compound present in chemical libraries will be able to bind and possibly modulate the biological target of interest. VS can be divided in ligand- and receptor-based VS. In the first approach the structural information available for ligands already described as being active for the target is used to feed a computational algorithm that by measuring the bi-dimensional (2D) or three-dimensional (3D) similarities between the active ligands and the one for which a prediction is requested will classify this latter as being active or inactive. On the other hand, in structure-based methods, the steric and electrostatic complementarity between the test ligand and the 3D structure of the biological target is measured through a computational method (i.e. docking calculations). This allows for predicting if a compound will be able to bind with a certain affinity to the given target.

3   Related Work

There have been numerous studies in the field of drug discovery that have used machine learning and deep learning techniques to predict the biological and physical properties of new chemical entities [1]. For example, several studies have used artificial neural networks (ANNs) [8] to predict the binding affinity of small molecules to a target protein, using a variety of molecular descriptors as input features. Another popular approach is the use of graph-based convolutional neural networks (GCNNs) for virtual screening [4]. These methods use the chemical structure of a molecule as a graph, with atoms as nodes and chemical bonds as edges, and apply convolutional filters to extract features from the graph [6]. These features are then used to predict the binding affinity of the molecule to a target protein. There have also been a number of studies that have used isomorphism and similarity between graphs as a method for virtual screening. For example, other studies used the maximum common subgraph (MCS) [3] algorithm to find structural similarities between a set of active compounds [5]. In general, these studies have shown that machine learning and graph-based methods can be effective for virtual screening, and have the potential to improve the efficiency and accuracy of drug discovery. However, the application of AI in drug discovery is still in its infancy, and much more research is needed to fully understand the potential of these methods in the drug discovery process.

4   Data Preprocessing

Data used within this work are originally in the SMILE format (acronym of Simplified Molecular Input Line Entry System). SMILE format is a method for the description of the molecular structure using a short ASCII string. In terms of graph-based computational procedures, SMILES is a string obtained by printing the symbols of the nodes on the graph representing the structure formula. Hydrogen atoms are first removed from the graph, then cycles are opened to convert the graph into an open tree. Where loops have been opened, numerical suffixes are added to indicate which nodes are connected. The branches of the tree are indicated through the use of parentheses. The SMILE extension for the structural description of molecules, as is evident from the previous paragraph, is a valuable aid and a powerful means of representing molecules in graphs, but it is not compatible, at least directly, with the tools used. Therefore, a first step was to convert the dataset entries in order to be able to use them in the graph manipulation tools. Smile files were initially converted to Spatial Data File (SDF) format. This is a single-user geodatabase file format developed by Autodesk. The design of the SDF format utilises low-level storage components from SQLite using flat binary serialisation. The result is a set of large binary objects (BLOB). This format is still incompatible with the Neo4j tool, which is the tool used for this work, so a second conversion is necessary. Through the SDFEater tool, therefore, the SDF files were converted into Neo4j-readable Cypher files. The entire data preprocessing process can be represented by means of the diagram in the Fig. 1. In addition, in order to create the dataset useful for the purposes of this work, the compounds were tagged with an identifier indicating whether they are active towards the first protein, active towards the second protein or part of the test dataset.

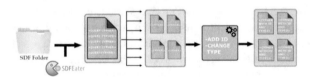

Fig. 1. Data preprocessing pipeline

5   Methodology

Once we have obtained a graph representation of the molecules, our problem reduces to finding all possible subgraphs common to N graphs stored in an array. The basic idea is to search the database of compounds active towards the first protein for common substructures and to repeat the process for the second dataset of compounds active towards the second protein. At this point, a list of substructures that are located in a certain percentage within the two datasets is generated, the objective being to search, within the third dataset, for molecules that present substructures with a high percentage of recurrence in both the first dataset and the second dataset. The second part of the problem is actually the computationally less demanding one, as in the literature there are algorithms for graphs that, given a pattern and a graph, search for the pattern within the graph and return a value indicating its presence or absence. On the other hand, the most computationally expensive part of the problem is the search for the most recurring subgraphs within the dataset since it requires:

1. Generate all possible subgraphs of a graph
2. Search within the other graphs in the dataset for the subgraphs found
3. Repeat steps 1 and 2 for all graphs
4. Sort the vector of common subgraphs found by occurrence (Fig. 2)

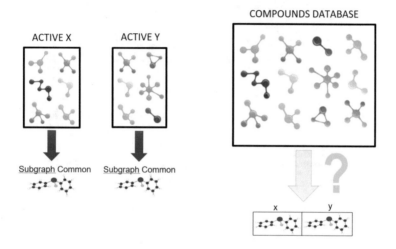

Fig. 2. Problem infographic

It is clear that without any kind of simplification of the problem and without any kind of parallelisation of the computation, solving this algorithm is very hard. We therefore focused on defining initial assumptions such as the maximum length of the graph pattern in order to make the proposed solution viable. Furthermore, we proceeded with the parallelisation of the code in order to speed up the computation. We define:

- G: Array containing the graphs to be compared
- i and k: indices of the pair of graphs to be compared
- F: Array of the node combinations
- j: index of the combinations (Fig. 3)

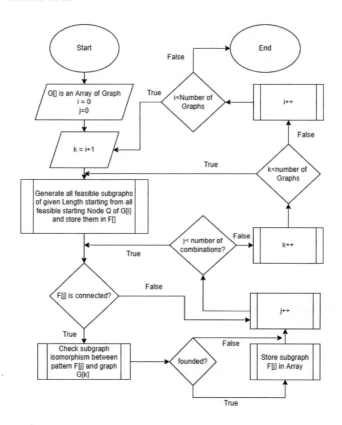

Fig. 3. Algorithm flowchart

Following the generation of the subgraphs of $G[i]$ and the matching of each of them with each possible subgraph of $G[k]$, each subgraph that verifies the condition of sub-isomorphism is saved in a three-dimensional matrix. This matrix is composed on the horizontal plane by the n and m indices representing the n–th and m–th graphs. On the vertical plane, for each pair of compounds n-m, all the subgraphs present in both $G[n]$ and $G[m]$ are saved. Of this matrix, the occurrences of the individual subgraphs are counted, this can only be done by applying again the exact isomorphism between all elements of the matrix. Another matrix is then generated, this time two-dimensional. Assuming we have applied the algorithm on a number of graphs equal to 100, the matrix will similar to Table 1.

Table 1. Number of occurrencies in Graphs

SubGraph	Occurrences	% of presence
Graph.Object	22	22%
Graph.Object	76	76%
Graph.Objcct	14	14%
Graph.Object	27	27%

Where, the number of occurrences is the number of times a subgraph appears in a pair of different graphs, the percentage occurrence is calculated as:

$$\%ofPresence = \frac{NumberOccurences}{NumberofGraphs} * 100 \tag{1}$$

Now apply the entire algorithm to both data sets:

- Active compounds towards P1
- Active compounds towards P2

Note that the result will be two matrices in the form of Table 1, one containing the information of all the most recurrent subgraphs in the first dataset and the other containing the most recurrent information of the second dataset. We now apply the function that check the isomorphism between a pattern and a graph using as arguments each element of the third dataset (containing the compounds to be tested) and each element of both two-dimensional matrices that has a % of recurrence higher than 50%. Compounds in the third dataset that have at least two matches, one in each matrix, are marked as possibly active on both proteins. Compounds that have matches in just one matrix are marked as possibly active on the protein corresponding to the matrix.

6 Algorithm Parallelization and Results

In this section, we will discuss a first implementation of the parallelized algorithm proposed in its Python version. This parallel algorithm was run on three high-performance machines that have the characteristics reported in Table 2.

Table 2. System specifications

System	Cores	RAM	HDD
Master	40	180 GB	500 GB
Slave-1	20	90 GB	50 GB
Slave-2	20	90 GB	50 GB

The parallelization is necessary because the algorithm is computationally very heavy, especially when increasing the maximum size of the subgraphs to be generated. Using a distributed approach, each node of the cluster takes care of generating the subgraphs of a single graph and comparing them to the other graphs in the belonging dataset, significantly speeding up the time required for the execution of the algorithm. The parallelization was carried out using the Ray library, written for Python, which allows for running code in a parallel manner both for data parallelism and task parallelism. In this case, the features of Ray were crucial in performing distributed computation in the described Cluster. In particular, Ray allowed for dividing the execution of the subgraph generation and isomorphism functions evenly among the various nodes of the cluster. Furthermore, subgraph generation was parallelized within the same core by using multiple processes in a single node. Thanks to the design decision described above, it was possible to use not only all the nodes of the cluster but also all the cores available in each node of the cluster by using two levels of parallelization. The first level is composed of the distributed distribution of tasks and data through the use of Ray on the various nodes of the cluster, the second level is composed of the parallelization on a single machine by dividing the subgraph generation work among the various cores of the single node.

The algorithm was tested with 2 pairs of compounds, about 100 subgraphs were found with an occurrence percentage within the pairs greater than 50%, limiting the maximum number of nodes per subgraph to 4. This example is only provided for illustrative purposes, clearly increasing the minimum and maximum size of the subgraphs to generate the number of subgraphs in common between two graphs is clearly smaller. The algorithm was launched for a greater number of pairs and has, as a function of this variable (Number of graphs), a polynomial complexity.

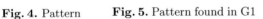

Fig. 4. Pattern   **Fig. 5.** Pattern found in G1   **Fig. 6.** Pattern found in G2

In Fig. 4, one of the subgraphs found in graph G1 is shown. In adjacent Figs. 5 and 6, it is respectively shown that the pattern has been found in both graph G1 and graph G2, highlighting it with respect to the rest of the graph.

7   Conclusions and Future Work

In conclusion, this paper presents a computational method for recognizing chemical entities that have the potential to act as polypharmacological compounds by binding and modulating two different proteins that have been demonstrated to be critical for cancer metastatic potential and aggressivity. The method aims to develop a parallel algorithm that is able to provide an exact solution to the problem using classical techniques of isomorphism and similarity between graphs. This approach is intended to be used in the sub-field of virtual screening in drug discovery pipeline. The paper highlights the importance of new computational methods in the drug discovery process and the potential utility of AI in drug discovery. As a future development of this work, we aim to increase the degree of parallelism of the algorithm in order to optimize its execution speed and to seek assumptions that simplify the problem without invalidating the solution. We will also investigate quantum computing techniques for the parallel computation of possible subgraphs. We will also investigate the possibility to use a Big Data Pipeline [2] in order to store and retrieve graphs in a more efficient, fast and structured way combined with the approach based on computational patters and automatic deployments based on primitives presented in [7].

References

1. Agatonovic-Kustrin, S., Beresford, R.: Basic concepts of artificial neural network (ANN) modeling and its application in pharmaceutical research. J. Pharm. Biomed. Anal. **22**(5), 717–727 (2000)
2. Branco, D., Di Martino, B., Venticinque, S.: A big data analysis and visualization pipeline for green and sustainable mobility. In: Barolli, L., Woungang, I., Enokido, T. (eds.) AINA 2021. LNNS, vol. 227, pp. 701–710. Springer, Cham (2021). https://doi.org/10.1007/978-3-030-75078-7_69
3. Ehrlich, H.-C., Rarey, M.: Maximum common subgraph isomorphism algorithms and their applications in molecular science: a review. WIREs Comput. Mol. Sci. **1**(1), 68–79 (2011)
4. Han, K., Lakshminarayanan, B., Liu, J.Z.: Reliable graph neural networks for drug discovery under distributional shift. CoRR, abs/2111.12951 (2021)
5. Jayaraj, P.B., Rahamathulla, K., Gopakumar, G.: A GPU based maximum common subgraph algorithm for drug discovery applications. In: 2016 IEEE International Parallel and Distributed Processing Symposium Workshops (IPDPSW), pp. 580–588 (2016)
6. Jing, Y., Bian, Y., Hu, Z., Wang, L., Xie, X.-Q.S.: Deep learning for drug design: an artificial intelligence paradigm for drug discovery in the big data era. AAPS J. **20**(3) (2018). Article number: 58. https://doi.org/10.1208/s12248-018-0210-0
7. Martinez, I., Montero, J., Pariente, T., Di Martino, B., D'Angelo, S., Esposito, A.: Parallelization and deployment of big data algorithms: the toreador approach. In: 2018 32nd International Conference on Advanced Information Networking and Applications Workshops (WAINA), pp. 408–412. IEEE (2018)
8. Yinqiu, X., Yao, H., Lin, K.: An overview of neural networks for drug discovery and the inputs used. Expert Opin. Drug Discov. **13**(12), 1091–1102 (2018). PMID: 30449189

Experiences in Architectural Design and Deployment of eHealth and Environmental Applications for Cloud-Edge Continuum

Atakan Aral[1], Antonio Esposito[2], Andrey Nagiyev[1], Siegfried Benkner[1], Beniamino Di Martino[2], and Mario A. Bochicchio[3,4(✉)]

[1] Faculty of Computer Science, University of Vienna, Vienna, Austria
{atakan.aral,andrey.nagiyev,siegfried.benkner}@univie.ac.at

[2] Università degli Studi della Campania Luigi Vanvitelli, Via Roma 29, 81031 Aversa, CE, Italy
{antonio.esposito,beniamino.dimartino}@unicampania.it

[3] Dipartimento di Informatica, Università degli Studi di Bari Aldo Moro, via E. Orabona, 4, 70125 Bari, Italy
mario.bochicchio@uniba.it

[4] CINI Consorzio Interuniversitario Nazionale per l'Informatica, via Ariosto 25, 00185 Roma, Italy
mario.bochicchio@consorzio-cini.it

Abstract. The Cloud-Edge continuum has lately exponentially grown, thanks to the increase in the availability of computational power in Edge Devices, and the better capabilities of communication networks. In this paper, two use cases, in eHealth and environmental domain, are presented in order to provide an application context to exemplify the approaches driving the analysis and selection of Cloud-Edge architectural solutions and patterns, the structural design, the allocation and deployment of distributed applications targeted to the Cloud Continuum. The main focus of this paper is the comparison of the architectural choices made for the two use cases, and how they have been driven by typical non-functional requirements, guiding the adoption of a Cloud Continuum solution.

1 Introduction

The Cloud-Edge continuum has lately exponentially grown, thanks to the increase in the availability of computational power in Edge Devices and the better capabilities of communication networks that, as of today, can support massive exchanges of data with high reliability.

There are still drawbacks to this approach, regarding the necessity to optimally manage the available computational resources, and to avoid traffic congestion that can always happen, even with modern networks. The main drive

---

A. Aral and A. Esposito—Contributed equally to this work as the first authors.

L. Barolli (Ed.): AINA 2023, LNNS 655, pp. 136–145, 2023.
https://doi.org/10.1007/978-3-031-28694-0_13

behind the use of the Cloud-Edge continuum paradigm resides in the necessity to keep data elaboration close to data sources, to avoid privacy and security issues, and to better organize resource management. However, this requires a clear decomposition of the applications' components that need to be re-thought in a completely distributed manner and having in mind that Edge Devices do not possess infinite computational power. The use of the Cloud-Edge continuum, which exploits centralized Cloud resources to satisfy the computational needs of the application, can resolve several resource scarcity problems but still requires to be accurately designed. Patterns to define Cloud-Edge architectures and to identify the optimal data workflows in such applications are currently being identified, studied, and applied, but there is still a strong need for standardization and general acceptance from programmers' communities.

Edge computing was originally proposed to satisfy four non-functional requirements, namely high responsiveness, scalability, privacy enforcement, and fault tolerance [1]. However, more recent, innovative applications of Cloud-Edge call for further requirements, including low network accessibility, deployment of personalized software configurations, computational offloading, and energy management.

In this paper, two use cases, in the eHealth and environmental domain, are presented in order to provide an application context to exemplify the approaches driving the analysis and selection of Cloud-Edge architectural solutions and patterns, the structural design, the allocation and deployment of distributed applications targeted to the Cloud Continuum. The main focus of the paper is the comparison of the architectural choices made for the two use cases and how they have been driven by the abovementioned non-functional requirements, guiding the adoption of a Cloud Continuum solution. Moreover, we discuss how these requirements entail unique hardware designs for the two use cases.

2   Reference Architectures for the Cloud-Edge Continuum

2.1   Multi-layer Architecture

The Cloud-Edge paradigm is still in full development, especially as regards computational, energetic and privacy requirements expressed by distributed applications. Therefore, there is the need to develop techniques for the efficient definition of architectures, deployment and management methodologies for algorithms, to be run on distributed devices, and schedule the computation offload.

Several architectural solutions are being developed to support the efficient development of Cloud-Edge platforms, with a particular interest in Mobile scenarios [2]. Patterns have been proposed for Cloud-Edge, also by private, commercial organizations that apply them in their everyday activities.

Cloud-Edge Patterns can be divided into four main categories:

Architectural Cloud-Edge Patterns that support the creation of Cloud-Edge Architectures, and that are strongly influenced by multi-layered approaches already in use in several distributed paradigms. Computational capabilities of the Edge Nodes are the main drive behind the selection of a specific Architectural Pattern.

Cloud-Edge Patterns for Data are less focused on architectural design, and more interested in discussing how data should be transferred among distributed components.

Deployment Cloud-Edge Patterns define how the deployment of distributed components should be handled in a distributed environment, especially when different versions of the same application may be running on the Edge Nodes.

Among the Architectural Patterns, one of the most complex is represented by the **Multi-tier Architecture Pattern with Edge Orchestrator** [3] shown in Fig. 1a. Such a Pattern describes a common situation in which the Cloud computational capabilities are separated from the Edge network, and they are both managed separately. The Edge layer is indeed organised through a Metropolitan Area Network (MAN), where and Edge Orchestrator specifically manages the computational offload among the Edge nodes.

Cloud-Edge Patterns for Data greatly differ according to the specific problem they focus on [4]. Common proposed solutions regard Synchronous and Asynchronous accesses to data produced by Edge Devices, or describe the exact workflow of the data streaming through the Edge framework. Figure 1b shows the Subsequent Data Retrieval Pattern, where the organization of the data flow is clearly presented.

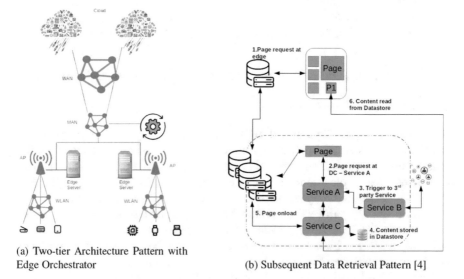

(a) Two-tier Architecture Pattern with Edge Orchestrator

(b) Subsequent Data Retrieval Pattern [4]

Fig. 1. Cloud Edge Patterns examples

Patterns for Data management in complex Cloud-Edge systems are necessary, in order to describe the correct workflow of information within such systems and to optimize the energy consumption and computational loads for all the collaborating devices. To stress a privacy-by-design development of distributed algorithms, Federated Learning techniques have arisen [5–7], which guarantee the absence of personal or identifiable data in the parameters exchanged over networks to train Machine and Deep learning algorithms. Federated Learning (FL) approaches are extremely useful for the efficient exploitation of computational nodes in distributed environments, and most importantly for the positive impact they have on privacy. Since several FL-oriented approaches are possible, being guided by architectural and computational patterns becomes fundamental to reduce risks. Indeed, FL Patterns and reference architectures have already been defined [8,9]. Applications of such patterns are available [10], but are still quite limited. Also, support for the application of such Patterns and the development of FL algorithms is still missing.

2.2 Event-Driven Architecture

Another highly connected complex of approaches and patterns for distributed applications is based on the event-driven architecture (EDA) [11]. The methods and techniques for using EDA are constantly changing and improving by adding new viewpoints and interaction mechanisms. Nevertheless, the immutable cornerstone of the architecture is the concept of the event, as an entity carrying a certain state throughout the system.

Considering the application of EDA for the Cloud-Edge paradigm, it operates with a sequence of events with a certain topic, forming a stream that transmits information from the edge devices as producers of the data, to clients consuming these events by a subscription of this topic. At the same instant, amongst the most fundamental capabilities of EDA is the availability of the data, which is achieved by the usage of mechanisms inside the brokers, acting as middleware nodes between producers and consumers. Brokers are combined into an event-driven cluster and are used as physical nodes, providing the mechanisms of replication, retention, partitioning and availability of data streams. Event-driven patterns can implement the approaches for the logical management under streams, organizing complex event processing and providing flexible procedures for the transmission of data from edge devices.

EDA has been implemented in commercial tools, e.g. [12], which are used in various organizations and activity areas, providing flexible loose coupling between different parts of systems, and supporting near-real-time interaction. Examples of the implementation of EDA to varying degrees might be such tools as Kafka or RabbitMQ. The application of the event-driven paradigm to FL and other Cloud-Edge application scenarios allows ample room for research and further improvements [13].

3   Applications to the Use Cases

3.1   E-Health Application

E-health applications have surely flourished thanks to the availability of advanced computational power, raw data to be analysed and, above all, distributed devices that can tackle part of the required computations locally.

The E-health use case that will be used here as an example regards the monitoring of maternal and fetal health status during pregnancy. In particular, we consider a monitoring system that includes wearable devices and series of sensors connected to it, that detect vital parameters (temperature, heart rate, blood oxygenation level, and blood pressure variances) of the mother and fetus and report any detected abnormalities to a central system. Considering that there were more than 4 million births in Europe in 2020 (about 140M worldwide) and that about one third of these are associated with health problems for the mother or the fetus, the potential user segment for such a device translates to more than one million mothers/year in Europe, and 35M mothers/years worldwide. These figures triple when considering that during the COVID-19 pandemic, remote monitoring proved essential to limit the risk of infection for pregnant women. Several companies in Europe and the U.S. are working towards the same goal, but none has yet developed clear leadership, in part because of the technical difficulties inherent in solving the problem. A first important challenge faced by this kind of Use Cases, indeed, is about the non-invasive monitoring of fetal health through sensors placed on the maternal abdomen. This requires the development and fine-tuning of sophisticated de-noising techniques and subsequent separation of signals originating from the maternal body (heartbeat, uterine contractions, muscle activity) from those originating from the fetus (mainly cardiac signal). To solve this problem, the most advanced and performing solutions make extensive use of machine learning techniques [14] that require a significant computational load and an energy expenditure that exceed the computational capacity and energy autonomy of processors currently used in wearable devices.

Figure 2a reports the envisioned target architecture for the implementation of the Use Case. Such an architecture follows the Two Tiers Pattern shown in Fig. 1a, by dividing the overall framework into two main layers: the Edge tier and the Cloud Tier. In particular, the Edge Tier comprehends:

- The **Data Ingestion Layer** represents the input section of the architecture, that is responsible for the acquisition of data. In particular, this architecture foresees the use of a generic Detector component, which represents the sensors acquiring the data. For the Use Case, the Detector represents the tools used to detect the signals from the mother and the fetus during the monitoring activities.
- The **Application Logic Layer** is in charge of elaborating the incoming data to obtain the expected results. Such a layer contains three sub-components, represented by: the Operator, that is the actual computational node; the MessageCompressor, acting on exchanged messages; the SecureAggregator that

focuses on security aspects. The Operator represents here the computational capabilities of the wearable devices used to monitor the Mother and Fetus vitals.

The Cloud Tier comprehends, instead:

- The **Resource Management Layer** is in charge of the critical aspect of managing computational and data resources within the envisioned framework. Load balancing, scaling and traffic management are the main responsibilities of such a layer. This layer acts as the Edge Orchestrator, as reported in the Architectural Pattern in Fig. 1a.
- The **Application Logic Layer** is in charge of elaborating the incoming data, to obtain the expected results. Such a layer contains five sub-components, represented by: the Processor, that is the computational node/cluster of nodes within the Cloud. A MessageCompressor and a SecureAggregator as in the Edge tier; a ClientRegistry and a ClientSelector that are in charge of deciding the target Cloud platform, when multiple ones are available.

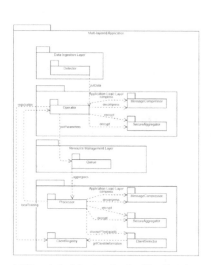

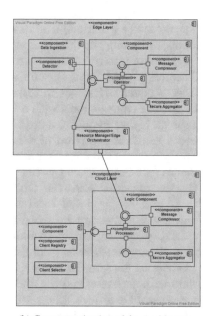

(a) Target Architecture for the Use Case implementation

(b) Componentization of the Architecture

Fig. 2. Cloud Edge Patterns examples

Figure 2b shows how the Architecture can be componentised and stresses the fact that communications between the layers only happen through the Resource management component.

The communication path followed by the application is better described through the Sequence Diagram in Fig. 3. All the information follow a linear flow

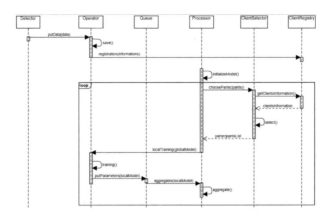

Fig. 3. Communications among components in the E-health use case

through the components of the application, with Detector and Operator communicating the data directly. The Processor retrieves the data from a common data structure (a generic Queue in the Architecture of Fig. 2a), and it operates in a loop, retrieving the data and elaborating them locally. Different Processors can participate into the elaboration of the data.

Theoretically speaking, the minimum sampling frequency for an optimal ECG recorder is equal 50 Hz, but in order to obtain precise measurements a typical ECG recorder samples data with a frequency of more 500 Hz [15]. In some situations even higher frequencies are required: this means that the Detector needs to feed a continuous stream of data to the local Operator, which in turn has to rapidly analyse the data and provide immediate feedback, in order to detect anomalies (*putData* method in the Sequence Diagram in Fig. 3).

On the opposite side, the Operator will not send the data (*registration* method) and the updated local model (*putParameters* method) frequently. Indeed, the updates can be scheduled on a regular basis, and their rate can vary a lot, according to the overall settings of the system. It is evident that dividing the system between the Edge Tier, with Detector and Operator working at strict contact, and a remote Cloud Tier, where the Processor can analyse the data without haste, and the update of the global model must be coordinated among several Operators, seems a feasible and suitable solution.

After extracting the fetal heart rate (FHR) from the maternal abdominal signal, this information can also be used to trigger alarms when its level is outside the standard range for the specific gestational age and context (e.g., walking, sleeping, etc.). Similar alarms can be associated with the other monitored vital signs (e.g., SPO2, body temperature, blood pressure, blood glucose level) based on the specific risks associated with the patient. This scheme of customized triggers and alarms fits very well with the EDA approach mentioned above.

3.2 Environmental Monitoring Application

Environmental monitoring and real-time decision-making are essential to environmental protection. Various applications towards pollution monitoring (air, water, soil, etc.) and disaster early warning (seismic activity, avalanches, etc.) already benefit from Cloud-Edge deployment [16]. In water quality monitoring, there exist monitoring systems for marine regions and freshwater bodies (both ground and surface water). SWAIN project (https://swain-project.eu/) focuses on surface waters, particularly rivers. The project aims to detect and locate pollutant sources (e.g., industrial leaks or failed wastewater treatment plants) through an unprecedented implementation of Edge Computing and IoT for the real-time analysis of water contamination data. Timely decision-making is crucial in this use case as the river water polluted upstream might be used downstream for irrigation or for municipal water intake.

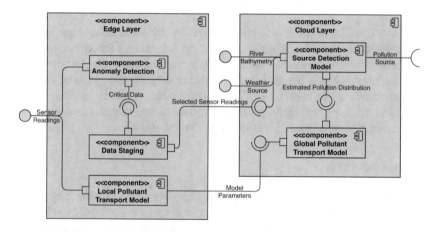

Fig. 4. Component diagram for the environmental monitoring use case.

Data analytics components of this application are deployed on the Cloud-Edge continuum in order to benefit from its responsiveness and network access advantages. Figure 4 visualizes these components and their interfaces. Since watersheds are located in remote areas, it is not always possible (due to the lack of reliable networks) to transmit sensor readings to the cloud, where complex river models can be executed. The edge subsystem, therefore, acts as intermediate data storage and pre-processing locations, which mitigates delayed decisions or data loss. As demonstrated in Fig. 4, the transport model of pollutants is decomposed into local and global components. Local components are less complex since they estimate the pollutant distribution only in their local geographical area. They transmit trained model parameters to the Cloud subsystem, where a global model incorporates the local updates. Based on the critical data readings selected by the Edge subsystem and the pollutant distribution estimated by the pollutant transport model, the source detection model is able to identify the source of the pollution.

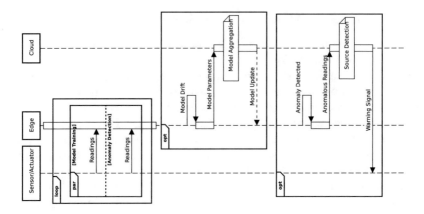

Fig. 5. Sequence diagram for the environmental monitoring use case.

Figure 5 illustrates the event communication between IoT, edge, and cloud components as a sequence diagram. The loop fragment demonstrates the synchronous data transmission between IoT sensors and the Edge component. The rest of the messaging is event-driven, as illustrated by the optional fragments. Model parameters are sent to the cloud only when a model drift is observed during local training, whereas pollution source detection at the cloud is only triggered when anomalous data is detected at the edge.

4 Conclusions

As exemplified by the two application use cases described in the paper, Multi-Layer and Event-Driven architectural solutions can be used effectively to capture various nonfunctional requirements that characterize numerous industrially and socially relevant application domains For instance, the low network access requirement of river monitoring entails the two-tier architecture and local training at the edge layer. Local training results in substantial data reduction compared to transmitting raw sensor data to the cloud. Similarly, EDA is motivated by the energy constraints of edge resources since processing is only triggered by rare events resulting in low power consumption.

In general, EDA is a powerful approach for building IoT-based environmental monitoring systems due to their particular non-functional requirements, such as (i) loose coupling between the tiers (intermittent network connectivity), (ii) real-time processing (delays might result in irreversible environmental impact), (iii) low energy consumption (no access to the electricity grid in remote areas) and (iv) reduced bandwidth utilization (the transfer is based only on changes in data).

Acknowledgement. This work was partially funded by the Digital Europe Programme, Project DANTE EDIH, ID: 101083913, as within the activities conducted by the "CINI - Consorzio Interuniversitario Nazionale per l'Informatica".

A. Aral was supported by the CHIST-ERA grant CHIST-ERA-19-CES-005 and by the Austrian Science Fund (FWF): I 5201-N.

References

1. Satyanarayanan, M.: The emergence of edge computing. Computer **50**(1), 30–39 (2017)
2. Baresi, L., Mendonça, D.F., Garriga, M., Guinea, S., Quattrocchi, G.: A unified model for the mobile-edge-cloud continuum. ACM Trans. Internet Technol. (TOIT) **19**(2), 1–21 (2019)
3. Zheng, T., Wan, J., Zhang, J., Jiang, C.: Deep reinforcement learning-based workload scheduling for edge computing. J. Cloud Comput. **11**(1), 1–13 (2022). https://doi.org/10.1186/s13677-021-00276-0
4. Koloth, A.: Data patterns for the edge: data localization, privacy laws, and performance (2022). https://www.infoq.com/articles/data-patterns-edge/. Accessed 25 Nov 2022
5. Xu, J., Glicksberg, B.S., Su, C., Walker, P., Bian, J., Wang, F.: Federated learning for healthcare informatics. J. Healthc. Inform. Res. **5**(1), 1–19 (2021). https://doi.org/10.1007/s41666-020-00082-4
6. Rieke, N., et al.: The future of digital health with federated learning. NPJ Digit. Med. **3**(1), 1–7 (2020)
7. Shyu, C.-R., et al.: A systematic review of federated learning in the healthcare area: from the perspective of data properties and applications. Appl. Sci. **11**(23), 11191 (2021)
8. Lo, S.K., Lu, Q., Zhu, L., Paik, H.-Y., Xu, X., Wang, C.: Architectural patterns for the design of federated learning systems. J. Syst. Softw. **191**, 111357 (2022)
9. Lo, S.K., Lu, Q., Paik, H.-Y., Zhu, L.: FLRA: a reference architecture for federated learning systems. In: Biffl, S., Navarro, E., Löwe, W., Sirjani, M., Mirandola, R., Weyns, D. (eds.) ECSA 2021. LNCS, vol. 12857, pp. 83–98. Springer, Cham (2021). https://doi.org/10.1007/978-3-030-86044-8_6
10. Di Martino, B., Graziano, M., Colucci Cante, L., Cascone, D.: Analysis of techniques for mapping convolutional neural networks onto cloud edge architectures using SplitFed learning method. In: Barolli, L., Hussain, F., Enokido, T. (eds.) AINA 2022. LNNS, vol. 451, pp. 163–172. Springer, Cham (2022). https://doi.org/10.1007/978-3-030-99619-2_16
11. Bellemare, A.: Building Event-Driven Microservices. O'Reilly Media Inc. (2020)
12. Shapira, G., Palino, T., Sivaram, R., Petty, K.: Kafka: The Definitive Guide: Real-Time Data and Stream Processing at Scale. O'Reilly Media Inc. (2020)
13. Martín, C., Langendoerfer, P., Zarrin, P.S., Díaz, M., Rubio, B.: Kafka-ML: connecting the data stream with ML/AI frameworks. Future Gener. Comput. Syst. **126**, 15–33 (2022)
14. Mohebbian, M.R., Vedaei, S.S., Wahid, K.A., Dinh, A., Marateb, H.R., Tavakolian, K.: Fetal ECG extraction from maternal ECG using attention-based CycleGAN. IEEE J. Biomed. Health Inform. **26**(2), 515–526 (2021)
15. Ajdaraga, E., Gusev, M.: Analysis of sampling frequency and resolution in ECG signals. In: 2017 25th Telecommunication Forum (TELFOR), pp. 1–4 (2017)
16. De Maio, V., Aral, A., Brandic, I.: A roadmap to post-moore era for distributed systems. In: Proceedings of the 2022 Workshop on Advanced Tools, Programming Languages, and PLatforms for Implementing and Evaluating Algorithms for Distributed Systems, pp. 30–34 (2022)

Programming Paradigms for the Cloud Continuum

Geir Horn[1]([✉]), Beniamino Di Martino[2], Salvatore D'Angelo[2], and Antonio Esposito[2]

[1] Department of Informatics, University of Oslo, P.O. Box 1080, Blindern, 0316 Oslo, Norway
Geir.Horn@mn.uio.no
[2] Università degli Studi della Campania Luigi Vanvitelli, Via Roma 29, 81031 Aversa, CE, Italy
{beniamino.dimartino,salvatore.dangelo,
antonio.esposito}@unicampania.it

Abstract. The Cloud continuum is a notion for the distributed infrastructure able to run third party software on heterogeneous hardware ranging from the high performance core Cloud data centres, through smaller fog data centres, down to resource constrained Edge servers. Beyond that, one will often have Internet of Things (IoT) devices on a wired or wireless infrastructure for data collection and actuation. The application components should be allocated along this continuum in order to optimize the application performance which is a combined measure of communication speed and computational power. However, most applications are not written with distribution in mind, and may not even be component based. This paper discusses various programming paradigms and how they can be extended to support distribution and embedding into the microservice architecture supporting tomorrow's Cloud applications.

1 Introduction

Cloud computing has transformed over the last decade from a model where the users rented Virtual Machines (VMs) in data centres as remote, hosted servers for the application, to today's model where the application components can additionally be hosted as 'micro-services', 'server less' as bare functions, or simply using third party applications as callable 'services'. Cloud computing applications are therefore, by definition, distributed systems consisting of a set of interacting *components*, where the term component is denoting any parts of the application refining input data to some output data used by other application components.

Recently, also network providers have started offering computing resources distributed in their networks to allow the processing of data closer to the sensors generating the data. This is driven by the ever increasing demand on mobile network bandwidth forcing the radio cells to become smaller, and handover and antenna management to be grouped on 'edge servers' for a set of geographically close antennas. However, these edge servers should have spare capacity that can be reused for user applications spanning from content caching for media streams consumed by locally attached users, to event processing of IoT sensor data and automatic control of Cyber-Physical Systems (CPSs).

Cloud computing was about renting computing resources as a utility when needed, and these new, dispersed resources available for rental adds a geographical dimension

L. Barolli (Ed.): AINA 2023, LNNS 655, pp. 146–156, 2023.
https://doi.org/10.1007/978-3-031-28694-0_14

to the distribution of the application's components as they can be deployed anywhere in the so called Cloud continuum ranging from the high performance computing data centres at the core of the network to the resource constrained servers at the edge of the network.

The problem is that software developers are generally not trained in developing distributed applications, and there are many legacy applications available that need complete re-implementation to work well in the distributed Cloud continuum. The core contribution of this paper is the identification of essential software engineering principles needed to make applications distributed and adaptive to changes in the application execution environment, and a vision for how the existing principles can better support the Cloud continuum application developers.

There are two aspects to the challenge: one is the application code development, and the second is the subsequent allocation and orchestration of the application's components over the available resources. The natural starting point for an application developer is probably to start with existing coding practices, and in Sect. 2 we will discuss compiler based techniques directly instructed by the application developer. In Sect. 3 we will outline how object oriented software should be developed in order to support the required distribution. Section 4 considers the extreme case where a huge number of application components may collaborate to realise jointly the goals of the application. The management of applications in the Cloud continuum is discussed in Sect. 5, and we will argue our vision in Sect. 6.

2   Code Annotation

Code annotation is the practice of adding additional information or metadata to code in order to improve its readability, organization, and functionality. Annotation has long been a common practice in software development, and it has become increasingly important with the rise of distributed computing systems.

One of the earliest forms of code annotation was the use of comments, which are lines of code that are ignored by the compiler but are intended to provide context or explanations for the rest of the code. Comments have been a standard feature of programming languages since the earliest days of computing, and they are still widely used today. As distributed computing systems became more common, developers began to use code annotations to help manage the complexity of these systems. For example, annotations could be used to specify which parts of a program should be run on which servers or to provide information about the dependencies between different pieces of code. More recently, code annotation has become an important tool for documenting Application Programming Interface (API) and other software systems. By providing clear and concise explanations of the purpose and behaviour of different parts of the code, annotations can help developers more easily understand and use these systems. Overall, code annotation has played a vital role in the evolution of distributed computing and continues to be an important tool for developers today.

When considering multi-core framework, Open Multi-Processing (OpenMP) [16] as a programming language has gained a lot of popularity and is becoming a de-facto standard. This is mainly because it offers very simple annotation extensions known as 'pragmas' for common programming languages such as C++ and FORTRAN. OpenMP can be seen as a set of compiler directives that can be used to specify the parallelism in a program. The directives are inserted in the source code of the program and are used by the compiler to generate the parallel code. However, the simplicity of the approach is counterbalanced by the need to deeply understand the implications of applying the different pragmas to obtain good optimisation. Optimising and tuning an OpenMP annotated program is never an easy task, and sometimes the overall performances are much lower than expected because of a non optimal use of the available annotations.

The TrustwOrthy model-awaRE Analytics Data platfORm (TOREADOR) Horizon 2020 project aimed at defining models and programming paradigms to enhance programmers' capabilities to develop Big Data analytics. One of the main contribution of the project is represented by the so-called "code-based approach". The idea behind the code-based approach is to provide the users with tools to annotate or instruct their code, in order to be able to decompose and distribute it on the Cloud continuum. The main targets of the decomposition were represented by Big Data analytics frameworks, such as Apache Spark[1] and Apache Hadoop[2]. The TOREADOR user is an expert programmer well aware of the flexibility and controllability of the framework, and the analytics developed from scratch or migration of legacy code. In particular, the approach is divided in two main phases:

1. The use of parallel computation of an existing coded algorithm, in terms of parallel primitives and directives.
2. The use of a Skeleton-Based Code Compiler, which is a source to source transformer [8,9]. The compiler is incarnating the Skeletons defined within the Project, which transforms the sequential code augmented with the parallel primitives into parallel versions, specific to the different platforms and technologies.

Within the TOREADOR project, this directive based approach was experimented on several algorithms, specifically to distribute Big Data analytics taking in consideration container-based execution environments. The programmer was provided with primitives and functions that expressed the available computation models, following a Decorator Pattern [10]. Such primitives were provided as comments to a compiler, that could ignore them and treat the code as a simple Python program, in case no parallelisation or code distribution was explicitly requested.

Two different kinds of primitives were identified:

- Task oriented primitives that guide the computation according to a task based decomposition, and follow specific computational patterns, *e.g.*, map reduce, bag of task, producer consumer.
- Data oriented primitive taking in consideration the nature and locality of data to be distributed among the computational nodes to identify the best code decomposition.

There are two main drawbacks when considering annotations approaches, like OpenMP and the one proposed by TOREADOR. First of all, they are strongly language dependent. The same annotation is not guaranteed to be applicable to different

---

[1] https://spark.apache.org/.

[2] https://hadoop.apache.org/.

programming languages, and in most cases it will not be. This is due to the fact that different syntax require different kinds of annotations, and not all compilers recognise comments in the same way. Also, some control structures may be unavailable in specific programming languages. It could be possible to provide several version of the same annotation, to accommodate the different specifics, but this would require a huge effort, and it would still not cover all possibilities. The second limitation regards the compiler: once a kind of annotation has been adopted, the programmer needs to stick with a compiler that can understand it. Otherwise, no practical effect would be noticed as no parallelization would be achieved in any way. The compiler depends on the native language chosen by the programmer, and needs to be frequently updated to consider the different, ever changing target platforms available on the market.

3   Active Objects

Even though early thoughts on software 'objects' combining the variables representing the state of the application with functions operating on this state dates back to the origin of computers, it was first formalised in the programming language SIMULA with the objects as 'classes' with functions or methods, and creating the concepts of inheritance, dynamic binding, and concurrent execution [14]. SIMULA was later followed by languages like SmallTalk, C++ and Java, where the latter two are still among the most used and requested languages for developers today[3].

The success of the object oriented paradigm can be ascribed to the partitioning of the data space and the localisation of variables to the objects. In well written object oriented software, data and methods belonging to an object are accessible uniquely for the methods belonging to the same object, and invisible for code outside of the object. In this way, the object defines a clear interface to its data space through its publicly callable methods. This greatly simplifies testing and debugging as variable value changes in one object will not affect any other object or state in the software system.

The object oriented programming paradigm is already a very good starting point for distributed applications since the objects can in principle be executed individually and concurrently as *Active Objects*, *i.e.*, an object behavioural pattern for concurrent programming [22]. Active objects that only communicate by messages are known as *actors* [5], and actors can freely be created and destroyed, and they will not deadlock or suffer from data race conditions [15]. There are several actor languages [15], and good actor libraries exists for most object oriented programming languages. A further benefit is that active object languages admit verification by formal methods [11].

On a single computer this enables safe multithreading as actors can safely be served concurrently by individual threads with no need for the programmer to manually implement memory locks and thread signalling: The memory is protected by the actor object, and the signalling is done by the messages. The actors with pending messages ready for processing will be scheduled as individual threads for execution by the local operating system. A communication and management system is needed when the actors are distributed across multiple nodes. There is no standard for this. However, within the High

---

[3] https://spectrum.ieee.org/top-programming-languages-2022.

Performance Computing (HPC) communities the Message Passing Interface[4] (MPI) is a popular communication standard with an open source implementation[5] that can also be used to create actors on the computing nodes, *e.g.*, a first actor on each node to bootstrap the distributed actor system.

This is a good foundation for writing distributed software for the Cloud continuum, yet two main issues remains: An actor registry, and a technology independent messaging layer. A registry of actors on remote execution platforms is needed given that actors can be created on demand and destroyed when no longer needed. The registry of actors must cover all the aspects of distributed resource discovery in a peer-to-peer network to ensure resilience against failures in the computing nodes or in the shared communication infrastructure.

The actor system messages are already software objects and the objects must be defined in a way that enables easy serialisation of the messages. However, a fault tolerant messaging layer that supports many different network technologies is needed to allow the actors to be distributed on resource constrained devices in the edge of the network and even to mobile IoT devices with limited wireless connectivity and battery bounded power restrictions [6]. The recently approved Internet Engineering Task Force[6] (IETF) Transport Services[7] (TAPS) standard could be integrated with an MPI based communication layer to provide a unified and flexible transport layer abstraction.

4   Agent Systems and Swarms

Several tools for programming and managing distributed infrastructures, based on the concept of actors or agents, have been proposed, as a possible solution to the lack of an edge oriented programming language or paradigm. Among these, we find JADE [3], JaCaMo [4], Jadex [21], SARL [23], PADE or SPADE [18], and many more. All these frameworks support the definition and execution of 'Agents', which can be employed to execute computations on dynamic edge devices and that can be synchronized with a centralized control. However, the resulting environment is too dependent on the proposed frameworks, which also act as an execution layer. At the same time, they do not provide support for existing, legacy software, which needs to be completely rethought and rebuilt from scratch.

Several Agent-Oriented Programming Languages (AOPs) have been proposed in recent years. However, the acceptance and adoption of AOP in the software engineering community remain limited. Current AOP application practices do not convince that it has fully exploited its advantages and technical potential. In recent years, other programming models for decentralized smart devices have also been presented such as, such as Protocol Over Thing [24] and the Microservice Orchestration LanguagE (MOLE) [25]. Other approaches for the management and programming of groups of smart devices are inspired by nature, where the behaviour of software components has been modelled against swarms or colonies of insects, in which a complex action

---

[4] https://www.mpi-forum.org/.

[5] https://www.open-mpi.org/.

[6] https://www.ietf.org/.

[7] https://datatracker.ietf.org/wg/taps/about/.

derives from the decision taken by a "collective intelligence". We, therefore, speak of swarm intelligence, which according to the definition of Beni and Watt can be defined as: *"Property of a system in which the collective behavior of (unsophisticated) agents interacting locally with the environment produces the emergence of global functional schemes in the system"*. All frameworks that implement this approach target Unmanned Aerial Vehicle (UAV) or Micro Air Vehicle (MAV) and robot devices, where some applications use general purpose languages and others have defined their own language, for example: Buzz [19] and Meld [2] as custom languages, and Karma [7] using Java and Voltron [17] using C++.

An interesting extension to the agent based frameworks is represented by Electronic Institutions (EIs). From a computational point of view, an EI realizes an 'environment for agents' [27], namely it provides *"the surrounding conditions for agents to exist and that mediates both the interaction among agents and the access to resources"*. Considering the levels of support that an environment may provide, EIs focus on two particular levels of support [27]: an abstraction level that shields agents from the low-level details of deployment; and the mediated interaction between agents. With this aim, EIs provide a computational infrastructure for agent environment design [27], and the mechanisms to enforce and monitor the norms and laws that apply to a multiagent system in a given environment. More precisely, an EI has implemented itself as a multi-agent system composed of two types of internal agents: the so-called staff agents (STA) and governor agents. While governor agents mediate the interactions of external agents within the environment, STAs manage the norms within the institution. The dynamics of the environment is restricted to those external agents that satisfy the social laws represented by the norms and enacted by the coordinated actions of governors and staff agents. With this aim, governors and staff agents employ synchronization mechanisms for interaction mediation [20].

5   Orchestration

Optimizing distributed applications for deployment along the cloud continuum can be challenging, as it involves ensuring that the application can run effectively and efficiently in various environments. Typically, the functionality of these applications is provided by a combination of heterogeneous distributed components. On the other hand, an approach based on code decomposition, where complex computations are decomposed into many small functions that are coded directly for the node that has to execute them, is becoming more popular. This decomposition has led to the emergence of a new model of 'server less' computation called Function as a Service (FaaS), where functions are executed on random nodes but this decomposition approach can be applied very well to the Cloud continuum deployment of an algorithm or application, where the functionality is coded and deployed directly on the designated edge nodes. However, currently there is no specific language and no automation in decomposition and deployment of FaaS applications.

Any architectural design with distributed communicating components poses difficulties for the application's overall management and operation and necessitates the use of fresh ideas for deployment, configuration, operation, and termination. One of the main challenges in optimizing distributed applications for the Cloud is ensuring that they can scale to meet the needs of a large number of users and requests. This requires careful design and planning, as well as the use of appropriate technologies and architectures. Another critical challenge is ensuring that the distributed application is highly available and able to remain functional even in the face of hardware or software failures. This can be achieved through techniques such as load balancing, redundant servers, and failover systems. Ensuring the security of a distributed application is critical, as it must protect sensitive data from unauthorized access or tampering. This can be achieved through the use of secure communication channels, encryption, and robust access controls. Finally, distributed applications must be able to work with a wide range of different systems and technologies, both within the organization and externally. This requires the use of standardized interfaces and protocols, as well as the ability to integrate with a variety of different systems and platforms.

Fortunately, several application management frameworks have appeared over the recent years. These targets different needs. The first is the deployment of the distributed application where the best known framework is probably the Topology and Orchestration Specification for Cloud Applications[8] (TOSCA) [26]. It is based on the concept of a topology and service template, which can be used to define the structure and behaviour of the application. TOSCA also includes a set of standard interfaces that can be used to integrate with other systems and platforms. OpenTOSCA[9] is an open-source implementation of TOSCA.

A micro-service architecture based on communicating containers can make it easier to scale the individual components of the application independently, and one may use a framework like Kubernetes[10], which also supports application micro-service scaling over a given set of computing resources. The Juju[11] framework can be used for deploying, scaling and managing distributed Kubernetes applications on a variety of different infrastructure platforms. It is based on the concept of a 'charm', which can be used to define the structure and behaviour of the application. Juju also includes a set of standard interfaces that can be used to integrate with other systems and platforms.

There are run-time management frameworks like the Multi-cloud Execution ware for Large scale Optimised Data Intensive Computing[12] (MELODIC) for optimised application management including resource management and variable application architectures [12]. It uses an application *model* based on the Cloud Application Modelling and Execution Language (CAMEL) [1] describing the application components and their resource requirements. CAMEL allows the users to define the automatic scaling behaviour of their application, and it includes features such as redundant servers and failover systems to help ensure the availability of the application and provides security features such as encryption and secure communication channels to protect sensitive data and prevent unauthorized access. Frameworks based on CAMEL support a wide range

---

[8] https://www.oasis-open.org/committees/tosca.

[9] https://www.opentosca.org/.

[10] https://kubernetes.io/.

[11] https://juju.is/.

[12] https://melodic.cloud/.

of different technologies and platforms and includes features to facilitate integration with other systems.

In addition to these application management tools, RedHat Ansible[13] provides support for configuration management. It can be used to automate the deployment and management of distributed applications. It is based on the concept of a 'playbook', which can be used to define the structure and behaviour of the application. Ansible also includes a set of standard interfaces that can be used to integrate with other systems and platforms.

Finally, there are life-cycle management tools like the Agent-based Cooperating Smart Objects (ACOSO) methodology that is a middleware for programming, implementation and simulation of agent based software, oriented to the Internet of Things [13]. The ACOSO methodology supports the Smart Objects (SO) development phases of analysis, design, and implementation using metamodels featured by different levels of abstraction. The ACOSO middleware that provides an effective agent model for SO-based systems and a hybrid JADE-based platform to program both basic SOs and more complex IoT systems at the edge.

6 Discussion

Despite the existence of several programming paradigms and frameworks that can be exploited for the development of distributed applications, the current state of the art lacks a dedicated framework to support Cloud and edge programmers explicitly.

It is evident that there is a need for flexible and versatile programming languages and development and execution platforms to address the complex management that characterise distributed, decentralised, and dynamic computational models and systems such as multi agents and swarms. These languages, tools, and framework should be easy to use for developers with little or no knowledge of decentralised computing paradigms and application deployment, execution, and orchestration in the real, physical environment. Such languages should also exploit the capabilities of smart IoT devices and mobile networks to facilitate the seamless smart deployment and dynamic reallocation and scheduling of decentralised software components on heterogeneous edge nodes.

This is currently an open challenge: the basic blocks of such a framework potentially exist, as agent oriented technologies, active object programming paradigms, and native parallel languages are already available, but they need to be specifically integrated and tailored for the development of Cloud continuum applications.

A possible solution could be represented by the integration of different programming paradigms, which could provide different development interfaces to users, according to their level of knowledge of the Cloud continuum, and to the degree of control they want to retain over the distributed applications. Exploiting agent and actor oriented frameworks would simplify the programming and deployment of Cloud continuum distributed solutions, but they would need precise scheduling and task allocation policies to be implemented. Code annotation approaches, relying on pragmas or primitives could enable users to retain a strong control over their applications, paying a higher price in terms of complexity of the solution and its optimisation.

---

[13] https://www.ansible.com/.

In the end, a hybrid solution, taking the best from these two approaches and meeting the needs of programmers with different levels of expertise, would address both the need for simplicity and control over applications distributed in the Cloud continuum.

7   Conclusion and Future Work

This paper has examined three approaches for engineering distributed applications useful for the Cloud continuum starting from the code annotations enabling good compilers to generate the application components for the distribution based on the application developer's guidance directly expressed in the code. Within the realm of the object oriented programming paradigm, we argued the use of active objects and message passing as a way to explicitly model the application components and their interaction, and representing this model in the application's code. Finally, considering the application as a swarm of collaborating components, it would be possible to use established agent system techniques, tools, and languages to define the system and component interactions. The better approach depends on the application's tasks and how they process and communicate data. However, all three development paradigms have their pros and cons, and it may well be situations where the optimal software engineering approach would be a combination of all three approaches.

Once the distributable application has been developed, it must be distributed, deployed, and managed. This is a separate, but linked, challenge since the application developer should consider the application's life cycle when deciding on the development approach. This includes considering the scalability, variability, adaptivity, and volatility of the various application components and their interactions.

Even after almost a century of computer science results, there is unfortunately still no unified best response to the challenge of developing distributed applications for the Cloud continuum, but we envision that efficient solutions may be constructed in the near future from the solid foundational results already available and presented in this paper.

Acknowledgements. This work has received funding from the European Union's research and innovation programmes Horizon 2020 under grant agreement No 871643 MORPHEMIC (http://morphemic.cloud): *Modelling and Orchestrating heterogeneous Resources and Polymorphic applications for Holistic Execution and adaptation of Models In the Cloud* and Horizon Europe grant agreement No 101070516 NebulOuS (https://www.nebulouscloud.eu/): *A Meta Operating System For Brokering Hyper-Distributed Applications On Cloud Computing Continuums.*

References

1. Rossini, A., Kritikos, K., Nikolov, N., et al.: The cloud application modelling and execution language (CAMEL). Open Access Repositorium der Universität Ulm (2017). https://doi.org/10.18725/OPARU-4339
2. Ashley-Rollman, M.P., Goldstein, S.C., Lee, P., Mowry, T.C., Pillai, P.: Meld: a declarative approach to programming ensembles. In: 2007 IEEE/RSJ International Conference on Intelligent Robots and Systems, pp. 2794–2800 (2007). https://doi.org/10.1109/IROS.2007.4399480

3. Bellifemine, F., Poggi, A., Rimassa, G.: JADE-a FIPA-compliant agent framework. In: Proceedings of PAAM, p. 33 (1999)
4. Boissier, O., Hübner, J.F., Ricci, A.: The JaCaMo framework. In: Aldewereld, H., Boissier, O., Dignum, V., Noriega, P., Padget, J. (eds.) Social Coordination Frameworks for Social Technical Systems. LGTS, vol. 30, pp. 125–151. Springer, Cham (2016). https://doi.org/10.1007/978-3-319-33570-4_7
5. Hewitt, C., Bishop, P., Steiger, R.: A universal modular ACTOR formalism for artificial intelligence. In: Proceedings of the 3rd International Joint Conference on Artificial Intelligence, IJCAI 1973, San Francisco, CA, USA, pp. 235–245. Morgan Kaufmann Publishers Inc. (1973)
6. Graff, D., Richling, J., Stupp, T.M., Werner, M.: Distributed active objects - a systemic approach to distributed mobile applications. In: Proceedings of the Eighth IEEE International Conference and Workshops on Engineering of Autonomic and Autonomous Systems, Las Vegas, NV, USA, pp. 10–19. IEEE (2011). https://doi.org/10.1109/EASe.2011.10
7. Dantu, K., Kate, B., Waterman, J., Bailis, P., Welsh, M.: Programming micro-aerial vehicle swarms with karma. In: Proceedings of the 9th ACM Conference on Embedded Networked Sensor Systems, pp. 121–134 (2011). https://doi.org/10.1145/2070942.2070956
8. Di Martino, B., D'Angelo, S., Esposito, A.: A platform for MBDAaaS based on patterns and skeletons: the python based algorithms compiler. In: 2017 IEEE 14th International Conference on Networking, Sensing and Control (ICNSC), pp. 400–405 (2017). https://doi.org/10.1109/ICNSC.2017.8000126
9. Di Martino, B., Esposito, A., D'Angelo, S., Maisto, S.A., Nacchia, S.: A compiler for agnostic programming and deployment of big data analytics on multiple platforms. IEEE Trans. Parallel Distrib. Syst. 30(9), 1920–1931 (2019). https://doi.org/10.1109/TPDS.2019.2901488
10. Gamma, E., Helm, R., Johnson, R., Vlissides, J., Booch, G.: Design Patterns: Elements of Reusable Object-Oriented Software. Addison-Wesley Professional, Reading (1994)
11. De Boer, F., Serbanescu, V., Hähnle, R., et al.: A survey of active object languages. ACM Comput. Surv. 50(5), 76:1–76:39 (2017). https://doi.org/10.1145/3122848
12. Horn, G., Skrzypek, P.: MELODIC: utility based cross cloud deployment optimisation. In: Proceedings of the 32nd International Conference on Advanced Information Networking and Applications Workshops (WAINA), Krakow, Poland, pp. 360–367. IEEE Computer Society (2018). https://doi.org/10.1109/WAINA.2018.00112
13. Fortino, G., Russo, W., Savaglio, C., Shen, W., Zhou, M.: Agent-oriented cooperative smart objects: from IoT system design to implementation. IEEE Trans. Syst. Man Cybern. Syst. 48(11), 1939–1956 (2018). https://doi.org/10.1109/TSMC.2017.2780618
14. Holmevik, J.R.: Compiling SIMULA: a historical study of technological genesis. IEEE Ann. Hist. Comput. 16(4), 25–37 (1994). https://doi.org/10.1109/85.329756
15. De Koster, J., Van Cutsem, T., De Meuter, W.: 43 years of actors: a taxonomy of actor models and their key properties. In: Proceedings of the 6th International Workshop on Programming Based on Actors, Agents, and Decentralized Control (AGERE 2016), Amsterdam, The Netherlands, pp. 31–40. ACM (2016). https://doi.org/10.1145/3001886.3001890
16. Dagum, L., Menon, R.: OpenMP: an industry standard API for shared-memory programming. IEEE Comput. Sci. Eng. 5(1), 46–55 (1998). https://doi.org/10.1109/99.660313
17. Mottola, L., Moretta, M., Whitehouse, K., Ghezzi, C.: Team-level programming of drone sensor networks. In: Proceedings of the 12th ACM Conference on Embedded Network Sensor Systems, pp. 177–190 (2014). https://doi.org/10.1145/2668332.2668353
18. Palanca, J., Terrasa, A., Julian, V., Carrascosa, C.: SPADE 3: supporting the new generation of multi-agent systems. IEEE Access 8, 182537–182549 (2020). https://doi.org/10.1109/ACCESS.2020.3027357

19. Pinciroli, C., Beltrame, G.: Buzz: a programming language for robot swarms. IEEE Softw. **33**(4), 97–100 (2016). https://doi.org/10.1109/MS.2016.95
20. Platon, E., Mamei, M., Sabouret, N., Honiden, S., Parunak, H.: Mechanisms for environments in multi-agent systems: survey and opportunities. Auton. Agents Multi-Agent Syst. **14**(1), 31–47 (2007). https://doi.org/10.1007/s10458-006-9000-7
21. Pokahr, A., Braubach, L., Jander, K.: The Jadex project: programming model. In: Ganzha, M., Jain, L. (eds.) Multiagent Systems and Applications. ISRL, vol. 51, pp. 21–53. Springer, Heidelberg (2013). https://doi.org/10.1007/978-3-642-33323-1_2
22. Lavender, R.G., Schmidt, D.C.: Active object: an object behavioral pattern for concurrent programming. In: Vlissides, J.M., Coplien, J.O., Kerth, N.L. (eds.) Pattern Languages of Program Design 2, USA, pp. 483–499. Addison-Wesley Longman Publishing Co., Inc. (1996)
23. Rodriguez, S., Gaud, N., Galland, S.: SARL: a general-purpose agent-oriented programming language. In: 2014 IEEE/WIC/ACM International Joint Conferences on Web Intelligence (WI) and Intelligent Agent Technologies (IAT), pp. 103–110 (2014). https://doi.org/10.1109/WI-IAT.2014.156
24. Smirnova, D., Chopra, A.K., Singh, M.P., et al.: Protocols over things: a decentralized programming model for the Internet of Things. Computer **53**(12), 60–68 (2020). https://doi.org/10.1109/MC.2020.3023887
25. Song, Z., Tilevich, E.: A programming model for reliable and efficient edge-based execution under resource variability. In: 2019 IEEE International Conference on Edge Computing (EDGE), pp. 64–71 (2019). https://doi.org/10.1109/EDGE.2019.00026
26. Binz, T., Breitenbücher, U., Kopp, O., Leymann, F.: TOSCA: portable automated deployment and management of cloud applications. In: Bouguettaya, A., Sheng, Q., Daniel, F. (eds.) Advanced Web Services, pp. 527–549. Springer, New York (2014). https://doi.org/10.1007/978-1-4614-7535-4_22
27. Weyns, D., Omicini, A., Odell, J.: Environment as a first class abstraction in multiagent systems. Auton. Agents Multi-Agent Syst. **14**(1), 5–30 (2007). https://doi.org/10.1007/s10458-006-0012-0

Worker-to-Task Skill-Based Assignment

Vlad Rochian[1,2(✉)], Cosmin Bonchis[1,2], and Ionut Tepeneu[3]

[1] Department of Computer Science, West University of Timisoara,
Timișoara, Romania
cosmin.bonchis@e-uvt.ro
[2] e-Austria Research Institute, Timișoara, Romania
vlad.rochian@e-uvt.ro
[3] ETA2U, Timișoara, Romania
itepeneu@eta2u.ro

Abstract. We know that any optimization problem with constraints can be solved through linear programming methods that are NP-hard and thus have (at the time of writing) only exponential solutions. In practice, though, there are methods that perform well. There are, however, particular cases of constrained optimization problems that allow polynomial solutions. In this article we study how some real world worker allocation problems can be modeled as optimization problems, whether they can be solved in polynomial time and whether a polynomial solution works better in practice than the naive one.

1 Introduction

1.1 Motivation and Problem Statement

In the context of industrial planning of task allocation somehow similar with [7] we are looking at the problem of skill based allocation of workers to tasks up on different scenario.

In [7], the authors propose an optimisation model that can be used by planners to perform some particular task allocation for the problem of delivering maintenance services.

Assume we have a set of tasks that require certain skills and have a fixed time range, and a set of employed workers that possess certain skills, there are multiple particular cases of this problem, some of which are studied in this article. We aim to find a suitable solution for each subproblem rather than a general one.

1.2 Theoretical Foundation

Constrained Optimization Problems

An optimization problem is, in general terms, the problem of finding the best solution out of all feasible solutions. Formally, if every solution has an associated score given by a formula (the goal function), the problem is finding the maximum

L. Barolli (Ed.): AINA 2023, LNNS 655, pp. 157–168, 2023.
https://doi.org/10.1007/978-3-031-28694-0_15

possible score and the variables that produce this score. Constraints may appear where variables' values must respect some equations or inequations that restrict the feasibility of solutions.

A constrained linear optimization problem is a subclass of the aforementioned problem type, where the goal function and the constraints are all linear functions with respect to the variables. The field aimed at solving these problems is linear programming. The linear programming methods are NP-hard, meaning that currently there is no known polynomial-time solution to them.

Flow Networks

A flow network is a directed graph where every edge has a capacity. There are two special nodes, the "source" and the "destination" (or "sink"). Flow units can be sent from the source to the sink, with the constraints that, for every other node, the number of incoming units must be equal to the number of outgoing units. Additionally, for every edge, the number of units cannot exceed the capacity.

The maximum flow problem consists of finding the largest number of flow units that can be sent through the network while respecting the constraints. As this description suggests, the max-flow problem is a particular case of constrained linear optimization problems. However, it can be solved in polynomial time [1].

Bipartite Matching

A bipartite graph is a graph where the set of nodes can be split into two partitions such that all edges are between nodes from different partitions. A matching in a bipartite graph is a subset of edges with the property that for every node in the graph, there is at most one adjacent edge in the subset.

The maximum bipartite matching problem requires finding a matching in a bipartite graph such that the cardinality of the edge subset is the maximum possible. The problem is a particular case of the maximum flow problem, where there is a virtual source, a virtual sink, all other nodes are on two layers (the partitions) and all edges have a capacity of 1. There is a faster algorithm than the max-flow one that is specialized on maximum bipartite matching [3].

1.3   Related Work

There are several studies regarding skill-based allocation of resources. A study from 2002 [5] uses genetic algorithms to solve an allocation of tasks to resources, in particular they have analyzed how tutors can be optimally assigned to tutorials in a company based on their skill. Or similar optimisation problem for minimize the production and the labor costs over the total production waste is defined in [6].

There are also more recent studies that analyze how workers can be assigned to certain location-specific tasks [2], where they are dealing with the spatial crowdsourcing for assigning workers to some location to increase the impact of fulfilling the tasks. There are multiple problem types considered and the real life skill-based task allocation problems are investigated.

In a paper from 2021 [7], the authors propose an optimisation model that can be used by planners to perform the some particular task allocation. In [8] the authors focused on illustrating and experimenting with the principles of skill based assessment problems.

On the other hand in [4], the authors have searched for a solution for a skill-based automation process for the assembly of high quality machines for the medical industry in the Human-Robot-Cooperation context.

Closer to our study, there are some papers that formalize resource allocation problems such as linear programming (LP) problems for selection of projects and the human resource allocation, like [9]. Because the LP formulation has been proven to be NP-complete, they introduce a meta-heuristic to compute a feasible solution in polynomial time.

However, not many studies that we know of deal with the representation of this specific type of problems as graph theory models. In our paper we analyze different skill-based problem versions. In Sect. 2 we define some specific problems:

- Single-worker tasks spanning whole period (**W2T**)
- Multi-worker tasks spanning whole period (***W2T**)
- Tasks span certain periods and workers are active during certain time frames (**t*W2T**)
- Multi-worker to tasks planning with minimizing the number of workers with respect the working time table and the deadlines (**Min-t*W2T**).

Another part (Sect. 3) we show some primary experimental results that we have obtained regarding the runtime of the particular scenario discussed in Subsect. 2.3.

2   Workers to Tasks Problem Considered

2.1   Workers Active for Whole Period. Single-Worker Tasks Spanning Whole Period (W2T)

Problem Statement: In the context that the workers of a company will be available during the period of interest. The worker will be active for the whole period regardless of whether they have an active task. Assume all tasks span the whole period of interest. A task requires a single worker.

Goal: Assign the available workers to tasks, according to their skill.

Problem 1 [W2T]. Let there be a set of workers W and a set of tasks T. Let S_w be the set of skills of worker $w \in W$. Let R_t be the set of required skills for task $t \in T$. A worker w can perform the task t if $R_t \subset S_w$. The goal is to find an assignment (w, t) for each tasks.

Solution: We build a bipartite graph with one vertex in the left-hand side partition for every worker and one vertex in the right-hand side partition for every task. There will be an edge between every pair of worker and task vertices if the worker possess all the skills required for the task.

Formally:

$$G = (V_W, V_T, E)$$
$$V_W = \{v_w | \forall w \in W\}$$
$$V_T = \{v_t | \forall t \in T\}$$
$$E = \{(v_w, v_t) | \forall w \in W, \forall t \in T, R_t \subset S_w\}$$

Definition 1. The **solution for W2T** is to build an assignment as a bipartite graph $G = (V_W, V_T, E)$ with one vertex in the left-hand side partition for every worker $V_W = \{v_w | \forall w \in W\}$ and one vertex in the right-hand side partition for every task $V_T = \{v_t | \forall t \in T\}$. There will be an edge between every pair of worker and task vertices if the worker possess all the skills required for the task $E = \{(v_w, v_t) | \forall w \in W, \forall t \in T, R_t \subset S_w\}$.

We compute the maximum matching M for the graph G. The problem instance will have a valid solution if $|M| = |T|$. The Hopcroft-Karp algorithm can be used for computing the maximum matching. The time complexity for this algorithm proved in [3] is $O(|E| \sqrt{|V_W + V_T|})$.

2.2 Workers Active for Whole Period. Multi-worker Tasks Spanning Whole Period (*W2T)

Problem Statement: The scenario is the same as for the previous Problem 1, with the exception that a task may require multiple workers for multiple reasons: from the spreading period of task, to the complexity that cannot be fulfilled by only one worker. All workers must cover the entire skill set.

Goal: Assign the available workers to tasks, according to their skill.

With W, T, S_w and R_t with the same meaning as for the previous Problem 1 we define:

Problem 2 [*W2T]. For a set of workers W and a set of tasks T, let S_w be the set of skills of worker $w \in W$ and R_t be the set of required skills for task $t \in T$. Let n_t be the number of workers required by task $t \in T$. The goal is to find a set of assignments $M = \{(t, w) | t \in T, w \in W, R_t \subset S_w\}$, such that $\forall t \in T$, $|\{m \in M | m = (t, w)\}| = n_t$.

2.2.1 Solution 1

We alter the model from the previous problem by creating multiple vertices in the right-hand side partition, one for each worker requirement.

Definition 2. The **solution of *W2T** is to build an assignment as a bipartite graph $G = (V_W, V_T, E)$ with one vertex in the left-hand side partition for every worker $V_W = \{v_w | \forall w \in W\}$ and one vertex in the right-hand side partition for every task $V_T = \{v_{t_k} | \forall t \in T, \forall k \in 1...n_t\}$. There will be an edge between every pair of worker and task vertices if the worker possess all the skills required for the task $E = \{(v_w, v_{t_k}) | \forall w \in W, \forall t \in T, \forall k \in 1...n_t, R_t \subset S_w\}$.

We will use the same algorithm as for Problem 1.

Observation: The running time scales with the total number of workers (and similarly with the number of workers required for each task), not taking into account the existence of several workers with the same skill set.

2.2.2   Solution 2

We group the workers into same skill set categories, i.e. one category for every possible combination of skills. Optionally, we can perform the same grouping on tasks into same-requirement categories and sum their worker requirements.

Definition 3. The second model is to build flow network with two main layers (1 and 2) and two additional ones (0 and 3) for the source and destination respectively. The layers will have the following structure edges:

- Layer 0: A single source vertex: $V_0 = \{v_s\}$;
- Layer 1: A set of vertices containing one for every worker skill combination: $V_1 = \{v_c | \forall c\, \text{representing a combination of skills}\}$; Every vertex is connected to the source by an incoming edge with a capacity equal to the number of workers with the corresponding skill set;
- Layer 2: A set of vertices containing one for every task: $V_2 = \{v_t | \forall t \in T\}$; Every vertex is connected by an incoming edge to all $v_c \in V_1$ where $R_t \subset c$. The capacities can be considered infinite, as the maximum flow will be controlled by outgoing edges;
- Layer 3: A single destination ("sink") vertex: $V_3 = \{v_d\}$; The destination will be connected to every $v_t \in V_2$ by an incoming edge of capacity n_t.

Every flow unit corresponds to the assignment of a worker to a task "slot". The limit of workers per skill set category is enforced by the capacities for $V_0 \to V_1$ edges. The limit of workers per task is enforced by the capacities for $V_2 \to V_3$ edges. The validity of worker-to-task assignments is enforced by $V_1 \to V_2$ edges.

Example 1. In Fig. 1 we present an small example of a network flow for an instance of the Problem 2. Suppose there are two skill categories (category 1 with 3 workers and category 2 with 5 workers). There are 3 tasks, requiring 3, 2 and 2 workers respectively. A worker from category 1 can solve tasks 1 or 2, while a worker from category 2 can solve tasks 2 or 3. The values on the edges represent the flow and the capacity.

Constructing the flow network by Definition 3 we compute the maximum flow F in the network. The solution is valid if $F = \sum_{\forall t \in T} n_t$, i.e. all destination edges are saturated. The Edmonds-Karp algorithm can be used for finding the maximum flow, its complexity being $O(|V||E|^2) = O((|V_1| + |V_2|)|V_1|^2|V_2|^2)$ [1]. This is an improvement of the previous method if there is a relatively large number of workers but few skill set combinations.

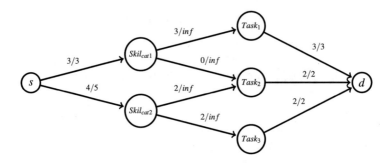

Fig. 1. Flow network for an instance of the worker-to-task assignment

2.3   Workers Active for Certain Periods. Tasks Span Certain Periods (t*W2T)

Problem Statement: This is a generalization of the previous problem. Instead of being available for the whole period, each worker has one or several time ranges when they are active. Similarly, each task has a fixed time range when it has to be performed.

Goal: Assign the workers to tasks, according to their skill and availability.

Problem 3 [t*W2T]. Let set of workers W and a set of tasks T. Let $A_W = \{a_w = (st_w, et_w) \| st_w/et_w$ are start/end time for worker $w\}$ be set of the active period for each worker and $A_T = \{a_t = (st_t, et_t) \| st_t/et_t$ are start/end time for task $t\}$. Let S_w be the set of skills of worker $w \in W$ and R_t be the set of required skills for task $t \in T$. Let n_t be the number of workers required by task $t \in T$. The goal is to find a set of assignments $M = \{(t, w, s, e) | t \in T, \ w \in W, R_t \subset S_w \ s \geq st_t, \ s \geq st_w, \ e \leq et_t, \ e \leq et_w, \ s < e\}$, such that for all moments in the timespan of the task, there are exactly n_t workers that perform it, formally $\forall t \in T, \forall q \in [st_t, et_t),$ $|\{m \in M | m = (t, w, s, e), q \in [s, e)\}| = n_t$.

Generalization of Solution 2 from the Previous Problem

Adaptation: The problem can be decomposed into several instances based on time slots. The naive approach is to consider the smallest unit of time taken into account, for example an hour. A more efficient split is to determine all moments when a "change" occurs, specifically when a task starts, a task ends, a worker clocks in and a worker clocks out. Then, an instance can be created for every time range between two such consecutive moments.

Every instance will generate a separate flow network that can be solved independently using the Edmonds-Karp algorithm [1]. This will give us a complexity of $O(I * C)$, where I is the number of instances and C is the complexity of the referenced solution.

2.4   Workers Active for Certain Periods. Tasks Span Certain Periods. Minimum Possible Number of Workers is Required (Min-t*W2T)

Problem Statement: This is an extension of Problem 3. We assume that every worker that has at least one assignment costs one unit (regardless of the number of assignments or the worked period). The assignment cost of a task is therefore the total number of unique workers used for the assignments.

Goal: Aside from assigning all tasks, the goal is to find a max assignment such that the total cost is minimal, i.e. a minimum size subset of the workers has to be found such that all tasks can be assigned.

Problem 4 [Min-t*W2T]. Let us consider the setting of the previous Problem 3 and a solution M of it. Let n_M be the number of unique workers used in the assignments from M. The goal is to find the minimum value n such that there exists a solution M_0 with $n_{M_0} = n$.

2.4.1   Solution 1

The logic behind the minimization process (Algorithm 1) is starting from a valid solution (a complete assignment), then trying to remove workers while solving the instance is still possible. A first, naive approach would be selecting randomly a subset of workers to remove and attempting to solve the instance with the remaining ones. If a solution is still possible, remove the subset. Otherwise, preserve the full set. Finally, repeat the process until an end condition. Multiple methods for subset selection were tested, the only remarkable one (but not remarkable enough) being a binary search-like algorithm:

Algorithm 1. Binary Search-like Algorithm

Require solve(t, w) - a function that finds a feasible assignment of workers to tasks
Input tasks, workers;
Output a minimal subset of workers;

    if $solve(tasks, workers)$ is null **then**
      return null
    end if
    $step = len(workers)/2$;
    while $step > 0$ **do**
      $s = subset(workers)$ such that $len(s) = step$;
      if $solve(tasks, workers - s)$ is not null **then**
        $workers = workers - s$;
      end if
      $step = step/2$.
    end while
    return $len(workers)$;

2.4.2   Solution 2

We apply a similar procedure to the one described in the previous section. We group the workers in category groups G defined by starting working time, ending working time and skill set.

Proposition 1. *If all workers are in the same category, then the solution described in the previous section works optimally. This also applies if there are multiple categories but it is only allowed to remove workers from a specific category.*

Proposition 2. *The maximum number of eliminated workers in a result obtained this way will be further referred to as $best_g$.*

Proposition 3. *Assuming we have the original problem and an optimal result defined by r_1, r_2, etc., r_k representing the number of removed workers from category g, we notice that $r_g \leq best_g, \forall g$, as $best_g$ is by definition the maximum number of eliminated workers.*

Based on Proposition 3, we can partition the workers into two groups, a "mandatory" group and an "optional" group. The optional group will contain $best_g$ workers from each category g.

Algorithm 2. Bounded Binary Search-like Algorithm

Require solve(t, w) - a function that finds a feasible assignment of workers to tasks
Input tasks, workers
Output a minimal subset of workers

    if $solve(tasks, workers)$ is null **then**
      return null
    end if
    $categories = getCategories(workers)$
    $optGroup = []$
    for all cat in $categories$ **do**
      $maxCut =$ binary search $maxLen(x$ from $cat)$ such that $solve(tasks, workers-x)$
      is not null
      $optGroup = optGroup + maxCut$
    end for
    $mandatory = workers - optGroup$
    $step = len(optGroup)/2$
    while $step > 0$ **do**
      $s = subset(optGroup)$ such that $len(s) = step$
      if $solve(tasks, mandatory + optGroup - s)$ is not null **then**
        $optGroup = optGroup - s$
      end if
      $step = step/2$
    end while
    return $mandatory + optGroup$

Proposition 4. *For any optimal solution, there is also one with the same cost with workers eliminated only from the optional group, as workers within a group are interchangeable.*

Considering these observations, we apply the solution implemented in Algorithm 2, with the mention that all elimination candidate subsets will be selected from the optional group.

3   Experimental Results

Multiple datasets of different sizes for problem described in Subsect. 2.3 were created. The size N of a dataset means the total number of necessary workers for the given tasks. The number of available workers was larger or equal than N (as a smaller number would cause an instant trivial negative resolution of the instance) but smaller than $2N$.

The Hopcroft-Karp algorithm for maximum bipartite matching [3] and the Edmonds-Karp algorithm for maximum flow [1] were applied on datasets with various sizes and were compared to a naive approach using linear programming (using the PuLP Python library).

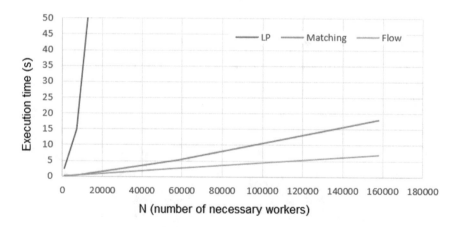

Fig. 2. Runtime for the linear programming method, the max-matching method and the max-flow method, in seconds.

The represented times in Fig. 2 were obtained by averaging the runtimes for the tests with the given size. It is worth mentioning that the existence of a solution did not impact the runtime for any of the algorithms. Lower times mean better solutions. For $N = 60000$, the runtime of the linear programming solution was approximately 6 min (364 s, to be specific). The point is not shown on the graph in order to not compress the figure.

The same time-range splitting method described in the previous section was used for all three methods.

Table 1. Percent of eliminated workers, expressed as the ratio between the number of workers eliminated by the algorithm and the total possible number of workers that can be eliminated

Algorithm	Elimination percent
Alg 1 - Binary search brute force	77%
Alg 2 - Bounded binary search	83%
Linear programming	100%

In a similar manner, the solutions for Problem 4 were compared to each other and to a linear programming method. In this case, the polynomial solution did not achieve a good performance. This is explainable due to the fact that they call a Problem 3 solving procedure multiple times, therefore increasing the theoretical complexity by a factor of $|W| * log|W|$. In Table 1 we show the ability to reduce redundancy. While it is guaranteed that the LP solution finds the optimal solution (a redundancy elimination percent of 100%), the polynomial solutions, Algorithm 2, (which use a metaheuristic approach) also achieve about 80%. We observe that the lowest redundancy elimination is done by binary search brute force algorithm with a good runtime, see Fig. 3.

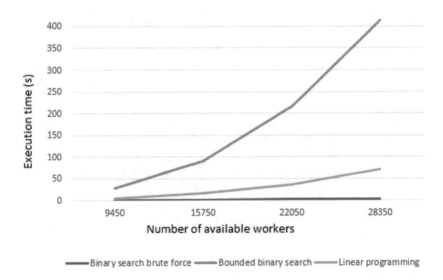

Fig. 3. Runtime for the worker minimization techniques

In Fig. 3, the runtimes of binary search brute force, bounded binary search and linear programming approaches for minimization of the workers of Problem 4 are compared. From our analysis, the LP algorithm achieves a good enough run time, but the best trade-off is achieved by the brute-force binary search

(Algorithm 1) minimizer, as it obtains almost 80% elimination in a much smaller time.

4   Conclusion

By analyzing the running time growth, one can notice that the matching and flow algorithms produce a running time that grows much slower than the naive linear programming solution. For smaller input sizes, the matching and the flow algorithms perform similarly. For larger input sizes, though, it can be easily noticed that the flow based algorithm performs better. The matching algorithm, despite having a better complexity, scales with the number of individual employees. On the other hand, the flow based algorithm scales with the number of skill set categories, thus allowing an arbitrary number of workers as long as they can be grouped into few categories.

However, if eliminating redundant workers is a concern, we notice that polynomial solutions have a large enough complexity that in practice it is not worth further improving based on greedy approaches.

This research should be continued naturally by analyzing some real word data based on the max-flow method proposed in this paper in order to get some real feedback from the industry partner to see if there are fits to our intuitions.

Acknowledgements. This paper is partially supported by the Competitiveness Operational Programme Romania under project number SMIS 120725 - SCAMP-ML (Advanced computational statistics for planning and tracking production environments).

References

1. Edmonds, J., Karp, R.M.: Theoretical improvements in algorithmic efficiency for network flow problems. J. ACM (JACM) **19**(2), 248–264 (1972)
2. Guo, B., Liu, Y., Wang, L., Li, V.O.K., Lam, J.C.K., Yu, Z.: Task allocation in spatial crowdsourcing: current state and future directions. IEEE Internet Things J. **5**(3), 1749–1764 (2018)
3. Hopcroft, J.E., Karp, R.M.: An $n^{5/2}$ algorithm for maximum matchings in bipartite graphs. SIAM J. Comput. **2**(4), 225–231 (1973)
4. Müller, R., Vette, M., Geenen, A.: Skill-based dynamic task allocation in human-robot-cooperation with the example of welding application. Procedia Manuf. **11**, 13–21 (2017)
5. Nammuni, K., Levine, J., Kingston, J.: Skill-based resource allocation using genetic algorithms and an ontology. In: Proceedings of the International Workshop on Intelligent Knowledge Management Techniques (I-KOMAT 2002) (2002)
6. Niakan, F., Baboli, A., Moyaux, T., Botta-Genoulaz, V.: A bi-objective model in sustainable dynamic cell formation problem with skill-based worker assignment. J. Manuf. Syst. **38**, 46–62 (2016)
7. Sala, R., Pirola, F., Pezzotta, G., Vernieri, M.: Improving maintenance service delivery through data and skill-based task allocation. In: Dolgui, A., Bernard, A., Lemoine, D., von Cieminski, G., Romero, D. (eds.) APMS 2021. IAICT, vol. 631, pp. 202–211. Springer, Cham (2021). https://doi.org/10.1007/978-3-030-85902-2_22

8. Smee, S.: Skill based assessment. BMJ **326**(7391), 703–706 (2003)
9. Zaraket, F.A., Olleik, M., Yassine, A.A.: Skill-based framework for optimal software project selection and resource allocation. Eur. J. Oper. Res. **234**(1), 308–318 (2014)

Prototype for Controlled Use of Social Media to Reduce Depression

Furqan Haider[✉], Hamna Aslam, Rabab Marouf, and Manuel Mazzara[✉]

Innopolis University, Universitetskaya St, 1,
420500 Innopolis, Respublika Tatarstan, Russia
f.haider@innopolis.university, {r.marouf,m.mazzara}@innopolis.ru

Abstract. This study examines social media addiction and prototypes a software application. This application can be integrated with any social media platform to alarm users when they are near the addiction stage. The work is ongoing, and this paper elaborates on the preliminary findings. We surveyed participants in the age range of 18–40 years to investigate the impact of social media on their lives. The participants are university students being monitored, and professionals from fields including information technology, medicine, education, and engineering. The prototype will not only help minimize social media addiction but also provide some productive activities as a replacement that will be entertaining simultaneously. The social media usage will be monitored, and users will be notified as per the customized settings of the application.

1 Introduction

Depression is one of the world's most serious psychological issues, particularly among young adults. Around 264 million people suffer from depression, according to the World Health Organization [19]. Depression can cause great suffering in one's life, primarily affecting daily routine activities such as studying, working, and household chores, while severe depression leads to suicidal attempts. More than 700,000 people commit suicide once a year, and suicide is identified as the fourth leading cause of death among children aged 15 to 19 years [20].

Several studies reported the correlation between the excessive use of social media among young adults and mental health problems, i.e., depression, anxiety, stress, and self-esteem issues. Social media is defined as [6] 'websites that enable profile creation and the visibility of relationships between users have become one of the most common leisure activities among their users'. Today, approximately half of the world's population actively uses social media, and the number of social media users is rapidly increasing day after day. Individuals use social media for a ramification of reasons, such as maintaining relationships and accessing information and entertainment. Hence, social media has become an inseparable component of individuals' ways of life [17]. Although social media provides several advantages and opportunities, its excessive usage raises concerns worldwide. Therefore, the excessive use of social media is defined as a behavioral addiction that is characterized as being overly concerned with social media, driven by an uncontrollable urge to travel, browsing, or use social media, and devoting plenty of time and energy to social media that impair other important life areas[1]. Although over the past decade, studies mainly explored the opportunities provided by the internet-social media, current research is exploring the adverse effects of the internet and social media among their users. Many prior studies [15] have explored the connection of social media usage with the users' mental state problems (e.g., depression, anxiety, stress, loneliness, and self-esteem) among the many cohorts of individuals in developed countries.

This work targets the general population of users (in a certain age range) to understand their addiction to social media and the impacts on their lives. The focus addresses possible solutions to control social media usage and proposes social media users a productive replacement. The ultimate objective is to enable users to efficiently time their social media activities and be more productive.

2   Background Work on Social Media Addiction

Haand and Shuwang [8] investigate the correlation between social media addiction and depression among college students in the Khost province of Afghanistan. Using stratified random sampling, a 46-item self-administered questionnaire was distributed to 384 students from the three universities, Shaikh Zayed, Ahmad Shah Abdali, and Pamir University. The Internet Addiction Test (IAT), created by Kimberly Young, was utilised to measure social media addiction, whereas the Centre for Epidemiologic Studies Depression Scale was used to measure depression (CES-D). Using the Pearson correlation coefficient, simple linear regression, and component analysis, the relationship between social media addiction and depression was investigated.

Further, Nguyen et al., [14] examine the connection between social media addiction and mental illness utilising current research among college students. A thorough document search was conducted based on six electronic databases, including PubMed, Scopus, ScienceDirect, Web of Science, JSTOR, and Pro-Quest Educationto find published research before the date of November 21, 2019. All the gathered research works investigate social media addiction and psychosis. Two reviewers used the Joanna Briggs Institute method to subjectively assess the study's quality. Through the screening procedure, five items were eliminated, leaving room for review. It was identified that college students have a significant prevalence of social addiction (9.7% to 41%). Positive for student health is the connection between social media addiction and mental illnesses. This article contributes to identifying these risk issues and locating solutions. The study backs up the association between eating disorders and online shopping addiction in social addicts, and it also covers the causes and consequences of social media addiction.

Social networking site addiction [9] is regarded as a concerning phenomenon where additional psycho-pathological issues may appear. Jasso-Medrano and Lopez-Rosales examine the connections between social media use, mobile device

---

[1] https://proorno.com/statement.php?id=13707.

use, compulsive behavior, depression, and suicide ideation. The surveys were given to a sample of 374 university students, with a gender split of 58.6% women to 41.4% males and an age range of 20.01 years (SD = 1.84) on average. Contrary to social media usage, addictive behavior was significantly linked to depressive symptoms and suicidal thoughts. This work proposes an explanatory model that was correctly adjusted and connected addicted behavior to daily hours, suicidal ideation, depression, and frequency of mobile phone use. It is determined that, in contrast to excessive use, addicted behavior is linked to unfavorable psychological traits.

However, when it comes to depression, addictive behavior can also be seen as a barrier to suicide ideation. Keles et al., [11] argue that a growth in young people's mental health difficulties is attributed to online social media. This systematic review compiles the data on the relationship between teen social media use and psychological distress, including depression, anxiety, and anxiousness. After scanning the PsycINFO, Medline, Embase, CINAHL, and SSCI databases, 13 pertinent papers were discovered, 12 of which were cross-sectional studies. The results were categorized using the four social media domains of time spent, activity, investment, and addiction. Across all aspects, there was a correlation between depression, anxiety, and psychological discomfort. However, because to methodological problems, the cross-sectional design, sample, and measures have major limitations. Long-term cohort studies and qualitative research are required to better comprehend the mechanisms underlying the reported detrimental effects of social media on mental health.

Further, J. Brailovskaia and J. Margraf [4] looked into the relationship between depression symptoms, exercise (such as jogging, cycling, and swimming), and compulsive social media usage (SMU). An online survey was used to evaluate depression symptoms, physical activity, and addictive SMU in a sample of 638 social media users [mean (M)age standard deviation (SD)age = 21.57 (4.89)]. The positive correlation between depressive symptoms and addictive SMU was dramatically attenuated by physical activity. The correlation between depressive symptoms and addictive inclinations weakens with increased physical exercise. People who are depressed tend to utilize social media heavily to escape their feelings and find solace. Thus, those users are more likely to exhibit addictive behaviors. This risk may be lowered and well-being may be promoted by physical activity. As a result, those with increasing depressive symptoms should be assessed for SMU problems and encouraged to exercise. I. Pantic [16] states that online social networking has significantly altered how individuals engage and communicate during the last ten years. However, some of these changes tend to make an unclear prediction of the potential impact of typical human behavior and result in psychiatric diseases.

According to numerous research, the regular use of social networking sites (SNS), such as Facebook, may be associated with depressed signs and symptoms. In addition, some academics have hypothesized that some SNS activities, especially those involving children and teenagers, may be connected to low self-esteem. In terms of the beneficial effects of social networking on self-esteem, some research has shown the opposite outcomes. There is still debate over the link between SNS use and mental health issues, and it is difficult to research this subject. This succinct review focuses on new research that suggests a link between social networking sites (SNS) and mental health conditions, including depression symptoms, changes in self-esteem, and Internet addiction. The study done by Banjanin et al., [3] looks at any possible links between adolescent sadness and internet addiction. 336 high school pupils in Belgrade, Serbia, were the subject of the cross-sectional observational study. Each student participated in responding to a questionnaire that included the Young Internet Addiction Test (IAT), the Center for Epidemiological Studies of Depression Scale for Children (CES-DC), and general inquiries on the usage of the internet and social networking sites (SNS). The findings reveal a positive correlation between depressive symptoms and internet use, as measured by the IAT scale, and internet addiction. In terms of preventive measures against social media addiction, Van Rooij et al., [18] focus on techniques to abandon excessive social media usage. The research was performed on Dutch adolescents, this research is significantly similar to the research paper under observation.

Social media [12] has become extremely popular, yet it is a double-edged sword that can have unfavorable outcomes like social media addiction. However, identifying the causes of social media addiction has received remarkably little attention. Artificial intelligence and expert systems were used to predict social media addiction in this study through a hybrid SEM-artificial neural network technique. Based on a sample of 615 Facebook users, an integrated model of the Big Five Model and the Uses and Gratification Theory was validated. In contrast to earlier social media research that employed SEM, we adopted a hybrid SEM-ANN technique with IPMA as an additional analytic in this work. The new SEM-IPMA-ANN study is a fresh methodological advancement that enables the drawing of insightful conclusions based on the construct's significance as well as its effectiveness in prioritizing managerial actions. The main emphasis will be on raising the performance of structures with significant importance but low performance. Based on the normalized importance of the ANN study using multi-layer perceptrons with the feed-forward-back propagation approach, the authors discovered nonlinear relationships between neuroticism and social media addiction.

Previously, only linear relationships had been discovered; therefore, this is an important discovery. Following agreeableness, neuroticism, hours spent, and gender as the strongest predictors are entertainment and entertainment value. With an accuracy rate of 86.67%, the artificial neural network can forecast social media addiction. The novel technique and study results will have significant effects on the existing literature on expert systems, artificial intelligence, and social media addiction. The authors talked about the study's methodological, theoretical, and practical contributions. The purpose of this study by Karakose et al., [10] is to look into the connections between depression, social media addiction, COVID-19-related burnout, and psychological distress. 332 school principals and instructors with doctoral degrees in the field of educational administration participated in the research, which was constructed using the relational survey paradigm. Online surveys were used to gather research data, which was then analyzed and tested using structural equation modeling (SEM). The results of the study showed that COVID-19-related psychological distress predicted COVID-19-related burnout quite substantially. In this situation, as COVID-19's psychological discomfort mounted, so did the feeling of burnout that went along with it. However, depression has shown a tendency to be significantly and positively predicted by burnout related to COVID-19.

According to SEM findings, psychological discomfort, caused by COVID-19, directly impacted social media addiction, sadness, and burnout. Additionally, a connection was identified, between COVID-19-related psychological distress and depression that is a result of COVID-19-related burnout and social media addiction. The study [1] of "addictive technology habits" has grown significantly over the past ten years. Research has also shown a high correlation between co-morbid psychiatric problems and technological addiction. In the current study [1], 23,533 adults (mean age 35.8 years, range 16 to 88 years) participated in an online cross-sectional study to examine the impact of demographic factors, signs of attention-deficit/hyperactivity disorder (ADHD), obsessive-compulsive disorder (OCD), anxiety, and depression, on differences in the addictive use (i.e., compulsive and excessive use associated with negative outcomes) of two types of modern online technologies. Social media and positive and substantial correlations were found between the symptoms of mental disorders and those of addictive technology use; even with the presence of only a tenuous connection between the two addictive technological practices. The addictive use of these technologies seemed to be inversely related to age. The addictive use of video games was significantly connected with being a male; whereas the addictive use of social media was considerably associated with being a female. Both social networking and video gaming were positively correlated with being single. Demographic characteristics accounted for 11 and 12% of the variance in the usage of addictive technologies, according to hierarchical regression studies. Between 7% and 15% of the variance was accounted for by the mental health variables. The study introduces Internet use disorder (often known as "Internet addiction") as a single, comprehensive construct that is not warranted and greatly advances the understanding of mental health symptoms on the one hand and their involvement in the addictive use of modern technologies on the other hand.

The number of adolescents and preadolescents using such sites has substantially increased during the last five years. A recent survey unfolded that more than half of teenagers log on to social media sites more than once each day, and 22% of teens visit their preferred social media site more than ten times per day. And today, 75% of teenagers have a cell phone, and 25% of them use it for social media, 54% for texting, and 24% for instant messaging [7]. Munene and Nyaribo [13] suggest that using social media can decrease productivity if it is not appropriately managed. Internet addiction [5] can lead to internet abuse, which can ruin not only a person's life but also a company's reputation.

2.1 Study Design

This paper conducts a qualitative study to investigate the research question:

The goal is to make the application appealing enough that the user will continue to use it. They will, however, not become addicted to it. When the user exceeds the daily limit for social media usage, the alarm activates to indicate potential risks. Furthermore, the proposed application recommends various useful courses to users that can engage them for the next 21 days. Specific rewards are integrated for completing each milestone.

The selected participants are in the age range of 18 to 40 years. We could reach 27 participants who are from different countries. A Google survey form has been used to collect the information from the participants.

2.2 Selections for Participants

The study participants are students and employees of various organizations. Participants' ages range from 18 to 40. The age range is selected with the intention to focus on individuals who just started university and people who have been in the working industry for quite some time. The qualitative analysis will identify the factors of social media that are common in both groups of students and employees. The current study is a work in progress and has a small sample size; however, in future work, we plan to reach a broader audience.

2.3 Survey Questions

The survey questions focus on collecting general impressions about social media use and a broader need analysis for an addiction prevention application prototype. As a result, the survey is purposefully kept short to concentrate on the main points that serve as the starting point for our requirements elicitation process [2]. The qualitative analysis of participants' responses is presented below and shown in the results section:

2.3.1 Gender of Participants
We surveyed a total of 27 participants, of whom 9 self-identified as female and 18 as male.

2.3.2 Age Range of Participants
The age of the participants was between 18 at the minimum and 46 at the maximum. The average age is 23 with a standard deviation of 50% is 19.

2.3.3 Residence and Nationality of Participants
The survey is performed on a global population, including Russia, Kazakhstan, Pakistan, Syria, and Turkey.

2.3.4   What Purposes are You Using Social Media Platforms For?

We aim to identify reasons (Fig. 1) for social media usage to formulate the problems and suggest the corresponding solution. About 48% (13 participants) stated that they use social media platforms for work purposes. This percentage increased to 74.14% for 20 respondents who use social media for work, leisure, and entertainment. However, 12 participants were using social media to stay up-to-date with current affairs. This also includes the work and entertainment category participants, making a total of 44.4%.

2.3.5   Does Using Social Media Cause Any Negative Feelings and Would you Like to have More Control Over that?

55.6% (15) of the participants stated that they have negative feelings because of social media; however, 44.4% (12) disagree with this.

2.3.6   How Much Time do You Spend per Day on Social Media?

For this research paper, understanding how much time the participants spend on social media daily is significantly essential. We found that 20 participants spent 2–4 hours, 5 spent around 5–6 hours, and almost 3–5 stated that they spent more than 10 h on social media, as shown in Fig. 2.

2.3.7   Is it Interesting for you to Know How Much Time You are Spending on Social Media?

The participants were asked if they wanted to track their social media usage. 66.7% (18) answered positively to that question, which indicates the absence of friendly monitoring of the social media platforms and participants who are interested in controlling their addiction to social media.

2.3.8   How do You Feel When Using Social Media?

We asked participants how they felt while using social, and they shared different feelings. Some even stated that they felt nothing, but the majority of the participants felt nervous, stressed, exhausted, and frustrated because of the use of social media. However, they do not know how to control the usage time. This identifies a strong need for the provision of a useful addition to replace the harmful addiction. 18.5% (5) of the participants stated that they feel nervous and stressed while using social media. 14.8% (4) felt exhausted and frustrated, as per the survey results. The rest of the participants stated mixed feelings regarding social media like 3.6% (1) felt themselves useless when they were using social media, 7.4% (2) felt nothing because of social media usage. 3.6% (1) stated it depends on the purpose and what news and impact social media can have. Sometimes, it enables them to communicate with friends and family. 3.6% (1) felt entertained. 3.6% (1) felt well informed. 3.6% (1) felt sociable. 3.6% (1) felt relaxed sometimes. 3.6% (1) stated the feeling of emptiness. 3.6% (1) stated neutral behavior, and the rest stated they felt nothing.

2.3.9   Do You Think You are an Addict of Social Media?

33.3% (9) of the respondents stated that they are addicted to social media; however, those who did not affirm the addiction point stated that they are just unaware of the fact that they are addicted.

2.3.10   How Often do You Post on Social Media in 48 h?

In the duration of three days, 29.6% (8) of the participants definitely posted on social media, and 7.4% (2) did not post at all. The rest of the participants are unaware of their posting schedule.

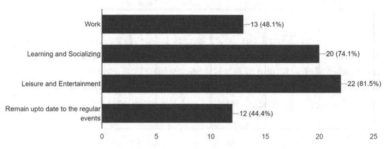

Fig. 1. Purpose

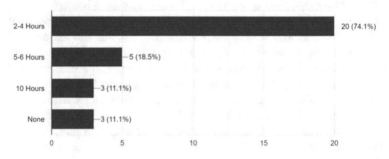

Fig. 2. Usage time

2.3.11 Does Social Media Causes Disturbance When you are Working or Focusing on Something Important?

Table 1 shows participants' responses where 55.6% of participants stated that they can not work with devotion and concentration because of the distraction caused by social media.

Table 1. Lack of Focus while working

Participants	Comments
15	Yes
7	No
5	Sometimes

Thus, our work aims at converting this disturbance or distraction into a productive activity alarm.

3 The Prototype of Addiction Control Application

Thus, our work is aimed at converting this disturbance or distraction into a productive activity alarm.

Therefore, this alarm has two features. One is *positive time* (targeted time or even less time spent on social media) and the other is *negative time* (time spent on social media more than the targeted time). This alarm is synced with each social media application and notifies the user of their situation.

The user will see a container of the most common social media applications as shown in Fig. 3 (d) (Facebook, Instagram, Twitter, TikTok etc.) available and can customize it if they want to exclude some social media from the addiction monitoring application. The application gives daily, weekly, biweekly, monthly, quarterly, bi-annually, and annual progress results Fig. 3 (e and f).

Once the user reaches the target time that is set by the specific social media platform, two features are activated. First, the user is notified about the fact that their *negative time* is starting. Second, a productive activity is recommended to the user, as a replacement for the specific social media. Productive time, such as learning from digital platforms, is also considered a *positive time* and is a constructive way to control social media addiction. If the user achieves the positive target for 21 days, they will be awarded some social activities, upon completing the first milestone of the controlled social media addiction. The final milestone award will be revealed upon the successful completion of the challenge, i.e., after 41 days.

Fig. 3. Prototype Design left to right screens: (a) login, (b) registration, (c) alarm settings, (d) app selection box, (e) app usage summary, (f) self-analysis

4   Conclusion

This paper proposes preliminary results of the work in progress, on preventive measures for social media addiction. We surveyed 27 participants to identify the general usage patterns of social media; such as how much time users spend on social media and what their reasons are for using certain platforms.

The qualitative findings suggest that users have an open attitude toward understanding their usage patterns on social media and are interested in learning about preventive measures for access usage. Hence, we prototype a software application that is required to be installed by users to monitor their social media usage time.

The application is simple to use and requires minimal intervention from the user. The user can customize this application; according to their preferred settings for time spent on social media. The proposed application can monitor the social media usage time, and an alarm can notify the user when they exceed their set time limit. The plan is to survey a broader audience to understand their rationale for social media usage. Hence, integrating preventive features into this application will be possible, based on the reasons identified for addiction.

This ongoing study is limited in its sample size and therefore requires targeting a diverse set of users. Furthermore, the application prototype will be tested to determine its applicability and impact on users' acceptance of it and social media usage patterns. We endeavor to achieve the study objective in the near future.

References

1. Andreassen, C.S., et al.: The relationship between addictive use of social media and video games and symptoms of psychiatric disorders: a large-scale cross-sectional study. Psychol. Addict. Behav. **30**(2), 252 (2016)
2. Aslam, H., Naumchev, A., Bruel, J.-M., Brown, J.: Examining requirements documentation through the focused conversation method. In: 29th International Conference on Information Systems Development (ISD 2021) (2021)
3. Banjanin, N., Banjanin, N., Dimitrijevic, I., Pantic, I.: Relationship between internet use and depression: focus on physiological mood oscillations, social networking and online addictive behavior. Comput. Hum. Behav. **43**, 308–312 (2015)
4. Brailovskaia, J., Margraf, J.: Relationship between depression symptoms, physical activity, and addictive social media use. Cyberpsychol. Behav. Soc. Netw. **23**(12), 818–822 (2020)
5. Chen, J.V., Chen, C.C., Yang, H.-H.: An empirical evaluation of key factors contributing to internet abuse in the workplace. Industr. Manage. Data Syst. **108**(1), 87–106 (2008)
6. Dollarhide, M.: Social media, Updated August 31, 2021 Reviewed by Amy Drury. Accessed 29 July 2022
7. Fernández, A.: Clinical report: the impact of social media on children, adolescents and families. Arch. Pediatr. Urug. **82**(1), 31–32 (2011)
8. Haand, R., Shuwang, Z.: The relationship between social media addiction and depression: a quantitative study among university students in Khost, Afghanistan. Int. J. Adolesc. Youth **25**(1), 780–786 (2020)
9. Jasso-Medrano, J.L., Lopez-Rosales, F.: Measuring the relationship between social media use and addictive behavior and depression and suicide ideation among university students. Comput. Hum. Behav. **87**, 183–191 (2018)
10. Karakose, T., Yirci, R., Papadakis, S.: Examining the associations between Covid-19-related psychological distress, social media addiction, Covid-19-related burnout, and depression among school principals and teachers through structural equation modeling. Int. J. Environ. Res. Public Health **19**(4), 1951 (2022)
11. Keles, B., McCrae, N., Grealish, A.: A systematic review: the influence of social media on depression, anxiety and psychological distress in adolescents. Int. J. Adolesc. Youth **25**(1), 79–93 (2020)
12. Leong, L.-Y., Hew, T.-S., Ooi, K.-B., Lee, V.-H., Hew, J.-J.: A hybrid SEM-neural network analysis of social media addiction. Expert Syst. Appl. **133**, 296–316 (2019)
13. Munene, A.G., Nyaribo, Y.M.: Effect of social media pertication in the workplace on employee productivity. Int. J. Adv. Manage. Econ. **2**(2), 141–150 (2013)
14. Nguyen, T.H., Lin, K.-H., Rahman, F.F., Ou, J.-P., Wong, W.-K.: Study of depression, anxiety, and social media addiction among undergraduate students. J. Manage. Inf. Decis. Sci. **23**, 4 (2020)

15. University of Pennsylvania: Social media use increases depression and loneliness, study finds, 8 November 2018. Accessed 29 July 2022
16. Pantic, I.: Online social networking and mental health. Cyberpsychol. Behav. Soc. Netw. **17**(10), 652–657 (2014)
17. Tutorials Point: Why has social media become an integral part of life? Updated on 30 July 2019 22 30 24 by Yashwanth Sitamraju. Accessed 29 July 2022
18. Van Rooij, A.J., Ferguson, C.J., Van de Mheen, D., Schoenmakers, T.M.: Time to abandon internet addiction? Predicting problematic internet, game, and social media use from psychosocial well-being and application use. Clin. Neuropsychiatry **14**(1), 113–121 (2017)
19. WHO: Depression, 13 September 2021. https://www.who.int/news-room/fact-sheets/detail/depression. Accessed 29 July 2022
20. WHO: Suicide, 13 September 2021. https://www.who.int/news-room/fact-sheets/detail/suicide. Accessed 29 July 2022

Why Zero Trust Framework Adoption has Emerged During and After Covid-19 Pandemic

Abeer Z. Alalmaie[1](✉), Priyadarsi Nanda[1], Xiangjian He[2], and Mohrah Saad Alayan[3]

[1] School of Electrical and Data Engineering, University of Technology Sydney, Sydney, Australia
abeer.z.alalmaie@student.uts.edu.au, priyadarsi.nanda@uts.edu.au
[2] School of Computer Science, University of Nottingham Ningbo, Ningbo, China
Sean.He@nottingham.edu.cn
[3] School of Software, University of Technology Sydney, Sydney, Australia
Mohrah.S.AlAlyan@student.uts.edu.au

Abstract. The rise of COVID-19 brought an unprecedented change in the way people lived. It left several people in a work-from-home situation. This Paper aims to investigate the recent works which applied Zero Trust and the reason that this framework adoption has emerged during and after the Pandemic. In this regard, a questionnaire was prepared, and its results are reported. According to its results, with Zero Trust Architecture (ZTA) gaining skyrocket popularity and trust, for around 60% corporates, ZT Access is planned for future, while for around 30% corporates, the project is in pipeline. None of the organizations surveyed have the ZTA in place. 14% of organizations are uninterested in adopting ZTA. Plus, in past 2 years, the percentage of north American organizations having a ZTA on the plans to establish one in the next 12–18 months has shot up.

1 Introduction

COVID-19 advanced at a very unprecedented time; leading a ceaseless work-from-home situation that organizations were prepared for. Businesses perpetually dealt with portable devises shortage, insufficient bandwidth, undersized infrastructure of Virtual Private Network (VPN) with IT department pushing the limits to maintain efficiency by enabling employee access to corporate resources/applications to work in full capacity. The pandemic caused profound re-modification and reorganization of human-to-human interaction [1]. Before the pandemic, interest in Zero Trust (ZT) was being driven by a need to modernize how the information security stack works. The traditional perimeter-centric security model is not compatible with the way businesses are working today. The pandemic forced the organizations looking at ZT because so many employees shifted to remote work that the organizations' networks were no longer a source of trust. Increased attack surface is a concern and increased mobility was already happening but accelerated in the pandemic. ZT offers encouraging solutions, but requires reasonable re-architecture, re-modification, and re-investment. That means work-from-home scenario is a continuous arrangement for job profiles who can fulfil professional obligations without commuting to and from their job site. These factors have seen the risk for cybercrime and cyberattacks increase. With older technologies like VPNs making headlines for security issues, a new approach to empowering distributed teams while ensuring optimal data security has emerged: ZT network architecture. In this regard, a survey is conducted that its results can help future works for developing an Intrusion Detection System (IDS) using Zero Trust Architecture (ZTA). Most of our respondents were around 40 + CISO's from various IT companies, banks and government in India and Saudi. The specific survey's objectives include:

1) Why businesses experienced similar pain points to enable secure remote work?
2) How a ZTA could assist business continuity in pandemic outbreaks?
3) Why are organizations committed to adopting a ZT security architecture?
4) Which are the key initiatives to enable ZT security adoption in their organization?

The rest of the paper is organized as follow: In Sects. 2 and 3 we review some related works and investigate the impact of COVID-19 on existing IT infrastructure, respectively. In Sect. 4, we explain how a ZT network functions. In Sects. 5 and 6 our survey objectives and results are mentioned, respectively. Eventually, in Sect. 7 we prepared a short conclusion.

2 Background

In [2], over 200 executive board members of 80 companies from 2014 to 2016 were interviewed, asking "How do we secure increasingly dynamic architecture in an environment without a perimeter?". The answers revealed Bring Your Own Devices (BYOD) were valuable opportunities but posed onerous risks. Setting up a centralized and scalable Mobile Device Management system using access controls (LDAP/AD[1]) was reported to be the most important challenge. This suggests a more-risk based approach to cybersecurity is needed in today's dynamic technological environment.

According to [3], ZT treated all network traffic as untrusted, continuously confirming users and endpoints by securing cloud data. The benefit of ZT is a highly flexible infrastructure that can be integrated with the cloud to enhance organizational security. To ensure the networks safety amid new cybersecurity threats, cybersecurity professionals should embrace additional philosophies alongside a ZT mindset.

According to [4], "Who we can trust with our data?" is one of the largest debates of our generation. It has never been more important to create security models that keep users safe. The author mentioned approximately 60% to 80% of network misuse comes from within the network. ZT therefore, offers a solution to both issues, with its ability to increase micro segmentation of a network offering more visibility of overall traffic through the inspection of users and devices which connect the network.

In [5], it was proved that working remotely, especially for employees with minimal cybersecurity resources, increased the risk for personal and organizational data to be compromised. In [6], authors elucidated the challenge pointing out the use of AI, and

---

[1] Lightweight Directory Access Protocol / Active Directory.

poorly secured technologies deployed in response to COVID-19 challenges increased risks for cybercrimes due to the high volume of data being generated and shared.

In [7], a two-stage ensemble classifier including hierarchical rotation forest and bagging classifiers, along with a hybrid evolutionary algorithm for feature selection was proposed for NID. In [8], a four-way ensemble classifier including Support Vector Machine, Linear Regression, Naïve Bayes, and Decision Tree was proposed which utilized a combination of feature selection methods.

Since ML methods require feature extraction and parameter tuning [9], DL methos have become a trend in AI problems like image, language, and speech processing and NID [10, 11]. A two-stage deep Neural Network (NN) was proposed for NIDS including a Deep Sparse AE as the feature extractor and a shallow NN classifier [12].

In [13], a Sparse Auto-Encoder was proposed for feature extraction, however, Support Vector Regression was used as the classifier instead of the shallow NN. The AE bottleneck features were shown to be effective in enhancing the NIDSs and giving the ability to feed any type of attributes to the NID model. Plus, bottleneck features were shown to be robust against noise.

In [14], a Recurrent Neural Network (RNN) was proposed for NID to consider the changes of the input in real-time applications. Also, deeper RNN models were used for NID which outperformed previous works. Since Long Short Term Memory (LSTM) cells hold the long term dependencies and prevent the vanishing gradient problem, in [15], extended the RNN models to LSTM and Bi-directional LSTM (BiLSTM) for NID.

In [16], a Convolutional Neural Network (CNN) classifier using a two-stage feature extraction including a PCA and a feature engineering method to select the most relevant have been proposed for NID. In [17], the CNN models have been used in combination of other classifier methods including RNN, LSTM, and Gated Recurrent Unit (GRU), which proved the power of CNN.

In [18], IGRF-RFE was introduced as a hybrid feature selection method for multi-class network anomalies using a Multi-Layer Perceptron network. It was a feature reduction model based on both the filter feature selection and the wrapper feature selection methods. The filter feature selection method was the combination of Information Gain (IG) and Random Forest (RF) importance, to reduce the feature subset search space. Recursive Feature Elimination (RFE) was a wrapper feature selection method to clear redundant features recursively on the reduced feature subsets.

In [19], a NID model was defined that fused a CNN and a gated recurrent unit. They tackled the low accuracy problems of existing ID models for multiple classification of intrusions and low accuracy of class imbalance data detection. They applied a hybrid sampling technique combining adaptive synthetic sampling and repeated edited nearest neighbors for sample processing to solve the positive and negative sample imbalance issue in the dataset. The feature selection was carried out by combining RF and Pearson correlation methods to address the feature redundancy problem.

In [20], an IDS was proposed to detect 5 categories in a network: Probe, Exploit, DOS, Generic and Normal. This system was based on misuse-based model, which acted as a firewall with some extra information added to it. Moreover, unlike most related works, they considered UNSW-NB15 as the offline dataset to design own integrated classification-based model for detecting malicious activities in the network.

3   Impact of COVID-19 on Existing IT Infrastructure

In [5], authors emphasized cybercrime is among the greatest threats for most organizations. The problem's magnitude was further elucidated by the financial burden, which they report was $3 trillion in 2015 and was projected to be over $6 trillion every year by 2021. The damages cybercrime cause is profound touching on data destruction, reputation attacks, shattering company progress, loss of intellectual property, embezzlement, increasing mitigation cost, and cost of damage control in case such attacks happen. Therefore, having a secure cyber for organizations is necessary.

The average ransom payment demanded by cybercriminals carrying out ransomware attacks went up by 33% in the first quarter of the year to $111,605, compared to the previous quarter. Phishing attempts have also exploded, with Google's Threat Analysis Group noting 18 million COVID-19-related phishing and malware Gmail messages each day in April.

Most companies with remote teams must address new cybersecurity concerns and points of vulnerability. They are working diligently to fortify their network perimeter, implementing the latest in hardened routers, next generation firewalls, and IDSs. However, a lot of IT departments weaken their own security by standing up websites, home banking systems, ERP/ERM systems that provide access to other networks and computers behind firewall; this enables the attackers to penetrate in the system. The major reason behind such diluted security is lack of acceptance and trial of different methods other than this traditional one.

Nowadays, as some users work from home and the network structures of all the organizations are changing, the traditional way of working is orienting towards the cloud and there is rise in the use of Software as a Service (SaaS). Meanwhile, many organizations are embracing flexible working, with staff connecting from multiple devices in various locations. Leading to declining traditional network cycle/perimeter which is causing decline in security. Additionally, hackers attempting to find unaddressed vulnerabilities in newly deployed remote work infrastructure. Hence, the businesses have felt a compelling need for advanced and dependable security solution.

Organizations are being driven by the stress the pandemic was putting on their infrastructure, particularly on VPNs. Before the pandemic, VPNs were good-enough to satisfy most companies' work-at-home demands, which were occasional. Though, it is difficult to cross VPN overnight, but factors like return on investment of traditional systems in present times and unanticipated cost of VPNs make businesses to find a suitable way. The legacy system model works on the principle of trust, where it considers the elements inside a particular network as harmless. Today, employees are working from home, which poses a huge threat to the security of network. With employees now connecting a lot, they are effectively creating a hacker's playground, with new, vulnerable endpoints and access points being exposed. However, all is not lost since the ever-advancing technology space has realized the ZTA importance, which can enhance cybersecurity and safety through its principles.

4 How a Zero Trust Network Functions

In prevailing times, with legacy model of security like VPNs witnessing non-success in security, ZTA is gaining popularity for its optimal data security and empowering redistributed teams. The idea behind ZT networks is hardly new. ZTA is operated through the basic philosophy that no technology user should be trusted [21]. It is self-explanatory, which means trust no network. ZTA is a security tenet that asks organizations not to automatically allow access from inside/outside into the network structure, but instead to verify each request trying to connect.

ZTA offers cybersecurity paradigms that are more focused on users, assets, and resources as opposed to traditional paradigms that were more static and network-based oriented. By evoking ZT principle in the restructuring workstation cyber-system security, ZTA seals the major loopholes exploited by hackers and malicious intruders. As in [22] authors noted the tactic to more the firewall from outside to inside the system architecture, ZTA makes it harder to target the system internally, which is the most used approach in cybercrimes. The checking and recording of traffic within a network allow for effective system monitoring, further securing the system. They cite the four main methods that help achieve a feasible and effective ZT strategy. The 4 steps emphasize the effectiveness of scalable security infrastructure:

1) Identity authentication, which validates credibility to be allowed in a network.
2) Access control regulates the layering degree an authenticated user can access.
3) Continuous encryption-based diagnosis to offer monitoring and feedback services that help trace a threat with ease to an origin point and potential damage.
4) Mitigation, which is critical in reducing the occurrence of damages through threat identification and preventions strategies.

ZTA goal is to prevent unauthorized access to data along with creating enforcement notions for control. To achieve this vision, several technical elements are necessary, and it is important to note a single commercial tool or technology will not be able to deliver all capabilities. As Per National Institute of Standards and Technology (NIST), the logical elements of ZT include policy engine, policy administrator, and policy enforcement point. Several data sources are necessary to provide input to these policy-based mechanisms which will feed the trust algorithm that ultimately determines whether to grant/deny access to information resources based on the level of evaluated trust of the endpoint/user combination. ZT models, according to the NIST, assumes an attacker is present on the network and an enterprise-owned network infrastructure is no different than any non-enterprise owned network. NIST categorizes the types of input as: access request; user identification, attributes, and privileges; asset database and observable status; resource access requirements; and threat intelligence (Fig. 1).

Authors in [23] noted that the scalable approach is thwarted threats to cybersecurity in a multistage strategy, hence more reliable than traditional network architecture.

Innate trust is removed from the network under ZTA principles which means one does not necessarily access to everything on network even though they are connected to that network. Inherent trust is removed and defied, so the devices and users are denied the access until they are verified on basis of pre-defined parameters. The pre-requisite for

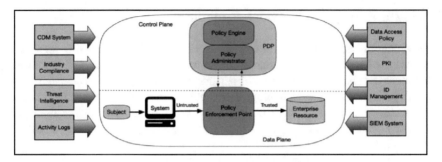

Fig. 1. NIST, 800–207, ZTA 2nd Draft

gaining access is authorization and crossing security set level. This works best in avoiding breachers who witness an attacker and move laterally into the network in cases where everything is trusted in the network. Treating the network as hostile has many advantages. By leveraging micro-segmentation and granular perimeter enforcement based on end-user characteristics like location, role, and permissions, a ZTA only gives people access to the specific resources they need. Therefore, the strategy secure layers with fine-grained segmentation, stringent system access controls, strict data retrieval management, and sophisticated data protection strategy [22]. All users access a system through gadgets must be authenticated while the information degree accessed is highly controlled and restricted to need-to-know bases [23]. The encryption ZTA strategy protects the information from internal and external intrusion and maintains a continuous monitoring and adjustment process that maintain proprietary interfaces in check. Thus, this study evaluates the effectiveness of adopting ZTA in the COVID-19 era to reduce the cyber threats risk and incidence.

5   Survey Objectives - Adoption of ZTA

We conducted a survey across multiple mediums (email, SMS, and web surveys). It incorporates qualitative and quantitative data to offer a qualifying and justifiable argument about the ZTA role in reducing cyber threats amid coronavirus pandemics.

5.1   Sampling and Sample Size

The survey involved 14 companies sampled through purposeful sampling method and then grouped into 2 categories. The 1st category of 7 companies, those using VPNs, is the primary architecture in cybersecurity. Number of them is based on accessible companies in the locality that can offer insights into the issue.

5.2   Data Collection and Analysis

The survey's data collection process entails asking the relevant question from selected individuals in the identified companies aimed to determine the degree of cybersecurity and safety offered by ZT networks compared to VPNs strategies. The data collection was done by sending the survey tool to the information technologists or persons responsible for maintaining cybersecurity in those organizations. Once they completed the questionnaire, they resent them via email for data extraction and analysis. The data was analyzed thematically for qualitative data while quantitative data was analyzed using SPSS version 22 to yield descriptive and appropriate inferential statistics.

6 Survey Results

With work-from-home, safeguarding company's network perimeter is more complex than logging in from a single location. Figures 2 and 3 show results of the survey's 1st and 2nd questions. In Fig. 2, at risk devices means unknown, unsanctioned or non-compliance endpoints, and the top challenge of companies in terms of securing the access to application and resources is the complex manual process. So, the ability to react quickly is an important feature for practical industry systems. In Fig. 3, application-specific access is based on a user's identity, device posture, and group membership, and most in most companies it was the chosen method.

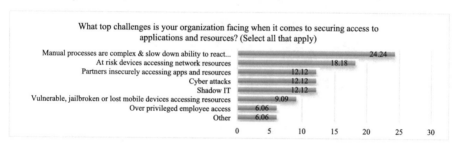

Fig. 2. Result for the first question in the survey

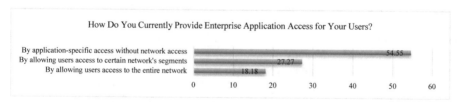

Fig. 3. Result for the second question in the survey

Recently, security was equal to the perimeter security model. It is built on its strength of outer defense. To make the network safe, the perimeter needs to be deemed impenetrable. Hence, there are ways incorporated like VPN, network segmentation and firewalls. Though, the model does not ensure security. Some attackers have demonstrated the complexity for large firms to avoid breach. The opportunity cost for security in perimeter-based security is operational agility. Plus, the network is kept secure via forming outer boundaries, what takes effort and efficiency is managing in the world micro-services and cloud computing where service communication requirements are changing frequently. Figure 4 illustrates the result of the 3rd question.

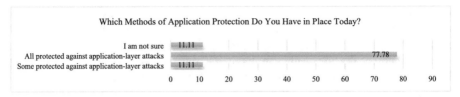

Fig. 4. Result for the third question in the survey

In Fig. 4, no one chose "None of our corporate applications are protected against application-layer attacks." option, and the majority mentioned their corporate applications are protected against application-layer attacks with a Web Application Firewall (WAF). Thus, WAF is the most-used current method for protection against application-layer attacks. In Fig. 5, the result for 4th question is shown, in which entity verification obviously means user, device, infrastructure, etc.

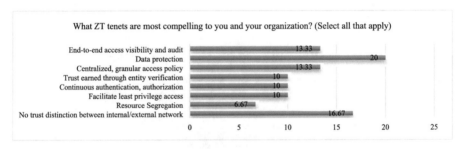

Fig. 5. Result for the fourth question in the survey

The idea is, when ZTA is detaching trust from the network, it simply amplifies the trust in the users, devices, and services. It is possible via undeterred authorization, authentication, and encryption. Its efficiency arises from its principle of authenticating each user connecting to the server regardless of where the access request is generated from. For effective use, the authentication and authorization levels and access policies should be well-defined, partaking all circumstances. The trust degree depends on the data value magnitude and impact of the performed action. Implementing ZTA on traditional systems is difficult. It needs to be installed in phases with iterations. Once the new approach foundation is laid over the legacy system, the establishment will be easier to further build on. Establishing a strong identity for users and devices or deploying modern authentication across the organization can be time-consuming.

According to the survey, 66% were neutral about adopting ZTA while 33% fall under the Satisfied to Extremely Satisfied spectrum. Around 60% plans to implement ZTA capabilities on-premises and SaaS. Figure 6 presents the 5th question's result.

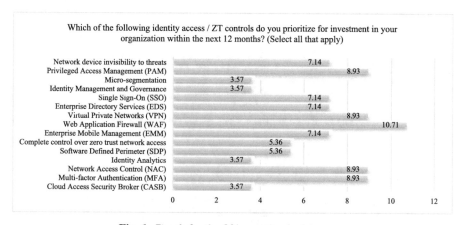

Fig. 6. Result for the fifth question in the survey

Many organizations that have implemented VPNs, with enterprise VPN usage may face data breaches in the absence of regular patching, updates, and the implementation of MFA for remote access accounts. For dealing with the challenges, IT teams attempted to make the access secured. Around 50% companies provided application access by allowing user access to only a certain network segment, however, other enacted by 38% organizations is by allowing application specific access, without network access post authorization scrutiny. Figure 7 shows the 6th question's result.

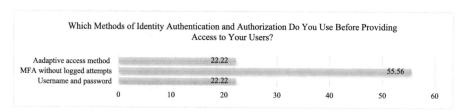

Fig. 7. Result for the sixth question in the survey

While transitioning to a new architecture, it is not suitable to start decommissioning traditional security controls before you have implemented and tested ZT controls. Due to the ZTA nature, it may leave systems exposed at considerable risk if they are not properly configured. Thus, it is vital not to dismantle the VPN establishment until the ZTA starts to perform satisfactorily. VPN can manage the potential threats if needed. Systems may be hosted using a traditional architecture or might not support the features of ZTA. So, some environments may need to manage both a traditional perimeter and a ZTA. This could involve using a split VPN tunnel to the legacy application or an authentication proxy. Figures 8, 9, 10, 11 present the results for 7th-10th questions.

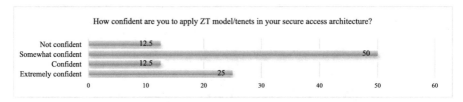

Fig. 8. Result for the seventh question in the survey

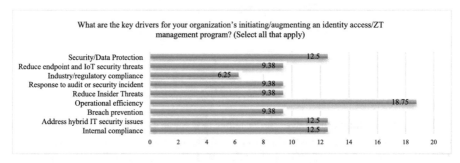

Fig. 9. Result for the eighth question in the survey

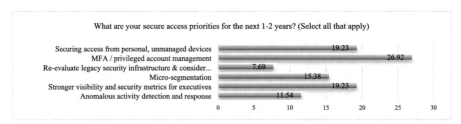

Fig. 10. Result for the ninth question in the survey

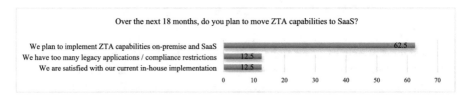

Fig. 11. Result for the tenth question in the survey

Figure 8 shows that more than 85% of the surveyed people have confidence to an extent about applying ZT model, which proves the problem of traditional approaches and the potential benefits of using this model. In Fig. 9, regulatory compliance could be HIPAA, GDRP, PCI DSS, etc. Plus, it shows that the most important management key driver for companies is operational efficiency. According to Fig. 10, the least appropriate secure access priority for companies is re-evaluating legacy security infrastructure and considering software-defined access. Furthermore, in Fig. 11, no one choose "Significant – we plan to solely use SaaS-based ZT access capabilities" option.

7 Conclusion

Employees are increasingly working from home. There is always potential risk while accessing corporate network or data even with strong arrangements. COVID-19 has taught businesses, resilience, and preparation for uncertainty. Hence reliability on ZTA in terms of cybersecurity will enhance the future alertness and adaptability. To summaries, the followings are a few disadvantages of perimeter-based security model:

1) Perimeter security largely ignores the insider threat.
2) The impenetrable fortress model fails in practice.
3) Network segmentation is time-consuming and difficult.
4) Defining network perimeter is difficult in a remote-work, BYOD multi-cloud world.
5) VPNs are often misused and exacerbate the further issues.

ZT attempts to mitigate these shortcomings by the following principles:

1) Trust flows from identity, device-state, and context; not network location.
2) Treat both internal and external networks as untrusted.
3) Act like you are already breached because you probably are.
4) Each device, user, and application must be authenticated, authorized, and encrypted.
5) Access policy should be dynamic and built from multiple sources.

The literature is replete with evidence supporting the superiority of ZTA over traditional VPNs in providing maximum possible cybersecurity and safety. However, ZTAs are not without challenges that can complicate their adoption and usage in securing personal and organization data against intrusions. At a time where cyberspace has attracted masses secondary to the pandemic complications, it is prudent to explore ZTAs feasibility in preventing and protecting cyber interactions and transactions from malicious attacks. Ultimately, implementing ZTA is the best choice for businesses which want to give remote access to users while maintaining the security posture.

References

1. Lallie, H.S., et al.: Cyber security in the age of COVID-19: a timeline and analysis of cyber-crime and cyber-attacks during the pandemic. Comput. Secur. **105**, 102248 (2021)
2. GriffyBrown, C., Lazarikos, D., Chun, M.: How do you secure an environment without a perimeter? Using emerging technology processes to support information security efforts in an agile data center. J. Appl. Bus. Econom. **18**(1), 90–102 (2016)
3. Puthal, D., Mohanty, S.P., Nanda, P., Choppali, U.: Building security perimeters to protect network systems against cyber threats [future directions]. IEEE Consum. Electron. Mag. **6**(4), 24–27 (2017)

4. Assunção, P.: A zero trust approach to network security. In: Proceedings of the Digital Privacy and Security Conference (2019)
5. Ahmad, T.: Corona virus (COVID-19) pandemic and work from home: challenges of cybercrimes and cybersecurity. SSRN Electron. J. (2020)
6. Al Hajj, A., Rony, M.: Cyber security in the age of COVID-19: an analysis of cyber-crime and attacks. Int. J. Res. Appl. Sci. Eng. Technol. (IJRASET). **8**(VIII), 1476–1480 (2020)
7. Tama, B.A., Comuzzi, M., Rhee, K.-H.: TSE-IDS: a two-stage classifier ensemble for intelligent anomaly-based intrusion detection system. IEEE Access **7**, 94497–94507 (2019). https://doi.org/10.1109/ACCESS.2019.2928048
8. Krishnaveni, S., Sivamohan, S., Sridhar, S., Prabhakaran, S.: Network intrusion detection based on ensemble classification and feature selection method for cloud computing. Concurr. Comput. Pract. Exp. **34**(11), e6838 (2022). https://doi.org/10.1002/cpe.6838
9. Tohidi, N., Rustamov, R.B.: Short overview of advanced metaheuristic methods. Int. J. Techn. Phys. Probl. Eng. (IJTPE) **14**(51), 84–97 (2022)
10. Tohidi, N., Rustamov, R.B.: A review of the machine learning in GIS for megacities application. In: Rustamov, R.B. (ed.) Geographic Information Systems in Geospatial Intelligence, pp. 29–53. IntechOpen, London (2020). https://doi.org/10.5772/intechopen.94033
11. Abolghasemi, M., Dadkhah, C., Tohidi, N.: HTS-DL: hybrid text summarization system using deep learning. In: The 27th International Computer Conference, the Computer Society of Iran, Tehran (2022)
12. Jiang, K., Wang, W., Wang, A., Wu, H.: Network intrusion detection combined hybrid sampling with deep hierarchical network. IEEE Access **8**, 32464–32476 (2020)
13. Preethi, D., Khare, N.: Sparse auto encoder driven support vector regression based deep learning model for predicting network intrusions. Peer-to-Peer Netw. Appl. **14**(4), 2419–2429 (2020). https://doi.org/10.1007/s12083-020-00986-3
14. Almiani, M., AbuGhazleh, A., AlRahayfeh, A., Atiewi, S., Razaque, A.: Deep recurrent neural network for IoT intrusion detection system. Simul. Model. Pract. Theory **101**, 102031 (2020). https://doi.org/10.1016/j.simpat.2019.102031
15. Lee, I., Kim, D., Lee, S.: 3-D human behavior understanding using generalized TS-LSTM networks. IEEE Trans. Multimed. **23**, 415–428 (2020)
16. AlTuraiki, I., Altwaijry, N.: A convolutional neural network for improved anomaly-based network intrusion detection. Big Data **9**(3), 233–252 (2021)
17. Vinayakumar, R., Soman, K.P., Poornachandran, P.: Applying convolutional neural network for network intrusion detection. In: International Conference on Advances in Computing, Communications and Informatics (ICACCI), Udupi, India (2017)
18. Yin, Y., et al.: IGRF-RFE: A Hybrid Feature Selection Method for MLP-based Network Intrusion Detection on UNSW-NB15 Dataset. arXiv:2203.16365 (2022)
19. Cao, B., Li, C., Song, Y., Qin, Y., Chen, C.: Network intrusion detection model based on CNN and GRU. Appl. Sci. **12**(9), 4184 (2022). https://doi.org/10.3390/app12094184
20. Kumar, V., Sinha, D., Das, A.K., Pandey, S.C., Goswami, R.T.: An integrated rule based intrusion detection system: analysis on UNSW-NB15 data set and the real time online dataset. Clust. Comput. **23**(2), 1397–1418 (2019). https://doi.org/10.1007/s10586-019-03008-x
21. Rose, S., Borchert, O., Mitchell, S., Connelly, S.: Zero trust architecture. In: National Institute of Standards and Technology (NIST) (2020)
22. Yan, X., Wang, H.: Survey on zero-trust network security. In: International Conference on Artificial Intelligence and Security, Singapore (2020)
23. Uttecht, K.: Zero Trust (ZT) Concepts for Federal Government Architectures. Lincoln Laboratory, Massachusetts Institute of Technology, Lexington, Massachusets (2020)

An Interoperable Microservices Architecture for Healthcare Data Exchange

Allender V. de Alencar[1(✉)], Marcus M. Bezerra[1], Dalton C. G. Valadares[2],
Danilo F. S. Santos[2], and Angelo Perkusich[2]

[1] Electrical Engineering Graduate Program, Federal University of Campina Grande,
Campina Grande, PB, Brazil
{allender.alencar,marcus.bezerra}@ee.ufcg.edu.br
[2] VIRTUS RDI Center, Federal University of Campina Grande, Campina Grande,
PB, Brazil
{dalton.valadares,danilo.santos,perkusic}@virtus.ufcg.edu.br

Abstract. Nowadays, the healthcare sector has stood out for offering the population a better quality of life. Under these circumstances, Information and Communication Technologies (ICT) are essential in mitigating the growing need for hospitals and care centers, providing computation and communication infrastructures to support healthcare services and applications. The introduction of new technologies in the healthcare industry, such as the Internet of Things (IoT), Artificial Intelligence (AI), and Big Data, improve service delivery vertically. However, a problem arises when introducing those new paradigms: the amount of data generated from healthcare devices and applications. Indeed, data sharing brings some challenges due to different types of healthcare standards, multiple protocols, systems diversity, and incompatibility of data. In this article, we introduce a microservices architecture for healthcare transcoding engines. Our solution show how transcoders can be encapsulated as independently functions using microservices. An application scenario that maps healthcare standards is also presented. We show that the use of microservices as transcoding engines improves scalability and flexibility features required by the healthcare sector.

1 Introduction

In the healthcare sector several systems need to interoperate to achieve an effective data exchange between professionals, institutions, and health information systems. This process involves correctly translating data among different standards to allow the systems to provide a better service and, with this, a cost reduction. Additionally, new challenges arise when it introduces paradigms like the Internet of Things, Artificial Intelligence, and Big Data in the healthcare sector. In particular, one of those problems is the amount of data generated from multiple sources. Indeed, data sharing brings some challenges due to different standards, data types, systems diversity, and incompatible data [3]. Besides,

L. Barolli (Ed.): AINA 2023, LNNS 655, pp. 193–205, 2023.
https://doi.org/10.1007/978-3-031-28694-0_18

most of these data are isolated into databases of systems unable to interoperate with each other [11].

To deal with this, some interoperability standards were created, including the Health Level 7 (HL7) [15] and the Fast Healthcare Interoperability Resources (FHIR) [5]. For example, while the HL7 standard focus on the representation and transferring of clinical and administrative data between health information systems, such as in clinics, offices, hospitals and public health systems, the FHIR standard describes data formats and an application programming interface (API) for exchanging Electronic Health Records (EHR). An EHR is a clinical document that contains information about the patient. We can notice that there is an intersection between both protocols related to the manipulation of EHR, but it is not an easy task to exchange data from different devices using different protocols. In this sense, the following question arises: how to enable dynamical, flexible and secure medical data exchanging between different devices and legacy protocols?

Systems can employ health interface mechanisms to solve this interoperability problem and keep the old implementation of standards in healthcare providers for more time. These mechanisms can format and, in some cases, reinterpret data during the transition of information from one system to another [6]. We call these mechanisms transcoding functions or, transcoders. These transcoders are usually integrated in the health-based service or application. However, based on their definition, requirements and purpose, one hypothesis is that transcoders could be encapsulated as independent functions into the health-based service or application. In this sense, a possible solution based on microservices technologies could be used.

In the context of health-based services, the aspect of horizontal scale and flexibility are essential due to the need to increase or decrease the transcoders based on a temporally demand of requests being made to them in on specific period [17]. For example, in one specific period of time is necessary to exchange data between two hospitals using HL7 and FHIR, but, few hours later is necessary to exchange data using FHIR and another healthcare standard.

As those technologies become increasingly sophisticated thanks to advances in computing power and communication technologies, the dynamism, flexibility, and criticality of digital healthcare solutions demand new requirements and architectures related to data exchange and usage. So, in this article, we propose a new interoperable microservices architecture for healthcare data exchange.

The main contributions of this paper are summarized as follows:

- demonstration of how transcoders can be encapsulated as independent functions in a healthcare system.
- description of a microservice-based architecture for healthcare data exchange considering different data standards/formats;
- validate our proposal in an application scenario considering two healthcare data standards, FHIR and HL7 v2.

The paper is structured as follows: In Sect. 2, we present basic concepts used for this work, such as monolithic services, microservices, and service mesh. In

Sect. 3, we expose the issues, assumptions, and constraints of our proposed architecture. We present the proposed solution in Sect. 4, including the system architecture and functionality and an application scenario, followed by the experimental results in Sect. 5. We also highlight some works related to the interoperability problem in the healthcare domain in Sect. 6. Lastly, we conclude this work in Sect. 7, highlighting future work.

2   Background

In the following subsections, we review the relevant concepts we used as the base for this work, including healthcare standards, monolithic services, microservices, and service mesh.

2.1   Healthcare Standards

This subsection introduces the three health standards used in this work: HL7 v2, FHIR, and Consolidated Clinical Document Architecture (C-CDA). They are summarized to briefly show the structure, characteristics, and data structure contained in the messages.

2.1.1   HL7 V2

HL7 v2 is a healthcare data format most used in hospitals, and clinics [15]. This broad adoption is the most crucial characteristic of this standard because, in a sizeable interoperable system, a connection with this standard will make it possible to access a significant part of the data in the database silos.

Version 2 of HL7 defines a set of electronic messages to support the exchange of messages between administrative, logistical, financial, and clinical processes. HL7 v2.x messages use a syntax for their encoding that is not based on XML but on segments (lines) and character delimiters.

The default delimiters are the carriage return <CR> as the segment separator, pipe (|) as the field separator, the character (^) as the component separator, and the ampersand (&) as the subcomponent separator. The tilde character (~) is the repetition separator. Every segment starts with a 3-character string that identifies its type. Each message segment contains a specific category of information. Every message has the Message Header (MSH) as the first segment and includes fields that describe the message type. The message type determines which other segments are expected in this message.

2.1.2   C-CDA

The Consolidated Clinical Document Architecture (C-CDA) is an architecture to achieve interoperability in healthcare data exchange using the extensible markup language (XML[1]) to encode the data. This standard support various scenarios of healthcare data [4].

The C-CDA, compared to the CDA, is a standard with stricter rules for the coding, structure, and semantics of clinical documents of CDA documents to make them more exchangeable. This standard can include structured information, such as medication lists, and unstructured information, such as images.

2.1.3   FHIR

Fast Healthcare Interoperability Resources (FHIR) is a standard for healthcare created by the Health Level Seven International (HL7) [5]. Based on the resources of old standards like HL7 v2, and HL7 CDA, this standard has two principal functions: the first is to define a universal standard for healthcare data; the second one is to create web resources to manipulate the data. Using this standard can get better integration between the EHRs than old versions [16].

FHIR is a modular architecture with well-defined resources. A resource defines every exchange of content. It has its specific data format, and it is possible to understand the data. In this way, FHIR enables communication between devices and domains. Technology clinicians, thus breaking a barrier in interoperability [2].

2.2   Application Architectures

This subsection introduces two architectural styles of software systems: monolithic and microservices.

2.2.1   Monolithic Services

A monolithic architecture is an implementation where all services are packed and coupled in a single application [10].

The monolith architecture is usually most effective on lower load [7]. The connection and integration of services have fewer problems than in microservices architecture [7].

2.2.2   Microservices

The microservices architecture divides the application into small and independent services applications, enabling greater agility and scalability in each one [10]. Microservices can lead to complex and increased network traffic [7]. This network complexity can be solved by a Service Mesh using observability, as described in Sect. 2.3.

One of the main advantages of microservices topology is the easy deployment in distinct infrastructures. Every microservice is independent of the others, which

---

[1] https://www.w3.org/XML/.

is helpful in distributed architectures. Shoumik et al. [17] developed an FHIR server based on microservices architecture. They also highlight some advantages of this implementation:

- *scalability:* Ability to horizontally scale specific services based on their load;
- *service deployment:* performing the deployment of services and their versions in a more practical way;
- *decoupling:* decreases the coupling between services. Since each service has its database, failure to run one of these services will not impact the others;
- *agility in development:* because microservices have REST-based access, each one has in its structure what is necessary to function without depending on others. So, it is possible to use different programming languages;
- *high availability:* as microservices manage to scale, then the availability of each service increases, as in addition to not depending on other services, there is more than one service of the same type running.

2.3   Service Mesh

A Service Mesh is a dedicated infrastructure layer coupled to applications. It allows to transparently add observability, traffic management, and security features without adding them to the application code. The term "service mesh" describes the type of software used to implement this pattern, and the security or network domain created when you use it.

Its features might include service discovery, load balancing, crash recovery, metrics, and monitoring. A service mesh often addresses more complex operational requirements such as A/B testing, continuous integration, rate limiting, access control, encryption, and end-to-end authentication. [9]

3   Architectural Issues, Constraints and Assumptions

This section provides analysis and justification for the choice of the microservices architecture in a healthcare integration service.

We defined the following requirements for a healthcare integration service:

- (a) transcode data: transform the data structure from one healthcare standard to another as independent functions, e.g., HL7 v2 into FHIR;
- (b) efficient and dynamic horizontal scalability: the architecture should enable the application to use available resources efficiently and dynamically, i.e., scale the necessary trasconding functions based on the demand;
- (c) flexible reconfiguration of standards: the solution must have a way to reconfigure the transcoding functions easily, e.g., add the CCDA-FHIR transcoder without stop the main service;
- (d) allow new clients, such as new Healthcare services, to connect on the fly: new providers must be added to the solution's without impacting the system.

In this sense, the use of microservices meets our healthcare service requirements since it is possible to run multiple instances of microservices, and keep
copies of essential services without the need for all application services to run,
fulfilling requirement (a) and partially requirement (b). Furthermore, we decided
that the solution would be controlled by a *Service Mesh* to enable scalability,
observability, and a secure communication channel between services, fulfilling
requirements (b), (c) and (d).

4   Proposed Solution

Our proposed architecture isolates transcoders as independent functions, where
their purpose is only to receive, transform and forward data. To make the system suitable for healthcare applications, we created two services, one for handle
received data and another to forward them. Usually, as in monolithic systems,
these functions are coupled. In our proposal, we decoupled them, as illustrated
in the block diagram of Fig. 1, which shows these functions in a microservices
architecture for transcoding healthcare data between a legacy system and a
FHIR server.

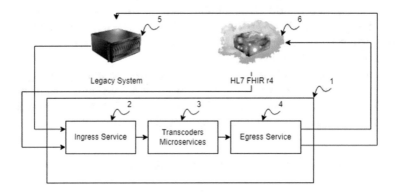

Fig. 1. Interface engine architecture.

The process in the diagram of Fig. 1 illustrates:

- (2) - Ingress service is the service that receives the request and sends the
  authorized requests to the corresponding transcoder microservice of healthcare data type;
- (3) - Transcoders microservices receive the data, transform and send them to
  the Egress service;
- (4) - Egress service sends the correct healthcare data type to the corresponding target of the message sender.

In the microservices block, it is important to notice that each function is
implemented as a independent function in different layers. Figure 2 illustrates an
example of transcoding microservices with two layers of depth:

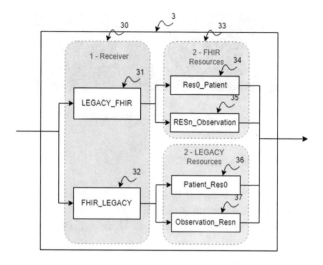

Fig. 2. Transcoders microservices.

- (30) - The first layer will receive the message from one format, check if it is a valid message, and send it to the second layer. It can be implemented using 2 microservices.
- (33) - The second layer process the message by the destination data type. For a specific resource on the destination, the message will be treated, transformed, and the requisition will be mounted to be sent. In our example it is composed by 4 microservices.
- More layers can be added by the implementation of the architecture, like in HL7 v.2 some messages contain the same types of "sub messages" (MSH, for example), and they can be treated by a specific microservice and returned to (33).

The algorithm of the Receiver (30) is presented in Fig. 3, and the algorithm of the Resources microservices (33) is presented in Fig. 4.

```
1   resource = receive_resource_data()
2   if(data_are_valid(resource)) do
3       resource_type = what_type_of_resource(resource)
4       if(type_of_resource_is_implemented(resource_type)) do
5           return request_to_resource(resource)
6       end else do
7           return not_implemented_resource_error
8   end else do
9       return invalid_data_error
10  end
```

Fig. 3. Algorithm of receiver.

```
1   resource = receive_resource_data()
2   data = transform_data_structure(resource)
3   data = adjust_data(data)
4   mapped = map_data_to_target_standard(data)
5   response = request_to_destination(mapped)
6   return response
```

Fig. 4. Algorithm of standard resources.

5   Experimental Results

This section presents and discusses experiments related to the proposed architecture. The goal of these experiments is to validate the requirements proposed
in Sect. 3. The microservices were implemented in Python with HL7[2] library.

5.1   Scalability of Transcoders

Our goal with this experiment is to validate the scalability, reconfiguration and
transcoding requirements. We prepared a setup to transcode data between a
HL7v2 compliant system and a FHIR complaint system. Figure 5 represents this
setup, since *System 1 HL7v2* (1) can communicate directly with *System 2 HL7v2*
(2), but *System 1 HL7v2* (1) cannot exchange information with *System 2 FHIR*
(4) directly. In our setup case, we inserted the *Transcoder HL7v2 <=> FHIR*
(5) between systems (1) and (4) to enable communication between them.

In this experiment, the proposed system was executed in the same host,
where all copies of the microservices share the same computer resources. We
generated copies of the transcoding microservices, always starting with no copy,
and scaling up to a number of copies (from 0 to 30 copies).

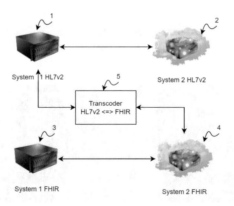

Fig. 5. Communication between HL7v2 and FHIR servers.

---

[2] https://python-hl7.readthedocs.io/en/latest/.

We validated that is possible to reconfigure or scaling up without restart the system. It is important to highlight that we only created copies of the necessary microservices for FHIR and HL7v2. In a monolithic system, if we need to reconfigure a service or change the implementation version, we must restart the system, but in the case of microservices, it is possible to start the execution of new service copies and remove old services from execution, so we do not have to restart the system.

In Fig. 6, we measured the startup time considering the time difference between creating the first copy and starting the last copy. As expected, although it is possible to dynamically start transcoders on-the-fly, there is a cost in the startup time, which should be considering when implementing this feature in a Service Mesh. It is important to highlight that the values presented in Fig. 6 should be analyzed in a relatively way, as the system was executed in the same computer host.

In the next experiment, we show how to use these features to guarantee that the system meets its quality-of-service (QoS) requirements when necessary.

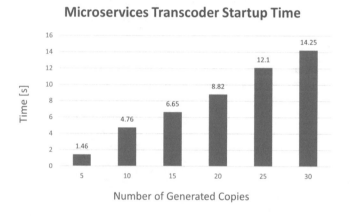

Fig. 6. Scalability of microservices transcoder - Startup time.

5.2 Response Time for QoS Requirements

The goal of this experiment is to validate that the use of microservices for transcoding can be used to guarantee that QoS requirements for health services and equipment, presented on standard IEEE 11073:00101 [1], are met. In this experiment, differently from the previous one, every microservice has 1 core and 512 MB of RAM in a virtual machine. Our goal is to guarantee a Qos requirement for real-time curves or alarms, which, for remote systems, has a maximum response time of 3 s [1].

The transcoder implemented was C-CDA to FHIR, and the data used is a C-CDA sample[3]. The requests were made by a script in Python using concurrent requests.

---

[3] https://github.com/microsoft/FHIR-Converter/blob/main/data/SampleData/
  Ccda/sample.ccda.

We compared three systems:

- (i) monolithic system: We used one copy of the transcoder service to simulate the monolithic system because the difference in technologies and implementation could interfere with the comparison. This system would represent a service that is running, but it is not prepared for an extemporaneous requests demand.
- (ii) multi-service system: This system has 5 copies of transcoding microservices, and is prepared for extemporaneous demand. However, it is not efficiently using its computer resources, as it uses 5 cores for microservices that are necessary.
- (iii) dynamic microservices: This system dynamically scale microservices and necessary.

In Fig. 7, we show the results of comparing monolithic and microservices. This result demonstrates that using dynamic microservices (iii) for interface engine can improve the response time to meet our QoS goal ($response_time < 3$ s) up to 800 parallel requests, without wasting computing resources.

It is important to notice that only microservices used in the demand were scaled.

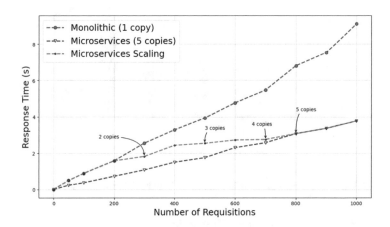

Fig. 7. Scalability of microservices transcoder - Response time.

6   Related Work

The *NextGen Connect Integration Engine* is an open solution that helps entities achieve interoperability, in which it is possible to integrate different formats such as XML, HL7 v 2.x and 3.x, CCD/C-CDA, FHIR, DICOM[4], NCPDP[5], X12[6],

---

[4] https://www.dicomstandard.org/.
[5] https://www.ncpdp.org/.
[6] https://x12.org/flow/health-care.

among others [8]. However, this solution does not use microservices topology, and it can only be used in architectures of the same type. In this solution, it is necessary to scale the whole system, and not only the transcoding functions.

Naveed et al. [14] created a *framework* to provide semantic interoperability, that is, when sending data from one system to another, keeping the same meaning in the receiver without the need for human intervention. Although promising, this solution still needs some type of correctness checking. On the other hand, in a transcoder solution, this is already defined by design. Shoumik et al. [17] developed a FHIR server based on microservices architectures. The advantages of their implementations are scalability, service deployment, decoupling, agility in development, high availability, and API. The problem with this solution is that even if FHIR servers are created, the servers that already exist with their standards, called legacy ones, still need to be directly included, leaving the need to change the legacy systems to FHIR.

Li et al. [12] designed and developed an interface engine that integrates IEEE 11073 DIM with FHIR using CoAP. This integration partially solves the interoperability problem between these two healthcare data types. However, even though the CoAP is faster than the HTTP protocol, the scalability for a considerable amount of data in a real scenario needs to be described, which means this architecture does not have the advantages of microservices architecture. Andersen et al. [2] created an architecture for mapping ICE (Integrated Clinical Environment) and FHIR. This work shows the viability of mapping IEEE 11073, but again the architecture needs to support a better service scale, and it is crucial in the context of multiple medical IoT devices and data exchange. Mukhiya et al. [13] presented a cloud-based architecture to achieve interoperability, and patients could interact directly with the Electronic Health Records. This architecture uses the cloud, which can slow the report's response and is not flexible to use in edge or hybrid environments.

7   Conclusion

This article presented an interoperable microservices architecture for interface engines applied to health data exchange. The novelty of this research work lies in the new architecture to provide interoperable healthcare services for data exchange based on independently transcoder functions implemented as microservices. The proposed architecture can run the interface engines in the cloud, on the edge of the network, or in a hybrid way. We also presented an Application Scenario that maps HL7v2 data to the FHIR standard. The experimental results showed that the architectural solution enhances the interoperability problem and fulfills our premises for health-data exchange.

For future work, we propose a testing platform to validate transcoders based on QoS requirements for healthcare services. i.e., the testing platform needs to ensure if a transcoder fulfills the QoS requirements based on the load of scenarios like pandemics, for example.

Acknowledgements. The authors would like to thank the Virtus Research, Development and Innovation Center and the Electrical Engineering Graduate Program (COPELE) from the Federal University of Campina Grande (UFCG) for supporting this research.

References

1. IEEE Health Informatics-PoC Medical Device Communication Part 00101: Guide-Guidelines for the Use of RF Wireless Technology. IEEE STD 11073-00101-2008, pp. 1–125 (2008)
2. Andersen, B., et al.: Point-of-care medical devices and systems interoperability: a mapping of ICE and FHIR. In: 2016 IEEE Conference on Standards for Communications and Networking (CSCN), pp. 1–5, Berlin, Germany. IEEE (2016)
3. Brauer, J.: Mirth: standards-based open source healthcare interface engine. Open Source Bus. Resour. **1**(205), 1 (2008)
4. D'Amore, J., et al.: Interoperability progress and remaining data quality barriers of certified health information technologies. AMIA Annu. Symp. Proc. **2018**, 358–367 (2018)
5. E. C. Q. I. (eCQI). Fast Healthcare Interoperability Resources (2020). https://ecqi.healthit.gov/fhir. Accessed 14 Oct 2022
6. Glaser, J.: Interoperability: the key to breaking down information silos in health care. Healthc. Financ. Manage. **65**(11), 44–49 (2011)
7. Gos, K., Zabierowski, W.: The comparison of microservice and monolithic architecture. In: 2020 IEEE XVIth International Conference on the Perspective Technologies and Methods in MEMS Design (MEMSTECH), pp. 150–153 (2020)
8. NextGen Healthcare. Seamless, scalable, and supported interoperability with NextGen connect integration engine (2021). Accessed 30 June 2021
9. Istio. The Istio service mesh (2021). https://istio.io/latest/about/service-mesh/. Accessed 14 Jan 2022
10. Kuryazov, D., Jabborov, D., Khujamuratov, B.: Towards decomposing monolithic applications into microservices. In: 2020 IEEE 14th International Conference on Application of Information and Communication Technologies (AICT), pp. 1–4 (2020)
11. Lehne, M., Sass, J., Essenwanger, A., Schepers, J., Thun, S.: Why digital medicine depends on interoperability. NPJ Digital Med. **2**(1), 79 (2019)
12. Li, W., Park, J.: Design and implementation of integration architecture of ISO 11073 DIM with FHIR resources using CoAP. In: 2017 International Conference on Information and Communications (ICIC), pp. 268–273, Hanoi, Vietnam. IEEE (2017)
13. Mukhiya, S.K., Rabbi, F., Pun, K.I., Lamo, Y.: An architectural design for self-reporting e-health systems. In: 2019 IEEE/ACM 1st International Workshop on Software Engineering for Healthcare (SEH), pp. 1–8, Montreal, QC, Canada. IEEE (2019)
14. Naveed, A., Sigwele, T., Hu, Y.F., Kamala, M., Susanto, M.: Addressing semantic interoperability, privacy and security concerns in electronic health records. J. Eng. Sci. Res. **2**(1), 31–38 (2020)
15. Noumeir, R.: Active learning of the HL7 medical standard. J. Digit. Imaging **32**(3), 354–361 (2018)

16. Petrakis, Y., Kouroubali, A., Katehakis, D.: A mobile app architecture for accessing EMRS using XDS and FHIR. In: 2019 IEEE 19th International Conference on Bioinformatics and Bioengineering (BIBE), pp. 278–283, Athens, Greece. IEEE (2019)
17. Shoumik, F.S., Talukder, M.I.M.M., Jami, A.I., Protik, N.W., Hoque, M.M.: Scalable micro-service based approach to FHIR server with golang and No-SQL. In: 2017 20th International Conference of Computer and Information Technology (ICCIT), pp. 1–6, Dhaka, Bangladesh. IEEE (2017)

D-insta: A Decentralized Image Sharing Platform

Yadagiri Shiva Sai Sashank, Ankit Agrawal$^{(\boxtimes)}$, Ritika Bhatia,
Ashutosh Bhatia, and Kamlesh Tiwari

Birla Institute of Technology and Science, Pilani, Rajasthan, India
{f20190068,p20190021,p20180022,ashutosh.bhatia,
kamlesh.tiwari}@pilani.bits-pilani.ac.in

Abstract. Due to the covid-19 pandemic, people have moved toward
digitization and using digital technologies in their daily life. For instance,
photographers and artists use social media platforms or stock photo websites to showcase their art to people to get recognition and credit. Since
social media platforms attract people more than stock photo websites,
we consider incorporating the stock photo website features into the social
media platforms. Currently, such platforms are running in a centralized
fashion where their proprietary algorithms mask most of the content to
which some users and advertisement posts are given more priority. Due to
the centralization, such hidden algorithms create trust issues among the
users along with other issues such as single point of failure, identity theft,
etc. This causes genuine artists and photographers to lose their interest
and motivation. Providing due credit to the authors and deserved recognition are significant concerns for photographers who share images on
stock photo websites or social media platforms. In this paper, we propose
a decentralized image-sharing platform/application utilizing blockchain
and a distributed file storage system to address all these issues. The proposed platform leverages Ethereum-based smart contracts to maintain
trust as deployed smart contracts are immutable, and the logic written in them is publicly available. We leverage a distributed file storage
system to solve the blockchain scalability issue in terms of storage.

1 Introduction

In the current era of the internet, many photographers and artists work hard to
be recognized in the vast sea of creators. They often post their works on popular
platforms like Instagram and Facebook to get the recognition they aspire to
get. The royalty-free and micro-stock business models on such websites have led
to intense competition among photographers. Everyone strives to upload more
images and get more recognition for their work. The centralized platforms use
their proprietary algorithms, which may not be fair and provide higher priority
to the post of some users than others.

The main issue with these platforms is that they are known for their propriety
algorithms, which are used to serve their user's requests; these algorithms mask
most of the content to which some users and advertisement posts are given

L. Barolli (Ed.): AINA 2023, LNNS 655, pp. 206–217, 2023.
https://doi.org/10.1007/978-3-031-28694-0_19

more priority [1]. Such hidden algorithms create trust issues among the users. Additionally, this causes genuine artists and photographers to lose interest and motivation. This is a massive loss to the art industry as many aspiring artists give up on their first trial, affecting them financially and mentally. Providing due credit to the authors and deserved recognition are significant concerns for photographers who share images on stock photo websites. The high commission rates on stock photo websites, which may run up to 55%, are also a common concern in the photography community [2].

The present applications/websites, like Instagram and Facebook, are based on web2 technology, which uses a central server to fulfill user requests. Using a central server may also give rise to a risk of spamming, fraud, and cyber attacks [3]. Due to the centralization, a single point of failure issue may occur. Additionally, such centralized systems are handled by a single authority that contains the user's personal (identity) and private information, creating a fear of data losses or mishappening, i.e., selling the private information. As social networks become increasingly integrated into daily life, social media identity theft is becoming more common [4]. Simple mistakes in centralized social media systems could expose your data, allowing scammers to steal your identity, exploit your personal information, or ruin your credit. Hence, there is a need to develop a Dapp to fulfill the above requirements built using decentralized blockchains such as Ethereum. Ethereum is a blockchain-based platform that helps build decentralized applications with no centralized server, and multiple users participate in the process through the blockchain platform [5].

Storing the social media post or images on the blockchain creates an issue of scalability in terms of storage [6]. Therefore, there is a need to store the data off-chain. Off-chain storage may include a database server or distributed file storage server. In general, distributed file storage system stores the data in a distributed manner and returns a unique address to fetch the data. Fetching data using the returned address in the blockchain through the smart contract may incur user costs. Therefore, there is a need to include a database server and the distributed storage server to store the data to reduce the cost and store the returned address onto the blockchain to provide verifiability and auditability. The fundamental motivation behind a decentralized image-sharing marketplace is to solve the centralization issues discussed above and correct image attribution, providing due credit to original image authors, and giving them their deserved recognition. This paper proposes a decentralized social media platform, especially for image sharing. This paper aims to create a platform where no malicious algorithms ruin the art form's genuineness. A transparent algorithm should be run that only shows the most appreciated posts on top of the feed.

The novel contributions of this paper include employing a programmable Blockchain, Ethereum, to create a decentralized image-sharing application, with the novel usage of perceptual hashes and Ethereum's robust smart contracts for encouraging the artists by tipping the posts. The images on the network are saved in a decentralized fashion through Inter-Planetary File System (IPFS).

Section 2 discusses the related works and their limitations. In Sect. 3, we discuss the key requirements of the proposed design and solution and the conceptual flow of the process of uploading and receiving the images. Section 4 discusses the implementation dependencies. In Sect. 5, we conclude this work.

2   Related Work

The authors in [8] proposed a blockchain-based framework to store and share the Neuro image medical data to solve transparency and security issues. The authors used Ethereum blockchain-based smart contracts to implement the framework and utilized IPFS as a storage layer to store and retrieve the image data. The authors in [9] proposed a framework that utilized blockchain technology to store the x-ray images to achieve availability, probability, and immutability. However, storing images in the blockchain creates an issue of scalability in terms of storage. Blockchain is a database, but it is not meant for storing huge amounts of data as it may impose a high cost to the user who uploads the images over the blockchain. Authors in [10] proposed a medical image-sharing framework that uses a blockchain to store the imaging metadata and the permissions to access the images defined by the patients and to share the images across domains. This framework utilized the existing imaging centers to store the images as they have invested a lot to set up the storage infrastructure and to use the existing infrastructure.

Authors in [11] proposed a blockchain-based framework to remove the hurdle of centralization while providing an image-sharing model for DICOM medical images. IPFS is used to store medical images. The article [12] is based on the application which enables secure medical image sharing based on zero trust principles and blockchain technology. The main idea behind this paper is that authors consider blockchain an effective tool for safeguarding sensitive information. However, to ensure the overall protection of medical data (images), several other security measures have to be taken at each step, from the beginning, during, and even after the transmission of medical images, which the zero trust security model ensures. This article proposed a decentralized and trustless framework combining these concepts for secured medical data, image transfer, and storage.

The authors in [2] address the major problem of correct image attribution and the malpractice of uploading duplicate images on stock photo websites and image-sharing marketplaces. This ensures that original photographers are rightfully recognized and credited for their works. A decentralized solution is proposed by utilizing the blockchain and IPFS to detect and reject the interfered images that are perceptually similar to images already present in the marketplace. This paper is very close to our problem statement but slightly deviates from the application's goal. Due to the inherent properties of Blockchain, our proposed framework has no central authority controlling it, no third-party interference, no single point of failure, zero censorship, and preserves online user privacy.

The authors in [13] proposed a blockchain-assisted social media network that allows data sharing among the users, and each user can claim ownership, trace, and control the content shared on social media. A personal certificate authority, not belonging to the blockchain, maintains each user's circle list of friends and family and issues a certificate for each circle that allows circle members to view the shared content. [14] provides additional capabilities to the decentralized social media network that takes advantage of several well-known and intriguing features of the current online social networks, including locality of access, predictability of access times, and others. The proposed system in [14] is particularly privacy-preserving since it enables users to manage content access more granularly. These articles [13,14] can be extended by incorporating our proposed framework to provide recognition to the genuine artist.

3   Proposed Design and Solution

Decentralized image sharing is an idea where the images posted on the application are stored in a decentralized manner using an Inter-Planetary file system (IPFS). IPFS ensures resistance to censorship and centralized point of failure, such as server issues or coordinated attacks. IPFS is a distributed file storage system that maintains a P2P network covering the whole world and comprising a list of nodes used to store and/or relay the information [7]. IPFS stores the uploaded images in a distributed fashion; the images are broken into small parts and stored by peers using IPFS. The images are retrieved using the hashes generated by IPFS; the hashes are also used to verify the correctness of the retrieved image. The retrieval process is possible because of the data structure of Merkle trees. The Merkle hash is stored on the blockchain; once it is stored on the blockchain(when the image is first uploaded), it is immutable. Hence, this provides censorship resistance. Any other user uploading the same images is denied the service because the image already exists in the blockchain. To eliminate the hefty commissions, we create an architecture where the user has to pay only for transaction charges related to the processing of the transaction by the Blockchain network.

We discuss all the key requirements of the proposed platform. All the requirements are listed at an abstract level to be expanded and implemented during the design phase. Detailed requirements of the proposed platform include:

- Create a user interface where a user can upload his post/image.
- Design a solution to secure the post over a blockchain to make everything decentralized.
- Formulating a way to identify the user accessing the platform.
- Design a way to store the data (images) in a decentralized manner.
- Design a tipping feature so that people can encourage the artwork on the platform.
- Design an algorithm that shows the most appreciated posts on top of the feed.
- Develop a smart contract that handles all the back-end logic.
- Create a back-end implementation such that the storage and front-end aspects of the application are synchronized.
- Deploy the smart contracts over the Ethereum blockchain.

Now we discuss the ways of handling these aspects one by one.

3.1   Designing the User Interface

The user interface has been designed with simplicity and functionality as its core features. The front end of the platform uses react and bootstrap, on which every user can see the photos that have been uploaded. The reason behind using react is that it is the most suitable option for building complex and robust front-end architectures through its various coding possibilities. Bootstrap is used to ease the design process by making the appearance more user-friendly; it was used to create some buttons and search bars that are important for users to interact with the application. The various options which are available for users to interact with include:

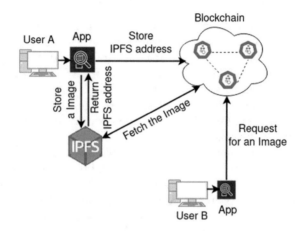

Fig. 1. Conceptual flow of the process of uploading and receiving the images

- Search-bar: is used to filter the posts based on tags.
- Upload-button: opens up a form where the user can upload the photo he/she desires along with the tag for the photo.
- Tip-button: is used by the user to tip the post that they like; this helps incentivize the content creators and keep their morale up.
- Delete-button: is used for deletion of posts (can only be done by the person who uploaded it).

3.2   Decentralizing the Application

Decentralization is introduced with the addition of blockchain into the platform. Ethereum blockchain is used to deploy the application. All the back-end logic, algorithms, and functionalities are put into the smart contracts and then deployed over the blockchain. Once a user requests some functionality (specified by some algorithm written in solidity), a transaction for that functionality is generated, mined, and added to the blockchain. This information is permanently stored on the blockchain and is immutable. Once the contract is deployed, we cannot make any changes, making it secure and immutable. In this way, decentralization of the application is made possible.

3.3   Identifying the User

The application is open to all users who have an Ethereum account. There is no login of any sort for the users to access the application. One must install all the dependencies and run the application to be able to access it. But there is a need for identification of the user for other purposes, such as for uploading a photo (to attach the identity of the author with the post) or for tipping a post (the tipped amount should be debited from the tipper and be credited to the author of the post). To accomplish this, the application senses the metamask wallet, a browser extension to identify the user. The application gets the user info (the user's public key) from this browser extension to identify the user.

3.4   Storing the Uploaded Images

Instead of having only a central database to store data, we utilize IPFS to store the images, and IPFS returned address (Hash) is stored on the blockchain to increase the scalability. IPFS is a decentralized peer-to-peer hypermedia distribution protocol, so there is no single point of failure, and all trust is not put on a single node. Moreover, it has been found to be more suitable for blockchain than any other off-chain storage (Swarm, StorJ, CoAP). IPFS is content addressable, uses cryptographic hashes to give unique fingerprints to blocks within files, removes redundancy across the network, and uses these unique hashes for information retrieval. The process of storing and retrieving the uploaded image is shown in Fig. 1. By storing images with the IPFS protocol, we make large savings in the size of the Ethereum test chain. Data stored on IPFS can be encrypted with any standard encryption scheme and retrieved. The encryption keys will only be shared with the appropriate user via smart contract. The data is replicated to all the participants in the form of hashes generated by IPFS to the Ethereum Blockchain, ensuring data validity despite having malicious users among the participants. Off-chain storage combined with the benefits of Blockchain-based architecture can put the power back in the hands of image contributors.

3.5   Enabling Tipping

One of our flagship functionality is to tip the authors of posts. This functionality enables the community to appreciate the artworks and encourages the art form. To enable this, first, we must hook every post with its author and store it permanently over the blockchain so that it cannot be changed (only paying the photo owner). Next, we must get the address of the tipper to create a transaction to request a tipped account from their wallet. This is done using the injected web3 object to sense the user from the browser wallet metamask. The smart contract requests the tip amount from the tipper address collected earlier. A metamask pop-up will open requesting the tip amount, and the user must validate the transaction. After validating the transaction, the tip amount is debited from the tipper and credited to the post's author by the smart contract.

3.6   Algorithm to Decide the Ordering of Posts

The application doesn't use any sophisticated algorithm and is not hidden from the users like the present image-sharing platforms such as Instagram, Facebook, etc. The posts are ordered based on the appreciation got from the community. The appreciation a post gets is assumed to be directly proportional to the amount of tip that particular post gets. The application orders all the posts in decreasing order of this tip amount. This would be the order of the posts for any user accessing the application and is not user dependent.

3.7   Designing the Smart Contracts

Ethereum provides a programming language called Solidity to write smart contracts. Solidity is the most used language to write smart contracts. The logic and programming constructs are incorporated and written in a solidity programming language. The logic and programming constructs include the tipping mechanism, the post-ordering algorithm, storing the photos uploaded by the user onto IPFS, and retrieving the photos stored on IPFS to display them on the user interface.

3.8   Back-end Implementation

The entire back end is developed using JavaScript; this is used to integrate our application with all of its components. The back-end code integrates our application with the react front-end user interface. The interface allows users to upload and store the images in IPFS and retrieve the photo from IPFS to show it in the feed. The back-end code includes some plugins and objects (JavaScript objects) that integrate the application to the web3 (a distributed web technology) to connect to metamask and indirectly to the blockchain.

3.9   Deploying the Application on Ethereum

Finally, our application is ready to be deployed over the blockchain. We have used truffle-suite to compile and deploy our contracts over the Ethereum blockchain. Once the application is deployed, any user with the URL can access it and its functionalities. Every interaction of the user will trigger a transaction over the blockchain. Some functions, like tipping and posting, will need some ether to go through, and others, like retrieving images from IPFS, will not charge any ether.

The block diagram of the proposed platform is shown in Fig. 2, which combines all its components. Four major entities are present in the platform: users, web clients, IPFS, and Ethereum blockchain. Users are the one who uploads the images on the platform to get recognition and credit. Web client includes a user interface to interact with the users. IPFS is an external entity used to store and fetch images. Ethereum blockchain provides decentralization, transparency, and security to the users. These properties create trust among the users. Smart contracts are designed to provide the following functionalities: 1) Tips are credited to the artworks' rightful owners. 2) The same photo cannot be uploaded twice, as IPFS will return the same image hash value.

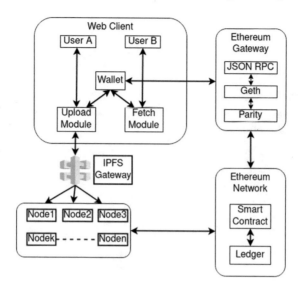

Fig. 2. Block diagram of the proposed platform

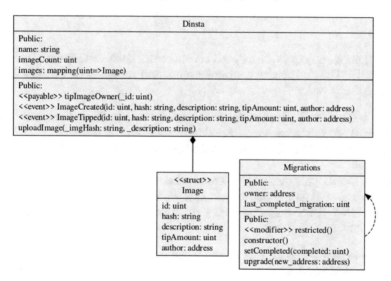

Fig. 3. The UML representation of the designed smart contracts

Figure 3 depicts the UML representation of the smart contract that handles the application's logic. The contract has three public attributes: *name, imageCount*, and *images*. The attribute *'name'* contains the application's name, i.e., D-insta. The attribute *'imageCount'* keeps track of the total number of images uploaded on the blockchain and acts as the index to that image. Finally, the 'images' attribute is a mapping from each image index obtained from the *'imageCount'* attribute and

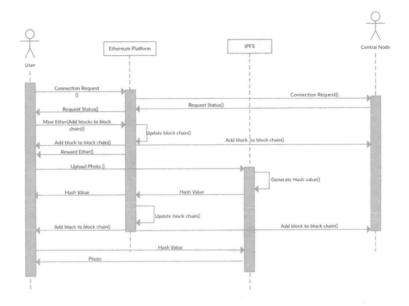

Fig. 4. The sequence diagram of the entire process in the proposed platform

the actual image object (a struct with attributes id, hash, description, tipAmount, and author). The Script also contains events and methods that are designed to perform specific tasks:

- *tipImageOwner*: A payable method that takes in the tip amount and transfers the tip ether to the image authors' address from the tipper.
- *ImageCreated*: This is an event triggered whenever a new image has been uploaded.
- *ImageTipped*: This event is triggered whenever an image has been tipped.
- *uploadImage*: This method is responsible for taking in the image hash generated from IPFS and the image description and uploading them to the blockchain.

Figure 4 depicts the flow of events when a new node tries to connect to the application and upload an image/retrieve an image. The sequence of events is described below:

1. The new node (User) sends a connection request to the blockchain application.
2. The request sends a notification to the central node, which manages the blockchain as a connection request.
3. The central node either accepts/rejects and sends the status to the user (If the central node chooses to reject the request, the process ends here).
4. Now, if the user's connection is successful, he can upload/retrieve images. The image first gets uploaded onto the IPFS, the hash of the image is generated and returned to the user, and the user uploads the hash to the blockchain.

5. If a user tries to retrieve an image, he has to provide the Hash of the image. This hash is used to pick up the corresponding image from IPFS and returns the image to the user.

4   Implementation Dependencies

To implement the application, we must have the following dependencies installed. Below is the list of dependencies used to develop the application. Detailed information about each can be found on their respective websites and corresponding GitHub repositories.

- **JavaScript:** JavaScript, often abbreviated JS, is a programming language that is one of the core technologies of the World Wide Web, alongside HTML and CSS. This is the application's main framework to integrate into web3, web2, IPFS, metamask, localhost, etc.
- **Metamask:** Metamask is a plugin for browsers that allows making Ethereum transactions. It allows running Ethereum DApps (Decentralised applications) as it manages identities and signs Blockchain transactions. The blockchain enables us to use our funds in the Dapp provided with proper authentication from the user.
- **Ganache:** Ganache is an Ethereum Blockchain emulator and an in-memory Ethereum node that allows testing all components locally. It also lets users deploy contracts, develop applications and run tests. This tool has been used to test the application's functionality over the blockchain environment and the proper working of smart contracts.
- **Geth:** Geth (Go-Ethereum) is a command line interface that lets users run and operate full Ethereum nodes. It is implemented in Go and allows to mine blocks to generate Ether, create smart contracts, send transactions to transfer funds between Ethereum addresses, inspect block history, and many other functions, used to test the application.
- **Truffle:** Truffle is used to compile, test, and build smart contracts and provides a development framework to increase speed in the development process. Used to compile the smart contracts and deploy them over the Ethereum blockchain.
- **IPFS:** The Inter-Planetary File System is a protocol and peer-to-peer network for storing and sharing data in a distributed file system. IPFS uses content-addressing to uniquely identify each file in a global namespace connecting all computing devices. IPFS stores all the photos uploaded by the users in a decentralized manner.
- **Node.js:** Node.js is an open-source, cross-platform, backend JavaScript runtime environment that runs on the V8 engine and executes JavaScript code outside a web browser. Used for running the application, it provides a runtime environment and executes the JavaScript code.

5   Conclusion

In this paper, we proposed a decentralized image-sharing platform that solves centralization and trust issues among users on social media or stock photo websites. Blockchain technology provides several key characteristics, such as decentralization, transparency, immutability, audibility, and others, that make the proposed platform works in an expected fashion, and the entire image-sharing happens in a decentralized manner. Blockchain-based solution for online stock photo marketplace provides many benefits over traditional marketplaces, such as a high level of security, permanent record of trade, custom licenses, the custom selling price for the image, higher revenue for contributors, and permits for distributed storage of images with the help of IPFS. Due to the absence of central authority, the marketplace is free from censorship and interference. Blockchain transactions only contain IPFS hashes for retrieving images. Information security is enforced by encrypting content beforehand in the smart contract with a choice of an encryption scheme and then transferring data to IPFS. Furthermore, our implementation is scalable. Thus the paper illustrates a novel use-case of proven technology to help fight a deep-rooted malicious practice in the stock photo or social media world.

References

1. Looft, R.: # girlgaze: photography, fourth wave feminism, and social media advocacy. Continuum **31**(6), 892–902 (2017)
2. Mehta, R., Kapoor, N., Sourav, S., Shorey, R.: Decentralised image sharing and copyright protection using blockchain and perceptual hashes. In 2019 11th International Conference on Communication Systems & Networks (COMSNETS), pp. 1–6. IEEE, January 2019
3. Yeung, C.M.A., Liccardi, I., Lu, K., Seneviratne, O., Berners-Lee, T.: Decentralization: the future of online social networking. In: W3C Workshop on the Future of Social Networking Position Papers, vol. 2, pp. 2–7, January 2009
4. Irshad, S., Soomro, T.R.: Identity theft and social media. Int. J. Comput. Sci. Netw. Secur. **18**(1), 43–55 (2018)
5. Mukhopadhyay, M.: Ethereum Smart Contract Development: Build blockchain-based decentralized applications using solidity. Packt Publishing Ltd. (2018)
6. Jayabalan, J., Jeyanthi, N.: Scalable blockchain model using off-chain IPFS storage for healthcare data security and privacy. J. Parallel Distrib. Comput. **164**, 152–167 (2022)
7. Benet, J.: IPFS-content addressed, versioned, p2p file system (2014). arXiv preprint arXiv:1407.3561
8. Batchu, S., Henry, O.S., Hakim, A.A.: A novel decentralized model for storing and sharing neuroimaging data using Ethereum blockchain and the interplanetary file system. Int. J. Inf. Technol. **13**(6), 2145–2151 (2021). https://doi.org/10.1007/s41870-021-00746-3
9. Li, M.M., Kuo, T.T.: Previewable contract-based on-chain X-ray image sharing framework for clinical research. Int. J. Med. Inform. **156**, 104599 (2021)
10. Patel, V.: A framework for secure and decentralized sharing of medical imaging data via blockchain consensus. Health Inform. J. **25**(4), 1398–1411 (2019)

11. Kumar, R., Tripathi, R.: Building an IPFS and blockchain-based decentralized storage model for medical imaging. In: Advancements in Security and Privacy Initiatives for Multimedia Images, pp. 19–40. IGI Global (2021)
12. Sultana, M., Hossain, A., Laila, F., Taher, K.A., Islam, M.N.: Towards developing a secure medical image sharing system based on zero trust principles and blockchain technology. BMC Med. Inform. Decis. Making **20**(1), 1 10 (2020)
13. Chakravorty, A., Rong, C.: UShare: user controlled social media based on blockchain. In: Proceedings of the 11th International Conference on Ubiquitous Information Management and Communication, pp. 1–6, January 2017
14. Narendula, R., Papaioannou, T. G., Aberer, K.: A decentralized online social network with efficient user-driven replication. In 2012 International Conference on Privacy, Security, Risk and Trust and 2012 International Conference on Social Computing, pp. 166–175. IEEE , September 2012

Ramification of Sentiments on Robot-Based Smart Agriculture: An Analysis Using Real-Time Tweets

Tajinder Singh[1], Amar Nath[1(✉)], and Rajdeep Niyogi[2]

[1] Sant Longowal Institute of Engineering and Technology,
Deemed-to-be-University, Sangrur, India
{tajindersingh,amarnath}@sliet.ac.in
[2] Indian Institute of Technology Roorkee, Roorkee, India
rajdeep.niyogi@cs.iitr.ac.in

Abstract. Social users and their sentiments on robot-based agriculture is an advanced area of research as demands for robots are increasing vividly in smart agriculture. Based on available studies, which usually depends on tweets, it helps the users to realize opinion on various aspects. Therefore, in this research work, a framework is designed to study users' sentiments, including their contextual behavior in terms of sentiment variations. The results show that the users have a positive attitude toward smart agriculture based on robots. Still, at the same time, they have a biased opinion also for various robot terms. Significant tweets based on the adoption of robots in agriculture are extracted in real-time using various event-based terms such as security, adoption rate, unemployment, and safety. Thus, this work will benefit the various business agencies, manufacturers, and technology-based organizations in understanding users' attitudes toward adopting robots in smart agriculture.

1 Introduction

Robot-based smart agriculture is a witness to a revolution. Robot-based smart agriculture is a domain introduced previously as numerous manufacturing units and technology-based industries are giving modern systems as required for smart agriculture. Due to this reason, social media users and public insight into robots for smart agriculture are becoming critical areas for smart agriculture research.

Therefore, it is assumed that social media can overcome such challenges by providing quality-based real-time data. Users share their sentiments on various social media platforms, among which Twitter is in huge demand. It helps the various social users to share information on time, which facilitates the users to re-tweet and propagate it among other users. Lack of information in tweets, such as missing socioeconomic behavior, is also a big challenge, including the noisy nature of data. Noise can be different, such as abbreviations, short forms, slang, spelling mistakes, special symbols, and emoticons [1]. Therefore, it is essential to address the noise before analyzing the data for sentiment analysis which is increasing steadily.

Yet, a significant position has to be given to the role of social media in the sentiment analysis of smart agriculture based on robots. Still, its significance in other areas, such as trend prediction [2], event detection and analysis [3], rumor detection [4], and Fake news analysis [5] on social media, is very much required. Therefore, the main motive of this research is to utilize social media (Twitter) for extracting real-time data for sentiment analysis on smart agriculture using robots. This work helps analyze the behavior of beings in robots-based smart agriculture and their influential role.

The remainder of the paper is structured as follows. In Sect. 2, related works are listed. In Sect. 3, methodologies s and notation are presented. The embedding-based sentiment analysis and polarity disambiguation are described in full in Sect. 4. In Sect. 5, conclusions are drawn.

2   Related Work

Robotics-based agriculture and their adoption rate with technology change have increased in the recent decade [6]. In the existing research, various studies have been conducted to analyze the ground acceptance rate from various facets. The major concerns include numerous benefits, safety concerns, and adoption processes, including social sentiments [2]. Including all these topics of concern, sentiment analysis has gained a gigantic amount of attention from various agricultural researchers. The main hurdle in such studies is getting a trustworthy platform or tool to extract genuine data to analyze [3]. From the study, it is observed that social media platforms are trustworthy and genuine for data extraction. Twitter is the most useful and major platform used by billions of users to post daily tweets [2,3]. Abundant platforms in the form of social media are available, which help users to extract feature-based data. Still, among them all platforms, Twitter is in huge demand due to its ease of use, and real-time analysis [1,9].

The basis of existing research work based on robotics-based smart agriculture in the context of sentiment analysis is quite interesting. Sentiment analysis based on time series is frequently used in various applications like rumor detection [7], event analysis [3], burst event analysis [2], fake news detection [4], and many more. Machine learning and deep learning approaches are usually used for sentiment analysis in various domains, which helps to categorize sentimental features. In [2], event analysis is based on various text streams of data, and machine learning approaches are used for classification. In [10], sentiment analysis using random forest is used from Twitter data sets for airline services in the US. To seek sentiments on the adoption of robots in smart agriculture is the same as identifying sentiment analysis as given in [3,4,10].

Delving further into the robotic-based smart agriculture adoption, [7] analyzes the IoT-based smart robotics system for detecting tomatoes, including picking point locations with machine learning. In [8], smart and emerging technologies for agriculture are explained, which helps the researchers to spread awareness to the people to get their way to adapt and enhance their usage. The facility provided by Twitter data helps to analyze data based on various tweets, which motivates the researchers to optimize the topic-based sentiments for classification. The societal impact can be computed on collected sentimental data. [13] gives detailed information on robotic-based grippers using sensor-based technology for agriculture purposes. This study explains that smart agriculture based on robotics is versatile in nature and interacts with the environment and objects to perform multiple recurring and monotonous tasks precisely.

Similarly, in [12], robots are used in various agricultural environments. In this study, the authors highlight the various challenges, including environmental, diversity, and biological variations, including chemical traits in environmental responsibility. Further, in [11], smart robotic-based agriculture environments are explained based on AI (Artificial intelligence). Thus, through this literature survey, the Robotics-based smart agriculture sentiment analysis is very important to analyze, which will be very helpful for the users to correct information at the right time. Based on the popularity index of robotic-based smart agriculture, adoption rate and benefits can be used to compute sentimental behavior.

3   Methodology

This research work mainly aims to analyze the sentiments on robot-based smart agriculture (framework is shown in Fig. 1). For this purpose, Twitter is used as a data extraction source, and the following research questions are answered through this research work. First, analyze the semantics of sentiments used by various users, which have the same sense referring to robot-based smart agriculture. Secondly, the context of sentiments has been studied, including the holistic observation of users with the change of time. Figure 1 is designed to explain the overall structure of the proposed work, and for this purpose machine learning approach is used.

3.1   Data Extraction and Pre-processing

Crawling data from Twitter: Using Twitter API, data is crawled using various timestamps, which helps to convert data into streams. Understanding the ground reality of data: The meaning of various short forms, slang, and out-of-vocabulary words (OOV) in terms of their contextual role is essential to analyze. Detection of sentimental words: In a real-time Twitter API-based data extraction scenario, sentimental query-based tweets are extracted. Relevant semantics and lexicons are selected from upcoming tweets based on the sentimental query

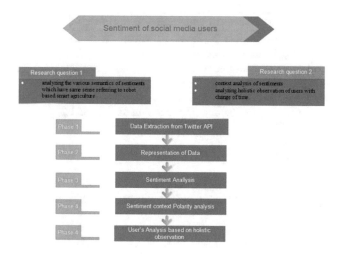

Fig. 1. Framework for sentiment analysis

theme. Contextual polarity disambiguation: The theme change from the actual topic is normal in real-time analysis. Therefore, it is essential to analyze the data from various periods to understand the context of various sentimental words. Every extracted tweet contains unwanted information. Therefore, it is necessary to filter the data to get useful information. Data pre-processing is a crucial phase to be performed dedicatedly to get better results. URLs, slags, short forms, hash (#) tags, @ tags, slang, and many more are the data that need to be filtered out. In [1] pre-processing mechanism is given, which helps to remove noise from the data. Similarly, in [14], authors also applied to pre-process collected tweets for signal detection. The BOW model is also beneficial for removing noise from data which is further converted into vector form to create features for dedicated machine learning models. Various steps are used for the pre-processing task [4,5], which helps to clean the crawled text, and in this study, we used the pre-processing mechanism as given in [3] for filtering unwanted data. Further, for contextual polarity disambiguation, Algorithm 1 is provided, in which the contextual behavior of sentimental words is computed.

3.2   Hand-Engineered Features for Text Pre-processing

In sentiment analysis, semantics and features used for representation vary from user to user. In this work, features are extracted and examined for sentiment analysis tasks which are given in Algorithm 1. The process for the same is depicted in Fig. 1

Algorithm 1. Text Stream pre-processing for sentiment analysis

---

Data: A continuous stream of tweets: Twitter Text stream $T_w = (t_{w_1}, t_{w_2}, t_{w_2}, \ldots t_{w_n})$, unidentified word w_{ui}.

Result: Collection of classified words W_i in terms of sentiments from text streams

1   **if** $(t_w \longrightarrow w_i)$ **then**

    /* Tweet is identified; update value of sentimental score   */

2      $S_{senti_{scar}} = S_i + 1$./* Update the sentiment score by 1   */

3      $t_w \longleftarrow t_w(url)$.

4      $t_w \longleftarrow t_w(@)$.

5      $t_w \longleftarrow t_w(url)$.

6      $t_w \longleftarrow t_w(splsymbol)$.

7      $T_w = (t_{w_1}, t_{w_2}, t_{w_2}, \ldots t_{w_n}) \in \{S\}$

    /* Collection of refined pre-processed Tweets, which belong to the Twitter text stream.   */

8   **while** *(not the end of stream)* **do**

    /* Compute indicator dataset of text stream corresponding to i */

9          $i \longrightarrow i + 1$ /* obtain the next stream from T_w */

10   .

11   **goto** line 1   /* for the collection of refined pre-processed Tweets, which belongs to the Twitter text stream for the new coming tweets   */

12   **for** *each each sentimental word* $S_{senti_{scar}} = S_i + 1$ **do**

13 | The similarity of each upcoming tweet is computed.

14   **end**

15   **end**

16 **else**

17   No change in the sentiment score is detected.

18      $senti_{score} = S_{senti_{score}}$.

19      $S_{score} \longleftarrow sentimentscore_{S_\alpha}$, where $S_{score} = S_{senti_{score}} \cup S_\alpha$

20 **end**

3.3   Impact of Pre-processing on Extracted Data

Crawling data from Twitter text streams is a complex task. In the Twitter stream, we extracted 23219 tweets in 120 min. After collecting data, we applied Algorithm 1 to clean unnecessary information. Figure 2 depicts the process of crawling text from Twitter API. It is observed that after applying the proposed Algorithm 1, we got 22478 labeled as a pre-annotated data set. A feature set is designed in the last in which embeddings are used by which tweets. The feature set is divided into training and test sets for evaluation purposes. The similarity of various sentimental words is computed using [2], which helps to obtain words closer to their sentimentally related words.

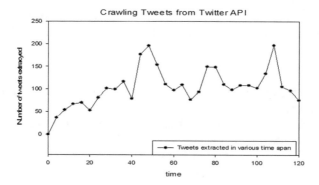

Fig. 2. Crawling tweets from the Twitter text stream

4 Embedding-Based Sentiment Analysis and Polarity Disambiguation

In the text, stream embedding plays a crucial role in representing each semantic-these representations present vector-based forms in a high-dimensional space. The contextual information and their synonyms can be represented accurately in such a single vector for context-based similar semantics. Such representations have a more expressive way of presenting semantics than the methods such as (BOW) bag-of-words. For the classification of sentiment and polarity disambiguation, we used SOTA methods of embedding, which are described below:

- *Word2Vec*-based embeddings: After pre-processing, embeddings can be applied to classify sentiment polarity. In our experiment, we use a feed-forward neural network. This neural network type is a simple solitary layer perceptron. Weights are connected and forwarded at output nodes in which input layers are directly linked and fed with lexicons. Input feature vectors are trained in comparison of sentimental words using the fit function.

$$f(x) = \frac{1}{1 + e^{-x}} \tag{1}$$

$$p(\alpha = v | O = o) = \frac{exp\left(v_v^\lambda \wp_o\right)}{\sum w \in stream\ exp(v_v^\lambda \wp_o)} \tag{2}$$

Equation 2 is used to compute continuous data in the text stream. Each word is represented as a set of vectors, including contextual information. The representation of the contextual and actual word is given in Eq. 2.

- GloVE: It is known as a log bi-linear. It consists of weighted least squares, which help to analyze the word's co-occurrence based on their probabilities as described in Eq. 3. It is a very simplistic approach in which we can encode various semantics in various forms, including their contextual behavior. The GLoVE is assumed to be superior to Word2vec in context analysis.

$$w_i, w_j = logP(i/j) \qquad (3)$$

As GloVe is an unsupervised learning algorithm. Therefore, training is performed on aggregated global word-word co-occurrence from the collected text stream to train this model. In this way, GLoVE considers whole data to understand contextual dependency among semantics.

- Fasttext: It is an open-source library that helps to create embeddings of collected data in a fast manner. Both supervised and unsupervised data can be converted into embeddings. In this model, each word is represented as an n-gram of characters.
- BERT: This is a transformer-based attention mechanism highly recommended for sentiment analysis to extract super-rich contextual information. The connection between different words is computed dynamically, which helps to obtain sentence prediction, including text. The representation of the BERT includes 12 layers, including 110M parameters which can be trained per the data requirement.

4.1   Comparison of Various SOTA Models

For evaluating the performance of real-time Twitter data, it is tested against various SOTA methods. The evaluation is measured and computed for analyzing the impact of contextual information on various features of sentiments Fig. 3. Metrics like precision, recall, and F1 as described in Table 3, whereas Table 1 represents the total number of sentimental words extracted for evaluation after implementing Algorithm 1. Features involved in pre-processed data sets are sensed by various SOTA methods, which are given in Table 2.

Table 1. Summary of Twitter text stream data for sentiment analysis

Data set	Source	Total	Selected	Training/Testing
Twitter Sentiment Analysis (API)	Twitter API	33219	22478	20000/2478

Table 2. Feature extraction using various SOTA methods

Data set	Word2Vec	GLoVE	Fasttext	BERT
Twitter Sentiment Analysis (API)	3374	3439	3472	3807

Table 3. Accuracy of sentiment classification in various SOTA methods

SOTA methods	Accuracy	Precision	Recall	F Score	References
Word2Vec	0.937	0.926	0.832	3807	[15]
GLoVE	0.924	0.881	0.783	0.850	[16]
Fasttext	0.936	0.923	0.814	0.862	[15,16]
BERT	0.946	0.932	0.821	0.851	[15]

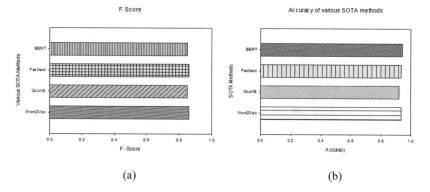

(a)   (b)

Fig. 3. F- score of various SOTA methods (a) and Accuracy in terms of sentiment context analysis (b)

4.2 User's Analysis and Impact on Society

It is also observed that when the extraction starts in continuous and discrete time intervals, the given query and contextual information are also evaluated by SOTA methods. The most significant features based on similarity context will be combined. Table 2 depicts the evolution of features in real-time data in which numerous features' occurrences, including their impact on the user, can be computed. As SOTA methods decently handle the contextual polarity, their impact changes with time and location can be predicted. Therefore, to predict the particular region where robot-based agriculture is highly appreciated at a high rate is analyzed from the feature value and user's sentimental score.

Moreover, to check the sentimental score in favor of robot-based agriculture, the Z score is computed for a set of features for a particular period for continuous intervals. It is observed that for a text stream, robot-based agriculture for a specific region is demanded. With an increase in the period, the involvement of various nodes, which participate with different means, also promotes robotics-based agriculture. The user's sentiments are computed for robot-based agriculture at various Z values. Five hundred random tweets from the pre-processed data are chosen to analyze their behavior.

The data sets analyzed for evaluation are based on their popularity and the attributes which provide ground reality. The limitation of contextual polarity disambiguation relies on matching contextual keywords with the exact context domain as a domain-specific relationship modeling between various sentimental-based features is demanded Fig. 4. Therefore, the SOTA methods detect user behavior, including contextual information, from time to time.

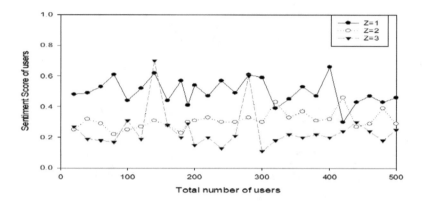

Fig. 4. Detection of various user's sentiments at different values of Z

5   Conclusion

Real-time data analysis from Twitter is quite challenging. Numerous methodologies are available in different areas. According to the above experimental study, various SOTA methods perform well in sentimental contextual analysis. BERT performs well among others for identifying context from various lexicons. It is observed that the design and dependency of various semantics in long sentences degrade the performance due to the complex structure of semantic dependency. In this work, we use four SOTA methods, and in terms of Accuracy, BERT gives a (0.946) maximum, including (0.937)in Word2Vec, 0.924 in GLoVE, and 0.936 in the case of Fast text. User analysis is also performed by selecting 500 random samples, showing that users favor robot-based agriculture. Significantly few negative sentiments are noticed from the collected data at different values of Z.

In the future, it can be extended further for classification tasks, including trust factor and combining it with other social media sources to compare users in terms of sentimental score. This is an open challenge for the data science community as only some methodologies have been explored in combining more than two social media platforms.

Acknowledgment. The authors thank the anonymous referees for their valuable comments that were helpful in improving the paper. The third author was in part supported by a research grant from Google.

References

1. Singh, T., Kumari, M.: Role of text pre-processing in Twitter sentiment analysis. Procedia Comput. Sci. **89**, 549–554 (2016)
2. Singh, T., Kumari, M.: Burst: real-time events burst detection in the social text stream. J. Supercomput. **77**(10), 11228–11256 (2021)
3. Singh, T., Kumari, M., Gupta, D.S.: Real-time event detection and classification in social text stream using embedding. Cluster (2022)
4. Kaliyar, R.K., Goswami, A., Narang, P.: FakeBERT: fake news detection in social media with a BERT-based deep learning approach. Multimedia Tools Appl. **80**(8), 11765–11788 (2021). https://doi.org/10.1007/s11042-020-10183-2
5. Choi, D., Oh, H., Chun, S., Kwon, T., Han, J.: Preventing rumor spread with deep learning. Expert Syst. Appl. **197**, 116688 (2022)
6. Hasan, M., Uddin, K.N.W., Sayeed, A., Tasneem, T.: Smart agriculture robotic system based on internet of things to boost crop production. In: 2021 2nd International Conference on Robotics, Electrical and Signal Processing Techniques (ICREST), pp. 157–162. IEEE (2021)
7. Bai, Y., Mao, S., Zhou, J., Zhang, B.: Clustered tomato detection and picking point location using machine learning-aided image analysis for automatic robotic harvesting. Precis. Agric. 1–17 (2022)
8. Friha, O., Ferrag, M.A., Shu, L., Maglaras, L., Wang, X.: Internet of things for the future of smart agriculture: a comprehensive survey of emerging technologies. IEEE/CAA J. Autom. Sinica **8**(4), 718–752 (2021)
9. Lukasik, M., Srijith, P.K., Vu, D., Bontcheva, K., Zubiaga, A., Cohn, T.: Hawkes processes for continuous time sequence classification: an application to rumour stance classification in twitter. In: Proceedings of the 54th Annual Meeting of the Association for Computational Linguistics (Volume 2: Short Papers), pp. 393–398 (2016)
10. Rane, A., Kumar, A.: Sentiment classification system of Twitter data for US airline service analysis. In: 2018 IEEE 42nd Annual Computer Software and Applications Conference (COMPSAC), vol. 1, pp. 769–773. IEEE (2018)
11. Pathan, M., Patel, N., Yagnik, H., Shah, M.: Artificial cognition for applications in smart agriculture: a comprehensive review. Artif. Intell. Agric. **4**, 81–95 (2020)
12. Vougioukas, S.: Annual review of control, robotics, and autonomous systems. Agric. Robot. **2**(1), 365–392 (2019)
13. Zhang, B., Xie, Y., Zhou, J., Wang, K., Zhang, Z.: State-of-the-art robotic grippers, grasping and control strategies, as well as their applications in agricultural robots: a review. Comput. Electron. Agric. **177**, 105694 (2020)
14. Nazir, F., Ghazanfar, M.A., Maqsood, M., Aadil, F., Rho, S., Mehmood, I.: Social media signal detection using tweets volume, hashtag, and sentiment analysis. Multimedia Tools Appl. **78**(3), 3553–3586 (2019)
15. Kolajo, T., Daramola, O., Adebiyi, A.A.: Real-time event detection in social media streams through semantic analysis of noisy terms. J. Big Data **9**(1), 1–36 (2022)
16. McMinn, A.J., Jose, J.M.: Real-time entity-based event detection for Twitter. In: Mothe, J., et al. (eds.) CLEF 2015. LNCS, vol. 9283, pp. 65–77. Springer, Cham (2015). https://doi.org/10.1007/978-3-319-24027-5_6

The Digital Humanities Trend in Chinese Film History: A Case Study of Filmmaker Lvban

Zitong Zhu[✉]

Peking University, Beijing, China
zztnikki@stu.pku.edu.cn

Abstract. The application of data algorithms improves the human perception of knowledge structures and provides interdisciplinary information storage and analysis tools for humanities-based academic research. Using computer-assisted tools, data visualization is applied to texts and images, which are then utilized to investigate film noumenon and film style. This paper, based on Lvban's films and their associated literary texts, found that *Before the Arrival of the New Director* was a week keyword in previous research; however, it can serve as a future point of research area for Lvban cluster analysis. Returning to research on Lvban's film style using the image itself, the author proposes a new experimental method for film research. Using computer-assisted tools CiteSpace and Darwin, this paper demonstrates that studies on Lvban are transitioning from the investigation of individual cases to the exploration of Lvban's identity, film style, and film aesthetic.

1 Introduction

Lvban was a Chinese actor and director whose artistic career encompasses films, dramas, and operas. He was known for his smart and amusing performance in the film *Cross Street*. Lvban's original name was Hao Enxing, and he joined the third squad of the Shanghai Salvation Drama Team in 1938 to work on propaganda on resistance and salvation. He studied comedic films in 1955. In the same year, Spring Comedy Club was established by the Film Bureau of the Ministry of Culture under the supervision of Lvban. In 1957, he was labeled a "rightist" for his play *The Unfinished Comedy* and was assaulted for his membership in the Xiaobailou Anti-Party Group, and he was finally vindicated in 1979, but he died in October that year. Throughout an extensive artistic career, Lvban directed a distinctive brand of sardonic comedy. Analyzing the characteristics of his style in conjunction with historical sources permits a more direct expression of the multiple ways in which the camera language is organized. The director's film shots, sets, and other expressive techniques have a distinctive personal style and traditional Chinese aesthetic.

According to the definition of digital humanities research, digital technologies are used to investigate traditional humanities artifacts. The humanities are currently facing several significant obstacles, including possible institutional extinction and cultural marginalization. Ideally, traditional and digital methods should be complementary to one another. The computational approach of digital humanities frequently makes use of interpretation, textualization, and other forms of qualitative analysis to transform cultural

L. Barolli (Ed.): AINA 2023, LNNS 655, pp. 228–240, 2023.
https://doi.org/10.1007/978-3-031-28694-0_21

works that are being studied into quantifiable datasets. These datasets are then analyzed, classified, and sorted using algorithms before being presented as visualizations, maps, and several other novel methods [1]. However, scholars in the field of digital humanities have reconsidered the gap between theory and practice, turning it into a debate that provides space to examine cultural values and elevates meaning-making beyond rigorous empiricism [2]. Research in the field of digital humanities pertaining to film and media has continued to concentrate on the quantification of moving picture texts, such as the calculation of shot lengths, timing of editing patterns, or comparison of production frames. It fails to offer a critique of the viewer's visual habits, but maintains conviction in the conventional ways of viewing things [3].

Comedies have flourished as satirical entertainment during China's "Seventeen Years" (1949–1966) period. The "Lvban Trilogy," consisting of *Before the Arrival of the New Director, The Unconventional Man, The Unfinished Comedy*, and each made outstanding contributions to the exploration of Chinese comedy. The investigation presented in this paper starts with an evaluation of previous studies on Lvban and then moves on to an examination of Lvban's film itself (both the text and the motion pictures). In many ways, new interpretations of the movie *Before the Arrival of the New Director* have been provided using several different methods, including textual analysis, video annotation, and word cloud analysis. The methodologies and tools that have already been created in film and media studies have been augmented by the addition of these approaches, which allow academics to examine films in new ways. These tools are used to forge a connection between quantitative film studies and the discipline of digital humanities. Similar to other computational methods, these methods have both benefits and drawbacks. Computers do not have the same insight into the meaning and structure of a film as directors and editors do, but they can help highlight subtle patterns in editing, composition, movement, and other aspects of the film and narrative that might otherwise be difficult to observe. Additionally, computers can compare any number of films, assisting researchers in understanding what is typical and unique in a given dataset and identify common features and similar patterns [4].

2 The Concentration and Lack of CiteSpace Methods

Through the utilization of data obtained from various computer-aided tools, the authors of this study intend to investigate movies using an original method. The data come from the narrative of the movie; however, it is impossible to pinpoint exactly what conclusions are drawn and what difficulties are brought to light by employing these quantitative measurements. The kind of question that is posed already influences the kind of data that will be produced; in other words, it is possible to anticipate hot future keywords and gaps in research.

CiteSpace has been selected to carry out the task of doing the textual analysis of the Lvban literature. This determines the pivotal points in the development of a field, particularly the points of knowledge, which are vital turning points in the evolution of the area. One of the ways that information is presented is by using a graphic known as "knowledge mapping," which analyzes and displays the patterns of structural and dynamic characteristics of an academic field on both macro and micro levels. The methodological foundation of co-occurrence analysis is the law of proximity linking and the principles of knowledge structure and mapping in psychology. This method can explain the hotspots of research on a particular topic over time by analyzing high-frequency terms. A total of 595 articles (including newspapers) regarding Lvban and his films were collected from the Internet regarding Lvban and his films, with 52 papers focusing on Lvban's film. Lvban's works will be examined from two distinct vantage points using CiteSpace.

A total of 122 keywords with 110 connecting lines were determined based on research that was conducted by utilizing Lvban as the primary study focus and selecting relevant material from the knowledge mapping (Fig. 1). The size of the nodes represents the frequency of the keywords, the links between the nodes represent connections made at various times, and the thickness of the linkages represents the intensity of keyword co-occurrence. The size of the nodes represents the frequency of the keywords, the links between the nodes represent connections made at various times, "comedy film," is the biggest nodes, followed by *"Before the Arrival of the New Director,"* and "new Chinese movies." If a node has a higher degree of centrality, which is proportional to its degree, it has greater relevance in the network.

According to the degree centrality index which demonstrates the significance of nodes (see Table 1), there is a high degree of co-occurrence between the terms "Guo Zhenqing," "comedy film," and "new Chinese film" and other widely used keywords. The following chronological order includes the terms "banker," "Taihang Mountains," "Beijing Film Studio," "Experimental Theater Group," and "Counter-Japanese Military and Political University Sing and Song Group." "Dongshan Shi" refers to both the "movie" and the "textbook." In recent years, there has been a decline in the amount of scholarly work done on "People's Artists," "Lvban," "Satirical Comedy," and "Tianmin Li." Over the past few years, fundamental concepts such as "Tianmin Li" and novel approaches like "performance," "comedic techniques," and "realism" realism have emerged. This is projected to become the main subject of Lvban's film studies in the future.

The intermediary centrality of keywords is a crucial indicator for identifying hotspots of study in this field and a significant criterion for evaluating where scholars' attention is concentrated. This can be achieved by examining the number of times specific keywords were used. The intermediary centrality indexes, which are used to reflect the promotion of nodes (see Table 2), indicate that "new Chinese film," "comedy film," "Zhenqing Guo," "new Chinese film" and *"Before the Arrival of the New Director"* and other top keywords offer better communication, suggesting that linked research relies on these keywords for content creation. As they can also be seen as research direction points for the future, "Konechuk," "Peilin Li," and "satirical humor" have a level of significance that is comparable to that of the three most important topics that preceded them. The significance of keywords is not only a major criterion for establishing where academics focus their attention, but is also an essential signal for determining the hotspots of study that are associated with this issue.

After analyzing 570 documents (excluding newspapers), 329 keywords and 391 related lines were discovered. The largest node was "Feature films," followed by "New Chinese Films" and "Comedy Films." The relative frequencies of the three terms did not vary significantly (Fig. 2). In terms of frequency (Table 3), "feature film," "new Chinese cinema," "comic film," "seventeen years," and "Changchun Film Studio" were significantly related to one another. The degrees of centrality of "film script," "feature

Fig. 1. Keyword co-occurrence knowledge mapping in CiteSpace (since the text analysis is in Chinese, the result will only show Chinese, so the author did the translation)

Table 1. Top ten keywords (in order of degree centrality)

Ranking	Keywords	Frequency degree	Degree centrality
1	Zhengqing Guo	3	9
2	Comedy Movies	7	7
3	New Chinese Film	4	6
4	Konechuk	1	6
5	Japanese	1	6
6	Peilin Li	1	6
7	Satirical Comedy	2	5
8	Changchun Film Studio	2	4
9	Geopolitics	1	4
10	Remake Film	1	4

film," and "new Chinese film" are not notably dissimilar (Table 4). The two tables of frequency and degree centralities revealed numerous comparable terms, indicating that they frequently co-occur with other phrases. In terms of time (Fig. 3), the terms "comic film," "comedy film," "feature film," "film policy," and "ideology" emerge early but span a broad range of time, and nearly all keywords are associated with it, as they are the most important terms in Lvban's work.

Through cluster analysis of two different literature selections, the terms "comedy film," "feature film," and "new Chinese film" new Chinese film appeared most frequently and ranked highest in the cluster analysis of the two distinct literary selections. It can be determined that these keywords will be Lvban's are the most popular. Simultaneously,

Table 2. Top 10 keywords (sorted by between centrality).

Raking	Keywords	Betweenness centrality	Ranking	Keywords	Betweenness centrality
1	New Chinese Film	0.17	6	Xiucen Yin	0.07
2	Comedy Movies	0.16	7	Konechuk	0.06
3	Zhenqing Guo	0.10	8	Japanese	0.06
4	Lvban	0.09	9	Before the arrival of the new director	0.06
5	Huangmei Chen	0.08	10	Jiayi Wang	0.06

the "film script," "satirical comedy," "Before the Arrival of the New Director," and *The Bridge*" are likely to be Lvban future study priorities. According to CiteSpace's rating of degree-granting institutions, Tables 5 and 6 illustrate the shortcomings of Lvban's recent film studies and the general scope of the study (mediated centrality is zero, ignored). Keywords such as "genre cinema," "ideology," "realism," "bureaucracy," and "directorial style" directorial style provide a starting point for further investigation. In accordance with the general trend, the degree of centrality and frequency of "Before the Arrival of the New Director" are weak, which may be utilized as an example for the next study in relation to Fig. 2.

Fig. 2. Keyword co-occurrence knowledge mapping in CiteSpace (since the text analysis is in Chinese, the result will only show Chinese, so the author did the translation)

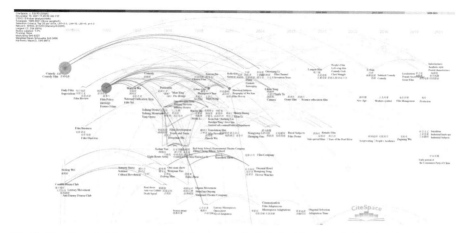

Fig. 3. Result sorted by time (since the text analysis is in Chinese, the result will only show Chinese, so the author did the translation)

Table 3. Result sorted by cluster

Cluster	Degree centrality	Betweenness centrality	Keyword	Year
33	10	0.17	Feature Film	1992
33	9	0.19	New Chinese Film	1987
32	7	0.07	Comedy Film	1993
15	3	0.02	Seventeen Years	2005
13	7	0.22	"Seventeen Years" Film	2001
13	3	0.2	Shi Dongshan	1984
12	1	0	Changchun Film Studio	2005
11	7	0.2	Beijing Film Studio	1957
11	1	0	Northeast Film Studio	1995
10	6	0.06	Film Director	1984
10	8	0.14	China Film	2001

Table 4. Result sorted by degree centrality

Degree centrality	Cluster	Betweenness centrality	Keyword	Year
12	8	0.06	Movie Scripts	1958
10	33	0.17	Feature Film	1992
10	2	0.07	National Unification Area	1981

(*continued*)

Table 4. *(continued)*

Degree centrality	Cluster	Betweenness centrality	Keyword	Year
9	33	0.19	New Chinese Film	1987
9	6	0.15	Film Art	1960
9	3	0.31	*The Red Flag*	1992
8	10	0.14	Chinese Film	2001
8	5	0.16	*The Lin Family Shop*	1999
7	32	0.07	Comedy Films	1993
7	13	0.22	"Seventeen Years" Film	2001

Table 5. Shortcomings of Lvban film studies (sorted by degree centrality)

Degree centrality	Cluster	Keyword
1	1	Realism
1	1	The Art of Film Performance
1	1	Directorial Exposition
1	1	Aesthetic Experience
1	1	Comedic Devices
1	1	Science Education Film
1	1	The Art of Theatre
1	1	Cultural Leadership
1	1	State Ideology
1	1	Bureaucracy
1	1	Ideology
1	1	Comedy Techniques

3 Research Expansion of Video Annotation and Word Cloud Methods

The relationship between text and images is inextricable in film studies. For this reason, I settled on the film *Before the Arrival of the New Director,* which is currently trending in co-occurrence but has a weak keyword. Nick Redfern has discussed that video annotation is still in its early stages in terms of methodology and software for humanities users [5]. Cinemetrics is currently the most well-known film measurement program, although it requires considerable manual labor from a single operator and the reference aid of a tool book. V7 Darwin [6] has been the program of choice in recent years owing to its stable and frequent updates. Automatic annotation and semantic segmentation tools can be applied to keyframes, making Darwin one of the most flexible picture and video

Table 6. General scope of Lvban film studies (sorted by degree centrality)

Degree centrality	Cluster	Keyword
0	1	The Aesthetics of "Seventeen Years" Film Performance
1	1	Tradition of Excellence
1	2	Rural Topic
1	2	Scientific and Educational Film
1	2	Genre Film
1	2	Class Struggle
2	1	Opposite Role
2	1	Directing Style
2	1	*Before the Arrival of the New Director*
2	4	Realism
3	1	"Cultural Weapons"
3	6	Ideology

annotation tools. This allows users to see how selected areas in intermediate frames are distorted from one another. Darwin is still evolving, but it is approaching the point where it can be used by those with little to no experience in annotating videos.

A total of 2791 photos were presented, as computed by Darwin's automatic measuring and compression method (one frame per second extracted, 60 images in total). It is important to manually trace and inspect the primary characters owing to the varying conversion speeds of the film shots. In Fig. 4, it can see that the chosen masks are in action. An intelligent selection analysis was carried out by automatically selecting and manually ticking character shots (including section chief Niu [Jingbo Li], Director Zhang [Ke Pu], Steward Cui [Guangting Chen], Suling [Jianfeng Su], and Lao Li [Yan] Han), scenery (non-character) shots, long shots, and key objects, an intelligent selection analysis was carried out (Fig. 5). Section chief Niu (the older man), Director Zhang (the film's protagonist), Steward Cui (the younger generation), Suling (the female lead), and Lao Li (the wisest character) received the most screen time. Clowns have made their way into the hall and have even become the primary characters in the making of comedies, especially satirical comedies in new Chinese films, dramatically altering the norms of character development [7]. The film features a large number of character-to-character dialogues and a stage-style approach to each other, precisely because it was adapted from a play, novel, and comic play with the same name, which demonstrates the close relationship between the creation of new Chinese comedy films, stage comedy, and comedy literature.

As Chinese films are influenced by classical drama, novels, and other narrative literature structure theories, including "beginning and end" and "first and last correspondence," a staircase was chosen as the focal point to punctuate the film with echoes of the beginning and the end. Scenes displaying a security room door, window, and stairway were used at the beginning and end of the film. The director primarily employed medium and close-up shots of solo, duo, and trio actors. More medium shots were used when the narrative function of the film was highlighted. It is understood that medium shots are crucial in "Seventeen Years" films due to the centrality of the actors' performances, especially in comedies where many dramatic crises and comedic gags depend on the performances of the actors. The lengthy take and stairwell viewpoints were also tools for the director's trade. Lvban employs a pictorial long-take style of expression, framing the film from far to near, to clearly show the activities of the people in the house, as well as the scenery and furnishings further away, in keeping with the idea that Yi is the way and it is a path of continuous movement, change, and circumstance. Push shots, for example, gradually close the lens and screen, shrink the outer frame, and enlarge the inside scene of the screen. This technique allows the audience to comprehend the inner actions of the characters more deeply, and the ironic and comedic atmosphere will be strengthened [8]. The viewer's gaze is guided as a horizontal shift shot in the composition by a continuous stream of vignettes of various sizes and forms [9].

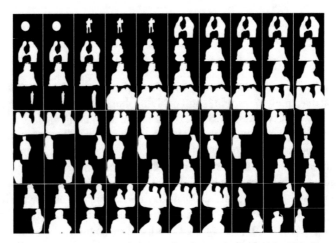

Fig. 4. Partial interception example of *Before the Arrival of the New Director*

2497	551	192	31	All
people old man	Very Overrepresented	456		
main	Overrepresented	397		
three people	Overrepresented	395		
shot	Medium Low Data	363		
two people	Medium Low Data	309		
young people	Medium Low Data	257		
people	Medium Low Data	243		
woman	Medium Low Data	221		
Stairs	Underrepresented	165		
longtake2	Underrepresented	149		
old one	Very Underrepresented	137		
object	Very Underrepresented	92		
longshot	Very Underrepresented	35		
window	Very Underrepresented	31		

Fig. 5. Footage intelligent selection analysis of *Before the Arrival of the New Director*

The term "data mining" refers to the process of automatically examining and obtaining information from various digital resources [10]. Although it is not limited to the analysis of so-called big data, it is highly effective on a large scale and is an integral part of the research approach for text, music, audio, image, and multimodal communication. Text analysis is a subfield of data mining that concentrates on linguistic examination such as NVivo and Voyant. Text processing can reveal elements of a text that are inaccessible to human readers, thus providing a new starting point for research that would otherwise not have been possible. The use of remote reading techniques allows for the detection of patterns of variation in vocabulary, nouns, terms, moods, themes, and an almost inexhaustible array of other topics.

Although "director" is still the most common word used in *Before the New Director Arrives*, "section chief" is not too far behind. The concept of "contradiction" serves as the primary humorous tension in this movie. This pairing or rivalry between characters, and other group interactions such as the "general affairs department," "general affairs," "office," "comradeship," are the primary focal points of the movie. The ideological clash between "boss" and "public," the bureaucracy of the department's head, and the public servant approach of the director, Cui's submission, and Suling's lack of fear of the government are shown in this passage. The actors are those who define the drama, while the actions (movements and behaviors), contradictions, and conflicts (including the narrative) of the objects most effectively convey the characters, ideas, and feelings of the characters. Without dramatic events, there would be no such thing as "film art," which refers to the feature film [11].

The change from satire to glorification is still apparent in new Chinese comedies because the comedic conflict shifts from "contradiction" to "misunderstanding". Figure 6 depicts the significance and subservience of the section leader, who is subordinate to the director. The phrases "misunderstanding," "public," and "old friend" indicate these characteristics. Addressing Steward Cui and Suling with authoritative phrases, such as "how dare you," "make a fuss," "do you understand?" and "no matter what" are used throughout the conversation. The speaker's perspective on Steward Cui and Suling can be inferred from the phrases presented here. The 'oppression' and 'upward mobility' of Steward Cui and Suling, who are currently under pressure, are part of the government. The audience observes how satirical comedy focuses on real-life as its central issue and how its "tears" apart numerous "useless" goods in contemporary society at the levels of ideology, lifestyle, and moral judgment. The lack of political, intellectual, and aesthetic leadership in cinematic art is the most flagrant example of bureaucracy in the film medium. Lvban is a perfect blend of bureaucrats and nationalists, and he flawlessly reflects the characteristics of a section chief Niu's bureaucracy and nativism.

The trend graph represents the lexical frequency of each document or segment within a document in the corpus. The exact nature of this representation is determined by the schema used. Each document in the corpus can be represented as a line on the horizontal plane in Voyant [12]. These lines are then cut into pieces of equal length (50 segments by default). The size of the bubble corresponding to a selected word reflects the number of times that the word appears in the text segment being analyzed. There is a correlation between the bubble size and the frequency with which the word appears. When one moves their mouse over a line on the page, a bubble appears that shows the number of times the term appears in that paragraph. When the mouse hovered over the label at the end of the document line, a breakdown of the term frequency for this document depending on the line content was presented. This label is located at the end of the document and represents the number of terms selected for this document. Figure 7 shows that the term 'director' and the section chief are paired throughout the play, establishing a 'pairing (contrast/opposition)' structure. However, Steward Cui and Suling formed a "contrast" structure. In conclusion, the play is centered on the term "director," even though the section chief is the most prominent character in the play. While Steward Cui and Suling form a separate group, Lao Li plays a more impartial role in the narrative of the novel.

Through image analysis, in the case of the movie *Before the Arrival of the New Director*, the findings obtained repeat the exact weak points (keywords) that the textual analysis embodies. Consequently, it is possible to hypothesize that there is also a relationship between the keywords of weaknesses. This means that Lvban's cinematic style mirrors the concept of seeing the huge with the little from the shots and keywords in a film), and this strategy can be applied to all his works.

Fig. 6. *Before the Arrival of the New Director* Word Frequency by NVivo

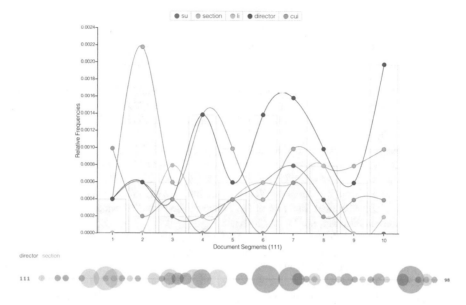

Fig. 7. *Before the Arrival of the New Director* Relative frequencies and bubblelines for main actor's line

4 Conclusion

Not only are digital humanities a source of information for other fields, but they are also a benchmark for acquiring new knowledge and implementing it in the actual world. In digital "post" film culture, new avenues of inquiry have opened as a result of digital aesthetics. However, this shift invariably creates theoretical difficulties and is related to alternative ways of conducting research. Considering the abundance of digital resources, academic fields within the humanities should integrate digital approaches into classroom instruction and theoretical critiques of digital culture in a more comprehensive manner. Initial surface-level experimental analysis of Lvban's films has the potential to make digital methods more accessible to film studies. Simultaneously, it highlights the necessity of algorithmic strategies grounded in the digital humanities, as well as the added value that these strategies provide. Due to this, the utilization of computer-aided tools helps in determining whether the theoretical assessments performed in the past were accurate. Within the fields of digital humanities and film studies there is still a great deal of work to be done to tackle issues that are unique to films. For example, when applied to films, dictionary vocabulary is too fashionable and rigid for the emotional analysis of script lines, and NLP (natural language processing) machine learning does not have models that are acceptable for film texts. Databases are not sufficient for conducting many studies, including film studies. In addition, fresh oral histories, fieldwork, and interviews are necessary. An ongoing practice of self-reflection is required to gain an understanding of the difference between the filmmaker in the database and the society.

References

1. Mittell, J.: Videographic criticism as a digital humanities method. In: Gold, M., Klein, L.F. (eds.) Debates in the Digital Humanities 2019, pp. 224–242. University of Minnesota Press (2019)
2. Drucker, J.: Humanistic theory and digital scholarship. In: Gold, M. (ed.) Debates in the Digital Humanities, pp. 85–95. Minnesota Scholarship Online (2012)
3. Ferguson, K.L.: Volumetric cinema. Transit. J. Videographic Film Moving Image Stud. **2**, 335–349 (2015)
4. Manovich, L.: Visualizing vertov. Russ. J. Commun. **5**, 44–55 (2013)
5. Heftberger, A.: Do computers dream of cinema? Film data for computer analysis and visualisation. In: Understanding Digital Humanities, pp. 210–223 (2012)
6. V7 - AI Data Platform for Computer Vision (n.d.). https://www.v7labs.com/
7. Daoxin, L.: The Historical Situation of New Chinese Comedy Cinema and Its Conceptual Transformation. Film Art, pp. 15–21 (2003)
8. Yan, X.: Film Art Dictionary, p. 261. China Film Press, Beijing (2005)
9. Niantong, L.: Discussions on Classical Aesthetics in the Study of Chinese Film Theory. Contemporary Film, pp. 102–109 (1984)
10. Underwood, T.: A genealogy of distant reading. Digit. Humanit. Q. **11**(2), 1–12 (2017)
11. Jin, J.: Action Center - Film Art Design Exploration 2. Film Art, pp. 60–68 (1963)
12. Sinclair, S., Geoffrey, R.: Voyant Tools Help. https://voyant-tools.org/docs/#!/guide/tools

A Tool for Creation of Virtual Exhibits Presented as IIIF Collections by Intelligent Agents

Dario Branco, Rocco Aversa, and Salvatore Venticinque[✉]

Department of Engineering, University of Campania "Luigi Vanvitelli",
via Roma 29, 81031 Aversa, Italy
{dario.branco,rocco.aversa,salvatore.venticinque}@unicampania.it

Abstract. In this paper we present a tool designed and developed to allow for the production of digital contents under the IIIF 3.0 standard. It provides an user-friendly web GUI to editors who just need to define and correlate entities uploading media and filling form. This software is part of the results delivered by the Cleopatra project, introduced later in this manuscript, which aims to use conversational agents alongside IIIF compliant contents and services in order to improve the immersiveness of the user experience. For these reasons, in addition to the organization of contents, the software is able to insert appropriate annotations in such a way as to increase the amount of information a conversational agent possesses in order to be able to answer a wider and more varied pool of questions (author of a work, year in which it was created, stylistic current, geographical location, etc.).

1 Introduction

The International Image Interoperability Framework (IIIF) [1] is a widely accepted interoperability standard for delivery of media contents trough the web. It allows to build digital repositories which can be accessed by open and uniform Application Programming Interfaces (APIs). Being able to create contents that comply with such a standard is fundamental for maximizing the popularity of media collections that can be reached trough the Internet and can be easily integrated into any applications for end-users. Thanks to the offered features and the extreme flexibility of this format, IIIF is becoming increasingly popular among museums and sites of cultural interest [2,3], contributing to its popularity and larger and larger acceptance. Production of IIIF contents consists of editing machine readable text files in a JSON[1] data format according to complex rules defined by the standard. Hence, despite the numerous strengths, creating contents that comply with the standard is often cumbersome and complicated, especially for those who are not computer literate and/or unfamiliar with this type of technologies. To this end, there is a need for tools that are able to generate content automatically, through a graphical interface that is easy to use,

---

[1] https://www.json.org/.

© The Author(s), under exclusive license to Springer Nature Switzerland AG 2023
L. Barolli (Ed.): AINA 2023, LNNS 655, pp. 241–250, 2023.
https://doi.org/10.1007/978-3-031-28694-0_22

even for the less geeky employees. In this paper we present a tool designed and developed to allow for the production of digital contents under the IIIF 3.0 standard. It provides an user-friendly web GUI to editors who just need to define and correlate entities uploading media and filling form. This software is part of the results delivered by the Cleopatra project, introduced later in this manuscript, which aims to use conversational agents alongside IIIF contents and services in order to improve the immersiveness of the user experience. For these reasons, in addition to the mere creation of contents, the software is able to insert appropriate annotations in such a way as to increase the amount of information the conversational agent possesses in order to be able to answer a wider and more varied pool of questions (author of a work, year in which it was created, stylistic current, geographical location, etc.).

2   Related Work

The IIIF standard is rapidly assuming a central position in the field of multimedia content delivery, and this is confirmed by the worldwide effort in the creation of tools related to the standard, from visualisers to annotators to libraries useful for the creation of the standard. Typically, a fundamental role for the delivery of multimedia contents is played by the content viewer. The Mirador viewer is one of the most known available IIIF compliant technology. A very thorough and precise review of the Mirador tool, carried out from an academic point of view, is presented in [4]. In [5] a Curation Platform is proposed for the creation of IIIF collections where both Mirador and Universal Viewer are used for the visualisation part. In reference [6], authors monitor the area of the images which are have been most frequently viewed by users counting the number of pixels. They create heat maps that highlight the most viewed areas. Since the first presentation of the IIIF standard, several tools have been proposed for the creation of contents that complies with the standard in terms of automatic generation of IIIF manifests. Many ho these have been designed to be used by people with a background in computer science and who can use Java, Python, PHP or similar libraries. Most of the proposed solutions involve the coding of software and do not provide an effective answer to users of the application field, who usually do not have programming skills. Some technological implementation for the generation of IIIF manifests are directly supported by the IIIF consortium and can be found in the official documentation. There are few documented tools that allow the generation of IIIF manifests by a GUI (Graphical User Interface) that is easy to use. For example, the Bodleian Manifest Editor [7], which, however, produces manifests in IIIF v. 2.1 and not in the recent IIIF v. 3. Another open source application, the biiif tool [8], allows the creation of IIIF v. 3 compliant manifests, but it has no GUI and requires the organization of contents in folder and text files according to a particular naming convention and a folder tree structure defined in their documentation. Our contribution represents an alternative technology that is capable of generating manifest in v. 3 in a simple and intuitive manner and that provides a web GUI that was easy to use even by novice users.

3   The IIIF Standard

The International Image Interoperability Framework is a standard for the description and distribution of multimedia contents [9]. The idea was born in 2011 from a collaboration between various museums, libraries and university research institutes, and the first official API was officially released in 2012. The framework is mainly composed of two main APIs plus other secondary APIs that complement it with secondary functionalities. The two main APIs deal with image retrieval and image display. As far as the IIIF Image API is concerned, it specifies a webservice that returns an image following an HTTP or HTTPS request. Within the request URI, various parameters can be specified, which are used to return the relevant image. Through the use of parameters, it is possible, for example, to select an image that is only a part of a larger image, change its size, rotate it, define the colour scale and the format in which we want the image to be saved. As far as the IIIF Presentation API is concerned, it deals with the display of digital objects that contain several multimedia contents within them, as if they were multimedia galleries. This API returns a structured JSON document describing the structure and layout of multimedia content or even other collections of images with their contents. The Presentation API thus allows a visualisation software to immersively display all the contents linked to the container and navigate through them. In addition to the two main APIs mentioned above, for the sake of completeness we list here the other APIs made available by the standard: Authentication API, Search API, Content State API and the Change Discovery API.

The creation of JSON files, on the other hand, can be overwhelming and uncomfortable, especially for those without an IT or Programmer background. From this problem arises the need for automatic software for the creation of multimedia galleries under the IIIF 3.0 standard (the latest available and released). The Cleopatra software platform utilizes several key concepts in order to augment the cultural experiences of an user that is enjoying a virtual or live exhibition. One of these concepts is the **manifest**, which is the main component within IIIF and defines all information relating to an IIIF object at various levels of detail. The manifest also defines which components of the object to display, such as a sequence of images, and how. In Fig. 1 a fragment of an IIIF manifest is shown. Within a manifest, the **canvas** serves as the reference frame for the spatial presentation of the content. **Annotation** represents any additional information, not foreseen by the standard, that can be added to the manifest. For example, the author of a work of art, its geographical or temporal location, etc. A **Collection** is a group of Manifests hierarchically linked or a set of related objects. An **Image server** that provides IIIF compliant objects and an **Image viewer**, a software capable of understanding, visualizing, and manipulating IIIF contents provided by the Image Server.

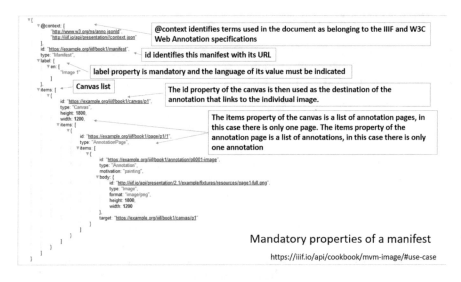

Fig. 1. An example of Manifest with mandatory elements.

4   Cleopatra Project

The Cleopatra software platform [10] enables the construction of a dynamic P2P overlay of users and software agents. This platform provides advanced services to users that aim to enhance their cultural experiences by using their smartphone or any other device installed in physical museums, or from their desktop if they prefer to enjoy a virtual tour from home. [11]. Intelligent software agents are responsible for delivering cultural knowledge, being aware of the user's information and the digital archives through ad-hoc knowledge base and digital archives, which expose an IIIF compliant interface. According to the concept platform represented in Fig. 2, a collection of IIIF manifests is used as a meta-representation of a museum. IIIF manifests include a sequence of media components that represent either a tangible artifact or an abstract cultural content. This can be a virtual depiction of a building, a monument, a work of art, a natural scene, a museum space or any other self-consistent information through a set of multimedia objects (photos, videos, drawings, etc.) that are semantically connected and have their own meta-data. The Cleopatra software agent guides the visitor along a cultural itinerary, complementing a proactive storytelling with an interactive conversation, similarly to what is described in [12], presenting the descriptive information while semantically linked multimedia are displayed. The user may search through the images associated with the cultural contents they wish to view and can also inquire about any pertinent details. The agent's awareness is not limited to the contents of the conversation engaged with the users [13]. As an example, a trivial query like "What is this picture?" could not be addressed by the agent if it were unsure about which canvas the Image viewer displays at that moment. To deliver a positive experience, during the conversation we make the agent aware about how the user is interacting with the viewer, by a strict integration between the web chat and the IIIF viewer. The user's interaction with the viewer is communicated to the agent in terms of IIIF events, which can help overcome this restriction, informing the agent on how the user is acting with respect to the standard.

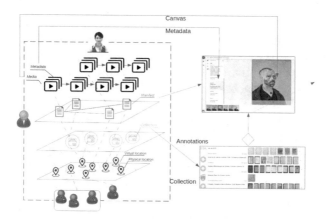

Fig. 2. The Cleopatra platform.

The set of information that need to be provided to represent the IIIF context are: the current manifest, the displayed canvas, the degree of zoom and any selected annotation. Moreover, in order to link the IIIF contents to the conversation in natural language with the agent, additional information can be provided by an original tool for IIIF manifest generation. This tool is specifically designed to allow the agent for providing the right answer when the topic of the conversation depends on the content selected by the viewer, but also to show an alternative or a related content for better understanding or for an effective presentation of the answer, or simply to select relevant recommendations. For example, the annotation for the predicted question "Who is the author of this painting?" could be a short presentation of the author's profile and the visualization of a secondary artwork.

5 IIIF Manifest Generator

Experts of the cultural field, who need to design a Cleopatra virtual museums, which complement or replace physical locations, need to focus on those aspects that enable the digital multi-user interaction. They needs to collect, annotate and link multi-media contents, in a way which can be presented to the users and handled by software agents. A platform to support this kind of activity has been designed and implemented as part of the Cleopatra project. The platform is based on laravel[2] that is a free, open-source PHP web framework. The management software is able, through an advanced ORM (Object-Relational Mapping) system, to communicate with a database that has been structured to facilitate its translation to the IIIF 3.0 standard supported by the most famous software for the use of multimedia content in virtual platforms for museums (such as Mirador). Eloquent ORM is an advanced PHP implementation of the active record pattern, providing at the same time internal methods for enforcing constraints on the relationships between database objects. Following the active record pattern, Eloquent ORM presents database tables as classes, with their object instances tied to single table rows. Figure 3 shows the ER diagram on which the management software is based. Proceeding to the description with a Bottom-Up approach, it can be seen how the relationships and the components refer to the hierarchical structure of the IIIF standard. At the lowest point of the hierarchy is the media content (work of art), which has a name, a description, a reference to the author and to the manifest it is associated with in addiction to a list of question and answer used by the conversational agent regarding the media content. Proceeding upwards we find the manifest which are nothing more than a set of media contents (artworks), each manifest refers to a collection which in turn refers to an exhibition (like museums or archaeological sites).

Figure 4 presents the component diagram of the web application. User requests are handled by routing software which dynamically redirects to the corresponding View. Once the view has been reached, it is populated by a Controller that manages the data to be displayed and the possible actions on them. The Routing component allows access to the Server IIIF Generator interface after logging in, which then saves the generated IIIF library within the FileServer, making it accessible to a Mirador instance on the WebServer.

Finally, the IIIF Generator server will expose a REST API that is able to receive the name of the collection whose IIIF is to be produced, to generate the corresponding IIIF and return the parent path that contains the generated files. It is possible to insert annotations in JSON format comprehensible by the Cleopatra framework at the time of object creation in order to increase the dialogue capacity of conversational agents within the system. Note how the creation of data in IIIF format occurs asynchronously with the addition of annotations. This design decision makes it possible to add, modify or delete annotations to an existing collection without having to generate it again from scratch [14]. An example of interface is provided in Fig. 5 where one of the windows of the IIIF

---

[2] https://laravel.com/.

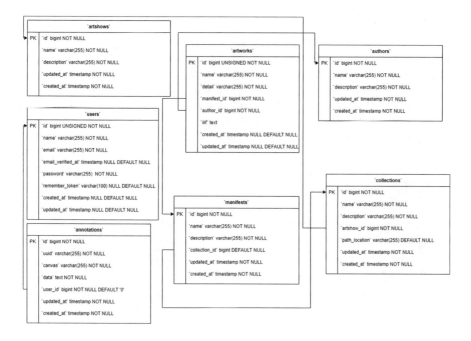

Fig. 3. ER diagram of Cleopatra IIIF Generator software

Generator web application is shown. In the window shown, it is possible to modify a previously loaded multimedia content (other views are similar). In addition to a name and a description, it is also possible to associate it with the author,

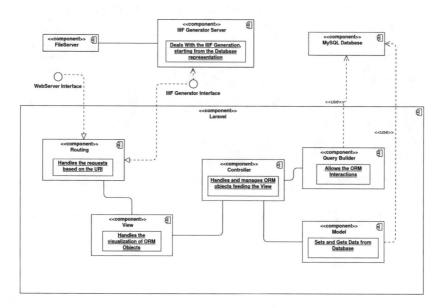

Fig. 4. Software architecture of the Cleopatra IIIF Generator

the collection to which it is associated and a list of questions which, in conjunction with the respective answers, are used to train the conversational agent to manage a conversation with the user about the artwork. In the example, the Conocchia is shown which is an ancient funerary monument built by the ancient Romans and located along the Ancient Appian Way. On the other hand, Fig. 6 shows the functioning of the Cleopatra application where, together with the IIIF Mirador viewer, the conversational agent and his virtual Avatar answer the user's questions previously foreseen in the IIIF Generator.

Fig. 5. Example of Media Content Editing

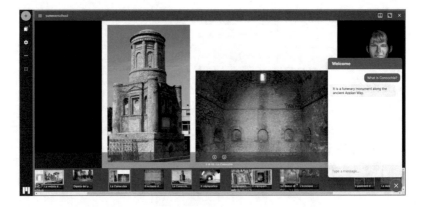

Fig. 6. Example Media Content shown in Mirador Viewer

6   Conclusions

This work presented a web tool for the design of collections of multimedia contents via an user-friendly GUI. The tools generates IIIF standard manifests, which can be handled by intelligent agents able to interact with the users and presents the collection. The objective is to let the expert of the cultural field focus on the creation of multimedia galleries and on the way they can be made accessible to all and on the way they can be presented. The proposed software infrastructure includes an annotator that allows for the use of the Cleopatra framework, which enhances user immersion and engagement through interactive interaction with a virtual assistant. On the other hand, the compliance with the IIIF standard, that is hidden to the user, guarantees the interoperability with many tools and the integration with public repository, providing large visibility.

Acknowledgment. The Cleopatra Project has been funded by the University of Campania "Luigi Vanvitelli" through the VALERE 2019 research program.

References

1. Salarelli, A.: International image interoperability framework (IIIF): a panoramic view. In: International Image Interoperability Framework (IIIF): A Panoramic View, pp. 50–66 (2017)
2. Manoni, P.: L'adozione del iiif nell'ecosistema digitale della biblioteca apostolica vaticana. DigItalia **15**, 96–105 (2020)
3. Robson, G.: Implementing the international image interoperability framework at the national library of wales. J. Digital Media Manag. **4**(4), 353–359 (2016)
4. Van Zundert, J.: On not writing a review about mirador: mirador, IIIF, and the epistemological gains of distributed digital scholarly resources. Digital Medievalist **11**, 5 (2018)
5. Kitamoto, A.: IIIF curation platform: creating and sharing virtual image collection on a global scale. In: International Conference: Glocal Humanities in the Era of Hyperconnectivity, vol. 9 (2019)
6. Nishioka, C., Nagasaki, K.: Measurement and visualization of IIIF image usages. In: Jatowt, A., Maeda, A., Syn, S.Y. (eds.) ICADL 2019. LNCS, vol. 11853, pp. 308–311. Springer, Cham (2019). https://doi.org/10.1007/978-3-030-34058-2_30
7. University of Oxford. Bodleian manifest editor (2022). https://digital.bodleian.ox.ac.uk/manifest-editor/#/?_k=uob5lo
8. Build IIIF (2022). https://github.com/IIIF-Commons/biiif
9. Snydman, S., Sanderson, R., Cramer, T.: The international image interoperability framework (IIIF): a community & technology approach for web-based images. In: Archiving Conference 2015 (2015)
10. Ambrisi, A., et al.: Intelligent agents for diffused cyber-physical museums. In: Camacho, D., Rosaci, D., Sarne, G.M.L., Versaci, M. (eds.) Intelligent Distributed Computing XIV, pp. 285–295. Springer, Cham (2022). https://doi.org/10.1007/978-3-030-96627-0_26
11. Kiourt, C., Pavlidis, G., Koutsoudis, A., Kalles, D.: Multi-agents based virtual environments for cultural heritage. In: Proceedings of the 26th International Conference on Information, Communication and Automation Technologies (ICAT), pp. 1–6 (2017)
12. Muratović, E., Prazina, I.: A web-based service-oriented solution for a mobile digital storytelling application. In: Proceedings of CESCG 2017: The 21st Central European Seminar on Computer Graphics (2017)

13. Riedl, M., Saretto, C.J., Young, R.M.: Managing interaction between users and agents in a multi-agent storytelling environment. In: Proceedings of the Second International Joint Conference on Autonomous Agents and Multiagent Systems, AAMAS 2003, pp. 741–748. Association for Computing Machinery, New York (2003)
14. Amato, A., Aversa, R., Branco, D., Venticinque, S., Renda, G., Mataluna, S.: Porting of Semantically Annotated and Geo-Located Images to an Interoperability Framework. In: Barolli, L. (ed.) CISIS 2022. LNNS, vol. 497, pp. 508–516. Springer, Cham (2022). https://doi.org/10.1007/978-3-031-08812-4_49

Recommender Systems in the Museum Sector: An Overview

Alba Amato[✉]

University of Campania "Luigi Vanvitelli", Caserta, Italy
alba.amato@unicampania.it

Abstract. Recommendation Systems are used in hundreds of applications, each with particular rules, with affinity and, at times, points of departure from pure recommendation. The applications we will analyze often exploit hybrid approaches, attempting to reduce or completely eliminate the negative effects of pure methods. The first factor to consider when designing a Recommendation System is the application domain, as it has an important effect on the algorithmic approach that should be adopted. In this paper we provide an overview of the typologies of recommender systems that are applied in the museum sector.

Keywords: Recommendation systems · Cultural heritage · Digital humanities

1 Introduction

Recommendation Systems are used in hundreds of applications, each with particular rules, with affinity and, at times, points of departure from pure recommendation. The applications we will analyze often exploit hybrid approaches, attempting to reduce or completely eliminate the negative effects of pure methods. The first factor to consider when designing a Recommendation System is the application domain, as it has an important effect on the algorithmic approach that should be adopted. There are many types of domains:

- entertainment: we take care to provide advice on films, music, games;
- containers: personalized news, web page recommendations, e-mail filters;
- e-commerce: advice to be given to consumers on the products to buy such as, for example, books, T-shirts, gadgets;
- services: recommendations from experts in a particular field, recommendations for rental houses;
- social: suggestions from possible people you might know and personalized recommendations related to particular social media content such as tweets and Facebook statuses.

The developer of a Recommendation System for a given domain should understand the specific aspects of that domain, its requirements, application challenges

L. Barolli (Ed.): AINA 2023, LNNS 655, pp. 251–260, 2023.
https://doi.org/10.1007/978-3-031-28694-0_23

and limitations. Only after analyzing these factors is it possible to select the optimal recommendation algorithm and design an effective human-machine interaction. In this paper we provide an overview of the typologies of recommender systems that are applied in the museum sector.

2   Typologies of Museum Recommendation Systems

The Recommendation Systems can be incorporated into various contexts, one of these is the museum environment [13]. Nowadays, there are huge museums all over the world, such as, for example, the Louvre in Paris, the Prado in Madrid or the Hermitage in St. Petersburg, which contain various collections of works of art from all over the world, made in different areas, with different styles and belonging to different periods, cultures or currents. Usually these art products can be spread across multiple rooms, multiple floors, and even multiple buildings. Even in small museums, sometimes the number of exhibitions is large enough to make it impossible to view all the collections at once, taking into account the fact that, as the visit progresses, the visitor's interest and attention decrease. Furthermore, it often happens that you have to face an art exhibition about which very little is known. Once inside a museum, one is therefore forced to make choices. It seems that questions like "where to go?", "Which exhibitions to see?" and "how to quickly find interesting works for us?" are not easy to answer, as visitors are faced with information overload induced by the richness of the museum's content. Since the visit to the museum itself is a leisure activity, an educational experience, in which different visitors have different needs, objectives and interests and follow their own paths at their own pace, it is important to acquire knowledge about users and understand their needs, interests and preferences, in order to satisfy them. In fact, most of the exhibits they see may not be of particular interest to them. This could make them lose works of art or other items that are more related to their personal preferences. It is therefore certainly important to visit all the highlights of the museum, the so-called "Hubs" (works so famous that they cannot be skipped, for example "La Gioconda" by Leonardo da Vinci in the Louvre), but also to expand the knowledge on the collection based on to their specific interests [10].

These challenges are addressed by the Recommendation Systems, which take into account the needs and preferences of users [1]. Recommendation Systems can be a winning choice for a museum as their primary motivation is to help visitors cope with this "information overload" by helping them find their way into the collection of works of art by providing them with the right information at the right time, increasing their awareness on the themes of art history, recommending them the works they might like and providing them with a personalized and pleasant experience in the visit, perhaps stimulating them and tempting them to visit the museum more often or to provide positive reviews, useful for other potential visitors. An optimal solution should, on the one hand, offer users a personalized selection of articles based on their preferences and, on the other hand, provide in-depth material on these, which extends their knowledge [3]. Some museums, such as the Louvre, already offer software that allows the user to personalize their visit

[4], thus promoting cultural heritage even in younger generations. For example, the Rijksmuseum in Amsterdam [17] provides various online services through its website, where visitors can preview the décor of the exhibition halls and introductions to the exhibitions. Since 2005, the Rijksemuseum has collaborated with researchers from the Eindhoven University of Technology in the Cultural Heritage Information Personalization (CHIP) project [7], part of the Continuous Access to Cultural Heritage in the Netherlands (CATCH) program, funded by the Dutch Science Foundation, which promotes the access to cultural heritage in the Netherlands. A web application has been developed which, in the form of an interactive quiz, helps users in finding potentially interesting works in the Rijksmuseum collection. This application initially presents the museum's works of art to customers, so that they can evaluate them [2]. On the basis of these evaluations, the system searches for other related items and topics, again among the works of the Rijksmuseum, that users might find interesting. To give an example, if a user gives a high score to portraits and low to landscapes, the system will deduce that the user is more interested in the "Portraiture" topic than in the "Landscape" one. The Recommendation System therefore recommends "Portraiture" as a Topic, along with other portraits found within the Rijksmuseum collection. The system also allows users to rate the recommendations provided by offering their personal feedback on the recommendation. As can be seen from Fig. 1, each user can also view his own profile with the choices already made (he can also modify them), and obtain explanations on why he has been recommended a particular work. It should be emphasized how this application exploits the works of the museum itself. Other applications instead prefer to use works related to them and not part of the collection, in order to be able to profile the user by recommending works within the museum that they might like, but without "spoilers". Some museums have patented audio guides with, inside, a Recommendation System based on the analysis of visitor behavior: trajectory (accurate measurement of the user's position, complete with time dedicated to each work of art) and interaction with the device ("I like" given to the works, searches for additional information) [9]. Other museums, on the other hand, have focused on creating applications for smartphones which, by profiling the user, not only suggest works that they might like, but also use algorithms

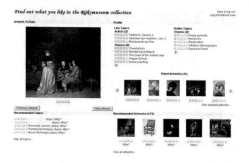

Fig. 1. Users profile

typical of operations research (for example the algorithm that solves the problem of the traveling salesman), in order to also allow him to view a map with the shortest route to be able to see them all [5].

Considerable research efforts have also focused on the development of recommendation applications that exploit virtual and augmented reality, already available as an addition to exhibitions in various museums, offering a more interesting and engaging experience for visitors [16].

The usefulness of a recommendation system lies in the ability to analyze the history of the interactions of the various users on a platform and, based on the characteristics of the objects with which the user interacted in the past, recommending other items that might suit user's interests or user's needs, considering [14]:

- User's history: Looking at the user's history, objects similar to those found interesting in past.
- Similar users: Considering users with similar tastes, objects considered interesting by them users.
- Context information: This is information surrounding the context of use of the system, such as demographic information about users or even temporal information.
- Domain information: Information specific to the application domain, which allows a more in-depth analysis of the objects. For example, a recommendation system for use on a video streaming service can use the percentage of video watched by the user to obtain an indirect measure of the interest of the user.

Recommendation Systems usually work with two types of data: user-item interactions (such as ratings or purchasing behavior) and information on user and item attributes (such as textual profiles or relevant keywords). Recommendation system models that use the first data types are referred to as Collaborative Filtering, while methods that use the others are referred to as Content-Based [11]. There are also Recommendation Systems based on "knowledge" (Knowledge-Based) [6], in which recommendations are based on requirements explicitly specified by the user. Some Recommendation Systems combine these different aspects to create hybrid systems. They can leverage the strengths of various types of Referral Systems to create techniques that offer better performance. Within specific areas, the recommendation plays a critical role and it could be important to provide suggestions using time data, location-based data or data from Social. Context-Based Recommendation Systems [8] are born from this idea. The models mentioned above will be described more fully below.

2.1   Collaborative Filtering Models

The Collaborative Filtering model is one of the most used. It exploits the evaluations provided by users to the items and the similarity between them (item-item, user-user) in order to generate new suggestions. The collaborative method generally uses rating matrices, in which each row represents a user profile, that is, a continuously updated vector of customer ratings for items. These ratings can be Boolean (I like, I don't like) or values that indicate the degree of preference. The rating matrices have a peculiar characteristic that, at times, can become a problem: they are usually very sparse. Let's consider, as an example, a music application in which users indicate their liking or not to the songs. Since there are billions of songs in the world, the majority of users will have heard only a small part of this vast universe, of as a result, most of the ratings will be missing. Collaborative Filtering Recommendation Systems take care to value these missing votes. This task can be difficult in the presence of very sparse matrices. Most of these models focus, as already mentioned, on the use of correlations between elements or correlations between users for the purposes of the forecasting process. There are two types of methods commonly used:

- Memory-Based Methods, more commonly called Neighborhood-Based. This category includes algorithms that rely on the history of ratings entered by users to predict missing votes. They are divided into two subfields:
  - User-Based Collaborative Filtering. Users are determined who are similar to each other according to certain criteria established by the application (creating sets of customers). Subsequently, for each evaluation not given by a customer to an item, the weighted average of the votes of the group of users similar to him on that item is calculated, all in order to predict a possible vote that the user himself would give to the item. Therefore, if two customers have rated songs in a similar way in the past, it is possible to use the ratings of one of the two on certain pieces in order to predict the missing ratings of the other on them.
  - Item-Based Collaborative Filtering. In order to predict the possible vote for an article by a user, we first determine a set of articles similar to the item in question. The votes given by the user to the latter are used to predict whether or not they will appreciate the item itself.
- Model-Based Methods. This category includes algorithms that exploit the history of the system in order to learn a model that is used to generate recommendations. The Recommendation Systems that are based on these techniques can encounter problems, as it could happen, for example, that all users "similar" to the customer have not provided any evaluation for a given item. In this case, the estimate of the possible evaluation of the article for that specific user becomes difficult. This often happens in the presence of scattered rating matrices, as already pointed out.

2.2 Content-Based Recommender Systems

In the Content-Based Recommendation Systems the attributes (features) of the items are used, i.e. a set of all the characteristics capable of summarizing the content (hence, precisely "Content-Based"), to formulate recommendations. Combined with item attributes, in these models, user ratings are also used. For example, consider a situation in which a user has positively evaluated a work of art, but we do not have access to the evaluations of other users. Excluding the use of a Collaborative Filing method for this very reason, you can decide to exploit the description of the work in question, expressed through keywords, which, if similar to those of other artistic works, allow the recommendation of the latter to the customer. To provide a more concrete example, a user who likes nude representations could be suggested, as an artistic work, "The Birth of Venus" by Botticelli, or the "Venus de Milo", as both are artistic nudes. For each user, the items that he has already purchased or evaluated are displayed, in order to predict whether or not he will like an item for which his evaluation or purchasing behavior is unknown. The Content-Based Recommendation Systems have, also for this reason, some disadvantages that must be, in some way, contained. These models, in many cases, provide obvious recommendations. For example, if a user has never rated an article with a certain set of keywords, those items are unlikely to be recommended. This phenomenon therefore tends to reduce the diversity of recommended articles. They are susceptible to the Cold-Start problem. While content-based methods are effective at providing recommendations for new articles, as you might imagine, they are not effective at providing recommendations for new users. This is because the model, for the user, must be able to use his past votes. Such content or knowledge, in fact, may not always be available. If these are not available, such as when a new customer arrives, it becomes more difficult to apply this method. Therefore, numerous specific methods have been designed to solve the cold start problem [15]. While the foregoing provides a conventional view of Content-Based techniques, more elaborate versions of this method are sometimes used. For example, users can specify relevant keywords in their profiles and these can be combined with item descriptions to make recommendations. Such an example approach could be useful in cold boot scenarios. However, such methods are often seen as a distinct class of Recommendation Systems, known as Knowledge-Based Systems.

2.3   Knowledge-Based Recommender Systems

Knowledge-Based Recommendation Systems work by trying to suggest items starting from the data available, from the needs and preferences of each user, which are often collected explicitly or deduced. The user profile can be any representation of knowledge that derives from this inference. Knowledge-Based Referral Systems are particularly useful in the context of items that are not purchased very often, such as real estate, cars, tourism inquiries, financial services, or luxury goods. In such cases, there may not be sufficient ratings available for the recommendation process, as items are rarely purchased and with different types of detailed options. Furthermore, the nature of consumer preferences can evolve over time when it comes to such items. For example, the model of a luxury car, which few people can afford, can evolve significantly over the course of a few years, following which user preferences can show a corresponding evolution. Attributes that correspond to its properties can be associated with a particular article and a user can only be interested in articles with specific characteristics. As an example, again in the case of cars, they can have different brands, models, colors, characteristics of the engine and user interests can be adjusted by a very specific combination of these options. The recommendation process is performed on the basis of similarity between customer requirements and item descriptions. Knowledge-based Recommendation Systems can be classified according to the type of interface (and corresponding knowledge) used to achieve the objectives mentioned above:

Constraint-Based Recommendation System. Users specify requirements or constraints (for example, lower or upper limits) on the attributes of the element. For example, a user wishing to purchase an item could specify the minimum price and the maximum price that he is willing to pay in the interface made available by the system.

Case-Based Recommendation System. The user directly specifies the specific characteristics that the item in question must have. For example, a user looking for a house will specify, in the dialog box provided by the system, the number of bathrooms she wishes to have, the price he is willing to pay, etc. Similarity metrics are defined on the attributes of the articles, in order to retrieve similar elements. Note that in both cases, the system offers the user the opportunity to change the specified requirements if he is not satisfied with the results of the suggested recommendation. Graphical interfaces are particularly useful for expressing feedback in such systems, where users iteratively change one or more attributes of an item at each iteration. Through an iterative sequence of change requests, it is possible to arrive at a desired article. Due to their use of attributes on content, knowledge-based systems inherit some of the same disadvantages as content-based systems. For example, recommendations in knowledge-based systems can sometimes be obvious.

Demographic Recommender Systems. The demographic model classifies the user on the basis of his personal attributes (age, height, weight, origin, ...) and elaborates the suggestion according to the demographic class to which he is assigned. Many websites have simple and effective personalization solutions based on demographics. For example, users are sent to certain websites based on language or country. A first Recommendation System, called Grundy [12], recommended books, exploiting the user characteristics collected with the use of an interactive form. In many cases, demographic information can be combined with additional context to guide the recommendation process. This approach is related to the methodology of Context-Based Recommendation Systems. In fact, although the demographic model does not usually provide the best results on a standalone basis, it significantly increases the power of other Referral Systems as a component of hybrid models. Demographic Recommendation Systems are not subject to the "new user" problem as they provide suggestions simply by evaluating the demographic information entered. The main limitation consists in the difficulty of finding the relevant information for the purpose of rating, both due to the growing awareness of privacy and the consequent skepticism of users in providing personal information in an online form.

2.4   Context-Based Recommender Systems

Context-based Recommendation Systems take into account various types of contextual information in order to make recommendations. This information could include the user's time, location or social data. For example, the types of clothing a clothing site recommends might depend on both the season and the customer's location. The context of the user, when looking for a recommendation, can therefore be used to better customize the system output. For example, in a time context, vacation recommendations in winter should be very different from those given in summer, and a restaurant recommendation for a Saturday night with your friends should be different than one suggested for a day's lunch. It has generally been observed that the use of this contextual information can greatly improve the effectiveness of the recommendation process, as it increases the degree of personalization of the suggestion.

2.5   Hybrid Recommender Systems

The recommendation methods described above have some weaknesses, for which it is sometimes necessary to combine them in order to minimize the disadvantages deriving from each model. The combination can take place in several ways, and the resulting method is called a hybrid. A Hybrid system, which combines two techniques, attempts to use the advantages of one in order to correct the disadvantages of the other, or limit them. The models already mentioned, as seen, can work well in different scenarios. For example, Collaborative-Filtering systems rely on community assessments, Content-Based methods leverage textual descriptions, Knowledge-Based systems focus on user interactions, establishing knowledge bases, and demographic systems use attributes personal data of users to make recommendations. Some Recommendation Systems, such as Knowledge-Based, are more effective in Cold-Start situations, where a significant amount of data is not available. Others, such as Collaborative-Filtering methods, are more effective when a lot of data is available. To give an example, coupling a collaborative system with one based on knowledge, in this case, would allow the former to solve the cold start problem. In conclusion, the various aspects of the different types of systems can be combined to achieve better performance, solving the weaknesses that such systems would have if they were stand-alone. The power of multiple types of models is thus combined, in order to create a better one.

3   Conclusions

Recommendation systems have now become a fundamental and necessary product of any company that offers a platform for the use and/or sharing of content. Where first the information retrieval techniques used through the search for content on the part of the user were the only method of interaction with the user, the systems take over recommendation that, through analysis techniques of the user's previous history, are successful to create a model of his tastes and become the primary intermediary for the use of contents, while direct research becomes a secondary link. Recommender systems have become particularly important for platforms based on multimedia content and, as we have seen, they adapt perfectly to any area, even reaching the museum one, suggesting to customers which works they might like in a museum, in order to constitute a point of reference for anyone wishing to make a visit, helping them in choosing works of art and improving his experience. The fast-paced world we live in leaves us little time to devote to the things we love most. Attending a museum could be an educational leisure activity for which the necessary time is not always available. Moreover, sometimes the museums are so large that it is impossible to view all the collections at once, taking into account the fact that, as the visit progresses, the visitor's interest and attention decrease. It also often happens that you have to face an art exhibition about which very little is known or a collection that is not of particular interest. This could cause us to lose works of art that are more related to our personal preferences. Museum Recommendation Systems can significantly improve the quality of the visit. Taking into account time needs and user preferences, Recommendation Systems can be a winning choice for a museum, helping visitors find their way through the collection of art by providing them with the right information at the right time, increasing their awareness on the themes of art history, recommending them the works they might like and, in essence, providing them with a personalized and pleasant experience in the visit.

References

1. Alves, P., et al.: Modeling tourists' personality in recommender systems: how does personality influence preferences for tourist attractions?, pp. 4–13. Association for Computing Machinery, New York (2020). https://doi.org/10.1145/3340631.3394843
2. Amato, A.: Procedural content generation in the game industry. In: Korn, O., Lee, N. (eds.) Game Dynamics, pp. 15–25. Springer, Cham (2017). https://doi.org/10.1007/978-3-319-53088-8_2
3. Amato, A., Di Martino, B., Scialdone, M., Venticinque, S.: Multi-agent negotiation of decentralized energy production in smart micro-grid. In: Camacho, D., Braubach, L., Venticinque, S., Badica, C. (eds.) Intelligent Distributed Computing VIII. SCI, vol. 570, pp. 155–160. Springer, Cham (2015). https://doi.org/10.1007/978-3-319-10422-5_17
4. Angelis, S., Kotis, K., Spiliotopoulos, D.: Semantic trajectory analytics and recommender systems in cultural spaces. Big Data Cogn. Comput. 5(4), 80 (2021). https://doi.org/10.3390/bdcc5040080
5. Berre, D.L., Marquis, P., Roussel, S.: Planning personalised museum visits. In: Proceedings of the Twenty-Third International Conference on International Conference on Automated Planning and Scheduling, ICAPS 2013, pp. 380–388. AAAI Press (2013)
6. Burke, R.: Knowledge-based recommender systems, vol. 69, pp. 180–200. Marcel Dekker, New York (2000)

7. van Hage, W.R., Stash, N., Wang, Y., Aroyo, L.: Finding your way through the rijksmuseum with an adaptive mobile museum guide. In: Aroyo, L., et al. (eds.) ESWC 2010. LNCS, vol. 6088, pp. 46–59. Springer, Heidelberg (2010). https://doi.org/10.1007/978-3-642-13486-9_4

8. Javed, U., Shaukat Dar, K., Hameed, I., Iqbal, F., Mahboob Alam, T., Luo, S.: A review of content-based and context-based recommendation systems. Int. J. Emerg. Technol. Learn. (iJET) **16**, 274–306 (2021)

9. Keller, I., Viennet, E.: Recommender systems for museums: evaluation on a real dataset. In: Fifth International Conference on Advances in Information Mining and Management, pp. 65–71 (2015)

10. Loboda, O., Nyhan, J., Mahony, S., Romano, D.M., Terras, M.: Content-based recommender systems for heritage: developing a personalised museum tour. In: Proceedings of the 1st International 'Alan Turing' Conference on Decision Support and Recommender Systems (DSRS-Turing 2019), p. 7 (2019)

11. Pazzani, M.J., Billsus, D.: Content-based recommendation systems. In: Brusilovsky, P., Kobsa, A., Nejdl, W. (eds.) The Adaptive Web. LNCS, vol. 4321, pp. 325–341. Springer, Heidelberg (2007). https://doi.org/10.1007/978-3-540-72079-9_10

12. Rich, E.: User modeling via stereotypes. Cogn. Sci. **3**(4), 329–354 (1979). https://www.sciencedirect.com/science/article/pii/S0364021379800129

13. Ruiz, M.T., Mata, F., Zagal, R., Guzmán, G., Quintero, R., Moreno-Ibarra, M.: A recommender system to generate museum itineraries applying augmented reality and social-sensor mining techniques. Virtual Real. **24**(1), 175–189 (2020)

14. Ryding, K., Spence, J., Løvlie, A.S., Benford, S.: Interpersonalizing intimate museum experiences. CoRR abs/2011.11386 (2020)

15. Son, L.H.: Dealing with the new user cold-start problem in recommender systems: a comparative review. Inf. Syst. **58**, 87–104 (2016)

16. Tavcar, A., Antonya, C., Butila, E.: Recommender system for virtual assistant supported museum tours. Informatica **40**(3) (2016). http://www.informatica.si/index.php/informatica/article/view/1433

17. Wang, Y., Stash, N., Aroyo, L., Hollink, L., Schreiber, G.: Semantic relations in content-based recommender systems (2009)

Towards the Enrichment of IIIF Framework with Semantically Annotated and Geo-Located images

Alba Amato[✉] and Giuseppe Cirillo

Department of Political Science, University of Campania Luigi Vanvitelli,
Caserta, Italy
{alba.amato,giuseppe.cirillo}@unicampania.it

Abstract. The fields of application of digital to cultural heritage are vast and constantly changing, as they are based on technology, a factor in constant evolution. An important element to take into consideration is the fact that digitization has a double key to understand: on the one hand, it becomes a production tool, while, on the other, it also covers the role of means of fruition with regard to the heritage taken into consideration. This paper presents the porting of a collection of semantic annotated images belonging to a content management system. To support the IIIF interoperability standard, we use a new technological stack made of open source independent software.

1 Introduction

The digital age can be defined as an historical era marked by the diffusion of digital products and by social, economic and even political changes, which occurred thanks to the coming of the digitization of access to information, resulting in the formation of the current information society. The basic objective of digitization consists in the reorganization of knowledge according to criteria of efficiency and effectiveness, implementing a simplification in the selection of news in a world that is now submerged by information of all kinds. Digital is also widely used in the archival and librarian field, as it is possible to preserve and make available digital documents, both those born as such and those subsequently transformed, all of which is then inventoried and managed electronically, like the various physical institutions. Libraries and digital archives provide the resources, including specialized personnel, to select, organize, intellectually access, insert, distribute, preserve the integrity, and ensure the persistence of digital collections over time, so that they can be readily accessible and affordable for a defined community or for a set of communities. Greater accessibility, especially at a technological level, for users entails, on the one hand, a major improvement in the service, while, on the other, a leveling of discrepancies and cultural barriers. Being able to expand the services also allows for the preparation of new functions, reaching a previously foreign target audience and consequently increasing one's catchment area. The use of digital at this juncture has to do mainly with the field

L. Barolli (Ed.): AINA 2023, LNNS 655, pp. 261–270, 2023.
https://doi.org/10.1007/978-3-031-28694-0_24

of information and communication. Libraries and digital archives manage to de-contextualize information to make it available to a varied audience and not just to the so-called sector insiders, widening the communication of knowledge to all interested parties and not just to the narrow circle of the community of experts originally foreseen: in this way the user no longer plays a passive role, but is able to share different conceptions and points of view, without falling back into the standardization of thought. An important element to take into consideration is the fact that digitization has a double key to understand: on the one hand, it becomes a production tool, while, on the other, it also covers the role of means of fruition with regard to the heritage taken into consideration [1]. The fields of application of digital to cultural heritage are vast and constantly changing, as they are based on technology that is in constant evolution. IIIF features make it perfect for publishing high-quality images on the web. In fact, with IIIF an image becomes a portable digital object and it is possible to see images in other digital libraries with very high definition, with possibility of zoom, image comparison, annotations, photo editing tools (color, contrast, etc.).

One of the most used functions of IIIF is the possibility of displaying, in the same viewer, different images, even from different institutions.

This paper presents the porting of a collection of semantic annotated images belonging to a content management system. To support the IIIF interoperability standard, we use a new technological stack made of open source independent software [2].

In Sect. 2 the literature review is presented. The case study is described in Sect. 4. Finally conclusions are drawn.

2   State of Art

Within museum institutions it is essential to recognize but even more understand what are the power and driving force of digital and as a consequence also of the media in the enhancement of cultural heritage in order to be able to implement innovative policies for protection, dissemination and sharing. Today's museums can no longer only cover the role of conservators of heritage but it is their obligation to pursue and implement constant renewal, using all the resources at their disposal in promotion, communication and in attracting to better affect society, feed the civil and political growth of the public, guide them in learning, help them re-establish their own unmistakable identity. Therefore, this is the juncture in which it is appropriate to use digital in the artistic-cultural field, so as to experiment with a new and alternative way of organizing the relationship between the public and art, suggesting an innovative interpretative reading. Digitization allows visits to museums and archaeological sites from a computer or any other mobile device, with the availability of information in real time.

A benefit that can be defined from an economic point of view can be found in the development of innovative technologies and digital services, given that the digitized material can be used as a bringer of innovation in various sectors, such as tourism and education. Several problems around the topic, referring to different areas of interest, are the following:

- The overall financial cost and public funding available to European cultural institutions for the digitization of their collections;
- The best maximization models for access and use of the material;
- The role and responsibility of public and private organizations for copyrighted works;
- Ensure the sustainability of digitized resources.

Moreover with the aim of ensuring greater accessibility to cultural heritage it is necessary to consider that Cultural heritage is part of present and future knowledge, imagination and creativity and all these factors tend to evolve constantly; Digitization must be seen as a moral obligation, in fact it is essential to bring culture online in an era in which goods are increasingly used in this way.

The International Image Interoperability Framework (IIIF) is a set of open standards for delivering high-quality, attributed digital objects online at scale. It's also an international community developing and implementing the IIIF APIs. IIIF is backed by a consortium of leading cultural institutions.

The library is the institution in which knowledge regardless of its form (traditional or electronic resources) is preserved, cataloged, and made available for users. The Internet has broken down traditional barriers to access brought about by geographic distance, economic circumstances, political boundaries, and cultural sensitivities, and today it is the market place for research, teaching, expression, publication and communication of information. As in the digital society traditional knowledge publications in the form of physical items are often replaced by an infinite knowledge space of dynamic networked data. The library, as the knowledge and memory institution, must now engage in complete flow of knowledge, i.e. in connection of knowledge generation with knowledge reception through collecting, providing access to, and preserving all types of documents published by wide range application of the newest technological solutions [5].

A growing community of the world's leading research libraries and image repositories have embarked on an effort to collaboratively produce an interoperable technology and community framework for image delivery [4].

IIIF is a way to standardize the delivery of images and audio/visual files from servers to different environments on the Web where they can then be viewed and interacted with in many ways.

Modern Web browsers understand how to display formats like .jpg and .mp4 at defined sizes, but cannot do much else. The IIIF specifications align with general Web standards that define how all browsers work to enable richer functionality beyond viewing an image or audio/visual files. For images, that means enabling deep zoom, comparison, structure (i.e., for an object such as a book, structure = page order) and annotation. For audio/visual materials, that means being able to deliver complex structures (such as several reels of film that make up a single movie) along with things like captions, transcriptions/translations, annotations, and more.

IIIF makes these objects work in a consistent way. That enables portability across viewers, the ability to connect and unite materials across institutional boundaries, and more.

The prime unit in IIIF is a Manifest. It can be seen as the package or envelope which contains links to all of the resources that make up a IIIF item. This IIIF Manifest is what is shown in a Viewer and is usually the thing that can be imported into viewers and other tools. A IIIF Manifest usually represents a physical object such as a book, an artwork, a newspaper issue, etc., but it does not have to.

This IIIF Manifest itself is accessible via a URL that points to a document online (in a format called JSON, or JavaScript Object Notation) which a IIIF tool can read and display.

IIIF has some limitations [6] such as the absence of support for semantic search and integration of geo-located images.

3 The Case Study

In this section is described a transformation of a digital archive making it compliant with the IIIF standard in order to guarantee interoperability and make semantic search even easier. The use of this standard therefore implies the transforming the annotations into the format it requires (JSON), keeping the information used by the framework to perform semantic search. We will focus on generating the IIIF manifest for the presentation of georeferenced and semantically annotated images. With this aim, the data and metadata of the repository used have been identified to support content indexing, semantic search and geolocation of the contents themselves.

These data were subsequently exported and entered in the format required by the IIIF standard.

The initial digital archive is a container of numbered folders (1,2,3...), each of which represents a Point of Interest (POI). Each folder has its own internal:

- a subfolder whose name represents the POI, an IRDF file and a file xml
- inside the subfolder there is another folder that contains the original image with its thumbnail and also a series of files, including the image of the POI and the IRDF file with the same name

Let us dwell on the file "nomeImmaginePOI.irdf". An IRDF file contains all the information relating to the annotation added by the annotator. The Fig. 4 shows a simple.irdf file containing a single annotation (Fig. 1).

As can be seen, an annotation of this type is based on the RDF (Resource Description Framework). It is a model for representing metadata. THE RDF documents allow you to establish relationships between various web resources, allowing thus the interoperability between applications that exchange information via the Web understandable. One of its main functions is to make the specification of the semantics of the data introduced by XML and also allowing to define a resource description mechanism independent from the domain of application and without defining the latter a priori. RDF doesn't just describe what you can find on the Web but also anything that can be associated with a URI. The data model can be illustrated by a graph representing the Triple Resource- Property-Value that we can define through Nodes and Arcs where the nodes are the resources described by RDF while the arcs are labeled with the name of the property related to the resource and are directed from node to value.

The first elements specify the namespaces, with which the property names are associated, in so that their semantics are unambiguous. More precisely, the first

```
xmlns:rdf=&quot;
http://www.w3.org/1999/02/22-rdf-syntax-ns#&quot;
```

indicates the namespace relative to RDF, refers to all concepts defined by the RDF syntax, while the others are the namespaces with which all other names are defined, especially those of property, whose definition is independent of RDF. As can be seen from Fig. 4, the

```
namespace xmlns:
j.1=http://www.dcs.shef.ac.uk/~ajay/image/annotation#
```

Fig. 1. Initial digital archive

indicates the annotation property. Through this information it can be structured annotation in json format. In addition, a ".csv" file was generated from the database with the information for the georeferencing. A CSV (Comma-Separated Values) file is a text file with a specific format that allows you to save data in a structured table format. The first line of a csv file defines how the information is divided. Such file looks like this (Fig. 2):

As we can see, each element, preceding a comma, matches to the following information: info; position; name; description. This data will be used for the "navPlace" structure that is contained in a specific section of the manifest as we can see in Fig. 3.

```
"info","position","name","description"
http://parsec2.unina2.it/~pristweb/contents/2/1_pagina_html_1POI/, 41.58850693414777 12.9657375365321
    Introduzione_topografica_e_cenni_storici,Introduzione Topografica e cenni storici
http://parsec2.unina2.it/~pristweb/contents/4/2_pagina_html_2POI/, 41.58856360321379 12.9655544310631
    Storia_degli_studi_e_degli_scavi,Storia degli studi e degli scavi
http://parsec2.unina2.it/~pristweb/contents/6/Html_POI_4/, 41.08591114627852 14.250052259659798,
    Macchinari,I macchinari
http://parsec2.unina2.it/~pristweb/contents/7/Html_POI_1/, 41.085536453320906 14.250108028859502,
    Sotterranei,Descrizione sotterranei
http://parsec2.unina2.it/~pristweb/contents/8/Html_POI_2/, 41.085942897565566 14.250101004416964,
    La_rete_idrica,La rete idrica dei sotterranei
http://parsec2.unina2.it/~pristweb/contents/9/Html_POI_3/,  41.08565320046599 14.250048317154487,
    Le_mensole,Le mensole alle pareti dei sotterranei
http://parsec2.unina2.it/~pristweb/contents/10/Html_POI_5/, 41.08600323097812 14.250071086765832,
    Oratorio,Oratorio di epoca medievale
http://parsec2.unina2.it/~pristweb/contents/12/Html_POI_9/, 41.086023840354684 14.249880216679891,
    Scala_di_accesso_ai_sotterranei,Scale di accesso ai sotterranei
http://parsec2.unina2.it/~pristweb/contents/13/Html_POI_8/, 41.0862543149471 14.250048149977882,
    Galleria_raffigurata_da_Alvino,Galleria raffigurata da Alvino
http://parsec2.unina2.it/~pristweb/contents/14/Html_POI_7/, 41.086187849503176 14.250194385917132,
    Galleria_raffigurata_da_Alvino,Una delle gallerie disegnate da Alvino
http://parsec2.unina2.it/~pristweb/contents/15/Html_POI_6/, 41.08594133862283 14.24997541518259,
    L_asse_minore_disegnato_da_Alvino,Asse minore dei sotterranei nelle vedute storiche
http://parsec2.unina2.it/~pristweb/contents/16/Html_POI_14/, 41.08576730567651 14.250152989357598,
    Copertura_delle_gallerie,Copertura delle gallerie
http://parsec2.unina2.it/~pristweb/contents/17/Htlm_2_POI_25/, 41.08500515719259 14.250111220618697,
    Anfiteatro_Campano_introduzione,Anfiteatro Campano vedute storiche
http://parsec2.unina2.it/~pristweb/contents/18/Htlm_3_POI_29/, 41.0859533047439 14.25010483807861,
    Podio_arena,Vedute storiche di podio e arena
http://parsec2.unina2.it/~pristweb/contents/19/Html_1_POI_1/, 41.084125064406045 14.250204924950244,
    Introduzione_generale,Il ninfeo sul piazzale
http://parsec2.unina2.it/~pristweb/contents/20/Html_1_POI_15/, 41.08510776203704 14.250196770537208,
    Area_di_rispetto_e_recinzione,Area di rispetto e recinzione
http://parsec2.unina2.it/~pristweb/contents/21/Html_1_POI_24/, 41.084601837017466 14.250188220796787,
    Anfiteatro_repubblicano,Anfiteatro repubblicano
http://parsec2.unina2.it/~pristweb/contents/22/Html_2_POI_1/, 41.084125064406045 14.250204924950244,
    Introduzione_generale,Notizie storiche su Capua antica
http://parsec2.unina2.it/~pristweb/contents/23/Html_2_POI_10/, 41.086014273805034 14.251023080284654,
    Ingresso_orientale,Vedute storiche del monumento da est
http://parsec2.unina2.it/~pristweb/contents/24/Html_2_POI_13/, 41.085648225381874 14.251060636310328,
    Vedute_storiche,Vedute storiche del monumento da sud est
http://parsec2.unina2.it/~pristweb/contents/25/Html_2_POI_15/, 41.08515966773119 14.250561641246412,
    Arco_inciso_e_pilastro_con_Eracle,Arco inciso sul pavimento
```

Fig. 2. CVS

4   IIIF Semantic Annotation

In the old system the digital archive is organized with a file containing the image and a file with *IRDF* extension containing semantic annotation in XML format.

The canvas that contains a specific image in the manifest file is searched by the annotator. Moreover, the annotator extracts from the file IRDF the semantic annotation and the information portion of the image that has been annotated. In Fig. 4 the coordinates of the top and bottom corners of the annotated area are shown. Those coordinates are identified by the *topX, topY, botomX and botomY* XML tags. The semantic concepts associated to the defined area is referred by the *annotationText* tag.

The list of annotations for each images are listed as a json object of the canvas including the selected image in the manifest. Figure 5 shows the *body* object of each annotation containing the class or the concept of the annotation. The id of the canvas and four values defining the annotated area (in terms of coordinates of the upper right corner, width and height of the rectangle) are contained in the target properties.

4.1 Geo-Localization in an IIIF Manifest

A GIS database contains geo-located information. The location is expressed as latitude and longitude. The geo-located information are composed of a link to the image, the indication of longitude and latitude of the Point Of Interest and a short description of the image. For each geo-located image, the *Annotator* adds a new json object described in the manifest using the *navPlace IIIF Extension* [3].

```
"navPlace": {
  "type": "FeatureCollection",
  "features": [
    {
      "type": "Feature",
      "properties": {
        "label": {
          "en": [
            "Camera location for picture of the front of the Laocöon sculpture."
          ]
        }
      },
      "geometry": {
        "type": "Point",
        "coordinates": [
          -118.4745559,
          34.0776376
        ]
      }
    }
  ]
}
```

Fig. 3. navPlace

Using a GeoJSON Feature Collection containing one or more Features, the navPlace property identifies single or multiple geographic areas related to a given resource. Figure 6 shows the *navPlace* json object of a canvas containing one *Point* identified by its latitude and longitude.

The manifest obtained at the end of the process can be used by any IIIF image viewer to visualize semantic annotations and geo referenced images.

Figure 7 shows the results visualized using Mirador viewer. The semantic annotations are shown in the left vertical tab and are highlighted on the selected image.

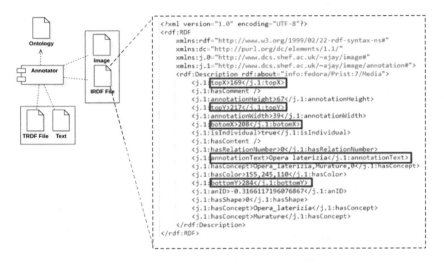

```xml
<?xml version="1.0" encoding="UTF-8"?>
<rdf:RDF
    xmlns:rdf="http://www.w3.org/1999/02/22-rdf-syntax-ns#"
    xmlns:dc="http://purl.org/dc/elements/1.1/"
    xmlns:j.0="http://www.dcs.shef.ac.uk/~ajay/image#"
    xmlns:j.1="http://www.dcs.shef.ac.uk/~ajay/image/annotation#">
    <rdf:Description rdf:about="info:fedora/Prist:7/Media">
        <j.1:topX>169</j.1:topX>
        <j.1:hasComment />
        <j.1:annotationHeight>67</j.1:annotationHeight>
        <j.1:topY>217</j.1:topY>
        <j.1:annotationWidth>39</j.1:annotationWidth>
        <j.1:botomX>208</j.1:botomX>
        <j.1:isIndividual>true</j.1:isIndividual>
        <j.1:hasContent />
        <j.1:hasRelationNumber>0</j.1:hasRelationNumber>
        <j.1:annotationText>Opera laterizia</j.1:annotationText>
        <j.1:hasConcept>Opera_laterizia,Murature,0</j.1:hasConcept>
        <j.1:hasColor>155,245,110</j.1:hasColor>
        <j.1:bottomY>284</j.1:bottomY>
        <j.1:anID>-0.3166117196076867</j.1:anID>
        <j.1:hasShape>0</j.1:hasShape>
        <j.1:hasConcept>Opera_laterizia</j.1:hasConcept>
        <j.1:hasConcept>Murature</j.1:hasConcept>
    </rdf:Description>
</rdf:RDF>
```

Fig. 4. IRDF annotation example.

```json
{
    "id": "http://cleopatra-project.cloud/anfiteatro/index.json/canvas/0",
    "type": "Canvas",
    "items": [-
    ],
    "label": {-
    },
    "width": 449,
    "height": 417,
    "thumbnail": [-
    ],
    "annotations": {
      "id": "http://cleopatra-project.cloud/anfiteatro/index.json/canvas/0/annotationpage/0",
      "type": "AnnotationPage",
      "items": [
          {
            "id": "http://cleopatra-project.cloud/anfiteatro/index.json/canvas/0/annotation/0-tag",
            "type": "Annotation",
            "motivation": "tagging",
            "body": {
              "type": "TextualBody",
              "value": "Opera laterizia",
              "language": "it",
              "format": "text/plain"
            },
            "target": "http://cleopatra-project.cloud/anfiteatro/index.json/canvas/0#xywh=169,217,39,67"
          },
          {
            "id": "http://cleopatra-project.cloud/anfiteatro/index.json/canvas/0/annotation/1-tag",
            "type": "Annotation",
            "motivation": "tagging",
            "body": {
              "type": "TextualBody",
              "value": "Sotterranei",
              "language": "it",
              "format": "text/plain"
            },
            "target": "http://cleopatra-project.cloud/anfiteatro/index.json/canvas/0#xywh=414,1,24,34"
          },
```

Fig. 5. Annotations into an IIIF manifest

```
"navPlace": {
 "type": "FeatureCollection",
 "features": [
   {
     "type": "Feature",
     "properties": {
       "label": {
         "en": [
           "Location of painting"
         ]
       }
     },
     "geometry": {
       "type": "Point",
       "coordinates": [
         41.08607557968985,
         14.25016008109071
       ]
     }
   }
 ]
},
```

Fig. 6. Geo-located information of a canvas into an IIIF manifest.

Fig. 7. Mirador visualizzation

5 Conclusion

Cultural heritage is an intrinsic part of shared values and cultural diversity of a population. So it is necessary to preserve it and in order to preserve it, it is necessary to ensure that it is fully embedded in the Digital Decade. Nowadays less than 20% of the collections at our museums, galleries or libraries are digitised. The situation is even more dramatic for sites and monuments. At the same time, these treasures are increasingly exposed to natural and man-made risks. In this paper the porting of a digital archive is described. The digital archive belongs to an old content management system and contains annotated images and text. Tn order to obtain interoperability with standard tools and standard viewer, the archive has been converted in IIIF.

References

1. Amato, A.: Procedural content generation in the game industry. In: Korn, O., Lee, N. (eds.) Game Dynamics, pp. 15–25. Springer, Cham (2017). https://doi.org/10.1007/978-3-319-53088-8_2
2. Amato, A., Di Martino, B., Scialdone, M., Venticinque, S.: Personalized recommendation of semantically annotated media contents. In: Zavoral, F., Jung, J.J., Badica, C. (eds.) Intelligent Distributed Computing VII, pp. 261–270. Springer, Cham (2014). https://doi.org/10.1007/978-3-319-01571-2_31
3. IIIF Community: navplace extention (2022). https://iiif.io/api/extension/navplace/
4. Manoni, P.: "discoverability" in the IIIF digital ecosystem. JLIS.it **13**(1), 312–320 (2022). https://doi.org/10.4403/jlis.it-12770. https://www.jlis.it/index.php/jlis/article/view/440
5. Piotrowicz, G.: Array(0x55cd383741a0) (2017). https://library.ifla.org/id/eprint/2073/
6. Salarelli, A.: International image interoperability framework (IIIF): a panoramic view. JLIS.it **8**(1), 50–66 (2017).https://doi.org/10.4403/jlis.it-385. https://www.jlis.it/index.php/jlis/article/view/385

Comparison of ML Solutions for HRIR Individualization Design in Binaural Audio

Simone Angelucci[✉], Claudia Rinaldi, Fabio Franchi, and Fabio Graziosi

Department of Information Engineering, Computer Science and Mathematics
(DISIM), University of L'Aquila, L'Aquila, Italy
simone.angelucci@graduate.univaq.it,
{claudia.rinaldi,fabio.franchi,fabio.graziosi}@univaq.it

Abstract. The exploitation of Machine Learning (ML) solutions is presented in this paper with the aim of obtain individualized Head Related Impulse Responses (HRIRs) without measuring them directly on the interested individuals. Different regression models have been explored in order to find the most accurate one in predicting the samples of the HRIRs, given a set of anthropometric features and given a specific virtual source position.

1 Introduction

Since the beginning of the digital era, scientists have wondered how to encode intelligence in computers, [1], and the research developed around this main purpose has been collected under the term Artificial Intelligence (AI). During the 80s new questions arise in this field, pushing the birth of Machine Learning (ML) paradigm, where the problem is now how to teach new tasks to a computer without explicit instructions, but exploiting data and inference, [2]. What happened next are various algorithms whose efficiency is constantly increasing and an incredible widespreading of ML applications in many fields, from computer vision to social media, from wildlife preservation to healthcare, from language translation to bank domain.

This paper mainly deals with exploitation of ML for audio, which is a wide context that has already seen many examples in literature going from speech recognition to audio signal processing, [3–6]. In particular, we focused our attention in ML solutions for improving spatial audio, which gained particular attention in recent years due to its capability in improving the immersivity of the applications which resort on it. For instance, the importance of exploiting spatial audio for Virtual Reality (VR) has been confirmed in various works such as [7].

Spatial audio allows to locate sound sources in a virtual 3D space, which can be listened through headphones or loudspeakers, resorting on the HRIRs, or Head Related Transfer Functions (HRTFs) in the frequency domain. HRIRs capture all the information related to the propagation of sound waveforms from

L. Barolli (Ed.): AINA 2023, LNNS 655, pp. 271–278, 2023.
https://doi.org/10.1007/978-3-031-28694-0_25

a sound source to the entrance of our ear canal. Such filters are typically measured in anechoic rooms on individuals by fixing a real sound source on a specific position and by placing microphones inside its ears. Thus, by filtering a sound source through a pair of HRIRs, one for the right ear and one for the left ear, such sound source will be perceived in the same position for which the filters were measured. However, the localization will be precise only if we use the HRIRs measured for that specific individual. HRIRs are indeed not only direction-dependent functions, since they are mostly related in a complex way to the anthropometric characteristics of the users on which they are measured. In order to achieve properly spatialization, one should measure the HRIRs for every possible individual, which is impractical.

The main purpose of ML for improving binaural audio experience in augmented reality contexts, is to obtain HRIRs (or HRTFs) for a specific user, as close as possible to his individual version. Authors in [8] focused toward cues affecting the perception of the source elevation and exploited CNN to identify the salient ones encoded in the HRTFs. In [9] ML is used to identify ears shape parameters from pictures, which are then used for numeric acoustic simulations identifying individualized HRTF. Other examples of ML for binaural audio can be found for instance in [10–12].

2   Scenario and Problem Definition

The final purpose of this research is to develop a binaural audio guide to visitors inside a museum or in an open space site, in order to improve the quality of their augmented experience while saving energy and computational resources. Such a scenario requires: a wireless communication between augmented device and network providing the service, real-time end user positioning and head tracking for providing the augmented content properly placed in the space and real time communication between end user device and service provider for guaranteeing a non- annoying experience. While the connection problem has been partially discussed in [13], in this paper the main focus is toward the optimisation of the binaural audio experience.

In our work, we decided to exploit some of the numerous ML algorithms available in literature with the main purpose of finding the most accurate one in predicting the samples of HRIR, given a set of anthropometric features belonging to a specific user and given the position of the source to be reproduced with respect to the user. This purpose can be pursued given the availability of many HRTF, HRIR datasets as for instance CPIC, by the University of California, LISTEN from IRCAM, KEMAR from MIT and many more, [14].

The data used in this work have been taken from the Center for Image Processing and Integrated Computing (CIPIC) dataset. This dataset contains the samples of individual HRIRs measured on 45 different subjects with different anthropometric parameters, which are provided too. Actually, only 37 subjects out of 45 possess both the anthropometric features and the samples of the HRIRs, so only the data related to these 37 subjects have been used. The anthropometric parameters contained in the dataset are related to the head, the torso and the pinna of the subjects, since it is known that these are the most influencing parameters affecting the HRIRs. Being more specific, the dataset contains 17 parameters related to the head and the torso, which can be viewed in Fig. 1, and 10 parameters related to the pinna of each ear, showed in Fig. 2. So, the total number of the anthropometric parameters is 37 since for each ear we have 10 pinna measurements. For each subject, the HRIRs have been measured in 1250 different positions and so, since for each position 2 HRIRs are measured because one is for the left ear and one is for the right ear, the dataset contains 2×1250 HRIRs for each subject. The 1250 positions are expressed in the Interaural-Polar Coordinate System, where the human head is considered to be the origin of such coordinate system and sound sources positions are completely determined by the azimuth θ and elevation ϕ angles.

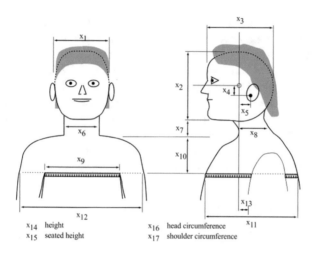

Fig. 1. Head and torso measurements.

Given the availability of such information, the main issue is related to the specificity of each person in terms of its physical structure (head, pinna, torso, ears distance), which strictly influences the perception of the direction of a sound. Various studies have been conducted along the years to understand how much individual characteristics influence the perception, and it is possible to state that an individualized HRIR is usually desired for the best binaural audio experience, [15, 16]. Here comes ML solutions to solve the impractical problem of measuring HRIRs for each individual given the long and expensive procedure generally required, [17]. This topic is not a novelty in literature of the recent years, since many experiments have already been conducted, e.g. [10–12]. Our purpose is to evaluate some of the available regression models for producing individualized HRIRs in terms of objective parameters as: Normalized Root Mean Square Error (NRMSE) and Root Mean Square Error (RMSE) expressed in dB. Since the

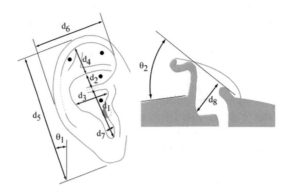

Fig. 2. Pinna measurements.

typical aim of a regression model is to understand the relations between the features and the targets of a training dataset, in order to exploit them on new features for predicting new targets, we choose the anthropometric parameters contained in the CIPIC dataset as features and the HRIRs samples as targets. Then, we trained several regression models on a training dataset, which consisted of a subportion of the CIPIC dataset and we exploited them in order to predict the new HRIRs samples providing to the models new input features, contained in the remaining subportion of the CIPIC dataset, namely the testing dataset. Then, we compared the predicted HRIRs with the ones contained in the testing dataset in order to evaluate the accuracy of the implemented models.

3   Problem Approaches

Two approaches have been considered in order to evaluate the previously described problem:

1. For each sound position, a dataset containing the anthropometric parameters of the 37 subjects as the features and the samples of the left and right ear HRIRs related to that position as the targets was created. Then, several machine learning algorithms were applied on each dataset. In order to speed up the training time of the models, we considered only the data related to 65 positions. In this way, each regression model was trained 65 times in order to consider all the datasets related to each position;
2. A unique dataset was created in which we included also the information related to the sound source position as additive feature. In this way the number of features were 39, since we added the information related to the azimuth and the elevation.

The first approach seemed to be the most natural one since, for each position, the relations between the anthropometric features and the samples of the HRIRs are different. On the other hand, due to the poor availability of the data, since in this case the datasets were composed by only 37 entries, thus the number of the individuals, bad performances were expected. A larger dataset composed by 37×65 entries was instead available for the second approach, thus giving the chance of properly training the models.

In both cases, the performance of the following models was investigated: (i) Linear Regression - Batch Approach, (ii) Kernel Regression with RBF kernel, (iii) Support Vector Machines with RBF kernel, (iv) Regression Tree, (v) Random Forest with 1, 10 and 100 trees, (vi) Deep Neural Network with 5 hidden layers, each of them with 64 neurons and using ReLU as activation function.

3.1 Training and Testing Datasets

Concerning approach (1) we considered 65 training datasets and 65 testing datasets where the training datasets contained the data related to 29 subjects, while the testing ones contained the data related to the other 8 subjects. This can be justified from the fact that if we had considered less subjects in the training dataset we would had worse performances since the training data were not enough, so it seemed to be a good trade off between training and testing data.

Concerning approach (2) we built a training dataset containing 29×65 entries and a testing dataset containing 8×65 entries.

3.2 Error Metrics

Two error metrics have been considered for performance evaluation. The Normalized Root Mean Square Error (NRMSE), defined as:

$$NRMSE = \frac{\sqrt{\frac{1}{N} \sum_{n=1}^{N} (y(n) - \hat{y}(n))^2}}{max(y(n)) - min(y(n))} \tag{1}$$

and the Root Mean Square Error expressed in dB (RMSEdB):

$$RMSEdB = 20 log_{10} \left(\sqrt{\frac{1}{N} \sum_{n=1}^{N} (y(n) - \hat{y}(n))^2} \right) \tag{2}$$

In both cases, N represents the length of the HRIR, so each error metric refers to the error between the "original" HRIR $(y(n))$ and the predicted one $(\hat{y}(n))$.

4 Results

Results are shown in Tables 1 and 2. Approach (1) is named "nopos" since the positions are not included in the dataset while approach (2) is named "pos". Both are expressed in terms of one of the error metrics and they refer to the prediction of the HRIRs for the left or right ear. In each table the performances are related to the training and the testing phase. The "nopos" results were

Table 1. Right ear HRIR prediction NRMSE.

Algorithm	NRMSE tr pos	NRMSE tr nopos	NRMSE tst pos	NRMSE tst nopos
LR Batch	8.6%	3.4%	19.1%	42.7%
Kernel RBF	4.9%	0.7%	17.2%	6.8%
SVM RBF	0.7%	0.8%	8.2%	6.7%
Tree	4.4%	3.2%	9%	9%
Forest1	7.2%	6.5%	8.9%	9%
Forest10	5.7%	4.5%	7.1%	6.8%
Forest100	5.4%	4.2%	6.8%	6.6%
DNN	5.7%	3.7%	8%	8.7%

obtained by averaging the error metrics of each subjects and then averaging for all the tested positions. The "pos" instead were obtained by averaging the error metric of all the subjects.

Obtained results related to the prediction of the HRIRs of the right ear expressed in terms of NRMSE are shown in Table 1.

First, an important trend can be observed in each considered algorithm: the performances in the testing phase are always worse than the ones in the training phase. In particular, this can be observed for the Linear Regression algorithm, in the "nopos" approach. In this case we can see that the error in the training phase is only about 3.4% while in the testing phase it is 42.7%, a very bad result. Indeed, bad results were expected for the Linear Regression model since it is well-known that a complex non-linear relation exists between the anthropometric features and the HRIRs.

The performance related to the SVM and the Kernel regression with the RBF are very similar in the testing phase of the "nopos" approach. We recall the fact that the advantage of the SVM is to avoid the problem of overfitting since it exploits only the important information to make the predictions, the so-called support vectors. However, in the "nopos" approach the datasets were very small, so we can think that basically most of the data were support vectors and so the performances were almost the same. The difference between the two approaches can be instead observed in the "pos" performance difference where SVM always performs better than the kernel regression algorithm.

Another interesting trend can be observed in the results related to the regression trees and the forests. It is possible to see that in both approaches, the performances of the regression trees are always better than the forest ones in the training phase, while are worse in the testing phase. Also here we can see the effect of the overfitting and how it is reduced with the random forest. We recall that the random forests are made of regression trees built on independent sub-datasets taken at random from the original one and that the final prediction is the average of all the predictions made by each single tree that composes the

Table 2. Right ear HRIR prediction RMSE.

Algorithm	RMSE tr pos	RMSE tr nopos	RMSE tst pos	RMSE tst nopos
LR Batch	−23.8	−32.5	−18.6	−10.7
Kernel RBF	−29.3	−45.4	−19.2	−25.8
SVM RBF	−47.3	−44.6	−24.1	−25.9
Tree	−30	−32.7	−23.5	−23.5
Forest1	−25.6	−26.5	−23.6	−23.5
Forest10	−27.5	−29.6	−25.4	−25.8
Forest100	−27.9	−30.1	−25.8	−26.1
DNN	−28.3	−31.5	−24.5	−24

forest. This also explains why the performances of the forest composed by only one tree are in general worse, especially in the training phase, than the single regression tree since the tree of the random forest is grown by considering only a random portion of the original dataset. Moreover, the advantage of increasing the number of the trees in the forest is clearly visible.

Good performances were obtained also with Neural Networks. Left ears performance are obviously similar to those for the right ears and are omitted for space saving.

The performances expressed in terms of the RMSE in dB, are shown in Table 2, which are basically the same results observed with the NRMSE metric but on another scale.

5   Conclusions and Future Works

In this paper we explored various ML algorithms for finding the most accurate one in predicting the samples of the HRIRs, given a set of anthropometric features and given a specific virtual source position. We can conclude that the approach that in general performed better was the "nopos" one, so the one in which we have trained different models for each position, and in particular the SVM with the RBF kernel, the kernel regression with the RBF kernel and the random forest with 100 trees. Future works are directed toward subjective evaluations of predicted HRIRs on real people. This would require collecting individuals anthropometric features, which in some cases are not straightforward to obtain. Moreover we will investigate the variation of the prediction accuracy when the number of features is reduced. Finally, in order to have a well-built model, the algorithms are going to be trained on larger datasets and this calls for the normalization of input parameters.

Acknowledgement. This work was supported by the Italian Government under MiSE "Programma di supporto tecnologie emergenti - Asse I (Casa delle Tecnologie Emergenti) Progetto SICURA" - CUP C19C20000520004.

References

1. Turing, A.M.: I.-computing machinery and intelligence. Mind **LIX**, 433–460 (1950)
2. Hopfield, J.J.: Neural networks and physical systems with emergent collective computational abilities. Proc. Natl. Acad. Sci. **79**(8), 2554–2558 (1982)
3. Deng, L., Li, X.: Machine learning paradigms for speech recognition: an overview. IEEE Trans. Audio Speech Lang. Process. **21**(5), 1060–1089 (2013)
4. Haeb-Umbach, R., et al.: Speech processing for digital home assistants: combining signal processing with deep-learning techniques. IEEE Sig. Process. Mag. **36**(6), 111–124 (2019)
5. Rong, F.: Audio classification method based on machine learning. In: 2016 International Conference on Intelligent Transportation, Big Data & Smart City (ICITBS), pp. 81–84 (2016)
6. Purwins, H., Li, B., Virtanen, T., Schlüter, J., Chang, S.-Y., Sainath, T.: Deep learning for audio signal processing. IEEE J. Sel. Top. Sig. Process. **13**(2), 206–219 (2019)
7. Hoeg, E.R., Gerry, L.J., Thomsen, L., Nilsson, N.C., Serafin, S.: Binaural sound reduces reaction time in a virtual reality search task. In: 2017 IEEE 3rd VR Workshop on Sonic Interactions for Virtual Environments (SIVE), pp. 1–4 (2017)
8. Thuillier, E., Gamper, H., Tashev, I.J.: Spatial audio feature discovery with convolutional neural networks. In: 2018 IEEE International Conference on Acoustics, Speech and Signal Processing (ICASSP), pp. 6797–6801 (2018)
9. Kaneko, S., Suenaga, T., Sekine, S.: DeepEarNet: individualizing spatial audio with photography, ear shape modeling, and neural networks. J. Audio Eng. Soc. (2016)
10. Zhang, M., Wang, J.-H., James, D.L.: Personalized HRTF modeling using DNN-augmented BEM. In: ICASSP 2021 - 2021 IEEE International Conference on Acoustics, Speech and Signal Processing (ICASSP), pp. 451–455 (2021)
11. Lee, G.W., Kim, H.K.: Personalized HRTF modeling based on deep neural network using anthropometric measurements and images of the ear. Appl. Sci. **8**(11), 2180 (2018)
12. Miccini, R., Spagnol, S.: HRTF individualization using deep learning. In: 2020 IEEE Conference on Virtual Reality and 3D User Interfaces Abstracts and Workshops (VRW), pp. 390–395 (2020)
13. Rinaldi, C., Franchi, F., Marotta, A., Graziosi, F., Centofanti, C.: On the exploitation of 5G multi-access edge computing for spatial audio in cultural heritage applications. IEEE Access **9**, 155197–155206 (2021)
14. McMullen, K., Wan, Y.: A machine learning tutorial for spatial auditory display using head-related transfer functions. J. Acoust. Soc. Am. **151**, 1277–1293 (2022)
15. Wenzel, E., Arruda, M., Kistler, D., Wightman, F.: Localization using nonindividualized head-related transfer functions. J. Acoust. Soc. Am. **94**, 111–123 (1993)
16. Xu, S., Li, Z., Salvendy, G.: Individualization of head-related transfer function for three-dimensional virtual auditory display: a review. In: Shumaker, R. (ed.) ICVR 2007. LNCS, vol. 4563, pp. 397–407. Springer, Heidelberg (2007). https://doi.org/10.1007/978-3-540-73335-5_44
17. Li, S., Peissig, J.: Measurement of head-related transfer functions: a review. Appl. Sci. **10**(14), 5014 (2020)

Performance Analysis of a BESU Permissioned Blockchain

Leonardo Mostarda[1], Andrea Pinna[2(✉)], Davide Sestili[1], and Roberto Tonelli[2]

[1] University of Camerino, IT, Camerino, Italy
{leonardo.mostarda,davide.sestili}@unicam.it
[2] University of Cagliari, IT, Cagliari, Italy
{pinna.andrea,roberto.tonelli}@unica.it

Abstract. We present a set of tests on a real permissioned blockchain where nodes are maintained by different independent public organizations in various geographic sites. Such configuration sets up real working conditions where a permissioned blockchain is not ruled and run by a single entity. Used platform is Ethereum-Hyperledger BESU implemented through docker technology. We compared standard "Caliper" tests against more detailed and customizable tests executed by launching different transactions to stimulate the answer of the entire network under typical working conditions. Results show that, unexpectedly, not all blockchain nodes work equivalently: under certain conditions, only some nodes contribute to validate transactions and to include them into blocks, while others only append empty blocks to the ledger. This work has a twofold purpose. First, behind the analysis of the specific permissioned blockchain, the aim is to investigate and detect general issues and pitfalls related to this kind of blockchain. Second, providing and improving a tool that can be customized for performance analysis.

1 Introduction

Permissioned blockchain have possible employments in private or consortium organizations [1–4]. A major difference with respect to public blockchain is that every node is controlled by the consortium giving the possibility to check and apply some level of control also on performances as well. In public blockchains every node can join and network load and performances are an emerging property. Delay in transaction approval are related to fees and can be unpredictable [5]. In a permissioned one bad performing nodes can be removed, their number can be tuned and transaction load can be set and ruled in order to optimize blockchain performance. This can depend not only on the internal configuration, but also from external features, mainly from network latency and nodes reachability. We analyze what are the main features influencing blockchain performances that can be taken care of, namely those variables that can be tuned and ruled by the blockchain consortium itself. Other research works tried to address this problem [6–11] which is a general issue in Blockchain Software Engineering [12, 14].

© The Author(s), under exclusive license to Springer Nature Switzerland AG 2023
L. Barolli (Ed.): AINA 2023, LNNS 655, pp. 279–291, 2023.
https://doi.org/10.1007/978-3-031-28694-0_26

We built a customizable tool and compared the result with *Caliper* [15], obtaining more details on performances. The performed tests can be useful for organizations or consortium aiming at implementing a permissioned blockchain by means of the docker container technology and the Hyperledger BESU blockchain under the IBFT2 consensus protocol. The is a first step forward for identifying the main features that is possible to work on for blockchain performance optimization, for identifying bottlenecks, critical issues and for configuring an overall testing and diagnostic framework for this kind of analysis. The approach is useful for consortium where blockchain nodes are maintained by independent authorities and single policies may differ from one organization to another or from one node to another.

2   The Blockchain Network

The blockchain network is built up as a private permissioned network. Permissioning is implemented by the Consensys [16,17] Dapp mechanism where new nodes can be onboarded if added into the permissioning smart contracts by administrators. Administrator can be onboarded and registered as well and can manage the addition of new nodes or new administrators with their cryptographic credentials.

The technology is Hyperledger BESU with IBFT2.0 consensus protocol where *validator nodes* in a Round Robin turn collect transactions into blocks and propose blocks to other validators. Empty blocks can be validated as well and added to the chain. Since the network is private and permissioned transactions are free and gas price has been set to zero.

Nodes are run by different and independent organizations which may adopt different access policies and firewalls and different machines and software. They are geographically distributed in different sites so that internet interaction can play a true role in the exchange of information among them. Every organization implements a node by means of the same docker-container technology but BESU versions run by the nodes can be different.

Network set-up required four bootnodes as validator nodes to start with the minting process. After setting up the four validators, two more standard nodes where added to the network. New validators can be added or existing validators can be removed as well through a voting process ruled by *curl* queries addressed to existing validator nodes. Smart contracts are written in Solidity 0.8 and only authorized addresses can deploy them while every address can query the network and send standard transactions. Some smart contracts have been deployed for testing purposes in order to analyze nodes performances against specific smart contracts calls.

3   Tools

In this work we aim not only at investigating our specific blockchain set-up, but also at getting hints and detect general issues and pitfalls related to permissioned/private blockchain and to the IBFT protocol more in general. Furthermore we aim at building and improving a tool that can be customized and fine tuned for this kind of analysis and can provide more details with respect to existing tools. To these purposes we used "Caliper" and compared the results with a custom built tool presented in the next sections.

3.1   Hyperledger Caliper

Hyperledger Caliper [15] is a blockchain performance benchmarking tool for Hyperledger Besu and other Ethereum-like blockchain technologies. Caliper allows to set up a workload specifying the nodes to which send requests (via web socket) and information about the transaction load. Transaction loads are completely customizable, allowing the user to define the fields of the transactions that will be sent, as well as the number of transaction in the load and the *send rate*: the number of transaction to submit per second.
The metrics measurable with Caliper are:

- *Read latency*: the time elapsed between a request is submitted and a reply is received;
- *Read througuput*: the total amount of read operations successfully submitted for which a reply has been received per second;
- *Transaction latency*: the amount of time a transaction requires to be part of the network from the moment it was submitted;
- *Transaction througuput*: the rate at which valid transactions are committed to the blockchain.

3.1.1   Custom Java Benchmark Tool

The custom java benchmark tool we developed allows us to send custom ethereum transactions to a node at a steady configurable rate and retrieve various metrics. The application allows us to configure various parameters to be used in a benchmark run, including:

- *Transaction send delay*: Number of milliseconds the application waits in order to send a new transaction after sending the previous one;
- *Transaction number*: number of transactions to be sent in a given run;
- *Timeout*: the number of milliseconds after which a transaction is considered to be not included in the blockchain if its transaction receipt hasn't been received;

- *Communication protocol*: The protocol employed for communicating with the node (either HTTP or Web socket);
- *Node address*: The node IP address and port number used to communicate with the blockchain node;
- *Sender key*: Ethereum account private key used for signing transactions.

The transactions that can be sent by the application are either simple value transfer transactions or smart contract method calls, more specifically, the smart contract method call that the application natively supports is the *transfer* method of the ERC20 smart contract. In case the application is run in the *smart contract method call mode*, the application will first commit a new ERC20 smart contract, wait for its inclusion in a block, and then will start sending the method call transactions.

After a benchmark run, the application outputs the *average transaction delay* –which is the average number of milliseconds elapsed from when a transaction has been sent to when it has been included in a valid block– and the number of blocks that have been generated during the run. Furthermore, for every transaction committed successfully, the application outputs information regarding the transaction and the block that contains the transaction. This information contains: block number; the timestamp when the transaction was sent to the node; the timestamp of the block; the ethereum address of the validator that proposed the block; the *transaction delay*.

4   Tests

The Smart Contract code we used for Caliper testing is a standard one and is reported in Listing 1.

Listing 1. Hyperledger Caliper benchmark smart contract deployed in the blockchain

```solidity
pragma solidity >=0.4.22 <0.6.0;

contract simple {
    mapping(string => int) private accounts;

    function open(string memory acc_id, int amount) public {
        accounts[acc_id] = amount;
    }

    function query(string memory acc_id) public view returns (int amount) {
        amount = accounts[acc_id];
    }

    function transfer(string memory acc_from, string memory acc_to, int amount)
        public {
            accounts[acc_from] -= amount;
            accounts[acc_to] += amount;
        }
}
```

The code we used for Customized testing is a standard ERC20 Token and can be found at "OpenZeppelin" [18]. Caliper allows to call the three functions depicted in the SC. One call can create accounts (Token), one is only a query, the third one is the most interesting and sends a transaction. Since simple queries do not consume gas, do not change blockchain state and are not included into blocks and since account (Token) creation is by far simpler and less interesting (there is not a reverting process), we choose to test performances using transactions to transfer Tokens. It must be noted that our blockchain set-up sets gas cost to zero and the used addresses does not have Ethers (the balance is zero). This means that transactions sending Ethers are reverted but nevertheless they consume gas, so that must be included in blocks. With this in mind we tested our tool against calls to the "Transfer" method of the ERC20 SC aimed at transferring a Token from sender's address to another address. Each transaction has been inserted into blocks but they triggered the revert procedure of the method since the address called the transfer method having no Tokens. As a consequence Smart Contract's balances cannot be updated but gas is consumed anyway.

Transactions were sent with different transaction request rates per second and with a different total number of requests spanning different configurations in order to test the blockchain answer under different workloads. The details of each configuration are reported in following section.

5   Results

Blokchain network performance metrics were first measured using the Hyperledger Caliper tool and then via the Custom Java benchmark tool. In particular, we will discuss the use of the custom Java benchmark tool as a diagnostic tool which provides more details useful to determine the causes of anomalies in network metrics.

5.1   Caliper Results

Hyperledger Caliper shows measured metric values for three different blockchain interactions: *open*, *query*, and *transfer*. The *open* and *transfer* tests are performed with gas-consuming transactions to the smart contract. The *query* test is a call to the view function of the smart contract. Transactions were sent for a programmed time of 100 s at the different send rates of 0.2, 1, 2, and 10 transactions per second (TPS). An additional test at the rate of 100 transactions per second was set up but was not carried out by the tool. Caliper metric results are summarized in Tab. 1 where we report the results of the "transfer" test.

The results show how the average latency time of the transactions tends to increase with the increase of the send rate. This value should not depend on the number of transactions and should be close to half the block time (which in our case is 6 s). However, it seems that the throughput is close to optimal. Caliper therefore allows us to detect an anomaly in latency metrics but does not provide further information to understand finer details.

Table 1. Results of the execution of Hyperledger Caliper tests at four different transaction send rates. Only the results relating to the *transfer* tests are reported.

Tx Send Rate (TPS)	0.2	1	2	10
Maximum latency (s)	5.20	6.66s	12.39	13.42
Minimum latency (s)	0.23	0.64	0.69	0.97
Average latency (s)	2.70	3.58	6.41	7.23
Throughput (TPS)	0.2	0.9	1.9	9.5

5.2   Custom Tool Results

We configured the Custom tool to send transactions using WebSocket as a protocol, to use the closest node (in our case, the local host) as destination, and to use a single sender key. We set proper values for the parameters in order to have

$$transaction\ number/transaction\ send\ delay = 100$$

in other words, transactions were sent for a programmed time of 100 s. Different values for *transaction send delay* were set in order to send transactions at the different send rates of 0.2, 1, 2, 10, and 100 transactions per second (TPS). We performed the test twice (i.e., test 1 and test 2 in the following) to verify the repeatability of the observations. After running the tests, our custom tool produces the results in the form of serialized data on a file that can be easily processed. All of the data shown below are the results of processing the custom tool's data without the need for any additional queries to the blockchain.

Tables 2 and 3 report the block metrics of the two tests performed with the same configuration of the tool. In these tables we can read for each send rate: the number of blocks created during the test; the total block time elapsed between the first and last block created; and the average block time.

Thanks to the data produced by the Custom Java Benchmark tool, we can understand how the blockchain network acquires transactions and how and when these are inserted into new blocks. For both tests, the average block time is stable for any TPS (6 s).

Table 2. The blocks metrics of the test 1 are reported in each column of the table for a different value of the transaction sending rate.

Tx send rate (TPS)	0.2	1	2	10	100
Number of Tx	20	100	200	1000	10000
N. of blocks	16	17	19	19	24
Total block time (s)	90	96	108	108	139
Average block time (s)	6.0	6.0	6.0	6.0	6.043

Table 3. The blocks metrics of the test 2 are reported in each column of the table for a different value of the transaction sending rate.

Tx send rate (TPS)	0.2	1	2	10	100
Number of Tx	20	100	200	1000	10000
N. of blocks	17	18	18	19	25
Total block time (s)	96	102	102	108	144
Average block time (s)	6.0	6.0	6.0	6.0	6.0

The tool allows the evaluation of latency metrics and to compare the results with the measurements made by Caliper. The transaction latency (i.e., the time interval between the instant of sending the transaction and the instant of confirmation) allows us to verify if the transactions are correctly inserted in the blocks. The latency is expected to be a local maximum if the transaction is requested immediately after creating a block. Conversely, minimal latency is expected if the transaction is sent close to the creation of a new block. Furthermore, an average latency of close to 3 s is expected for each test (or half of the block time established during the blockchain's genesis, which in our case is 6 s). As for the number of transactions per block, each block is expected to contain the same number of transactions, except for the first block and the last block. Any noticeable difference between expected and measured data is worth further investigation and can be traced back to network node operational anomalies.

The Tables 4 and 5 show the latency measures of test 1 and 2 respectively. The results confirm the values of the Hyperledger Caliper metrics obtained for the transaction send rates up to 10 TPS.

In both tables, for rates up to 10 TPS, the minimum delay remains below one second. At a rate of 100 TPS the minimum delay increases considerably up to 4.7 and 2.8 s for test 1 and test 2 respectively. For the rates of 2, 10, and 100 TPS, a significantly higher than expected average latency is observed, and, for both test1 and test2, the average latency increases with higher send rates.

Table 4. Test 1. The values of the transaction confirmation delays are reported in each column of the table for a different value of the sending transaction frequency parameter.

Tx send rate (TPS)	0.2	1	2	10	100
Maximum latency (s)	5.99	6.289	12.235	13.001	22.112
Minimum latency (s)	0.897	0.190	0.402	0.337	4.720
Average latency (s)	3.405	3.206	5.246	6.759	12.638
Throughput (TPS)	0.208	0.997	1.779	8.927	69.136

Therefore, an anomaly is found for the transaction send rates of 2 transaction per second and above. The data produced by our Custom benchmark tool allows the examination, transaction by transaction, of the latency, the block number, and the block's miner. This allows for delving into the reasons for the increasing in the latency. To investigate the causes of this anomaly, we consider the smallest sending rate that causes anomalies and the highest send rate, i.e. 2 TPS and 100 TPS.

Table 5. Test 2. The values of the transaction confirmation delays are reported in each column of the table for a different value of the sending transaction frequency parameter.

Tx send rate (TPS)	0.2	1	2	10	100
Maximum latency (s)	6.917	6.664	12.196	12.774	18.528
Minimum latency (s)	0.833	0.567	0.289	0.377	2.835
Average latency (s)	3.781	3.596	5.092	6.580	10.712
Throughput (TPS)	0.199	0.960	1.855	8.879	66.454

5.2.1   Investigation at 2 TPS

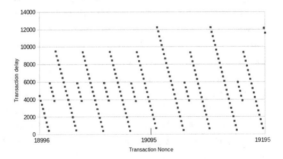

Fig. 1. Transaction delay (latency) measured in the test 1 at a transaction send rate of 2 TPS.

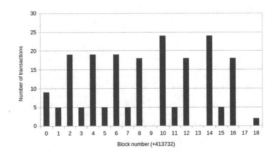

Fig. 2. Number of transactions per block in the test 1 at a transaction send rate of 2 TPS.

The first investigation is based on the data from both tests, at a transaction request rate of 2 TPS. We will show below the graphs relating to the transaction latency and to the number of transactions per block. Latencies (i.e. the confirmation delays) of transaction sent in test 1 at 2 TPS are shown in the graph of Fig. 1. In this figure, each point in the graph represents a transaction and its confirmation latency. Transactions are sorted by nonce. The y-axis represents the latency time in milliseconds. An expected pattern would show the points forming a transversal alignment, and each alignment would be formed by the same number of points (except the first and last). For each alignment, the upper-left point corresponds with the transactions with higher latency (sent just after the creation of a block, so that they need to wait the next block to be included), and the bottom-right point corresponds with the transactions with lower latency (sent close to the next block creation).

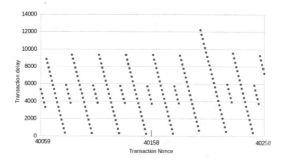

Fig. 3. Transaction delay (latency) measured in the test 2 at a transaction send rate of 2 TPS.

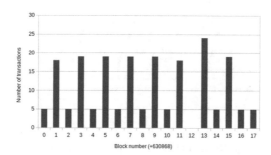

Fig. 4. Number of transactions per block measured in the test 2 at a transaction send rate of 2 TPS.

Table 6. Validators' addresses starting from the first block created during the test 1 at 2 TPS.

Block number	0	1	2	3	4	5	6	7
Validator	$0 \times 5a$	$0 \times 7c$	0xb3	$0 \times 2d$	$0 \times 5a$	$0 \times 7c$	0xb3	$0 \times 2d$

Table 7. Validators' addresses starting from the first block created during the test 2 at 2 TPS.

Block number	0	1	2	3	4	5	6	7
Validator	$0 \times 7c$	0xb3	$0 \times 2d$	$0 \times 5a$	$0 \times 7c$	0xb3	$0 \times 2d$	$0 \times 5a$

Alignments of different lengths are shown in Fig. 1 for test 1 at 2 TPS. Longer alignments are interspersed with shorter alignments, and two alignments appear missing. This is due to an uneven distribution of transactions in the blocks. Proof of this can be found in the graph in Fig. 2 where each bar represents the number of transactions per block. Blocks are ordered by their number (or height), starting at zero for the first block created during the execution of the test. The y-axis represents the number of transactions in each block. We can see that odd blocks contain only 5 or zero transactions sent during the experiment. In particular, two blocks (9 and 13) are empty.

Similar results are showed in Fig 3 and Fig. 4 for the test 2 at 2 TPS, at the same configuration. In this second test, even blocks contains only five or zero transactions sent during the test 2 at 2 TPS.

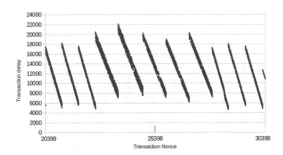

Fig. 5. Transaction delay (latency) measured after the test 1 at a transaction send rate of 100 TPS.

The test results confirmed that the validators are selected for the creation of the next block via a simple rotation. Table 6 provides the order in which the validators create new blocks, as determined in test 1 at 2 TPS, and Table 7 as determined in test 2. For the sake of simplicity, only the first byte of each validator's address is reported. Given the regularity of the validators' sequence,

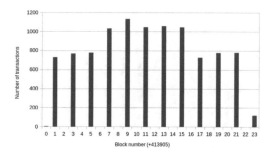

Fig. 6. Number of transactions per block measured after the test 1 at a transaction send rate of 100 TPS.

the nodes that mine blocks with few or zero transactions are always the same two: $0 \times 7c$ and $0 \times 2d$.

5.2.2 Investigation at 100 TPS

Similar examinations were made on the results of test 1 and test 2 at the rate of 100 transactions per second. A more peculiar behavior can be observed at this rate. As regards the latency, in Fig. 5 and Fig. 7 it can be seen that the distribution of the points in the graph does not have the same slope in each group. Also, narrow groupings are no longer present.

At this transaction send rate, there is a significant increase in the minimum, maximum, and average latency, with respect to the results of lower send rate. The reason of the increasing of the latency can be seen in the graph in Fig. 6 for the test 1 at 100 TPS, where every odd block with hundreds of transactions is followed by an even block with zero transactions. Similar results are reported in the Fig. 8 for the test 2 but this time the blocks without transactions are the even blocks. This means that transactions could wait two blocks or more until they are included in a block. At this rate, the same circular order of the validators already determined in the 2 TPS rate is confirmed. We can assert that the validator "$0 \times 7c$" and the validator "$0 \times 2d$" are working differently than

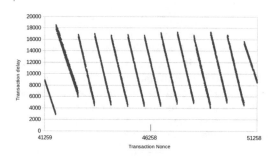

Fig. 7. Transaction delay (latency) measured after the test 2 at a transaction send rate of 100 TPS.

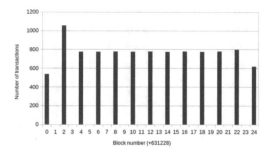

Fig. 8. Number of transactions per block measured after the test 2 at a transaction send rate of 100 TPS.

expected. These validators systematically enter a limited number of transactions per block, and do not confirm transactions in blocks at rates of 10 and 100 TPS.

6 Conclusions

In this work we presented a tool for the analysis of blockchain performances under various and customizable loading conditions. Tool's results has been briefly compared with those of "Hyperledger Caliper", used as a benckmark tool for analyzing blockchain performances. While further analysis is quite limited with Caliper, our tool is able to provide more details on the role of each blockchain node, of every block and on every transaction that can be traced by the sending time and the nonce. This allowed us to detect validator behaviours that resulted in anomalies when adding, or missing to add, transactions to the due blocks. The tool can be further improved and is virtually able to test any kind of possible behaviour or misbehaviour of validator nodes.

References

1. Baliga, A., Subhod, I., Kamat, P., Chatterjee, S.: Performance evaluation of the quorum blockchain platform (2018). arXiv preprint arXiv:1809.03421
2. Shapiro, G., Natoli, C., Gramoli, V.: The performance of Byzantine fault tolerant blockchains. In: 2020 IEEE 19th International Symposium on Network Computing and Applications (NCA), pp. 1–8. IEEE, November 2020
3. Helliar, C.V., Crawford, L., Rocca, L., Teodori, C., Veneziani, M.: Permissionless and permissioned blockchain diffusion. Int. J. Inf. Manage. **54**, 102136 (2020)
4. Vukolić, M.: Rethinking permissioned blockchains. In: Proceedings of the ACM Workshop on Blockchain, Cryptocurrencies and Contracts, pp. 3–7, April 2017
5. Pierro, G.A., et al.: A user-oriented model for Oracles' Gas price prediction. Future Gener. Comput. Syst. **128**, 142–157 (2022)
6. Dabbagh, M., Choo, K.K.R., Beheshti, A., Tahir, M., Safa, N.S.: A survey of empirical performance evaluation of permissioned blockchain platforms: challenges and opportunities. Comput. Secur. **100**, 102078 (2021)

7. Leal, F., Chis, A.E., González-Vélez, H.: Performance evaluation of private Ethereum networks. SN Comput. Sci. **1**(5), 1–17 (2020)
8. Choi, W., Hong, J.W.-K.: Performance evaluation of Ethereum private and Testnet networks using hyperledger caliper. In: 2021 22nd Asia-Pacific Network Operations and Management Symposium (APNOMS). IEEE (2021)
9. Fan, C., Lin, C., Khazaei, H., Musilek, P.: Performance analysis of hyperledger besu in private blockchain. In: 2022 IEEE International Conference on Decentralized Applications and Infrastructures (DAPPS), pp. 64–73. IEEE, August 2022
10. Abdella, J., Tari, Z., Anwar, A., Mahmood, A., Han, F.: An architecture and performance evaluation of blockchain-based peer-to-peer energy trading. IEEE Trans. Smart Grid **12**(4), 3364–3378 (2021)
11. Mazzoni, M., Corradi, A., Di Nicola, V.: Performance evaluation of permissioned blockchains for financial applications: the ConsenSys quorum case study. Blockchain Res. Appl. **3**(1), 100026 (2022)
12. Porru, S., Pinna, A., Marchesi, M., Tonelli, R.: Blockchain-oriented software engineering: challenges and new directions. In 2017 IEEE/ACM 39th International Conference on Software Engineering Companion (ICSE-C), pp. 169–171. IEEE, May 2017
13. Bistarelli, S., Mazzante, G., Micheletti, M., Mostarda, L., Sestili, D., Tiezzi, F.: Ethereum smart contracts: analysis and statistics of their source code and opcodes. Internet Things. **11**, 100198 (2020)
14. Marchesi, L., Marchesi, M., Tonelli, R.: ABCDE-Agile block chain DApp engineering. Blockchain Res. Appl. **1**(1–2), 100002 (2020)
15. https://hyperledger.github.io/caliper/
16. https://consensys.net/
17. https://github.com/ConsenSys/permissioning-smart-contracts
18. https://github.com/OpenZeppelin/openzeppelin-contracts/blob/master/contracts/token/ERC20/ERC20.sol

AI-Powered Drone to Address Smart City Security Issues

Ramiz Salama[1(✉)], Fadi Al-Turjman[2,3], and Rosario Culmone[4]

[1] Department of Computer Engineering, Research Center for AI and IoT, AI and Robotics Institute, Near East University Nicosia, Mersin 10, Nicosia, Turkey
ramiz.salama@neu.edu.tr

[2] Artificial Intelligence Engineering Department, AI and Robotics Institute, Near East University, Mersin 10, Nicosia, Turkey
Fadi.alturjman@neu.edu.tr

[3] Research Center for AI and IoT, Faculty of Engineering, University of Kyrenia, Mersin 10, Kyrenia, Turkey
Fadi.alturjman@kyrenia.edu.tr

[4] Computer Science Division, School of Science and Technology, University of Camerino, 62032 Camerino, Italy

Abstract. The idea of a dazzling metropolis has drawn interest from all across the world. New innovations like blockchain, IoT, artificial intelligence, robots, and many other things were added to it. Security is one of the top issues for people living in big cities, and everyone wants to feel completely secure when traveling around every single day. In this research, we will look into how CEOs in affluent cities use and value robots, especially in terms of security. To understand the robot security the board stream, some approaches and intricacies are used. Following that is a discussion of issues with urban security and the application of artificial intelligence to drones as a management tool. The use of cutting-edge technology, such as blockchain, to support smart urban community management is covered in the last part. The smart city idea and all of its benefits for local inspection are supported by robotic use. The idea of thriving urban areas is spreading around the world and is crucial in the context of developing economies. We examined how to leverage cutting-edge technologies in this project to make it feasible. Artificial intelligence, robotics breakthroughs, and blockchain technologies all have significant effects. This research highlights their significance to analysts and how they consider them when assessing prospects. This project will be very beneficial for analysts and experts in a relevant subject.

Keywords: Drones · Blockchain · Environment exploration · Security concerns · Artificial Intelligence

1 Introduction

One of today's quickly developing and highly embraced technologies is artificial intelligence. AI uses simulations of human brain activity to address practical issues. Robots,

L. Barolli (Ed.): AINA 2023, LNNS 655, pp. 292–300, 2023.
https://doi.org/10.1007/978-3-031-28694-0_27

smart cars, prediction, e-commerce, navigation, human resources, healthcare, agriculture, gaming, automobiles, social media, and marketing are just a few of the application areas for AI. One of the main issues for every person, business, or community is security, which is being handled with the use of artificial intelligence. Artificial intelligence (AI) security refers to technologies and methods that use AI to automatically detect and/or respond to possible cyberthreats based on similar or prior activities. When it comes to AI security, there are many different aspects. To detect malware, execute pattern recognition, and identify even the smallest behavior of malware even before it enters the system, for instance, advanced AI algorithms are frequently used. AI and machine learning can assist in keeping up with hackers, automating threat detection, and responding more efficiently than traditional software-driven or manual procedures in today's world of quickly developing cyberattacks and rapidly multiplying gadgets. AI is used to detect threats. Artificial intelligence approaches can greatly aid in preventing cyberthreats and malicious behaviors in the cyberworld. AI can assist in extracting patterns from textual material in conjunction with NLP to discover any potential threats. Another intriguing topic is the fight against bots, which are dangerous programs that can create phony accounts using stolen credentials [1, 2]. In addition to the previously mentioned AI application fields.

2 Background

A smart city is made up of a number of integrated parts that work together to allow it to run and provide automated services to its residents. The ICT network is made available for the transfer of data produced by smart city components to the next hop devices of the communication network topology. Large-scale data collecting is referred to as data volume. Analytics of acquired data depend on the urgency, processing power, and physical closeness of edge devices to the main cloud platform. They also depend on the level of urgency. For instance, IoT data collected from curbside sensors must be transmitted to the next hop edge device, where it can then be processed further in the cloud. Large volumes of data are continuously transmitted from devices with limited resources, such as drones, to a central cloud for processing and subsequent decision-making [3, 4]. In a platform for smart cities, artificial intelligence (AI) is a key component of accurate decision-making based on data collection.

3 Use Cases

For taking photos and shooting movies at great altitudes, drones or autonomous flying machines were frequently used. However, these robots have the potential to improve business processes when they are led by artificial intelligence (AI), computer vision, machine learning (ML), and other cutting-edge innovations. They are capable of continuously gathering, storing, and cycling information. This convergence of developments enables organizations to enhance work processes and execution while concentrating on yield. A CAGR of 28.8% is predicted for the global business drone market, which will increase from $8.15 billion in 2022 to $47.38 billion in 2029. Artificial intelligence drones can use PC vision for a variety of purposes, including observing and inspecting. Drones can be used to capture continuous images or movies. Robots might be programmed to use ML algorithms to assess data at the edge and deliver diagnostics. Associations can utilize these insights to help them make smart business decisions. In addition to autonomous driving, robots may also be employed for other purposes, such as swarming CEOs, monitoring and regulating disasters, and more. These drones operated by artificial intelligence can also be used by police personnel to monitor the crowd at any event. They might even send out robots to search for offenders in crowded public spaces like markets. Furthermore, during common calamities, these cutting-edge innovation-based robots may access locations that humans are impossible to get and can gather the vital information for the professionals to take crucial action [5, 6]. Labeled picture data is carefully managed to identify unidentified information in order to prepare AI computations.

4 Materials and Method

Drones are ethereal, robotic machines used for a multitude of purposes. When these gadgets were first created, their physical characteristics were well restricted. In any case, drones today frequently incorporate artificial intelligence, automating some or all of their duties. The merging of simulated intelligence allows the drone industry to collect and utilize natural and visual data from sensors attached to the robot. This information makes action simpler, makes flying autonomous or aided possible, and makes it more accessible. Drones now form a part of the smart portable products that are now offered for sale to consumers and enterprises. Computer vision is crucial to the artificial intelligence of robots. Thanks to this development, robots can now detect objects while flying, allowing for the inspection and collection of data on the ground. PC vision, which is applied picture handling carried out with a brain organization, is in charge of managing elite execution. Artificial intelligence uses a brain network, which is a multilayer design, to perform calculations. Robots powered by neural networks are able to identify, describe, and track objects. Robots can detect and track objects while avoiding collisions because to continuous data joining. In order to imitate brain networks in rambles, scientists must first set up the AI algorithms to accurately classify and organize objects in a number of scenarios. This is done by including closely examined photos into the algorithm. The characteristics that various item classes share as well as how to differentiate one type of article from another are communicated by these images to the brain's neural network [7.8]. More advanced brain networks learn continuously while engaged in activities, advancing the growth of identification and research.

Drones have Three Main Applications

Observation
A number of reconnaissance tools that can take HD video and still photos continually may be attached to robots. Robots might be able to record phone calls, select GPS locations, and collect tag data thanks to technology. Numerous research frameworks, such as lidar scanners, multi- and hyperspectral devices, and many others, can be used

constantly at low cost and with little staff because of the high payload similarity. Drone observation is the practice of using unmanned aerial vehicles (UAV) to take still pictures and videos of certain targets, such as people, groups, or situations, in order to gather information [9, 11]. Drone surveillance makes it possible to stealthily gather data on a target as observed from a significant distance or elevation.

Conjecture about the weather
In reality, drones are not as effective as satellite symbolism at forecasting bad weather conditions. When things go awry, they are ready to offer substantial support. Government experts and guarantors are becoming more conscious of how they could be used to evaluate post-debacle damages, particularly in areas that are not clearly indicated as being accessible by people. Robotic data collection represents a major improvement over current information gathering techniques and has the potential to greatly increase the accuracy of weather prediction models. What does it matter in the end? Both the common perspective and the higher perspective are impacted by more accurate models. As a result, meteorologists can provide us more accurate forecasts for the coming days as well as more detailed warnings for storms like tornadoes and the location and timing of typhoon landfalls [12, 14]. The limit layer, which is the lowest part of the atmosphere, is where most of our meteorological conditions take place.

Conveyance
A multitude of uses exist for drones. Robots can transport therapeutic products like blood components, antibodies, and other supplies as well as medical supplies like medications and clinical tests. During the Coronavirus crisis, drones started offering clinical delivery of personalized protective gear and Coronavirus testing. By the end of October 2020, Zipline has flown more than 70,000 drone-delivered medical packages [15, 16]. Robots are used in the clinical component of drone delivery to transport therapeutic supplies such as organ transplants, immunizations, pharmaceuticals, counter-agent toxin combinations, and blood donations.

5   Findings and Discussions

The ability of robots and the IoPST to create a compelling and effective correspondence network for public security is their most important feature. Drones provide a number of benefits, like quick setup, LoS, reprogram capability, and others. The organization's implementation of the IoPST and robot cooperation before, during, and after a disaster should therefore greatly improve. OPNET 14.5 is used to create the recreation network. Three robots and three SAR responders are present in the scenario. It is suggested that a flexible impromptu organization (MANET) connect the robots and responders to efficiently carry out tasks. The AODV directing convention is also taken into account for unicast and multicast steering services. The reproduction time is 500 s as well. As seen in Fig. 2, one robot is moving in closer to the events. In accordance with the actionable information provided by drone-2, this robot will take a photo of the event and offer it to other robots and SAR on the ground for doing relevant tasks. Robot 3 and Drone-1 are used to transport the provisions that the SAR or people in a certain area are expecting. SAR-1, SAR-2, and SAR-3 are carrying out their duties via controlling robots, and their wearable technology helps drones determine where they are needed. For evaluating the presentation of the robots and SARs specifically instances, QoS boundaries like postponement and throughput are taken into consideration. To verify collaborative execution during a crisis, the throughput and delay are evaluated for every device linked to an organization. Additionally, a Matlab function determines and recreates the misfortune probability.

With the increase in bundle size across all hubs, the throughput increased. As shown in Fig. 3, the difference between each number of hubs also increases throughout the increasing parcel size over time in this manner. When a remote hub increases, throughput decreases. Throughput is defined as the total amount of information that is effectively received and transmitted to the robot in bits per second. The highest throughput has increased from 2814.4 pieces per second due to robot 1 to 7908.8 pieces per second due to SAR-1. Similar to how the typical throughput changes from a limit of 4016.3 due to SAR-1 to 1827.2 due to robot (Fig. 1).

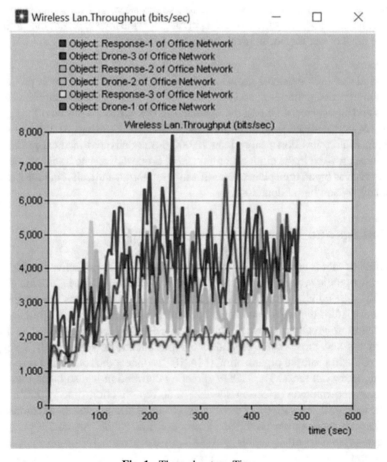

Fig. 1. Throughput vs. Time.

Table 1. Throughput for different nodes.

	Time to taken to peak (sec)
Throughput blue	250
Throughput cyan	100
Throughput lime	470
Throughput pink	160

Each hue represents either a drone or a receiver transferring data to another drone or receiver; the shorter the peak time, the faster the transmission of data from the drones to the receiver. As shown in Fig. 2, the time delay will increase as the number of remote hubs increases and will decrease as bundle size increases across all possible hub counts. SAR-2 is the least reduced in terms of postponement, whereas drone-2 has the most notable postponement, at 0.8 ms. Drone-1, in any event, has no postponement. Similarly, it is the same due to the most extreme and typical postponement for each hub, as shown in Table 1.

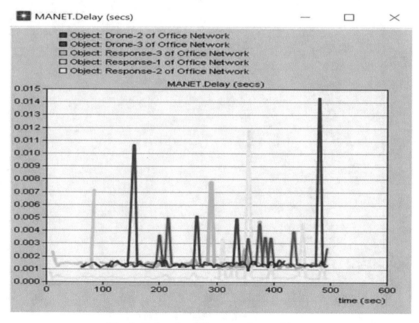

Fig. 2. Delay for different nodes vs. Time.

Table 2 shows that the faster the data transfer from the drones to the office network, the less time it takes to peak. Each color represents a drone or an office network relaying data to another drone or office network. Figure 3 shows the LoS likelihood vs rise point for metropolitan, rural, and thick metropolitan regions. The S bend, which is an

Table 2. Delay for different node.

	Time taken to peak(sec)
Node blue	490
Node cyan	85
Node lime	390
Node red	220
Node yellow	350

instantaneous estimate of ITU-R P.1410–2, is shown in Fig. 4. Whatever the case, the forecast is more accurate. In most cases, the height point expands as the LoS likelihood increases. At 30° heights in the metro area, LoS likelihood reached its maximum. However, the LoS chance reaches its maximum around 60° and 70° due to rural and dense metropolitan zones, respectively.

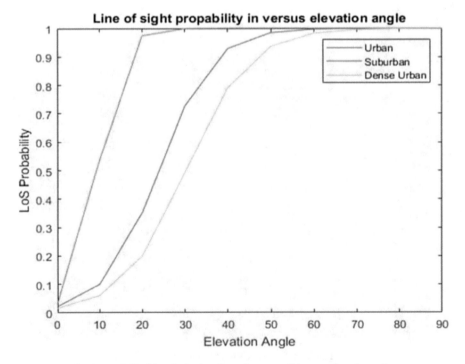

Fig. 3. LoS likelihood of robot in various conditions versus rise point.

Figure 4 illustrates the LoS likelihood for various scenarios using a different height point from the smart robot. The climate of a metropolitan, rural, or densely populated urban area can influence the LoS likelihood. However, in metropolitan areas, the rise point should not be more prominent than 70° and the height point should be small to maintain the LoS likelihood. In either case, the association will be lost. Although the LoS likely has similarities to the LoS that was first used, this forecast was more accurate because the LoS likelihood at each location and height point was evaluated.

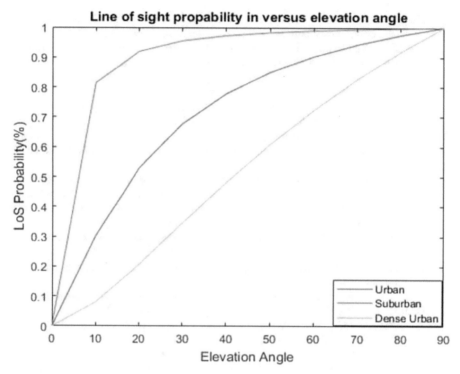

Fig. 4. Depicts the LoS likelihood for various scenarios using a different height point from the intelligent robot.

6   Conclusion

Given the backdrop of developing economies, the idea of "smart urban zones" is becoming more and more popular worldwide. We examined how to leverage cutting-edge technologies to make it possible in this work. Artificial intelligence, blockchain, and robotics/drone advancements are significant. This study illustrates the importance of these elements and how analysts consider them when assessing prospects. This work will be very helpful for analysts and specialists working in a connected field.

References

1. AlTurjman, F., Salama, R.: An overview about the cyberattacks in grid and like systems. In: AlTurjman, F. (ed.) Smart Grid in IoT-Enabled Spaces: The Road to Intelligence in Power, pp. 233–247. CRC Press, Boca Raton (2020). https://doi.org/10.1201/9781003055235-11

2. AlTurjman, F., Salama, R.: Cyber security in mobile social networks. In: Security in IoT Social Networks, pp. 55–81. Elsevier, Amsterdam (2021). https://doi.org/10.1016/B978-0-12-821599-9.00003-0

3. Salama, R., Al-Turjman, F.: AI in blockchain towards realizing cyber security. In: 2022 International Conference on Artificial Intelligence in Everything (AIE), pp. 471–475. IEEE, August 2022

4. AlTurjman, F., Salama, R.: Security in social networks. In: Security in IoT Social Networks, pp. 1–27. Elsevier, Amsterdam (2021). https://doi.org/10.1016/B978-0-12-821599-9.00001-7

5. Parvaresh, N., Kulhandjian, M., Kulhandjian, H., DAmours, C., Kantarci, B.: A tutorial on AI-powered 3D deployment of drone base stations: state of the art, applications and challenges. Veh. Commun. **36**, 100474 (2022). https://doi.org/10.1016/j.vehcom.2022.100474

6. Heidari, A., Navimipour, N.J., Unal, M.: Applications of ML/DL in the management of smart cities and societies based on new trends in information technologies: a systematic literature review. Sustain. Cities Soc. **85**, 104089 (2022). https://doi.org/10.1016/j.scs.2022.104089

7. Alsamhi, S.H., et al.: Blockchain-empowered security and energy efficiency of drone swarm consensus for environment exploration. IEEE Trans. Green Commun. Netw. **7**, 328–338 (2022)

8. Alam, T., Gupta, R., Qamar, S., Ullah, A.: recent applications of artificial intelligence for sustainable development in smart cities. In: AlEmran, M., Shaalan, K. (eds.) Recent Innovations in Artificial Intelligence and Smart Applications, pp. 135–154. Springer International Publishing, Cham (2022). https://doi.org/10.1007/978-3-031-14748-7_8

9. Pirotta, V., Hocking, D.P., Iggleden, J., Harcourt, R.: Drone observations of marine life and human-wildlife interactions off Sydney. Australia. Drones **6**(3), 75 (2022)

10. Ivanova, S., Prosekov, A., Kaledin, A.: A survey on monitoring of wild animals during fires using drones. Fire **5**(3), 60 (2022)

11. Giusti, E., et al.: A compact drone based multisensory system for maritime observation. In: 2022 IEEE Radar Conference (RadarConf 2022), pp. 1–6. IEEE (2022)

12. Xu, A., Wang, F., Li, L.: Vegetation information extraction in karst area based on UAV remote sensing in visible light band. Optik. **272**, 170355 (2022)

13. Job, I., Essien, E.N., Ododo, E.: A review application of machine learning in agriculture supply chain to improve food and agriculture industry. Asia-Af. J. Agric. **1**, 66–78 (2022)

14. Puangpontip, S., Hewett, R.: Towards intelligent management of internet of modern drones. In: Neri, F., Du, K.-L., Varadarajan, V.K., AngelAntonio, S.-B., Jiang, Z. (eds.) Computer and Communication Engineering: 2nd International Conference, CCCE 2022, Rome, Italy, March 11–13, 2022, Revised Selected Papers, pp. 3–13. Springer International Publishing, Cham (2022). https://doi.org/10.1007/978-3-031-17422-3_1

15. Solanki, A., Tarar, S., Singh, S.P., Tayal, A. (Eds.).: The Internet of Drones: AI Applications for Smart Solutions. CRC Press, Boca Raton (2022)

16. Setyawan, A.A., Taftazani, M.I., Bahri, S., Noviana, E.D., Faridatunnisa, M.: Drone LiDAR application for 3D city model. J. Appl. Geospat. Inf. **6**(1), 572–576 (2022)

Range Proofs with Constant Size and Trustless Setup

Emanuele Scala$^{(\boxtimes)}$ and Leonardo Mostarda

Computer Science, University of Camerino, Camerino, Italy
{emanuele.scala,leonardo.mostarda}@unicam.it

Abstract. Range proofs are widely adopted in practice in many privacy-preserving cryptographic protocols in the public blockchain. The performances known in the literature for range proofs are logarithmic-sized proofs and linear verification time. In contexts where the proof verification is left to the ledger maintainers and proofs are stored in blocks, one might expect higher transaction fees and blockchain space when the size of the relation over the proof grows. With this paper, we improve Bulletproofs, a zero-knowledge argument of knowledge for range proofs, by modifying its Inner Product Argument (IPA) subroutine. In particular, we adopt a new relation from the polynomial commitment scheme of Halo, based on standard groups and assumptions (DLOG and RO) with a trustless setup. We design a *two-step reduction* algorithm and we obtain a constant number of two rounds in the IPA and a constant-sized proof composed of 5 $\mathbb{G}_1$ points and 2 $\mathbb{Z}_p$ scalars.

1 Introduction

Bootle et al. [3] develop an Inner Product Argument (IPA) system, in which computational soundness relies on discrete logarithm (DLOG) assumption in standard groups. The IPA consists of an argument of knowledge of the openings of Pedersen commitments satisfying an inner product relation. Bünz et al. [7] adopt the system of Bootle and propose Bulletproofs, a zero-knowledge proof system optimized for *range proofs*. Such proofs are useful in confidential transactions where a sender wants to prove that a value is in a particular range, without revealing the value to the receiver of the transaction. Bulletproofs optimizes the communication complexity through a logarithmic number of rounds in the IPA protocol used as a subroutine in the range proof. From these results, many cryptographic protocols have been applied with range proofs in blockchain contexts: Quisquis [13] and Zether [6] are privacy-preserving payment schemes using range proofs to prove that transfer amounts and balances over homomorphic encryptions are non-negatives; ZeroMT [10] extends the Zether's relation to many transfer amounts and balances, proving that a batch of aggregated values are non-negatives; Lelantus [14] and Monero [1] are private cryptocurrencies that hide the coin values through Pedersen commitments, and prove that output commitments in a spend transaction are in the range of admissible values.

L. Barolli (Ed.): AINA 2023, LNNS 655, pp. 301–310, 2023.
https://doi.org/10.1007/978-3-031-28694-0_28

However, due to the complexity of the IPA protocol, range proofs are logarithmically sized and proof verification time is linear in the bit length of the range. It follows that many works try to optimize the IPA protocol with solutions from standard groups or pairing-friendly groups, e.g., the inner-pairing products (a complete description can be found in the related works Sect. 4). In our work, we consider the optimizations proposed by Bowe et al. in [5], and we show how Halo's modified IPA can be applied to Bulletproofs, keeping the trustless setup and avoiding expensive pairing checks.

Our Contribution. We present a new *two-step reduction* algorithm for the IPA of Bulletproofs. The reduction exploits the structure of the polynomial commitment of Bulletproofs and a new IPA relation presented in Sect. 3. Surprisingly, this adaptation yields a constant number of two rounds in the IPA and a constant-sized proof composed of 5 $\mathbb{G}_1$ points and 2 $\mathbb{Z}_p$ scalars. As a part of the contribution, we implement and evaluate concretely our solution in the arkworks [2] Rust ecosystem.

2   Preliminaries

Groups. Let $(\mathbb{G}, p, g)$ be a description of a cyclic group $\mathbb{G}$, where p is the order of the group and is a prime number, $g \in \mathbb{G}$ is a generator of the group, we consider groups in which the *discrete logarithm problem* is computationally hard. In particular, we refer to the *Discrete Logarithm* (DLOG) and *Decisional Diffie-Hellman* (DDH) security assumptions for such groups.

Pedersen Commitment. A Pedersen commitment can be defined over a cyclic group $\mathbb{G}$ of prime order. A *binding* and *hiding* commitment for a message $m \in \mathbb{Z}_p$, from the set of integers modulo p, can be generated by applying the Commit function such that: $\text{Commit}(m;r) = (g^m h^r) \in \mathbb{G}$, where g and h are two distinct generators of the group $\mathbb{G}$, and r is a randomly chosen blinding factor. One variant is *Pedersen vector commitment* which allows multiple messages to be committed at once. Pedersen commitments are *homomorphic additive* when the group operator $\cdot$ is applied between commitments.

Zero-Knowledge Proofs. Let $\mathcal{R}$ be a relation between an instance x and a witness w such that $(x, w) \in \mathcal{R}$ and $\mathcal{L}$ be the language for that relation such that $\mathcal{L} = \{x \mid \exists w : (x, w) \in \mathcal{R}\}$. An interactive zero-knowledge proof is a protocol between a prover $\mathcal{P}$ and a verifier $\mathcal{V}$ in which $\mathcal{P}$ convinces $\mathcal{V}$ that $x \in \mathcal{L}$ for the given relation $\mathcal{R}$ without revealing the witness. From the transcript of the protocol, the verifier can accept or reject the proof, which essentially reveals nothing beyond the validity of the proof. A proof system is *honest-verifier perfect zero-knowledge* (HVZK) if it has the properties of *perfect completeness*, *special soundness* and *honest-verifier perfect zero-knowledge*. The HVZK protocol is defined *public coin* if the messages from the verifier are uniformly random and are independent of the messages of the prover.

Bulletproofs Notation. The notations we use are those of Bulletproofs [7]. In summary, we denote:

- $\mathbb{Z}_p$ is a ring of integers modulo p prime and $\mathbb{G}_p$ is a cyclic group of prime order p.
- g and h are generators of $\mathbb{G}_p$.
- *Pedersen commitment* to the value a with blinding factor α is: $A = g^a h^\alpha$.
- Bold letters are vectors, e.g., $\mathbf{a} = (a_1, ..., a_n)$ with $\mathbf{a} \in \mathbb{Z}_p^n$.
- The *inner-product* of two vectors is $\langle \mathbf{a}, \mathbf{b} \rangle = \sum_{i=1}^n a_i \cdot b_i$.
- *Pedersen vector commitment* to a vector $\mathbf{a} \in \mathbb{Z}_p^n$: $A = \mathbf{g^a} = \prod_{i=1}^n g_i^{a_i}$ is *binding* (but not *hiding*) commitment, where $\mathbf{g} = (g_1, ..., g_n) \in \mathbb{G}^n$ is a vector of generators.
- *Vector polynomial* is defined as $p(X) = \sum_{i=0}^d \mathbf{p_i} \cdot X^i \in \mathbb{Z}_p^n[X]$, meaning that each coefficient of the polynomial p is a vector of field elements in $\mathbb{Z}_p^n$.

For the full notation of Bulletproofs, refer to Sect. 2.3 of [7].

Bulletproofs Proof System. Bulletproofs is zero-knowledge argument of knowledge in which a prover demonstrates that a value v is in a specific range, between zero and 2^{n-1}, where n is the range domain. Given as public parameters the tuple $(g, h, V = g^v h^\gamma)$, where g and h are generators of a group $\mathbb{G}_p$ and V is a Pedersen commitment of the value v, with a hiding factor from the randomness γ, the system ends by proving the equality $\hat{t} = \langle \mathbf{l}, \mathbf{r} \rangle$, i.e., that the inner-product of two committed vectors $\mathbf{l}$, $\mathbf{r}$ is a certain $\hat{t}$. In what follows, we first present the steps of the range proof prior to the inner-product. The prover generates two vectors $\mathbf{a}_L$ and $\mathbf{a}_R$ where $\langle \mathbf{a}_L, 2^n \rangle = v$ and $\mathbf{a}_R = \mathbf{a}_L - \mathbf{1}^n$, and commits to these vectors producing one commitment A. Further, the prover generates a second commitment S to blinding terms $\mathbf{s}_L$ and $\mathbf{s}_R$. The verifier generates and sends to the prover two random challenges y and z. The prover defines a polynomial $t(X)$ from the inner product of two vector polynomials $l(X)$ and $r(X)$, which in turn are derived from a linear combination of the vectors $\mathbf{a}_L$ and $\mathbf{a}_R$, the blinding vectors $\mathbf{s}_L$ and $\mathbf{s}_R$ and the two verifier challenges y and z. This results in a degree-two polynomial $t(X)$ with coefficients t_0, t_1 and t_2, where t_0 is the constant term, t_1 is the degree-one term and t_2 is the degree-two term. Then, the prover does not commit to the coefficient t_0, instead creates and sends to the verifier the commitments T_1 and T_2 to the coefficients t_1 and t_2. The prover convinces the verifier that it has the knowledge of the coefficients by proving that the polynomial $t(X)$ evaluates to a specific value $\hat{t}$ at a random point x. After receiving the challenge x, the prover sends to the verifier a blinding term τ_x for $\hat{t}$, a blinding factor μ for the commitments A and S and the two blinded vectors $\mathbf{l} = l(x)$ and $\mathbf{r} = r(x)$.

Prover	Verifier
$A, S \rightarrow$	
	$\leftarrow y, z$
$T_1, T_2 \rightarrow$	
	$\leftarrow x$
$\tau_x, \mu, \hat{t}, \mathbf{l}, \mathbf{r} \rightarrow$	
	verify

Finally, the verifier can check the commitment V (which is public) of the value v, that the two vectors $\mathbf{l}$ and $\mathbf{r}$ are valid and that $\hat{t} = \langle \mathbf{l}, \mathbf{r} \rangle$. In order to reduce the size of the range proof from linear to logarithmic size in n (bits of the range), instead of transmitting the vectors $\mathbf{l}$ and $\mathbf{r}$, the prover and verifier engage in an Inner-Product-Argument (IPA) protocol with the two vectors becoming witnesses. In the next section, instead, we present how we modify the IPA protocol following the IPA relation of Halo [5], to reduce the size of the proof to a constant size.

3   Two-Step Reduction Inner-Product-Argument

Bulletproofs [7] implements an Inner-Product-Argument in which the prover proves the knowledge of two vectors $\mathbf{a}$ and $\mathbf{b}$ to the verifier for the relation:

$$\{(\mathbf{g}, \mathbf{h} \in \mathbb{G}^n, \ u, T \in \mathbb{G} \ ; \ \mathbf{a}, \mathbf{b} \in \mathbb{Z}_p^n) \ : \ T = \mathbf{g}^{\mathbf{a}} \mathbf{h}^{\mathbf{b}} \cdot u^{\langle \mathbf{a}, \mathbf{b} \rangle}\} \tag{1}$$

Halo [5] introduce a new relation considering the following intuitions: by fixing the vector $\mathbf{b} = (1, x, x^2, ..., x^{d-1})$, where d is a fixed polynomial degree, we can claim that an evaluation v of a polynomial $t(x) = \langle \mathbf{a}, \mathbf{b} \rangle = v$ at random point x, where a is the vector with the coefficients of the polynomial t. With this variant, the vector $\mathbf{h}$ is no longer necessary, and we rewrite the new relation:

$$\{(\mathbf{g} \in \mathbb{G}^n, \ h \in \mathbb{G}, \ u, T \in \mathbb{G}, \ x, v \in \mathbb{Z}_p \ ; \ \mathbf{a} \in \mathbb{Z}_p^n, \ r \in \mathbb{Z}_p) \ : \ T = \mathbf{g}^{\mathbf{a}} h^r \cdot u^{\langle \mathbf{a}, \mathbf{b} \rangle}\} \tag{2}$$

where the additional generator h serves for the purpose of blinding the commitment T through the randomness r, x is the evaluation point used to construct the vector $\mathbf{b}$, and $v = \langle \mathbf{a}, \mathbf{b} \rangle$. Given the relation (2), in the following we design a *two-step reduction* IPA, adding the new relation to the range proof protocol of Bulletproofs.

From the definition of the polynomial $t(X) = t_0 + t_1 X + t_2 X^2$, we oberve that its evaluation at point x is

$$t(x) = \langle \mathbf{t}, \mathbf{b} \rangle = \hat{t}$$

where $\mathbf{t} = (t_0, t_1, t_2)$ is the vector of coefficients of $t(X)$ and $\mathbf{b} = (1, x, x^2)$. This means that if the prover proves the knowledge of $\mathbf{a} = \mathbf{t}$ and r for relation (2) also the relation (1) holds given

$$t(x) = \langle \mathbf{l}, \mathbf{r} \rangle = \hat{t}$$

with $\mathbf{a} = \mathbf{l}$ and $\mathbf{b} = \mathbf{r}$ for relation (1).

However, the length of $\mathbf{t}$ and $\mathbf{b}$ vectors is not a power of two, and so we cannot use them directly inside the Halo IPA. Then, we add an extra round into the protocol before the actual reduction step occurs.

In the first move, prover $\mathcal{P}$ and verifier $\mathcal{V}$ initialize a commitment

$$T' = T \cdot u^{\hat{t}}$$

where $T = \mathbf{g^t} h^r$, with $\mathbf{t} = (t_0, t_1, t_2)$, and $u \in \mathbb{G}$ is a random group element sent by $\mathcal{V}$. Then, $\mathcal{P}$ and $\mathcal{V}$ engage in an IPA for relation (2). Assuming $d = 2$ the degree of $t(X)$, the protocol proceeds in $k = 2$ rounds, from one extra round (at $j = k - 1$) to one reduction step (at $j = 0$).

In round $j = 1$, the prover sets three vector:

$$\mathbf{t}^{(1)} = (t_0, t_1) \ , \ \mathbf{b}^{(1)} = (1, x) \ , \ \mathbf{g}^{(1)} = (g_0, g_1)$$

Then, the prover samples at random $l_1, r_1 \in \mathbb{Z}_p$ and computes and sends to the verifier:

$$L_1 = g_1^{t_0} \cdot h^{l_1} \cdot u^{t_0 \cdot x}$$

$$R_1 = g_0^{t_1} \cdot h^{r_1} \cdot u^{t_1 \cdot 1}$$

The verifier samples and sends to the prover a random challange $\mu_1 \in \mathbb{Z}_p$. Then, the prover computes $t^{(1)} \in \mathbb{Z}_p$, $b^{(1)} \in \mathbb{Z}_p$ and $g^{(1)} \in \mathbb{G}$, such that:

$$t^{(1)} = t_1 \cdot \mu_1^{-1} + t_0 \cdot \mu_1$$

$$b^{(1)} = 1 \cdot \mu_1^{-1} + x \cdot \mu_1$$

$$g^{(1)} = g_0^{\mu_1^{-1}} \cdot g_1^{\mu_1}$$

Now the prover prepares for the next round ($j = 0$) three other vectors (note that in this way an effective reduction step does not occur):

$$\mathbf{t}^{(0)} = (t^{(1)}, t_2) \ , \ \mathbf{b}^{(0)} = (b^{(1)}, x^2) \ , \ \mathbf{g}^{(0)} = (g^{(1)}, g_2)$$

In round $j = 0$, the prover samples at random $l_0, r_0 \in \mathbb{Z}_p$ and computes and sends to the verifier:

$$L_0 = g_2^{t^{(1)}} \cdot h^{l_0} \cdot u^{t^{(1)} \cdot x^2}$$

$$R_0 = g^{(1)t_2} \cdot h^{r_0} \cdot u^{t_2 \cdot b^{(1)}}$$

The verifier samples and sends to the prover a random challenge $\mu_0 \in \mathbb{Z}_p$. Then, the prover computes $t^{(0)} \in \mathbb{Z}_p$, $b^{(0)} \in \mathbb{Z}_p$ and $g^{(0)} \in \mathbb{G}$, such that:

$$t^{(0)} = t_2 \cdot \mu_0^{-1} + t^{(1)} \cdot \mu_0$$

$$b^{(0)} = b^{(1)} \cdot \mu_0^{-1} + x^2 \cdot \mu_0$$

$$g^{(0)} = g^{(1)\mu_0^{-1}} \cdot g_2^{\mu_0}$$

After this final round, the verifier computes:

$$T^{(0)} = \prod_{j=0}^{k-1}(L_j^{\mu_j^2}) \cdot T' \cdot \prod_{j=0}^{k-1}(R_j^{\mu_j^{-2}})$$

And the verifier wants to check that:

$$T^{(0)} \overset{?}{=} g^{(0)t^{(0)}} \cdot h^{r'} \cdot u^{(t^{(0)} \cdot b^{(0)})} \tag{3}$$

where $r' = \sum_{j=0}^{k-1}(l_j\mu_j^2) + r + \sum_{j=0}^{k-1}(r_j\mu_j^{-2})$.

Note that the verifier can compute $g^{(0)}$ and $b^{(0)}$ by itself from the following inner-products:

$$g^{(0)} = \langle \mathbf{s}, \mathbf{g} \rangle = \langle (\mu_0^{-1}, \mu_0), (g^{(1)}, g_2) \rangle \text{ with } g^{(1)} = \langle (\mu_1^{-1}, \mu_1), (g_0, g_1) \rangle \tag{4}$$

$$b^{(0)} = \langle \mathbf{s}, \mathbf{b} \rangle = \langle (\mu_0^{-1}, \mu_0), (b^{(1)}, x^2) \rangle \text{ with } b^{(1)} = \langle (\mu_1^{-1}, \mu_1), (1, x) \rangle \tag{5}$$

To check the equality (3), we first rewrite the right side:

$$T^{(0)} = (g^{(0)} \cdot u^{b^{(0)}})^{t^{(0)}} \cdot h^{r'} \tag{6}$$

Hence, the prover and verifier engage in a Schnorr protocol in which the prover proves to the verifier the knowledge of $t^{(0)}$ and r'.

The prover samples at random $d,s \in \mathbb{Z}_p$, computes and sends to the verifier a new commitment R:

$$R = (g^{(0)} \cdot u^{b^{(0)}})^d \cdot h^s$$

The verifier samples and sends to the prover a random $c \in \mathbb{Z}_p$.
The prover computes and sends to the verifier the scalars z_1 and z_2:

$$z_1 = t^{(0)}c + d$$

$$z_2 = r'c + s$$

Finally, the verifier accepts or rejects the proof if and only if:

$$T^{(0)c} \cdot R \overset{?}{=} (g^{(0)} \cdot u^{b^{(0)}})^{z_1} \cdot h^{z_2}$$

Two-Step Reduction IPA Proof Size. A zero-knowledge proof is composed of all the scalars (elements in $\mathbb{Z}_p$), and elliptic curve points (elements in $\mathbb{G}$) that the prover forwards to the verifier. Our *two-step reduction* IPA generates two collections of elliptic curve points (L_1, L_0) and (R_1, R_0) at each j-th round, one group element R and two scalar field elements z_1, z_2 in the Schnorr protocol. The proof size is constant given the constant number of rounds in the IPA and the total proof consists of 5 $\mathbb{G}_1$ points and 2 $\mathbb{Z}_p$ scalars.

4   Related Work

Bowe et al. [5] propose Halo, a recursive proof composition from the notion of Incrementally Verifiable Computation (IVC), i.e. a method to inductively prove within a single proof the validity of past proofs. The recursion is made via a cycle of normal prime order elliptic curves, such that proofs over one curve can verify proofs over the other curve. An interesting technique is the *amortized succinctness* for *polynomial commitments*: by the structure of the two vectors behind the IPA, the linear-time work of the verifier is amortized across many proofs. This is done by an untrusted third party who executes the linear-time operations for each step proof and then proves the correctness of a batch of that proofs to the verifier. The verifier performs the same operations once for the entire batch. This batch of proofs is handled via an *accumulator* which does not grow in size with each step proof. With this amortization strategy, the IPA verifier results in a logarithmic cost barring the single linear time check.

Bünz et al. [8] establish an exciting result that generalizes the Halo's recursive composition to a class of non-interactive arguments which do not necessarily have succinct verification. The authors provide theoretical efficiency and security proofs for constructing *accumulation schemes* for any SNARK, which yields to Proof-Carrying-Data (PCD) scheme. Moreover, the authors prove a second theorem stating that if the SNARK verifier is succinct except for a specific predicate, and has an accumulation scheme for that predicate, it is possible to derive an accumulation scheme for the SNARK. Further, the authors prove that two polynomial commitment schemes have accumulation schemes in Random Oracle: (i) PC_{DL}, polynomial commitment scheme based on discrete logarithm assumption; (ii) PC_{AGM}, polynomial commitment based on knowledge assumption in bilinear groups. From this follows an interesting open question of whether constructions exist in the standard assumption instead of in the "trivial" knowledge assumption. From the efficiency perspective, PC_{DL} achieves an asymptotic logarithmic cost to check accumulation steps and a linear cost in the polynomial degree during the final opening check. Instead, PC_{AGM} has an asymptotic linear cost to check accumulation steps, while only one pairing is required in the final check.

Xiong et al. [19] propose VERI-ZEXE, an improvement of Zexe's Decentralized Private Computation (DPC) scheme [4], translating the circuit-specific trusted setup into a universal setup where a structured reference string (SRS) is reused for different circuits. Universal SNARKs built on Polynomial Interactive Oracle Proofs (PIOP) are often instantiated with pairing-based Polynomial Commitment Schemes (PCS) that require expensive pairing operations in the verifier circuit. To lighten the cost of pairing checks, VERI-ZEXE relies on the generalized *accumulation scheme* of Bünz et al. in PCD [8], designing a *two-step* IVC. Hence, with this algorithm, their goal is to delay the final bilinear pairing check, by attaching $2\mathbb{G}_1$ points to the transaction validity proof to be verified by the ledger maintainers.

Daza et al. [11] propose an optimization of the IPA protocol of Bootle et al. [3] on the verifier side. In particular, the authors try to achieve a logarithmic verification complexity in the circuit size. Their scheme is based on bilinear groups, secure under the standard assumption and Random Oracle, with an updatable and universal setup.

Bünz et al. [9] present a Generalized Inner Product Argument (GIPA) in pairing-based groups. With GIPA, the authors achieve a logarithmic-time verifier for a polynomial commitment scheme with a universal setup. This comes at the cost of square root complexity for prover time bounded to the polynomial degree and square root SRS size.

Lee [15] proposes Dory, an argument of knowledge system from inner-pairing products with a transparent setup. This result is established in the standard SXDH (Symmetric eXternal Diffie-Hellman) assumption which implies DLOG. The verifier work has an asymptotic logarithmic cost of n multi-exponentiation with respect to the length n of the IPA vectors, plus a constant number of pairings.

5   Implementation and Evaluation

In this section, we present an implementation and proof size evaluation of our *two-step reduction* IPA presented in Sect. 3, compared to the IPA of Bulletproofs [7]. The source code, available on GitHub [12], is written in Rust and is based on the **arkworks** [2] libraries. The elliptic curve we use for all group operations is the Barreto-Naehrig curve (BN-254). In Table 1, we report the evaluations in bytes of the size of the proofs, considering a fixed range domain of $n = 16$ bits and a variable number of aggregate range values m, from 2 up to 64 values. Measurements are executed on a machine running the Rust compiler with an Intel Core i7-10750H CPU and 16 GB of RAM.

Table 1. Inner-Product-Argument (IPA) proof size comparison. **BP-IPA** is the IPA of Bulletproofs [7]. **TS-IPA** is our *two-step* reduction IPA presented in this work. **n** and **m** are respectively the bit-range domain and the number of aggregate range values.

n	m	BP-IPA proof size (bytes)	TS-IPA proof size (bytes)
16	2	720	400
16	4	848	400
16	8	976	400
16	16	1,104	400
16	32	1,232	400
16	64	1,360	400

The results in Table 1 highlight that as the number of aggregated values increases, hence m, the proof size of BP-IPA grows logarithmically while in our TS-IPA the proof size is clearly constant. This is in line with the theoretical results: the aggregated Bulletproofs (in [7], Sect. 4.3) shows a logarithmic proof size asymptotically equal to $O(log_2(m \cdot n))$, considering that at each

IPA round there are two collections of group elements $(L_1, \ldots, L_{log_2(m \cdot n)})$ and $(R_1, \ldots, R_{log_2(m \cdot n)})$. Instead, in TS-IPA the two collections end up having only 4 group elements (L_1, L_0) and (R_1, R_0). Hence, the proof size is constant and the total proof consists of 5 $\mathbb{G}_1$ points and 2 $\mathbb{Z}_p$ scalars. Figure 1 shows the asymptotic sizes of the BP-IPA and TS-IPA proofs.

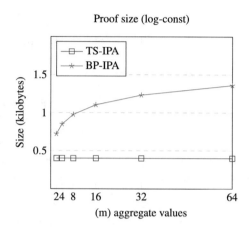

Fig. 1. Proof size comparison TS-IPA and BP-IPA

6 Conclusion and Future Work

Range proofs in standard security assumptions, standard groups and without trusted setup are attractive in confidential transaction protocols. However, range proofs lack succinct verification and proof size. We presented a modified Inner-Product-Argument protocol for range proof systems such as Bulletproofs, and our two-step reduction algorithm keeps the size of the proof constant. Moreover, we also reduce the communication complexity since the proof size is in the order of bytes. In this work, we assumed that the new relation for IPA introduced by Halo is sound and has zero-knowledge, however, further investigations are needed. As future work, we will validate our approach in real case studies involving streams of sensor data [16–18].

References

1. Alonso, K.M., et al.: Zero to Monero (2020)
2. arkworks rs. arkworks
3. Bootle, J., Cerulli, A., Chaidos, P., Groth, J., Petit, C.: Efficient zero-knowledge arguments for arithmetic circuits in the discrete log setting. In: Fischlin, M., Coron, J.-S. (eds.) EUROCRYPT 2016. LNCS, vol. 9666, pp. 327–357. Springer, Heidelberg (2016). https://doi.org/10.1007/978-3-662-49896-5_12

4. Bowe, S., Chiesa, A., Green, M., Miers,I., Mishra, P., Wu, H.: Zexe: enabling decentralized private computation. In: 2020 IEEE Symposium on Security and Privacy (SP), pp. 947–964. IEEE (2020)
5. Bowe, S., Grigg, J., Hopwood, D.: Recursive proof composition without a trusted setup. *Cryptology ePrint Archive* (2019)
6. Bünz, B., Agrawal, S., Zamani, M., Boneh, D.: Zether: towards privacy in a smart contract world. In: Bonneau, J., Heninger, N. (eds.) FC 2020. LNCS, vol. 12059, pp. 423–443. Springer, Cham (2020). https://doi.org/10.1007/978-3-030-51280-4_23
7. Bünz, B., Bootle, J., Boneh, D., Poelstra, A., Wuille, P., Maxwell. G.: Bulletproofs: short proofs for confidential transactions and more. In: 2018 IEEE Symposium on Security and Privacy (SP), pp. 315–334. IEEE (2018)
8. Bünz, B., Chiesa, A., Mishra, P., Spooner, N.: Proof-carrying data from accumulation schemes. *Cryptology ePrint Archive* (2020)
9. Bünz, B., Maller, M., Mishra, P., Tyagi, N., Vesely, P.: Proofs for inner pairing products and applications. In: Tibouchi, M., Wang, H. (eds.) ASIACRYPT 2021. LNCS, vol. 13092, pp. 65–97. Springer, Cham (2021). https://doi.org/10.1007/978-3-030-92078-4_3
10. Corradini, F., Mostarda, L., Scala, E.: ZeroMT: multi-transfer protocol for enabling privacy in off-chain payments. In: Barolli, L., Hussain, F., Enokido, T. (eds.) AINA 2022. LNNS, vol. 450, pp. 611–623. Springer, Cham (2022). https://doi.org/10.1007/978-3-030-99587-4_52
11. Daza, V., Ràfols, C., Zacharakis, A.: Updateable inner product argument with logarithmic verifier and applications. In: Kiayias, A., Kohlweiss, M., Wallden, P., Zikas, V. (eds.) PKC 2020. LNCS, vol. 12110, pp. 527–557. Springer, Cham (2020). https://doi.org/10.1007/978-3-030-45374-9_18
12. EmanueleSc. Zeromt
13. Fauzi, P., Meiklejohn, S., Mercer, R., Orlandi, C.: Quisquis: a new design for anonymous cryptocurrencies. In: Galbraith, S.D., Moriai, S. (eds.) ASIACRYPT 2019. LNCS, vol. 11921, pp. 649–678. Springer, Cham (2019). https://doi.org/10.1007/978-3-030-34578-5_23
14. Jivanyan, A.: Lelantus: towards confidentiality and anonymity of blockchain transactions from standard assumptions. IACR Cryptol. ePrint Arch. **2019**, 373 (2019)
15. Lee, J.: Dory: efficient, transparent arguments for generalised inner products and polynomial commitments. In: Nissim, K., Waters, B. (eds.) TCC 2021. LNCS, vol. 13043, pp. 1–34. Springer, Cham (2021). https://doi.org/10.1007/978-3-030-90453-1_1
16. Mehmood, N.Q., Culmone, R., Mostarda, L.: Modeling temporal aspects of sensor data for MongoDB NoSQL database. J. Big Data **4**(1), (2017)
17. Russello, G., Mostarda, L., Dulay, N.: A policy-based publish/subscribe middleware for sense-and-react applications. J. Syst. Softw. **84**(4), 638–654 (2011)
18. Vannucch, C., et al.: Symbolic verification of event–condition–action rules in intelligent environments. J. Reliable Intell. Environ. **3**(2), 117–130 (2017)
19. Xiong, A.I., Chen, B., Zhang, Bünz, B., Fisch, B., Krell, F., Camacho. P.: Verizexe: decentralized private computation with universal setup. *Cryptology ePrint Archive* (2022)

Sensorless Predictive Maintenance: An Example on a 'Not 4.0' Coffee Machine Production Process

Diletta Cacciagrano, Flavio Corradini, and Marco Piangerelli[✉]

Computer Science, University of Camerino, Camerino, Italy
{diletta.cacciagrano,flavio.corradini,marco.piangerelli}@unicam.it

Abstract. In this paper, we present a new method based on Network Analysis for implementing what we called Sensorless Predictive Maintenance (PdM) on production lines of coffee machines of an Italian manufacturing company, *Nuova Simonelli S.p.A.*. To the best of our knowledge, such an approach is an innovative one in the field of PdM; its final goal is to implement a way for performing PdM without installing sensors on machines or production lines. The results are very promising and allow us to gain knowledge about the assembly process after it took place. Then, such knowledge can be used for improving the process and putting into practice the Sensorless PdM on the new coffee machines.

1 Introduction

Back in 2017, The Economist published a story titled "The world's most valuable resource is no longer oil, but data" [3]. The cover featured an image of major companies, such as Microsoft, Google, Facebook and Tesla, as oil platforms intent on extracting data from the sea. The purpose of that piece was to point out that big tech companies used data from their users as a source of wealth (in addition to that derived from their services).

As the big companies exploits data provided by their customers for improving their own commercial services and increasing the turnover, so manufacturing companies, by collecting data from their activities or from sensors installed in their products, could increase their turnover and improve the quality of their products. For example, sensors on production chains can be used for predicting potential breakdowns: this approach is known as Predictive Maintenance (PdM). According to a McKinsey Global Institute report, "The Internet of Things: Mapping the Value Beyond the Hype", manufacturers' savings from PdM could globally total between 240 and 630 bn by 2025 [7].

The Authors thank Nuova Simonelli S.p.A. for sharing data.

From a more general perspective, the monitoring and analysis of production chains, or produced devices, are fundamental aspects in the economy of modern manufacturing companies and allow rapid decisions to be made for maintaining efficient and high quality production and related information systems.

One way to "easily" perform the above task is based on the application of sophisticated Machine Learning (ML) techniques. The use of these techniques is not straightforward and requires several steps before and after, such as the selection of relevant data from the ones the company produces, the definition of the most suitable data engineering/preparation strategy and the choice of the most appropriate computational resources.

As mentioned above, to implement this type of analysis, companies should equipped their machines with sensors [8] and collect data to feed to the various algorithms. What happens, however, if companies do not have this capability or more simply have no way to equip their machines and/or production lines? Should they abandon the idea to implement any PdM technique, or could they still use such a methodology perhaps by tackling the problem form another perspective?

In this paper we used data collected during testing in the production line and provided from an Italian leading manufacturer of high-quality coffee machines, *Nuova Simonelli S.p.A.*. We developed a methodology based on *networks analysis* for trying to answer the above questions and showing how data collected during testing in the production line can be used to improve the quality of the coffee machine production process and putting into practice what we called *Sensorless PdM*. Our contribution is twofold:

- our analysis is extendable to any production line
- our analysis is not based on any sensors thus allowing companies to implement PdM techniques without equipping their machines with sensors.

The paper is organized as follows. In Sect. 2 the used approach is described; in Sect. 3 we describe the data-set and the model we used to describe it; then, in Sect. 4 the obtained graphs are described. Furthermore, in Sect. 5 the results are presented and finally in Sect. 6 we discuss the results.

2 Our Approach

The main idea in our approach relies on the use of network theory. Network analysis utilizes the mathematical language of graph theory. Therefore, graphs are the starting point of network analysis. A graph, $\mathcal{G}$, is a structure made by a set of vertices, V, and edges, E, that connect couples of vertices; mathematically, $\mathcal{G}(V, E)$, with $E \subseteq V \times V$. Graphs can be of several types but the main reason for their widespread use is due to their capability to represent relations (edges) between entities (nodes). Most of the time such relations are symmetric between the nodes and represented by a number, a weight, describing how strong they are. Such graphs are called *weighted undirected graphs*. The process to produce a network graph answering a business question is listed below:

1. Formulate the exact business question.
2. Specify the representation of the nodes and edges including their weight concerning the question.
3. Identify the entities, facts, and measures in the data answering the question.
4. Map the identified data from step 3 to the representation of step 2.
5. Extract the nodes and edges accordingly from the data.
6. Filter the edges and nodes to the most important to set the focus.
7. Calculate statistics on the graph to gather additional information.
8. Colouring and sizing the nodes and edges accordingly to facts, measures, and statistic calculations

Steps 1 and 2 are fundamental and they are related in a very strong way. A clear business question is very helpful for extracting how links among the nodes should be weighted. Once a graph is available (Steps 1–5), it is possible to compute some ad-hoc statistical quantities, or attributes, to have insights about the entities and the links (Step 6). For example, referring to Fig. 1 we can identify two very important attributes:

- Node Degree Frequency Distribution
- Edge Weight Frequency Distribution

In a weighted undirected graph, the weighted degree of a node is the sum of the weights of edges incident in the node. For example in Fig. 1 node A has a degree of 12, C a degree of 10, B a degree of 7 and D a degree of 4, finally, E has a degree of 3.

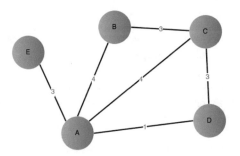

Fig. 1. Example graph. The graph has 5 nodes (A, B, C, D, E) connected by weighted edges. Each weight is specified on the edge. For example, the weight of the edge between A and E is 3.

Step 7 consists in using more statistical indexes such as *Betweenness Centrality*, *Modularity*, or *Eccentrity* in order to have insight into the structures inside the network.

3 Meeting the Data-Set

The data-set provided by *Nuova Simonelli S.p.A.* does not contain any numerical values. On the contrary, a lot of categorical values and fields filled with natural language notes are present. The data results from a series of measurements that are made to reflect the outcome or other quantitative characteristics of the assembly process of a coffee machine.

3.1 Multidimensional Model

It is vital to understand the structure, relations, and meaning of the data-set received from *Nuova Simonelli S.p.A.*. Multidimensional UML (mUML) [6] is the notation chosen to model the data-set of. mUML is a standard and extends the Meta-Modelling Language (MML), a sub-set of the United Modelling Language (UML) [10], via stereotypes that give a special meaning to certain UML constructs:

- Special classes: <<fact-class>> and <<dimension-class>> for modelling facts and dimensions
- Special relations: <<dimension>> to connect a dimension to a fact element and <<roll-up>> to connect adjacent hierarchy levels

The multidimensional model allows the modeling and viewing of the data in multiple dimensions. This is helpful to fully understand the entities and relations within the data-set. We used the star schema to model the data-set provided by *Nuova Simonelli S.p.A.* for two main reasons:

- **Performance**:
  The star schema performs in queries much faster because lesser joins are required, which is crucial for huge amounts of data
- **Simplicity**:
  Because all of the data connects through the fact table, the multiple dimension tables are treated as one large table of information, and thus makes queries simpler and easier to perform

In Fig. 2 the mUML is depicted. INTERVENTI builds the central <<fact-class>>, while COMPONENTI, MACCHINE_PRODOTTE, DIFETTI, and AZIONI are the <<dimension-classes>>. While for understanding and cleaning procedure, all fields are useful, this is not true for analysis. As illustrated, attributes and dimensions of the <<fact-class>> were discarded to create the dashboards, reports, and graphs. Also, some attributes in the <<dimension-classes>> were not used to extract information. All the used dimensions and attributes are highlighted in blue, while the others are greyed out.

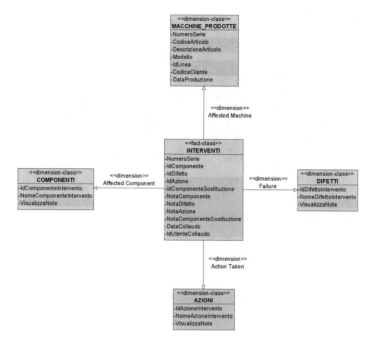

Fig. 2. mUML of the data-set of Nuova Simonelli as Star Schema. INTERVENTI builds the central <<fact- class>>, while COMPONENTI, MACCHINE PRODOTTE, DIFETTI, and AZIONI are the <<dimension-classes>>. All the used dimensions and attributes are highlighted in blue, while the others are greyed out.

3.2 Data Cleaning

As an example of data cleaning, we report the case of the "Altro". On a total of 22104 interventions recorded, 523 (2%) presented the intervention code ID 46, which stands for component code "Altro" ("Other"). For the failure components, 1349 (6%) entries listed the intervention code number 2, which also corresponds to the defect ID "Altro". Lastly, 1130 (5%) records presented the intervention code ID 7, which also points towards the action "Altro". The proportion of "Altro" in regard to the overall data is visualized in Fig. 3.

To face such an issue, all duplicate records were removed, which resulted in a reduction of 82 records. Then the notes field was used for gathering some information about the components, defects, and actions that are precisely classified as "Altro" but with different IDs. As soon as the required information was determined, the components, defects, or actions, the IDs of the records were changed, thus pointing towards the related component, defect, or action.

After performing the manual cleaning, a reduction of the IDs related to "Altro" was observed. As shown in Fig. 3, we observed the following results:

- "Altro" component from 523 records to 134 (<1%)
- "Altro" defect from 1349 records to 393 (2%)
- "Altro" action from 1130 records to 202 (1%)

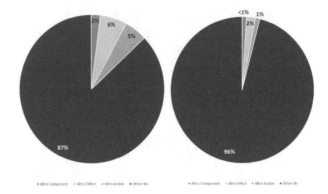

Fig. 3. "Altro" frequency in the initial data (not the left) and "Altro" in the cleaned data (on the right)

4   Graph Extraction

After the data cleaning, we used the business question for building our graphs. This phase can be considered a sort of pre-processing phase. In this paper, the main business question is *What are the correlations between failures?*

For answering this question we decided to split it into two sub-questions and build a graph for each of them. The sub-questions are:

1 Which components failed together within the same machine?
2 Which failure types occurred together within the same model?

In the following, the focus was set on the data for the year 2019. Gephi was used to create the visualization of the graphs and conclude the analysis of the found correlations. Gephi is a java based open source software tool for visualization and network analysis. It uses a render engine to visualize large networks in real-time [4].

4.1   Components Failed Within the Same Machine

In the following analysis, the focus is on the failed components per machine. In Fig. 4, the graph shows the components that failed within the same machine in 2019. The graph is represented as shown in Table 1:

Table 1. Description of how the graph in Fig. 4 is built

Nodes:	Components
Edges:	Links between two components that failed within the same machine
Edge-Weights:	Frequency the two components failed together within the same machine

To set the focus on the most important components and the most important correlations, filters on the nodes and edges were applied. We filtered out all the links with an edge weight lower than 6.

4.2 Failure Types Within the Same Model

In the following analysis, the focus is on the failure type per machine model. The aim is to find correlations between failure types and machine models. Thus, a correlation is defined whenever a failure type frequently occurred in the same model. Such a graph is built as described below:

Table 2. Description of how the graph in Fig. 5 is built

Nodes:	Failures & Models
Edges:	Links from failures to models when the failure occurred within the same model
Edge-Weights:	Frequency of failures within the same model

Again we filtered out all the edges with an edge weight lower than 29.

5 Results

In the following, the results are presented.

5.1 Components Failed Within the Same Machine

At first glance, the coloring makes it very easy to understand the different clusters. The graph in Fig. 4 shows the differentiation of the graph's nodes into two different clusters: the green one and the red one. The colors are obtained by computing the *Modularity*. It is a statistical measure of the structure of a graph representing the connectivity of the graph's nodes. When nodes have similar connectivity between each other they are said to have the same Modularity [1].

Moreover, the size of the nodes represents the impact of another measurement: *Betweenness Centrality*. Betweenness centrality is an indicator of the centrality of a node in a graph. It is equal to the number of the shortest paths from all vertices to all others that pass through that node. A node with high betweenness centrality has a large influence on the other nodes and therefore for the whole

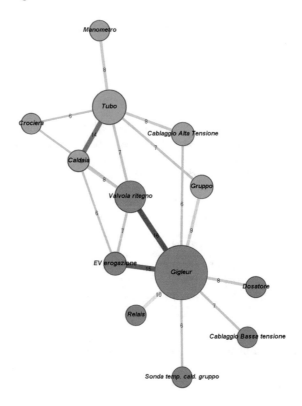

Fig. 4. Failed Components within the same Machine in 2019

graph [1]. For example, in Fig. 4 the node "Gigleur" is very connected and has therefore a high centrality. It is expected that a component that failed the most in the year is also the one that fails the most with other components together. Also "Tubo" and "Valvola Ritegno" are rather big and central, which indicates that they also often fail together with other components. Even if "Tubo" and "Gigleur" are very central, these two components never fail together; i.e. no edge connecting the two nodes. Indeed, "Tubo" and "Gigleur" are the "leaders" of the two clusters and have the biggest impact on the assembly process.

5.2   Failure Types Within the Same Mode

In the following analysis depicted in Fig. 5, the focus is on failures per model. Nodes colored in red represent failures, while nodes colored in blue represent the models. The edges mean that a failure type occurred in a model. Such a graph is a very particular graph and it is called a *bipartite graph*.

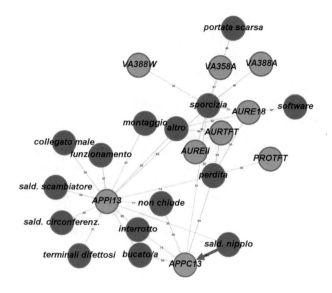

Fig. 5. Failure type within the same Model in 2019

In the graph illustrated in Fig. 5, the model "APPI13" has a lot of incoming arrows. The model "APPI13" is by far affected by the most different failures. It is difficult to pinpoint the main problem, but the issues can be categorized. It seems that there are welding problems "saldature" but also wiring defects "collegato male", and technical defects such as "non chiude", "bucato", "funzionamento", and "montaggio". The model "APPC13", on the other hand, is very frequently affected only by one failure, namely "sald. nipplo" illustrated by the red edge. This failure occurred 186 times within 2019 in the model "APPC13" and should be a focus to fix. In regards of the failure types, "sporcizia" and "perdita" are the most central type of failure in the graph. These defects happen with 6 different models and also rather frequently with a weight of over 60. The isolated failure "software" is interesting, because it happens 97 times within the model "AURE18", but not that often with others. Important to remember is that some of the edges are filtered out. It is to assume that this model has a serious problem with the software or that the software of this model is not very failure tolerant.

6   Discussion and Conclusions

The results presented in this paper intend to find a way to perform sensorless PdM. The final goal of this approach is to support *Nuova Simonelli S.p.A..* in improving quality management while producing coffee machines. In the literature there are contributions on sensorless anomaly detection for PdM (for further details please refer to [5] but, to the best of our knowledge, this is one of the first works to use a graph-based approach to perform PdM. It is worth mentioning that using graphs to model relations between machine models, components, and types of failures, or in general correlations between attributes is a robust method with well-established mathematical foundations.

We have proven the potential and power of network analysis by applying it in a non-conventional way in the data analysis of a business process in the industry. In particular, we were able to obtain new insights about the correlation of failures occurring during the assembly process of coffee machines at *Nuova Simonelli S.p.A.*. This work shows a lot of potential by applying network analysis beyond the conventional application on social networks, neuroscience, and finance. Furthermore, we are planning to extend and improve this work by adopting advanced techniques on graphs such as Topological Data Analysis (TDA) [9], clustering techniques [2] and machine learning techniques that could provide further advances in modeling industrial processes. Finally, we also see much more potential for *Nuova Simonelli S.p.A.* to further invest in the methods and tools used in this project to gain more insights into the failure correlations within their assembly processes.

References

1. Boccaletti, S., Latora, V., Moreno, Y., Chavez, M., Hwang, D.-U.: Complex networks: structure and dynamics. Phys. Rep. **424**(4), 175–308 (2006)

2. Corradini, F., Luciani, C., Morichetta, A., Piangerelli, M., Polini, A.: $TLV\text{-}diss_\gamma$: a dissimilarity measure for public administration process logs. In: Scholl, H.J., Gil-García, J.R., Janssen, M., Kalampokis, E., Lindgren, I., Rodríguez Bolívar, M.P. (eds.) EGOV 2021. LNCS, vol. 12850, pp. 301–314. Springer, Cham (2021). https://doi.org/10.1007/978-3-030-84789-0_22

3. Jossen, S.: The world's most valuable resource is no longer oil, but data. The Economist (2017)

4. Khokhar, D.: Gephi Cookbook. Packt Publishing, Birmingham (2015)

5. Li, X., Liu, X., Yue, C., Liang, S.Y., Wang, L.: Systematic review on tool breakage monitoring techniques in machining operations. Int. J. Mach. Tools Manuf. **176**, 103882 (2022)

6. Luján-Mora, S., Trujillo, J., Song, I.-Y.: Extending the UML for multidimensional modeling. In: Jézéquel, J.-M., Hussmann, H., Cook, S. (eds.) UML 2002. LNCS, vol. 2460, pp. 290–304. Springer, Heidelberg (2002). https://doi.org/10.1007/3-540-45800-X_23

7. Manyika, J., Chui, M., Bisson, P., Woetzel, J.R., Dobbs, R., Bughin, J., Aharon, D.: The internet of things: mapping the value beyond the hype. Technological innovation (2015)

8. Mehmood, N.Q., Culmone, R., Mostarda, L.: Modeling temporal aspects of sensor data for MongoDB NoSQL database. J. Big Data **4**(1), (2017)

9. Merelli, E., Piangerelli, M., Rucco, M., Toller, D.: A topological approach for multivariate time series characterization: the epileptic brain. In: Proceedings of the 9th EAI International Conference on Bio-inspired Information and Communications Technologies (formerly BIONETICS) (2016)

10. Object Managemt Group. About the Unified Modeling Language Specification Version 2.5.1 (2017)

Attendance System via Internet of Things, Blockchain and Artificial Intelligence Technology: Literature Review

Sarumi Usman Abidemi[1]([✉]), Auwalu Saleh Mubarak[1], Olukayode Akanni[1], Zubaida Said Ameen[1], Diletta Cacciagrano[2], and Fadi Al-turjman[1,3]

[1] Artifiical Intelligence Engineering Department, AI and Robotics Institue, Near East University, Mersin 10, Nicosia, Turkey
20214000@std.neu.edu.tr, {awalusaleh.mubarak,zubaida.saidameen, Fadi.alturjman}@neu.edu.tr
[2] Computer Science Division, School of Science and Technology, Camerino University, 62032 Camerino, Italy
diletta.cacciagrano@unicam.it
[3] Research Center for AI and IoT, Faculty of Engineering, University of Kyrenia, Mersin 10, Kyrenia, Turkey

Abstract. Many researchers have make efforts on creating several attendance systems to keep track of student attendance in school which is also part of academic curriculum that will have great impact in student academic performance. There have been different technologies such as RFID, face recognition, biometric, web-based attendance system etc. To capture attendance in a way that's better than the traditional method of pen and paper or roll call system. Making this attendance record available on blockchain network and applying artificial intelligence algorithm for prediction or classification in correlation with the data captured from student attendance purpose.

Several articles, papers and journals has been reviewed from different research plethora of research publishers' database through internet such as Google scholar, IEEExplore, Springerslink etc. This will make it easy for comparing several attendance systems with respect to internet of things, blockchain and artificial intelligence to address the research gap in this aspect.

Keywords: Attendance system record · Blockchain · Internet of things · Artificial Intelligence

1 Introduction

Student attendance in academic system is one of the basic and very important factors that cannot be overlooked, it's very necessary for the academic staff, lecturers, students and parents to keep and monitor the record of student attendance safe secure and it should accessible based on request by the authorize personnel [1]. Making uses of internet of things technology to capture the student attendance safely where by every student can

L. Barolli (Ed.): AINA 2023, LNNS 655, pp. 321–330, 2023.
https://doi.org/10.1007/978-3-031-28694-0_30

mark attendance with their respective phone or computer without sharing any item or making physical contact to carry out the attendance task due to the effect of spreading virus such as COVID-19 virus in the university during the pandemic era [2]. It has been a great benefit and rapid development of many wired and wireless connecting devices, which have make exchange of information's and monitoring to be easier and faster since every student can achieve this task with their phone or laptop, while the record will be kept secured [3]. In recent year. Researchers are focused on integrating this three technology, internet of things, blockchain and machine learning algorithm in educational sectors, to checkmate, appraise the students learning and lecturer teaching activities. Many study have shown several applications of attendance system using internet of things IoT technology, but they couldn't be able handle the issue of data privacy and making it available with unique features [4] and some could be able to achieve this but couldn't be able to use this for predicting or checkmating the student performance and lecturer teaching appraisal, by integrating artificial intelligence with the internet of things (AIoT) and also blockchain decentralized technology due to its excellent features such as reliability, immutability, audibility, and security.

2   Research Methodology

In this section of methodology there is extensive survey of several attendance system ranging from attendance system without internet application, the attendance system with internet of thing, attendance system with internet of things and blockchain technology, internet of things and machine learning, internet of things (IoT), blockchain and artificial intelligence (Fig. 1).

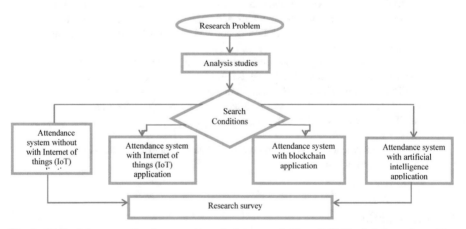

Fig. 1. Methodology on attendance system via Internet of things IoT, blockchain and machine learning application.

2.1   Digital Libraries & Respositories Used in the Research Survey

Springers Link (1 paper), IEEE Xplore (6 papers), Elsevier (1 paper), SDI Article (1 paper), CiteseerX (2 papers), UMP Institutional Repository (1 paper), NHS Digital (1 paper), Scholars archive (1 paper), Archivo Digital UPM (1 paper), Korea science (1 paper), Academia (2 papers), Researchgate (2 papers), Google scholar (3 papers).

2.2 Attendance System Without Internet of Things (IoT) Application

2.2.1   This project report [5] an online attendance system was designed that captured student attendance record in real time. It has an example of a generic architecture for an intra connected of RFID readers in a network at any institution e.g. A school. Additionally, it offers more services to simplify educational institution activities and has a fully automated system for tracking students' attendance. It is simple to integrate with other campus activities that make use of RFID readers like the canteen's automatic payment system and RFID base book library. This system just needs to be purchased once and has a lengthy lifespan; no additional costs are necessary.

2.2.2   In This article [6] describes the designs and construction of a portable, GSM-based fingerprint student attendance system. The terminal fingerprint acquisition module and attendance module are included in the system. This attendance system can receive fingerprint of individual, it will process this unique fingerprint by matching it with the fingerprints saved on the database and automatically use this information to create attendance record. After taking attendance, this system uses GSM to send each student's attendance to their parent's mobile device. It Access student's attendance in a certain class and this made it easier to serve as an attendance system. Deploying this system will eliminate the need for acquiring and uses of stationary items to maintain attendance records of the staffs.

Although, the primary purpose of this project is to track student attendance in lectures, tutorials, and laboratory practice exercise effectively and to inform parents of the student attendance. By allowing parents to receive SMS messages, this technology prevents students from skipping class. For the time and attendance system, biometrics has been employed successfully for more than ten years. The fingerprint attendance system is a straightforward, cost-effective method of identifying. Each person's fingerprint is distinct from the others and cannot be shared which make it effective for eliminating proxy attendance.

2.2.3   The authors [7] of this study have demonstrated how an RFID-based system may be created. Since this system is adaptable, other modules could be added to make it larger. The cards utilized for this system are RFID cards as a means of identification which represent the students on the attendance list and the programming codes. In addition, this algorithm programming codes has offer a good protection to the sensitive information they have placed on these cards. These cards can be used on campus and might take the role of student ID cards. As was seen, both staff and students can use these cards for a variety of functions; other features can always be added to the system and the cards can be given more protection. The moment has come for us to take advantage of RFID technology's ease and potential as it continues to advance. The major goal of this project was to develop an RFID-based system and show potential applications for the technology.

2.2.4   This paper [8] discussed the conception and creation of a attendance system fingerprint recognition based. This system reduced a number of problems, including the possibility of cheating when initiating attendance, and also it make it easier for lecturers to keep track of students' attendance records, to ensure security by applying encryption, preventing an anonymous fingerprint from having access to the data, and make it faster to take attendance rather than in a queue method. Features like indicator that will show the battery life, more security enhancements, a backup are all planned for the future. In summary, By integrating biometric fingerprint technology and an Arduino microcontroller, the authors constructed an attendance monitoring and management system that was both secure and portable.

2.2.5   This project report [9] has discussed the conception and creation of a portable attendance system based on fingerprint recognition. The system assisted in reducing a number of issues, including the possibility of cheating when recording attendance, ease of use for lecturers in maintaining student attendance data, increased security provided by the encryption technique, which prevents an anonymous fingerprint from being able to tamper with the data, and time savings from taking attendance on the go rather than in a line. Making this system wireless and utilizing the IOT (internet of things) idea are future projects.

2.3   Attendance System with Internet of Thing (IoT) Application

2.3.1   The goal of this project write-up [1] is to create a smart attendance system that effectively tracks and maintains student attendance in a setting on an automated basis. An Arduino ESP8266 Wi-Fi microcontroller that is installed in the classroom is used to construct the system. Additionally, Wi-Fi communication module was employed to facilitate connection between the ESP8266 Wi-Fi microcontroller and the student's devices, such as their phones or laptops. A database has been made where the supplied attendance will be recorded and saved. Internet is used to send data from the database to the front-end web page GUI graphical user interface and from the database to the information gathered by the microcontroller. The administration and students will complete a one-time registration process before connecting the student's device to the classroom microcontroller over Wi-Fi to record attendance. This system will compile a list of attendees with a real-time timestamp, save it to a database, and make it accessible to the instructor and admin at any time through the internet using a web browser like Google Chrome (GUI).

2.3.2   In this paper [10] NFC technology was utilized to build and execute an autonomous student attendance tracking system. Around 1100 first-year students and 8 courses were part of the monitoring that began at the beginning of the previous semester in 7 distinct lecture rooms. Almost without a hitch, the system ran the entire semester. The system accepted more than 200.000 card identifications with more than 20.000 biometric identifications during the semester. Out of the 1100 students, about 100 were barred from completing particular courses. There were no successful attacks against the monitoring system throughout the initial phases of its development. Students got used to using the system adequately two weeks following installation, and registration turned into a necessary task.

2.3.3   In this paper [11] the authors designed an automated attendance monitoring system that tracks students' attendance using the location data from their mobile devices. The created system was made to track the whereabouts of each student and professor, examine the distance distribution, and determine whether a student shows up for class, arrives late, departs early, or is missing. Some devices didn't display the desired outcome in accordance with the test results in this study, however the issue can be solved by applying error information and improving the dithering algorithm. The test results showed that the designed system could be used for attendance check if something needed to be improved upon was put into place and more study was done. Additionally, because the system may be used without any additional equipment, it will help produce high cost value.

In nutshell, identifying the geographic locations of smartphones, a mobile-based attendance system for students was created and built.

2.3.4   In this paper [12], the researcher presented a system for tracking student attendance that allows users to check their attendance records by scanning a QR code with their cellphones by deploying a novel mechanism that is All data will be organized and stored by the suggested system in a database server, making the data safe and secure from loss. Students will be able to correctly track their monthly attendance using the suggested approach by scanning a QR code with their smartphones. Additionally, because the entire system is computerized, the suggested solution will require less paper and administrative manual labor.

2.3.5   In this research report [13] a web-based Laboratory Attendance System (LAS) is a project that is basically designed for taking attendance records in the laboratory most especially in the university. This system has a lot of usefulness which make it more convenient for the laboratory instructor to grade student's assignments, allocate laboratory tool and materials, and also to evaluate student attendance because these tasks can all be done online. Additionally, this system makes use of RFID sensors with Arduino microcontroller which also benefits university administration in a number of ways, such as low cost, very fast and low energy consumption the users. It might persuade people to use information and communication technology (ICT) more frequently in their daily lives. Further study and application of this RFID Arduino strategy in the field of wireless transmitters and receivers for system integration is possible.

2.4   Attendance System with Blockchain Application

2.3.1   This paper [2] discussed about making attendance in the classroom with the help of their phone or laptop, students will be able to mark their attendance using the Wi-Fi attendance system that will be designed for this project. The record will be maintained on a distributed and decentralized block chain. This attendance system is set up so that after initiating connection with the Wi-Fi microcontroller module mounted in the classroom, students can register their attendance by entering their specific student number on a webpage. Every registered student has a copy of this record, which is stored and organized on a server database with time in and time out, the student's name, student number, date, and course ID. The generated attendance can be transferred through the internet to a node in a block chain. With the use of peer-to-peer capabilities of the distributer ledger, the generated attendance record can be easily accessible, the data can be trusted, and it can be transparent. Other parties, such as parents, students, or anyone on the network, can access the generated attendance record.

2.4.2   In this study's [14] the author stated that its challenging to exploit student attendance data because maintaining it is a very critical duty. Data forgery is quite widespread. Data modification is possible when using attendance data that is kept in traditional databases or on paper. Consequently, blockchain technology is applicable. With blockchain, data manipulation is impossible.

Most methods employed by attendance systems to store or retrieve data are traditional. The legitimacy of databases, papers, etc. cannot be verified by the traditional procedures. As a result, the attendance data needs to be kept up to date and shielded from outside interference. Blockchain is thus employed.

A web application will be our final product.

- they were able to store and safeguard attendance data on a blockchain that they built from the ground up.
- Using Python, they created blockchain from scratch

2.4.3   This study's [15] stated that blockchain technology can have effect how records are kept and maintained. Blockchain technology is a reliable invention. Blockchain aids in the decentralized and immutable maintenance of records and data. It also keeps the records accurate and transparent. It might alter the way we think about using third-party entities and relying on records. Additionally, it will lessen paperwork.

This study offers a plan for creating a fully functional student management and registration system that includes a system for recording attendance and storing student records.

2.4.4   This research paper [16] make discussion for Each and every student places a high value on attendance. Students at FSKKP will use a blockchain-based student attendance system to ensure that their attendance is protected from outside parties. Emulating the olden method with the uses of pen and paper for is said to consume more, especially when there are more pupils. This system's first problem statement is to limit the time wasting procedure. Lack of security is the second issue, as attendance sheets may not be as secure as computerized solutions. Because so many parties are involved, the data is easily lost. It is possible to abuse the attendance system, another issue is of space to keep record for all attendance of the student over a semester. In order to address the problem stated in the problem description, the objective has been identified. The first is to avoid the need for laborious procedures to track student attendance. The second is to use blockchain technology to produce a device that can track student attendance. Lastly establishing a system with efficient security.

2.4.5   In this report [17] the project was designed to be tied to wage payment and the imposition of employee discipline, employee attendance data is crucial to evaluating an employee's performance. As a result, the data must be accurately maintained and shielded against third-party data modification. To manage attendance transactions and ensure the accuracy and integrity of the information saved, a blockchain-staffs attendance system has been developed. Only people that have been granted right to access network will be aware of any changes to data there.

2.4.6   The goal of the project [18] is to create a complete Internet of Things to generate, gathering, store, and receiving reports on attendance records of students. This attendance system makes use of an Arduino NodeMCU board with a fingerprint sensor to produce data and transmit instantly gathered information to an Edge node through Wi-Fi. The data is generated with additional information at the edge node and can be saved inside a conventional SQL database after its has been filtered. The everyday attendance report is produced via a conventional web page in Excel format.

Finally, a decentralized application was created to make the daily report immutable and auditable by using smart contracts from the Ethereum blockchain and peer-to-peer InterPlanetary File System (IPFS) to store it.

The solution has been put to the test, and the results show that the blockchain's features make it impossible to change the daily record.

2.5   Attendance System with Artificail Intelligence Application

2.5.1   In this paper [19], an automated attendances system was design with monitoring features. it can show the live attendance activity of individual in a location, this attendance system will monitoring a student, staff, member or any personnel with is phone GPS location and google real time data storage, after the attendance has been recorded, this data record can be can be pushed to etheruim block chain network, with the help of JavaScript machine learning library, worldwide web network and blockchain network while the instant time-stamp monitoring is going on live, this record will be used for testing. Meanwhile the training dataset was retrieve from kaggle. At the end of the vent /semester/session the candidates performance can be evaluated using neural network artificial intelligence algorithm for overall performance predictions with the accuracy of 82%. This system can also act as a personal automated reminder.

2.5.2   In this research study [20], a facial recognition-based student attendance system was designed, the system can recognizes the face for unique identification of student taking an attendance by using Convolutional Neural Network to identify faces, Dlib's CNN or deep metric learning for facial embedding, and K-NN to classify faces. This attendance system records student attendance using personally identifiable information about the student, such as the student's ID number and attached it with, date, and timestamp. This attendance system is to bring a solution to some drawback of old or manual attendance procedure, which will be replaced with this technology, to automates the student attendance process. In order to speed up face identification for upcoming projects, cloud-based face recognition is being considered. Using a different, more advanced technology In order to compare performance, another, more advanced face recognition technique will be utilized. This technique, which in this case is the Convolutional Neural Network, should perform better in terms of speed and accuracy [20].

2.5.3   This project report [21] showed an artificial intelligence deep learning for facial recognition with web-based attendance system. The advanced deep learning face recognition was done with pre-trained models and get connected to the database through web application for storage of features in this proposed attendance system. The facial dataset have to be registered in other to make face matching process to be achievable to implement facial recognition which is embedded into this design. In order to replicate the actual situation, performance testing for facial recognition was carried out and measured in the production facility. It's very vital for All of the users face in the datasets must have high resolution and efficient lighting. Moreso, a number of test scenarios must be carried out to identify the appropriate threshold value to obtain with accuracy of about

92%, precision of 100% and recall of 90% which is a good result. This test was done by asking the user to make a testing verification on the attendance system.

2.5.4   This study [22] contains information of online signature attendance systems and a suggested process for creating one utilizing an online verification mechanism. Artificial Neural Networks-based online signature classification and verification method was examined with accuracy of 74%. Also a signature enrolment platform that would interact with an electronic signature pad for the purposes of enrollment and verification was considered. A relational attendance database was also suggested for compiling, analyzing, storing, and reporting the data gathered regarding lecture attendance. Additionally, a unit integration system that houses both the system's hardware and software components was suggested. The waterfall software development lifecycle model is suggested as a last step. The suggested system is able to offer remedies for the many recognized problems with lecture attendance.

2.5.5   In this article [23], the authors measure the impact of applying sophisticated models which is highly dimension but nonlinear with a random sample of 22,318 scheduled magnetic resonance imaging appointments at two different hospitals served as the basis for the training and evaluation of models with systematically varying levels of complexity based on several machine learning algorithm. The best performance was achieved by Machine learning based models with high dimensional gradient boosting features, which had an average precision of 0.511, and area under curve of 0.852 with 81 factors were necessary for ideal prediction performance that achieve 0.852 for the area under curve. Simulations indicated a net potential benefit that peaked at particular fee every appointment make during present prevalence as well as call efficiency, over a large volume of attendance features that occur within some patient factors necessitates more complicated models than have previously been used in the area for optimal attendance result prediction.

3   Conclusion

This research provide as extensive study on how Integration of Internet of things, Artificial intelligence and block chain technology has given the opportunity for large amount of critical data generated to be available on a decentralized and more secure network, which enhance a reliable dataset availability to training models, the model algorithm can be used to carry out some task such as prediction or classification, showing the advantages and limitations in those technologies in order to find a solution to the research gap with them.

References

1. Abidemi, S.U., Oghenetega, O.G., Daniel, S.A., Al-Turjman, F.: Wi-Fi attendance system in the IoT era. Lect. Notes Data Eng. Commun. Technol. **130**, 19–29 (2022). https://doi.org/10.1007/978-3-030-99581-2_3/COVER
2. Sarumi, U.A., Ameen, Z.S., Al-Turjman, F., Altrjman, C., Mubarak, A.S.: A novel attendance system via integrated WIFI and blockchain technologies. In: Proceedings - 2022 International Conference on Artificial Intelligence Everything, AIE 2022, pp. 209–215 (2022). https://doi.org/10.1109/AIE57029.2022.00046

3. Atzori, L., Iera, A., Morabito, G.: The Internet of Things: a survey. Comput. Netw. **54**(15), 2787–2805 (2010). https://doi.org/10.1016/J.COMNET.2010.05.010
4. Hassan, R.J., et al.: Editor(s): (1) Dr. Dariusz Jacek Jakóbczak. Asian J. Res. Comput. Sci. **8**(3), 32–48 (2021). https://doi.org/10.9734/AJRCOS/2021/v8i330202
5. Www, W., Patel, R., Patel, N., Gajjar, M.: Online students' attendance monitoring system in classroom using radio frequency identification technology: a proposed system framework. Int. J. Emerg. Technol. Adv. Eng. **2**(2), (2012). https://ijetae.com/. Accessed 02 Jan 2023
6. Verma, P., Gupta, N.: Fingerprint based student attendance system using GSM. Int. J. Sci. Res. *(2023)*. [Online]. http://wiki.answers.com/Q/How_does_a_fingerprint_sens
7. Saparkhojayev, N., Guvercin, S.: Faculty attendance control system based on RFID-technology. Int. J. Comput. Sci. Iss. **9**(3), 227–230 (2012)
8. Zainal, N.I., Sidek, K.A., Gunawan, T.S.. Manser, H., Kartiwi, M.: Design and development of portable classroom attendance system based on Arduino and fingerprint biometric. In: 2014 5th International Conference on Information and Communication Technology. Muslim World, ICT4M 2014, January. 2014. https://doi.org/10.1109/ICT4M.2014.7020601
9. Kumar Yadav, D.: Fingerprint based attendance system using microcontroller and Labview development of TDR based wireless system for slope monitoring in opencast mines View project Computer Science Department of IMA View project Pragyan Mishra (2015). https://doi.org/10.15662/ijareeie.2015.0406029
10. Benyo, B., Sódor, B., Doktor, T., Fördös, G.: Student attendance monitoring at the university using NFC. In: Wireless Telecommunications Symposium (2012). https://doi.org/10.1109/WTS.2012.6266137
11. Yong Hui, L., Hwan Seok, K., Byung Hwan, K.: A Study on the location based automatic attendance check system with smart devices (2014). https://doi.org/10.14257/astl.2014.60.18
12. Baban, M., Hikmat, M., Baban, M.: Attendance checking system using quick response code for students at the University of Sulaimaniyah a web based system for online student transaction among Sulaimaniyah City School Application view project using concrete strength to predict the amount of mixed material and lowest price estimation view project attendance checking system using quick response code for students at the University of Sulaimaniyah. J. Math. Comput. Sci. **10**, 189–198 (2014). https://doi.org/10.22436/jmcs.010.03.04
13. Arbain, N., Nordin, N.F., Isa, M., Saaidin, S.: LAS: web-based laboratory attendance system by integrating RFID-ARDUINO technology. In: *International Conference on Electrical, Electronics and System Engineering,* ICEESE 2014, pp. 89–94 (2014). https://doi.org/10.1109/ICEESE.2014.7154601
14. Prathyusha, O., Sushma Chowdary, K., Vaishnavi, N.P., Professor, A.: Securing attendance data using blockchain. *Int. Res. J. Mod. Eng. Technol. Sci.* www.irjmets.com @*International Res. J. Mod. Eng.* **3561**, 2582–5208 (2023), [Online]. www.irjmets.com. Accessed 03 Jan 2023
15. Joshi, S., Sharma, R., Sharma, S., Adgaonkar, A.: Blockchain in student registration system. Int. Res. J. Eng. Tecnol. **8**(5), 3652–3656 (2021)
16. [object Object]: Student attendance system using blockchain technology"
17. Ardina, H., Gusti Bagus Baskara Nugraha, I.: Design of a blockchain-based employee attendance system. In: 2019 International Conference on ICT for Smart Society (ICISS): Innovation and Transformation Toward Smart Region, November 2019. https://doi.org/10.1109/ICISS48059.2019.8969840
18. Padrón Núñez, J.: Biometric work attendance management and logging with a blockchain system. Thesis (Master thesis) (2021)
19. Vubangsi, M., Al-Turjman, F.: Design and Implementation of a conference attendance monitoring system using blockchain and AI technologies. In: Proceedings - 2022 IEEE International Conference on Artificial Intelligence in Everything (AIE), AIE 2022, pp. 197–202 (2022). https://doi.org/10.1109/AIE57029.2022.00044

20. Sutabri, T., Pamungkur, P., Kurniawan, A., Saragih, R.E.: Automatic attendance system for university student using face recognition based on deep learning. Int. J. Mach. Learn. Comput. **9**(5), 668–674 (2019). https://doi.org/10.18178/ijmlc.2019.9.5.856
21. Ismail, N.A., et al.: Web-based university classroom attendance system based on deep learning face recognition. KSII Trans. Internet Inf. Syst. **16**(2), 503–523 (2022). https://doi.org/10.3837/tiis.2022.02.008
22. "III. Online and Offline Verification Techniques"
23. Nelson, A., Herron, D., Rees, G.: Predicting scheduled hospital attendance with artificial intelligence. NPJ Digit. Med. **2**, 1–7 (2019). https://doi.org/10.1038/s41746-019-0103-3

An Overview and Current Status of Blockchains Performance

Hamza Salem[✉], Manuel Mazzara, and Siham Hattab

Software and Service Engineering Lab, Innopolis University, Innopolis, Russia
{h.salem,s.hattab}@innopolis.university, m.mazzara@innopolis.ru

Abstract. Since the introduction of Bitcoin and Ethereum, blockchain introduced new ways to handle payment and dealing with financial transactions. From 2009 till now, several blockchain systems were developed. However, the main issue for each blockchain is to know where to use it and what the best use case fits with it. Transaction per second (TPS) and confirmation time (CT) give the possibility to evaluate the speed and performance of each blockchain. It demonstrates the speed of blockchain in processing any transaction and saving it to the ledger. This work reviews and presents the current status of the state-of-the-art blockchain performance.

1 Introduction

Since almost ten years ago Satoshi Nakamoto introduced Bitcoin as the first distributed peer-to-peer financial system, that could be used to solve the double-spending problem in transactions [1]. Double-spending is the risk when the same unit of a digital currency is spent more than one time. It is unique problem to digital currencies in blockchain because digital information can be copied by individuals or groups who have skills such as blockchain network knowledge and the understanding of how much computing power to manipulate the blockchain [2]. On the other side, physical centralized financial systems do not have this problem. Also, decentralization limits blockchain speed in transactions processing [3]. It is a problem for all blockchain applications in the production environment.

Moreover, comparing blockchain financial applications' performance with other systems can not be fair because of the huge gap between both. For example, bitcoin can handle 6 transactions per minute and Visa transactions average 200,000 per minute. That's why most of the blockchain loses the performance race without even starting [4]. Therefore, transaction per second (TPS) is a very important topic for all blockchain systems. TPS represents the performance benchmark that the financial system measures the speed of transactions processing. Unfortunately, TPS is number declared by the blockchain founding team or tested for other blockchain systems using test networks. Also, some Blockchain provides simulators but all these numbers are not reliable until it will be seen the performance on production environment [5]. The main motivation was to discover the relationship between each chain and performance and provide an

© The Author(s), under exclusive license to Springer Nature Switzerland AG 2023
L. Barolli (Ed.): AINA 2023, LNNS 655, pp. 331–339, 2023.
https://doi.org/10.1007/978-3-031-28694-0_31

overview for scholars and researchers about the current state of blockchain performance. In this paper, we will provide an overview and the current status of blockchains performance. The paper is organized as follows. Section 2 presents an overview and comparison of Blockchains performance, and Sect. 3 concludes our work.

2   Overview and Comparison of Blockchains Performance

There are several factors that influence the performance of blockchain networks some of those factors are consensus algorithms, latency, node infrastructure, and transaction per second (TPS), this paper will be focused on the TPS and confirmation time (CT) as it is essential to identify how rapidly the network can confirm a trade or an exchange as well as it indicates the network's current capacity to process transactions.

To identify the material for our work we have performed a literature search in scientific databases such as Google Scholar, IEEE Xplore, WoS, Scopus databases, and the original white paper for some blockchains. All searches were conducted in December 2021. At the end of this process, we identified 38 blockchains that resource and white paper and already working at least as a test network.

In Table 1, we summarize the performance for 38 blockchain cryptocurrencies. In the next columns, we provide confirmation time for each transaction in seconds (CT) and the performance shown in Eq. 1. The performance is the benchmark to evaluate all blockchain. All TPS and CT coming from white paper or results from test network already running.

$$Performance = TPS/CT \tag{1}$$

As seen in Table 1 the values are not in the same range. For example, Solana (SOL) is counted as the fastest blockchain so far, it is faster 5 million times than bitcoin. Because of the big difference between the average of performance selected blockchain currencies, we have categorized blockchains regarding performance into four categories:

- Section A: Fast;
- Section B: Normal;
- Section C: Slow;
- Section D: Very Slow;

Section A is shown in Fig. 1 and this section contains 9 blockchains the performance range is 1700-12000 transactions per second including Solana, Nano, Avalanche, Cosmos, RED, STEEM, Hedera, EOS, and BitShares.

Table 1. Summary for 38 blockchain transaction per second, confirmation time in second and performance

Symbol	TPS	CT	Performance
SOL [6]	29000	2.575	11262.136
XNO [7]	1000	0.14	7142.858
AVAX [8]	5000	1	5000
ATOM [9]	10000	2	5000
RED [10]	14000	4	3500
STEEM [11]	10000	3	3333.334
HBAR [12]	10000	3	3333.334
EOS [13]	4000	1.5	2666.667
BTS [14]	3400	2	1700
NEO [15]	10000	15	666.667
XRP [16]	1500	4	375
XLM [17]	1000	4	250
WAVES [18]	100	2	50
MIOTA [19]	1500	60	25
ALGO [20]	1000	45	22.223
ICON [21]	9000	600	15
ONT [22]	5300	600	8.834
TRX [23]	2000	300	6.667
DGB [24]	560	120	4.667
XMR [25]	1000	1800	0.556
ADA [26]	250	600	0.417
BCN [27]	500	1200	0.417
DASH [28]	35	360	0.098
AE [29]	100	1080	0.093
DOGE [30]	33	360	0.092
BCH [31]	300	3600	0.084
ETH [32]	25	360	0.07
XEM [33]	2	30	0.067
Qtum [34]	70	1200	0.059
LTC [35]	53	1800	0.03
XTZ [36]	40	1800	0.023
BTC [1]	7	360	0.02
RVN [37]	116	6000	0.02
BCD [38]	56	3600	0.016
LSK [39]	25	3060	0.009
DCR [40]	14	1800	0.008
ZEC [41]	27	3600	0.008
BTG [42]	7	3600	0.002

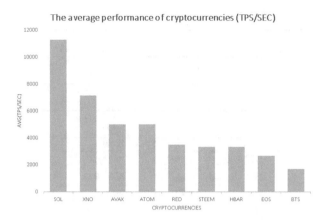

Fig. 1. Section A: fast chains

Section B is shown in Fig. 2 and this section contains 7 blockchains the performance range is 15–700 transactions per second including NEO, Ripple, Stellar, WAVES, MIOTA, Alogrand, and ICON.

Section C is shown in Fig. 3 and this section contains 6 blockchains the performance range is 0.4–15 transactions per second including Ontology, TRON, DigiByte, Monero, Cardano, and Bytecoin.

Section D is shown in Fig. 3 and this section contains 16 blockchains the performance range is 0.002–0.1 transactions per second including DashCoin, Aternity, DogeCoin, Bitcoin Cash, Ethereum, NEM, QUTM, Litecoin, Tezos, Bitcoin, RevenCoin, Bitcoin Diamond, LISK, Decred, Zcash and Bitcoin Gold.

Note that has been added the current and minimum TPS and CT for each chain, for example, Solana chain can run up to 65,000 transactions per second, however, we include the minimum that is shown in the tests and white paper. Also, Ethereum promise in Ethereum 2.0 will have higher TPS, and we include the current only on the chain. To sum up the observations, the first observation shows that every section can be used in specific use cases. For example, all chains in section A had a confirmation time of fewer than 5 s, and that shows why such as blockchain can work with instant transactions payment use-cases [43]. On the other hand, other chains can support transactions types that do not need to be instant. Also, our methodology can be shown in Fig. 5.

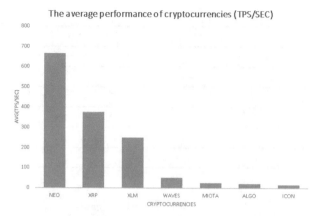

Fig. 2. Section B: normal chains

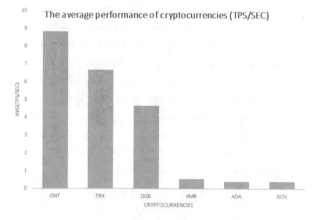

Fig. 3. Section C: slow chains

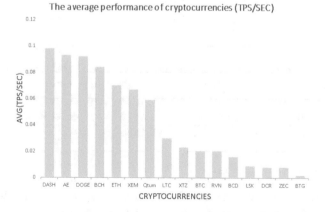

Fig. 4. Section D: very slow chains

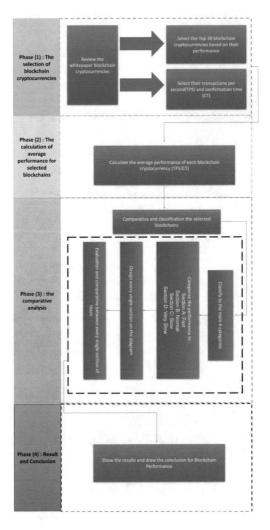

Fig. 5. Research methodology

The Next observation is that most chains still need to be simulated on several blockchain simulators [44] to confirm TPS and CT to be sure about the performance. Moreover, our investigations showed all chains still can not beat 24,000 TPS that VISA claim so far [45], however, the closest is Solana chain and all numbers should be verified in both systems.

3   Conclusion

In this paper, we have identified that the performance on 38 blockchains is still needed more verification methods such as a general simulator to measure Transaction per second (TPS) and confirmation time (CT) for transactions to be valid.

Most of the numbers found in all white papers are not valid by simulators. Moreover, Some blockchains already have test chains but still, it is not like measuring real production environment as well as undoubtedly there are other factors that affect the average speed of cryptocurrency performance, as mentioned during the research paper. Finally, we have provided a comparative analysis to identify blockchain performance using TPS and CT as main parameters to measure performance and we have categorized the 38 blockchains into 4 sections from the fastest to slowest regarding the current information that is available from the current state.

References

1. Nakamoto, S.: Bitcoin: a peer-to-peer electronic cash system (2008). [Online]. https://bitcoin.org/bitcoin.pdf
2. Karame, G.O., Androulaki, E., Capkun, S.: Double-spending fast payments in bitcoin. In: Proceedings of the 2012 ACM Conference on Computer and Communications Security, pp. 906–917 (October 2002)
3. Zhou, Q., Huang, H., Zheng, Z., Bian, J.: Solutions to scalability of blockchain: a survey. IEEE Access **8**, 16 440-16 455 (2020)
4. Velde, F.: Bitcoin: a primer (2013)
5. Mechkaroska, D., Dimitrova, V., Popovska-Mitrovikj, A.: Analysis of the possibilities for improvement of blockchain technology. In: 2018 26th Telecommunications Forum (TELFOR), pp. 1–4. IEEE (2013)
6. Yakovenko, A.: Solana: a new architecture for a high performance blockchain v0. 8.13. Whitepaper (2018)
7. Original Raiblocks/Nano Whitepaper - Nano Documentation. Nano Docs. https://docs.nano.org/whitepaper/english/
8. Tanana, D.: Avalanche blockchain protocol for distributed computing security. In: 2019 IEEE International Black Sea Conference on Communications and Networking (BlackSeaCom), pp. 1-3. IEEE (June 2019)
9. Cosmos Network. https://cosmos.network. Accessed 2019
10. Crain, T., Natoli, C., Gramoli, V.: Evaluating the red belly blockchain. arXiv preprint arXiv:1812.11747 (2018)
11. Guidi, B., Michienzi, A., Ricci, L.: Steem blockchain: mining the inner structure of the graph. IEEE Access **8**, 210251–210266 (2020)
12. Baird, L., Gross, B., Donald, T.: Hedera Consensus Service (2020)
13. Xu, B., Luthra, D., Cole, Z., Blakely, N.: Eos: An architectural, performance, and economic analysis (2018). Accessed 11 June 2019
14. Schuh, F., Larimer, D.: Bitshares 2.0: general overview. http://docs.bitshares.org/downloads/bitshares-general.pdf. Accessed June 2017
15. Elrom, E.: NEO blockchain and smart contracts. In: The Blockchain Developer, pp. 257–298. Apress, Berkeley (2019)
16. Qiu, T., Zhang, R., Gao, Y.: Ripple vs. SWIFT: transforming cross border remittance using blockchain technology. Procedia Comput. Sci. **147**, 428–434 (2019)
17. Lokhava, M., et al.: Fast and secure global payments with Stellar. In: Proceedings of the 27th ACM Symposium on Operating Systems Principles, pp. 80-96 (October 2019)
18. wavesenterprise. We-Whitepaper-En.Pdf (2021). https://wavesenterprise.com/we-whitepaper-en.pdf

19. Silvano, W.F., Marcelino, R.: Iota tangle: a cryptocurrency to communicate Internet-of-Things data. Futur. Gener. Comput. Syst. **112**, 307–319 (2020)
20. Gilad, Y., Hemo, R., Micali, S., Vlachos, G., Zeldovich, N.: Algorand: scaling byzantine agreements for cryptocurrencies. In: Proceedings of the 26th Symposium on Operating Systems Principles, pp. 51–68 (October 2017)
21. ICON-Whitepaper (n.d.). http://docs.icon.foundation/ICON-Whitepaper-EN-Draft.pdf
22. Introduction - Ontology Developer Center. https://docs.ont.io/
23. Li, H., Li, Z., Tian, N.: Resource bottleneck analysis of the blockchain based on tron's TPS. In: Liu, Y., Wang, L., Zhao, L., Yu, Z. (eds.) ICNC-FSKD 2019. AISC, vol. 1075, pp. 944–950. Springer, Cham (2020). https://doi.org/10.1007/978-3-030-32591-6_103
24. Kelly, B.: The Bitcoin Big Bang: How Alternative Currencies Are About to Change the World. John Wiley & Sons, New York (2014)
25. Miller, A., Möser, M., Lee, K., Narayanan, A.: An empirical analysis of linkability in the monero blockchain. arXiv preprint arXiv:1704.04299 (2017)
26. Kondratiuk, D., Seijas, P.L., Nemish, A., Thompson, S.: Standardized crypto-loans on the Cardano blockchain. In: Bernhard, M., et al. (eds.) FC 2021. LNCS, vol. 12676, pp. 579–594. Springer, Heidelberg (2021). https://doi.org/10.1007/978-3-662-63958-0_41
27. Bytecoin Whitepaper Doc.' Bytecoin Whitepaper. https://bytecoin.org/old/whitepaper.pdf
28. Duffield, E., Diaz, D.: Dash: a privacycentric cryptocurrency (2005)
29. Hess, Z., Malahov, Y., Pettersson, J.: Æternity blockchain (2017). https://aeternity.com/aeternity-blockchainwhitepaper.pdf
30. Dogecoin: Dogecoin: https://github.com/dogecoin/dogecoin/blob/master/README.md (7 November 2021)
31. Webb, N.: A fork in the blockchain: Income tax and the Bitcoin/Bitcoin Cash hard fork. NC J. Law Technol. **19**(4), 283 (2018)
32. Rouhani, S., Deters, R.: Performance analysis of ethereum transactions in private blockchain. In: 2017 8th IEEE International Conference on Software Engineering and Service Science (ICSESS), pp. 70–74. IEEE (November 2017)
33. Morgadas Palau, A.: Development of a distributed social network using NEM blockchain. Bachelor's thesis, Universitat Politècnica de Catalunya (2019)
34. Chavady, M.G.J.: A study of oracle systems for the QTUM blockchain eco-system. Doctoral dissertation, University of Dublin (2020)
35. Heilman, E., Lipmann, S., Goldberg, S.: The Arwen trading protocols. In: Bonneau, J., Heninger, N. (eds.) FC 2020. LNCS, vol. 12059, pp. 156–173. Springer, Cham (2020). https://doi.org/10.1007/978-3-030-51280-4_10
36. Allombert, V., Bourgoin, M., Tesson, J.: Introduction to the tezos blockchain. In: 2019 International Conference on High Performance Computing & Simulation (HPCS), pp. 1–10. IEEE (July 2019)
37. Fenton, B., Black, T.: Ravencoin: a peer to peer electronic system for the creation and transfer of assets (2019, July)
38. Bitcoin-Diamond-Whitepaper (n.d.). https://www.bitcoindiamond.org/wp-content/uploads/2018/08/Bitcoin-Diamond-Whitepaper-1.pdf
39. Alves, D.: A strategy for mitigating denial of service attacks on nodes with delegate account of LISK Blockchain. In: Proceedings of the 2020 the 2nd International Conference on Blockchain Technology, pp. 7–12 (March 2020)

40. Jepson, C.: DTB001: Decred Technical Brief. https://coss.io/documents/white-papers/first-blood.pdf. Additional information available at https://www.decred.org

41. Kappos, G., Yousaf, H., Maller, M., Meiklejohn, S.: An empirical analysis of anonymity in zcash. In: 27th USENIX Security Symposium (USENIX Security 2018), pp. 463–477 (2018)

42. Btcgpu (n.d.). Technical Spec · BTCGPU/BTCGPU Wiki. https://github.com/BTCGPU/BTCGPU/wiki/Technical-Spec

43. Zhong, L., Wu, Q., Xie, J., Guan, Z., Qin, B.: A secure large-scale instant payment system based on blockchain. Comput. Secur. **84**, 349–364 (2019)

44. Paulavičius, R., Grigaitis, S., Filatovas, E.: An overview and current status of blockchain simulators. In: 2021 IEEE International Conference on Blockchain and Cryptocurrency (ICBC), pp. 1–3. IEEE (May 2021)

45. Kiayias, A., Panagiotakos, G.: Speed-Security tradeoffs in blockchain protocols. IACR Cryptol. ePrint Arch. **2015**, 1019 (2015)

AgriBIoT: A Blockchain-Based IoT Architecture for Crop Insurance

Oumayma Jouini[1(✉)] and Kaouthar Sethom[2]

[1] Innov'COM Laboratory, University of Carthage, SUP'COM,
University of Tunis El Manar, ENIT, Ariana, Tunis, Tunisia
oumayma.jouini@enit.utm.tn
[2] Innov'COM Laboratory, SUP'COM, ENICarthage, University of Carthage,
Ariana, Tunisia
k_sethombr@yahoo.fr

Abstract. The climate has always played a central role in the lives of farmers. This close link is reinforced by climate change. Faced with increasingly significant consequences and growing uncertainties, they must find strategies to anticipate and protect themselves. Parametric agricultural insurance, which has recently appeared on the market, can help them. No need to send an expert on-site to see the damage, it is based on a climatic index (temperature, rainfall, humidity rate, etc.). When a certain threshold is exceeded on the index scale, compensation is automatically triggered. Around the world, several innovative start-ups have recently invested in this niche, with the ambition of combining their know-how in insurance with technology to offer solutions in response to climate change. In this paper, we propose to mix the IoT technology with the Blockchain to enhance and simplify the insurance process...

Keywords: Blockchain · Internet of Things · Smart agriculture · Wireless Sensor Networks · Smart contracts

1 Introduction

Agriculture is the oldest and most important human activity still practiced today. Due to its great importance for human beings, this area has become the object of a lot of studies, to improve it and resolve its current challenges. With the exponential growth of the world population, we will have to produce 70% more food by 2050, according to FAO estimates [1]. With the reduction of agricultural land and the depletion of natural resources, the need to improve agricultural yield has become critical. The limited availability of natural resources such as fresh water and arable land, as well as the slowdown in yield trends of tapping types of staple crops, have further aggravated this phenomenon. On the other hand, the changing structure and shrinking labor force are other concerns that are hampering the agricultural industry. Due to the decline in agricultural labor, the adoption of internet connectivity solutions in agricultural practices has

© The Author(s), under exclusive license to Springer Nature Switzerland AG 2023
L. Barolli (Ed.): AINA 2023, LNNS 655, pp. 340–350, 2023.
https://doi.org/10.1007/978-3-031-28694-0_32

become crucial in order to reduce the need for manual labor. IoT solutions aim to help farmers bridge the gap between supply and demand, improving yields and profitability with a positive impact on the environment. Agriculture has been one of the precursor sectors in the use of new technologies. Sensors, connected consoles, precision weather stations, and even drones have experienced strong growth in recent years [2–5].

This development makes it possible to gain precision, by obtaining weather data at the plot, for example, or even by determining the specific needs of the plant. The challenge today is to combine data from different sensors and connected objects to be able to carry out precise and predictive analysis of the meteorological, health, or even economic situation. Synergies are created between the different data recovery methods, thus generating high-added value in the use of digital technology. Associated with the IoT, the Blockchain presents new opportunities for agriculture 4.0. Blockchain technology was initially invented for the Bitcoin cryptocurrency and then used for other financial transactions. However, a great interest is made in this new decentralized and secure architecture, and is more and more reported as a possible solution in new areas. This popularity can be assessed by the high number of scientific articles dealing with blockchain in both Scopus and Web of Science databases. Blockchain may represent a possible answer to agriculture 4.0, because it offers deep security and scalability issues.

The paper is organized as follows: Sect. 2 is devoted to a literature review, Sect. 3 is where we discuss our AgriBiot architecture and its main components, Sect. 4 shows the registration and authentification of the IoT devices, the process of storing IoT devices data in the Blockchain is detailed in Sect. 5, Sect. 6 is dedicated to the AgriBIoT implementation, and Sect. 7 is where we wrap things up.

2 Blockchain to the Farmer's Rescue

Global warming, climate change, and climate anomalies... increasingly used terms reflecting reality of the present but above all the future. Climate anomalies, such as cyclonic phenomena, droughts, or extreme precipitation, are more and more frequent worldwide. To counter the unpredictability of farming, agricultural insurance schemes are a go-to option for most farmers. Farmers pay a reasonable insurance premium at the start of their cropping cycle and are eligible to receive a payout in case they experience any losses in their farms due to weather abnormalities. This gives farmers a fair bit of buffer room to tackle the uncertainties caused by weather changes.

There are various types of insurance policies available to farmers that differ based on how losses are calculated and payouts are executed. Indemnity-based insurance is a popular type of agricultural insurance that indemnifies farmers based on the assessment of an expert who visits the farm to examine the damages. Indemnity-based insurance, however, has numerous drawbacks related to damage assessment and lack of insurer information, and this negatively affects the farmers and the insurance agency as well in the long run. To provide a better alternative to indemnity-based insurance, blockchain technology enables the use of index-based insurance. Parametric insurance, also called index or index insurance, is mainly based on meteorological data: as soon as a climatic anomaly is observed (based on a rainfall index, temperature, or other criteria selected), compensation for the claim is triggered. The principle of parametric insurance is the construction of an index correlated with the variable of interest (i.e. sales, agricultural yields, crop quality, etc.). This index must however respect certain properties and must be:

- Observable and easily measured.
- Objective.
- Independently verifiable.
- Communicated in a timely manner.
- Consistent over time.

Blockchain can contribute to the improvement of index-based insurance in two major ways. Firstly, the basis of payments can be switched to a timely and automated parameter such as weather data. This parameter can be used to trigger the final payout based on the clearly defined terms of a smart contract. Secondly, all data sources including weather information and plant growth information can be automatically utilized in the insurance scheme with the help of a smart contract. This helps to improve the pay-out process and index determination substantially.

3   AgriBiot Architecture

In this section, we describe the main components of the proposed architecture as depicted in Fig. 1. It's divided into four main components as follows.

3.1   Perception Layer

Remote sensing is a potential tool for monitoring crop health and condition. This layer is composed by IoT sensors and an IoT node. Sensors are responsible for providing the essence of the Internet of Things, which is environmental monitoring and plant growth. These can either be embedded in devices or implemented as stand-alone objects to measure and collect data. IoT sensors are typically small in size, low in cost, low in power consumption, and easy to deploy. The monitored data is about environmental conditions including weather, wind speed, temperature, and plants growth conditions such as soil humidity, chemical and physical properties of soil like the pH level,... [2]. They can be powered by batteries, which can be recharged from some source of renewable energy like solar panels and wind turbines. The data sent and the commands received by this layer pass throw the IoT node. It consists of an IoT bridge with better capacity. It is used to expand the wireless range of the system and save IoT devices energy as they don't need more power consumption for higher range transmission to the far Gateway. The system relies on hop-by-hop transmission.

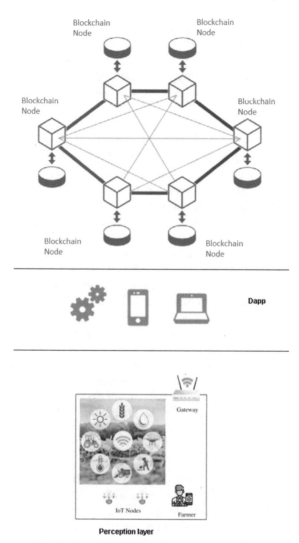

Fig. 1. AgriBiot architecture layers

3.2 Gateway

It facilitates communication between sensors and the rest of the system by converting sensor data into formats that are easily transferable and usable by other downstream system components. In addition, they are able to control, filter, and select data to minimize the volume of information that must be transmitted to the Dapp layer, which positively affects network transmission costs and response times. Thus, gateways provide a place for local pre-processing of sensor data ready for further processing. Message Queuing Telemetry Transport (MQTT) is used as a communication protocol between the IoT gateway and the logical layer. The IoT gateway can use long-range radio technology to connect to IoT nodes, but a link-distance link is required to connect the IoT node to sensors. The choice of technologies used is based on requirements in terms of required bandwidth and network availability in the considered area requirements. We use in our case LoRA technology for gateway wireless communications.

3.3   Dapp

All web applications consist of two parts: the front end and the back end. The particularity of our use case is that the back-end part exists on a blockchain. We opt for blockchain technology because it provides more security and reliability to the system while respecting the decentralized nature of the architecture. Web applications running on Blockchain are called distributed applications (Dapps). The code of a decentralized application (DApp) is known as the smart contract. The Dapp is designed to offer an easy-to-use, secure interface to manage insurance policies, farm details, and other information through a website and/or mobile app as shown in Fig. 2. The smart contract should consult an Oracle, and if the insurance policy is in effect (dated between the start date and expiry date), check if the geo-coordinates of the gateway are within the areas marked as under drought by the Oracle service (the Oracle service will return an entire set of polygonal vertices of areas under drought). If everything is OK, the smart contract writes the sensors updates on the blockchain. It's the responsibility of the gateway to authenticate the IoT devices before forwarding any request or event trigger to the Dapp layer (see Sect. 4 for more details). The code for the smart contract was written in Solidity language [6].

3.4   Blockchain Layer

There are many Blockchain frameworks for Distributed application deployment such as Hyperledger fabric [7], Quorum, Corda [8], Ethereum [9] etc. In our testbed, we decide to use Hyperledger fabric because it provides a private Blockchain. In fact, a public blockchain allows everyone to become a member and actor of the network unlike a private blockchain which only accepts members selected by the consortium or organization. In a public blockchain, any member can submit a transaction, they can read, write, or modify the database. This is not the case for a private blockchain where, to be able to submit a transaction, you must have the necessary permissions. This offers more confidentiality by prohibiting access to unknown users. The Hyperledger Fabric project is led by IBM and established under the Linux Foundation. Moreover, Hyperledger relies on a consensus protocol that does not require a native cryptocurrency. This reduces some risk/attack vectors and removes a resource-costly mining operation, which allows deploying a platform of the cost like any other distributed system.

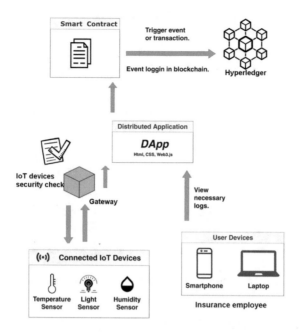

Fig. 2. Solution implementation logic

4   IoT Devices Registration and Authentification

First, to protect privacy all over the air transmissions, we assume secure communication channels between the IoT node, IoT devices, and the gateway. Some pre-established keys for ciphering exchanges are used. The keys can be refreshed by any classical algorithm such as A5. The farmer must register any new sensor using a frontend application provided to him by the insurance firm (Fig. 3). He logs into his account and adds all-new device information such as device name, MAC address, and description. He receives a code on his phone number to confirm the sensor added to the console. The system then generates a pair of keys (private and public) that will be used later by the gateway for the device's authentification and send it back to the device through the gateway and the IoT node.

Fig. 3. Interface for farmer's sensor registration

5   Recording IoT Devices Data Inside Blockchain

The simplest way to use a blockchain is to write data directly inside a transaction. Every transaction is then signed, timestamped, and mined by the blockchain's nodes, and finally stored permanently in a tamper-proof way. Any data within the transaction will therefore be stored identically but independently by every node, along with proof of who wrote it and when. The users can retrieve this information at any future time. Because IoT devices can generate important data volume, storing them directly in the blockchain is not a good choice [10]. We can solve data scalability in blockchains by embedding the hashes of data within transactions, instead of the data itself. Each hash acts as a "commitment" to its input data, with the data itself being stored outside of the blockchain or "off-chain". For example, using the popular SHA256 hash function, any updates received from the IoT sensor can be represented by only a 32-byte number, this puts us comfortably back in the territory of feasible bandwidth and storage requirements, in terms of the data stored on the chain itself. In our case, we use an off-chain platform a MongoDB database. Any insurance employee or application that needs the off-chain data can retrieve it inside the CouchDB database, then the on-chain hash serves to confirm who created it and when.

6   AgriBiot Implementation

For the physical layer, we used a temperature and humidity monitoring system using the DHT11 sensor connected to an Arduino UNO R3 board (Fig. 4). We used an MQTT Broker called Mosquitto [5]: the data is sent from the

Fig. 4. MQTT initialization and sensor monitoring

sensor to the MQTT server from the Arduino board. The IoT messages will trigger data analytics functions, which will extract farmer-relevant information and plants states. Only pertinent results will be saved in the distributed ledger on Hyperledger Blockchain thanks to the gateway. The stored information is then immutable, transparent, and traceable. The pre-treatment of IoT messages made by the gateway is necessary to avoid multiple updates inside the Blockchain. Each node among organizations will represent Farmers, insurance companies, and government regulators. The smart contract (chain code in Hyperledger) will be triggered automatically when a new farmer node event is submitted. A farmer node is composed of an SQL database to store IoT devices messages received from the MQTT broker, Express server is used for information analysis. The Fabric SDK is the module needed for communication with the Blockchain. The front end is the main part of any Blockchain application which is seen by the user. This part is developed using Nodejs and react frameworks. For the backend part, we have Hyperledger fabric and spring boot.

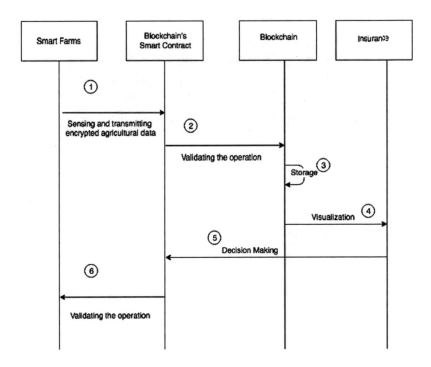

Fig. 5. Node's interaction in Hyperledger

A series of tests were executed. During each time, we simply send the value of a sensor through an MQTT broker to the IoT gateway and then write a transaction in the blockchain. For each test, we measured the time necessary to set the value in the blockchain (write time) and the time to read the information; the average values are summarized in Fig. 6. The number of IoT devices varies from 20 to 100. As we can see blockchain technology is not a good candidate for emergency applications as some time is needed to write and retrieve information (around 1 min). This is mainly due to the mining process. As our application is delay-tolerant, this will not affect the needed performance. Moreover, AgriBIoT scales well for an important number of transactions and nodes.

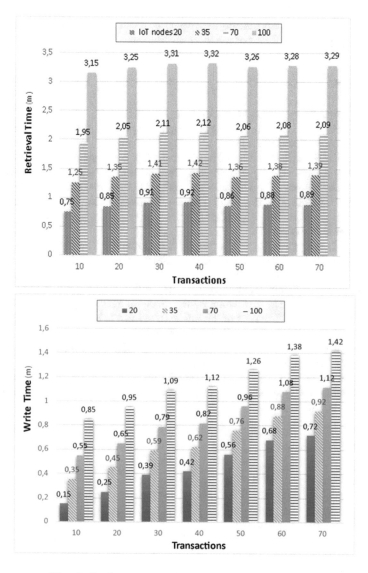

Fig. 6. Performance evaluation, time is in minutes

7   Conclusion

Conventional crop insurance systems are complex and often not economically feasible. The use of technologies like IoT and blockchain will facilitate the improvement of the current scenario in agriculture. This paper explores how the integration of IoT-based sensing and blockchains, can improve crop insurance for farmers. Hyperledger Fabric was used as the implementation blockchain. Our AgriBIoT solution benefits from a set of unique properties including immutability and transparency of cryptographically secured and peer-recorded architecture.

References

1. https://www.fao.org/news/story/en/item/35571/icode
2. Pinter, P.J., et al.: Remote sensing for crop management. Photogram. Eng. Remote Sens. **69**(6), 647–664 (2003)
3. Kamble, P., Shirsath, D., Mane, R., More, R.: IoT based smart greenhouse automation using Arduino. Int. J. Innov. Res. Comput. Sci. Technol. **5**, 347–5552 (2017)
4. Thakur, D., Kumar, Y., Kumar, A., Singh, P.K.: Applicability of wireless sensor networks in precision agriculture: a review. Wireless Pers. Commun. **107**(1), 471–512 (2019)
5. https://mosquitto.org
6. Dannen,, C.: Cryptoeconomics survey,. In: Introducing Ethereum and Solidity. pp. 139–147. Springer, Cham (2017), https://doi.org/10.1007/978-1-4842-2535-6_7
7. https://www.hyperledger.org
8. Quasim, M.T., Khan, M.A., Algarni, F., Alharthy, A., Alshmrani, G.M.M.: Blockchain frameworks. In: Khan, M.A., Quasim, M.T., Algarni, F., Alharthi, A. (eds.) Decentralised Internet of Things. SBD, vol. 71, pp. 75–89. Springer, Cham (2020). https://doi.org/10.1007/978-3-030-38677-1_4
9. Wood, G., et al.: Ethereum: a secure decentralised generalised transaction ledger. Ethereum Project Yellow Paper **151**(2014), 1–32 (2014)
10. Reyna, A., Martín, C., Chen, J., Soler, E., Díaz, M.: On blockchain and its integration with IoT challenges and opportunities. Futur. Gener. Comput. Syst. **88**, 173–190 (2018)

Distribution of the Training Data Over the Shortest Path Between the Servers

Ibrahim Dahaoui[1]([✉]), Mohamed Mosbah[2], and Akka Zemmari[1]

[1] LaBRI, Univ. Bordeaux, 33400 Talence, France
ibrahim.dahaoui@u-bordeaux.fr, zemamri@labri.fr
[2] CNRS, Bordeaux INP, LaBRI, UMR 5800, Univ. Bordeaux, 33400 Talence, France
Mosbah@labri.fr

Abstract. In this paper, we present a method that supports multi-server learning with distributed data, by collecting selected data from the neighbors of training servers, the proposed training scheme can train multiple models on multiple servers located at specific points. By collecting selected data from the neighbors of the training servers, the proposed training scheme can train multiple models on multiple servers located at specific points, then we will create a method to share the learning tasks by exchanging the data on the shortest path between the servers.

1 Introduction

Data collection tends to be thought about as the most expensive part of deep learning, with discovering efficiency generally boosting with the size of the information collection. Big data, such as photos and videos come from sites spread throughout different areas and also need substantial expense to collect in one area for learning, which can drastically interfere with the applicability of deep knowing. There are several types of computation to enhance the accuracy performance of deep learning, and the volume of input data affects the training performance. One example of a dispersed information learning approach is federated learning, which iteratively exchanges updates to the semantic network model criteria in between a main area as well as getting involved gadgets. A choice approach is to intelligently pick the most useful training examples for transmission. It has been observed that deep discovering scalability in terms of the number of training examples, the variety of version criteria, or both, can dramatically enhance the precision of the resulting classification.

We have developed a method to distribute the tasks between the connected servers so that each server with a good configuration (memory and CPU) takes the role of a learning server and the neighbors collect and send the distributed data. In this article, we highlighted the servers that finish the learning and wait for the others to finish sending their models. As we detail in Sect. 3 after the learning phase of each server that takes care of this task, we have noticed that there are servers that finish the learning before others, and it takes a lot of time

L. Barolli (Ed.): AINA 2023, LNNS 655, pp. 351–356, 2023.
https://doi.org/10.1007/978-3-031-28694-0_33

to finish the learning part of all servers. So we will propose a method that consists in restarting the learning by the available servers, i.e., they have completed this task with their data collected by sending it the data of the servers that are in the process of learning and it still has data to minimize the learning time. This choice will be based on the amount of data left from each server.

We experimentally validate the proposed method with several common image classification datasets. MNIST and CIFAR-10 based image classification results show that our proposed distributed training method achieves near-optimal time compared to the ideal centralized scheme with access to all data.

The rest of the paper is organized as follows. We review related work in Sect. 2. Sections 3 and 4 describe our proposed method in detail. Evaluation results are presented in Sect. 4 for several image classification datasets. Finally, Sect. 5 concludes the paper and discusses future work.

2   Related Work

In [1], they study the problem of efficient estimation of mean values for distributed data but did not consider the impact of the bandwidth constraint for machine learning [2]. While other authors study machine learning for distributed data, the approach is only valid for convex optimization problems and does not apply to general deep neural networks with rectified linear layers and maximum pooling layersb14.

As we did for [3], we have used a set of models trained on multiple sites and applied majority voting on all results to have a single model with an accurate prediction. Federated learning [4] gradient update methods to be communication efficient for distributed data, Alternative model-based methods, e.g., ensemble learning and model distillation. [5], can be used to recover a high-quality model by combining several models. The succession-based convex training approximation has been used for distributed neural network learning, although it does not apply to general deep architectures.

There is other work that compares deep distributed learning, such as Pytorch package [6], a tensor library optimized for deep distributed learning using GPUs and CPUs, and provides parallel distributed data as a *nn.Module* class, where applications provide their model at build time as a sub-module. Many techniques are built into the design to provide high-performance training, including compartment gradients, communication overlapping with computation, and synchronization hopping. This can be used later for each type of data if we want to achieve good training performance. The method we use is to distribute the data to a set of servers.

The most frequent problem in distributed learning is loss of time when we have very large data to use for training, so this prompts us to further investigate strategies for distributing learning and data across connected sites. We focus on the choice of the servers available for training and how to provide them with data from other sites that have more data during training.

2.1 Incremental Learning

As the data grows more and more, the training is more and more complicated and takes longer with less precision. In this article, we propose a method to do trench learning which means that a server which takes care of learning instead of waiting to collect all the data from connected servers starts learning with the data it has, and whenever it receives new data, it continues training and updates the model.

3 Experiments

In this section, we have implemented our method on a set of containers on docker [7], so we created and configured ten containers using docker-compose witch is a tool for defining and running multi-container Docker applications. With Compose, you use a *YAML* file to configure your application's services. Then, with a single command, you create and start all the services from your configuration. For the first version, we use 5k from the training set to initialize the common model, and organize the remaining into 10 folds of 5k images each (Fig. 1).

We are going to implement the proposed method in a multi-site setting. Our evaluations include two common image classification datasets, CIFAR-10 [8] and MNIST [9]. We have resized all images to 32×32. We randomly chose an equal number of training images for each of the 10 sites, we have resulted in 60,000 images in the training set and 10,000 images in the test set. We are going to work on data transfer (images, text...) so for that we have established a network configuration between a number of containers on dockers. For the communication between the users and the containers, we used a *flask* Api.

First of all, before directly starting the shortest path computation, we have already developed a method to select the servers that will handle the learning. This selection is based on each server's performance (CPU, memory) and its position.

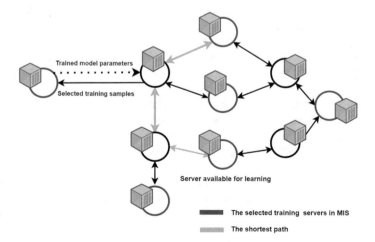

Fig. 1. A framework for continuous training of deep neural networks using selected server

Once the data is distributed, the servers start to collect the data from their neighbors and then start the learning process. When a server finishes its learning task, it sends a broadcast to all servers and if a server is still learning, it sends the remaining data to the nearest available server to help it learn what is left.

To do this we used the famous algorithm which is mostly used in road networks, but it is also used in traffic information systems and the Open Shortest Path First (OSPF), used in Internet routing. So how does a Dijkstra Algorithm work, I will explain this using four simple steps:

Step 1 Initially assign $Node(A) = 0$ as the weight of the initial node and $w(x) = \infty$ to all other nodes, where x represents the other nodes.

Step 2 Search x node for which it has the smallest temporary value of $w(x)$. Stop the algorithm if $w(x) = \infty$ or there are no temporary nodes. The node x is now labeled as permanent and as the current node, meaning parent of x and $w(x)$ will stay fixed.

Step 3 For each node adjacent to x labeled y which are also temporary, apply the following comparison: if $w(x) + Wxy < w(y)$, then $w(y)$ is updated to $w(x) + Wxy$, where W is the cost of the adjacent node. Now assign y to have parent x.

Step 4 Repeat the process from Step 2, doing as many iterations as required until the shortest path is found.

As shown above in the figure, we have configured all servers to be connected with their neighbors, our strategy is to reduce the overload of training data on the training servers by finding other servers that are available to take the training or contribute to the training in order to minimize the training time and also the energy provided for the servers.

All servers will go through the following steps to establish our strategy:

- **Start of the training:** During this phase each server after the collection of data it launches the learning.
- **Send a broadcast:** When a server finishes its training, it updates everyone by sending a broadcast.
- **Sending request:** The servers that are currently learning send a request to all available servers to ask for authorization to send them the remaining data. This request contains the learning progress status and a number that corresponds to the remaining data rate.
- **Selection of server:** Once the available servers receive the requests from the others, it processes and it makes the choice of the good servers in the course of learning. This choice is based on the distance parameter and the remaining data rate, the one with the shortest path will have priority to allow it to send the remaining data.
- **Data distribution:** Once the available server makes the choice, it allows him to send either all the data or a part of the data and it depends if the server being trained has found several servers available and very close, so in this case it will send a part of these data to each available server. Knowing that we use the algorithm of *Dijkstra* to calculate the closest distance between the servers.

4   Evaluation Results

We focus on the choice of the shortest path between training sites. Then, we evaluate the proposed models with and without the help of other servers to find out if we have reduced the time or perhaps also the performance. We tested two types of data on the learning servers, and noticed a **4.7x** faster learning time on the configuration we proposed compared to a single site centralized learning and **1.4x** compared to classic learning this means that each server does its task until the end, using the same resources (containers, configuration). The advantage of our method is not only the speed of learning but also the fact that we select the most efficient training servers from all the servers (Fig. 2).

(a) Before the intervention of available training servers

(b) After the intervention of available training servers

Fig. 2. Simulation of the transit time diagram for each epoch.

You can see in the figure that with the five training servers we have improved the learning time. For both types of data, we have the same result in terms of optimization but it differs in terms of the data rate.

5   Conclusions and Future Work

This paper presents an approach for distributing untrained data to available servers by the shortest path between them using Dijkstra's algorithm. Through this experiment using two popular datasets (MNIST and CIFAR-10), we show that the right choice of training sites and data accelerates learning and results in good performance in terms of time. We can apply this method using an IoT network to facilitate large-scale learning.

We have certainly used known and well-structured data types on machines with standard configurations to demonstrate the usefulness of the algorithms used. We can then develop a smarter system for selecting the shortest paths before starting to learn all the servers to see if we can also save more time. A larger scale study is needed to further demonstrate the benefits and limitations of our experience. It will also be interesting to examine the impact of machine performance on larger datasets.

References

1. Suresh, A.T., Yu, F.X., McMahan, H.B., Kumar, S.: Distributed mean estimation with limited communication. arXiv preprint arXiv:1611.00429 (2016)
2. Konecn'y, J., Richt'arik, P.: Randomized distributed mean estimation: Accuracy vs communication. arXiv preprint arXiv:1611.07555 (2016)
3. Dahaoui, I., Mosbah, M., Zemmari, A.: Distributed training from multi-sourced data. In: Barolli, L., Hussain, F., Enokido, T. (eds.) AINA 2022. LNNS, vol. 450, pp. 339–347. Springer, Cham (2022). https://doi.org/10.1007/978-3-030-99587-4_29
4. Konecn'y, J., McMahan, H.B., Ramage, D., Richtárik, P.: Federated optimization: Distributed machine learning for on-device intelligence. arXiv preprint arXiv:1610.02527. October 2016
5. Hinton, G., Vinyals, O., Dean, J.: Distilling the knowledge in a neural network. arXiv preprint arXiv:1503.02531. March 2015
6. Li, S., et al.: PyTorch distributed: experiences on accelerating data parallel training (2020)
7. Merkel, D.: Docker: lightweight Linux containers for consistent development and deployment. Linux J. **2014**(239), 2 (2014)
8. Krizhevsky, A., Hinton, G.: Learning multiple layers of features from tiny images (2009)
9. LeCun, Y., Bottou, L., Bengio, Y., Haffner, P.: Gradientbased learning applied to document recognition. Proc. IEEE **86**(11), 2278-2324 (1998). http://yann.lecun.com/exdb/mnist/

A Decentralized Architecture for Electric Vehicle Charging Platform

Marlon Rodrigues Martin and Fabiano Hessel[(✉)]

Pontifícia Universidade Católica do Rio Grande do Sul - PUCRS, Porto Alegre, Brazil
marlon.martin@edu.pucrs.br, fabiano.hessel@pucrs.br

Abstract. Hybrid and fully electric vehicles (EV's) have been drawing the attention of consumers and manufacturers due to the increase in fuel prices, the environmental impact that the burning of fossil fuels causes and the low volume of maintenance that vehicles in this category present. However, there are some issues that still keep consumers away from this type of vehicle, among them is the sales value, which is still much higher compared to combustion models (impacted by the cost of components such as batteries and other related items) and, the battery recharge time, which in the best case scenario, can take around 30 min, which is much higher than the time it takes to refuel a combustion vehicle. The scenario presented, where the driver needs to wait a certain period for the recharge of his EV's, presents a business opportunity where personalized products and services can be offered through data sharing. In this work, the architecture of a platform is proposed, which allows the connection between the driver and companies interested in offering personalized services while he waits for the electric vehicle to be recharged. The architecture in question is based on the concept of web 3.0 where the driver has full control over the data that will be shared, unlike other architectures where user data is centralized and under the domain of companies with no guarantee of security, traceability, or immutability of this information.

1 Introduction

They are classified as electric vehicles (EV's), automobiles, motorcycles and other vehicles, whose main characteristic is the presence of one or more electric motors responsible for their propulsion, either in full or in part (when the vehicle also has a combustion for the same purpose).

Parallel to this, issues related to the widespread use of fossil fuels, global warming and environmental legislation around the world, make not only car manufacturers, but also consumers in general, start to look for EV's as an alternative to road transport because, this category of vehicles allows the use of electricity produced from renewable energy sources, reducing the impact on the environment while road transport, based on engines powered by fossil fuels, is responsible for more than 70% of greenhouse gas emissions [8].

Even with the growing demand, the representation of EV's running on the roads is still small, this is due to issues such as the low availability of charging stations, limited autonomy, and the high cost of acquiring vehicles in this category, since it depends on entirely from batteries for energy storage and currently, they have a high development cost. For this reason, consumers sometimes opt for hybrid models, that is, they can count on a combustion engine associated with the electric one and, consequently, containing smaller (and cheaper) batteries [8].

We can also consider the waiting time to recharge the car battery as an obstacle to the popularization of EVs, which, depending on the type/model of the charger, can last between 4 and 8 h. It is important to point out that improvements in this area are constantly being evaluated, and it is already possible to observe results where improved station models that allow the driver to recharge approximately 80% of the vehicle's battery in 30 min. When comparing the improved recharge time with previous models, the evolution of results is evident, however, it is still a very distant scenario for users of combustion vehicles that need a few minutes for refueling [6]. In Brazil, it is possible to observe movements (public and private) that seek to increase the supply of EV charging stations. An example is in the city of São Paulo, which implemented legislation requiring new residential and commercial buildings to have internal charging stations and individualized electricity consumption [14].

Due to the fact that the most optimized scenario for recharging an EV's determines that its driver remains with the vehicle for a period of 30 min to obtain a considerable level of autonomy, an opportunity for new business arises with the objective of adding value to the driver during the time it waits for replenishment. This opportunity consists of a platform capable of offering personalized services and products, to be consumed while the driver waits for a recharge. For the best performance of an offer and recommendation application, data sharing is required between driver, vehicle, charging station and platform in a traceable format.

With the objective of providing greater security, transparency of use, traceability and immutability of data during the offer of personalized products and services, this study proposes an architecture based on the concepts of web 3.0, for the construction of a solution that allows the driver of a electric vehicle perform actions such as station reservation, availability check and also obtain commercial benefits from establishments close to that station, which registered on the platform, can provide a personalized experience during the recharge period. Unlike current solution architectures, where data and application are centralized on a server hosted in a data center under the domain of large technology companies, the proposed architecture takes advantage of the concept of decentralization of data and processing provided by Blockchain technology, where data are effectively the property of the user, and it is up to him to share (or not) this information [1].

The next Sects. 2 and 3 present background and related work respectively. Section 4 describes the main features of our architecture based on web 3.0 concepts and Sect. 5 discusses the architecture properties and its implementation strategy. Final considerations and future work are presented in the last Sect. 6.

2 Background

Motor vehicles were introduced to the market more than 100 years ago and, at the time, they were already offered with different propulsion formats such as electricity, steam and combustion, the latter being consolidated as the main propulsion format due to the high availability of fossil fuel on the market. Recently, electric vehicles have become more popular in the market because they provide their drivers with new experiences, such as being quieter (and pleasant to drive), economical (low cost for refueling and maintenance) and, as they do not harm the environment, since they do not emit polluting gases like combustion vehicles [12]. It is important to highlight that EVs that operate exclusively with electric motors have a zero-emission coefficient of pollutants through their exhaust, however, depending on the form and source of electricity used to recharge the vehicle, pollutant emissions may occur (on a smaller scale). And in this way, generate impact on the environment [6].

Currently, electric vehicles are classified into three different categories, which are basically defined by their electric propulsion format (full or partial, when combined with a combustion engine) and, the possibility of external energy recharging, they are: Hybrid Electric Vehicles (HEV's), Plug-in Hybrid Electric Vehicles (PHEV's) and Battery Electric Vehicles (BEV's). Table 1 presents characteristics that distinguish the electric vehicles classified in each category.

Table 1. Features that distinguish electric vehicle categories.

Feature	HEV's	PHEV's	BEV's
Auxiliary Combustion Engine	✓	✓	X
Regenerative Breaking	✓	✓	✓
Connector for external battery charging	X	✓	✓
Battery Life	Low	Medium	High

As we can see in Table 1, the PHEV's and BEV's categories have connectors that allow the battery to be recharged by agents and external sources such as outlets or specialized stations. These stations can be found in homes, commercial buildings, hospitals, shopping malls and gas stations, and it is possible to find them with two charging systems, they are alternating current (AC) or direct current (DC) [11].

The AC charging system is most found in homes and commercial buildings, because it can be installed in different electrical circuits, being able to provide a charging speed that can vary between 2 and 8 h and deliver to the vehicle a good level of displacement autonomy. On the other hand, fast charging stations use the DC system, which has the ability to supply a flow of energy continuous capable of charging the vehicle between 30 min and 2 h [11]. It is important to point out that fast charging stations require a considerable investment for their installation, which ends up alienating investors and making their distribution difficult.

Studies and initiatives that seek to find alternatives to the charging waiting time for EVs are being conducted, however, the results presented still maintain the time necessary to perform this task, still very far from what was observed for refueling a combustion vehicle.

There are also other initiatives, which seek to improve the experience and use of charging stations, an example is the Open Charge Point Protocol (OCPP), an open protocol for communication between the charging point (station) and a central charge management system. Seasons. This initiative was initially developed by the ELaadNL foundation in the Netherlands, which in essence allows the exchange of information (availability and status) between the station and its management center, in addition to allowing the execution of operations and configurations remotely [2]. OCPP allows central management or applications to perform functions such as:

- *Core*: Section that allows the execution of operations such as authorization, notification, station status change, general settings, communication analysis (heartbeat) and obtaining station meter values.
- *Firmware Management*: Section that allows to update the firmware of the charging station remotely.
- *Local Auth List Management*: Local file that allows transactions to be carried out even if they are offline in the environment.
- *Remote Trigger*: Section responsible for providing the current state of charging stations for an improved diagnosis.
- *Reservation*; Section that allows the reservation of stations for recharging vehicles.
- *Smart Charging*: Load profile configuration.

Present in all fast-charging stations for electric vehicles, OCPP is a fundamental protocol for the development of the proposed architecture, some manufacturers also provide an application layer or API's with new functionalities and functions that facilitate the integration of the station with different consumers [two].

The possibility of extending the connectivity between charging stations and vehicles through a solution that allows the exchange of information according to the user's interest, directed us to evaluate the architecture models of existing web solutions, the current model allows people to collaborate, communicate and create content in exchange for sharing their data (personal and behavioral), which are centralized in large servers and data centers, being held by companies that can determine how and when to use this information [1].

As the preservation of data privacy is becoming a major issue, due to recent regulations (in Brazil, LGPD and in Europe, the GPDR), aimed at protecting users' privacy and providing secure data, the concept of Web 3.0 leverages greater control over the use and sharing of your data because, among its essences are concepts of possession and ownership of well-defined data [1], where possession can be conceptualized in this context, as when a person or company has authorization from the data owner to use this information to build a personalized offer for him.

Blockchain networks are the key to the success of the Web 3.0 concept, since the issues with ownership and ownership of data are only possible due to the use of a shared data layer and the token model [1]. Blockchain technology enables the implementation of DLs (Distributed Ledger), storing the set of transactions in a block and chaining it using cryptography techniques [7]. As it is a P2P (peer-to-peer) network, Blockchain is based on transactions with trust where each node (computer) connected to it maintains a shared copy of the ledger and, when a transaction is not executed, until it is added to a block and linked to Blockchain. This link happens only after several sources validate this entry [1].

For a transaction to be added to Blockchain, they must be grouped into blocks by miners (or validators) and thus added to the network [7]. For this processing, a collective computational power of the miners is necessary through the resolution of a complex mathematical problem, fundamental to insert a new block in the network [5].

The mining or validation process allows participating nodes to reach an unbreakable consensus, even if some of the nodes are untrustworthy. Over the past few years, many different consensus algorithms have been used by Blockchain such as proof-of-work (PoW), where a miner must complete a certain processing to insert a new block into the chain and earn processing fees like reward; the second is proof-of-stake (PoS) where a randomly chosen validator can insert a new block into the network [7].

Once inserted, the block cannot be changed anymore and the record is distributed to several points in the network, being linked to each other forming a chain [5]. This is possible since the Blockchain uses an Elliptic Curve Digital Signature (ECDSA) algorithm, which assigns an encrypted hash to each block on the network. As you can see in Fig. 1, the signature, combined with the signature of the previous block and its timestamp, creates an immutable link between the blocks and the chain sequence, in this way the blockchain prevents new blocks from being added out of order and preventing that data can be modified or changed after the fact [1].

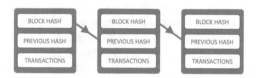

Fig. 1. Block chaining in blockchain technology. Adapted from [5].

This combination of encryption and traceability mechanisms, together with the need for a large computational power for processing, makes the probability of solving blocks simultaneously very low. Another essential factor for this to occur is that for a person to be able to hack the blockchain and insert an illegal transaction into a block, he would need to acquire 51% of the network's computational power and compete with the entire network to generate a longer blockchain so that the corrupted block could be added. [1] states that there is such a possibility, but the probability of happening is remote.

For the execution of some type of action associated with transactions registered on the Blockchain, it is possible to use smart contracts that provide an automation mechanism for code execution, according to rules and implemented situations that were agreed by the nodes participating in the network, [7] providing the ability to build decentralized applications, with integrated data and eliminating the need for centralized data centers [3]. Smart contracts were included in Blockchain 2.0 not supported by Bitcoin protocol, currently Neo and Ethereum protocols are the most used to configure these contracts [10].

Smart contacts are programmable digital contracts, which are stored on the network in a decentralized way and have the ability to automatically execute when one (or more) conditions are met by the parties involved. It is important to note that once published, it is not possible to prevent or censor the information contained in this contract. In this way, smart contracts eliminate intermediaries using the Blockchain network as a guarantee of immutability [10]. Other features that we can highlight is that with their structures, smart contracts can control which institutions, people or applications can have access to the contract, including reading and modification privileges, thus ensuring their privacy and confidentiality [13].

Each protocol (Bitcoin, Neo and Ethereum) has a shared layer of data, where they are stored on each node of the network and, consequently, used by different dApps (Decentralized Applications) that were built on these protocols. The Web 2.0 we know today works differently, platforms such as Google and Facebook offer a centralized data layer, where the user is allowed to maintain the state after a transaction and not effectively exercise their right of ownership over that data. [3]. Therefore, Blockchain returns the property right over the data to each entity, eliminating the intermediary (centralized server), since all communication will be P2P, a situation that we are already seeing in the financial market with Bitcoin.

There are numerous challenges for solutions based on Web 3.0, we can list in addition to the processing capacity of Blockchains, the limitation of the network for the storage of large volumes of data, since this generates a high cost for execution and storage. The concept of Web 3.0 suggests that for situations like this, a P2P distributed file system protocol such as IPFS, Storj, Skynet and Filebase should be used. This type of system aims to replace HTTP and the IP address as a way to locate and access content on the network, the access is performed through an immutable and permanent hash [4].

The distributed file system like IPFS solves the storage problem for Blockchains since it is responsible for storing large amounts of data and the immutable link generated for its access is included in the transaction within the Blockchain. Another advantage in using this file system is related to bandwidth savings, because as the files are distributed on the network, when trying to access a certain content from the hash, it will be directed to the node closest to the requester and this, will be responsible for delivering the content to him [1].

3   Related Work

We are following another moment of evolution of the internet and studies follow how the concept of Web 3.0 works with factors such as decentralization using Blockchain networks. Jennings [9] deals with the difference between decentralization models during the evolution of Web 3.0, which can be complete or open.

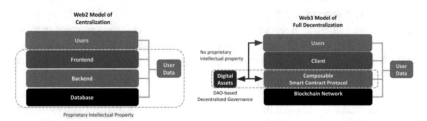

Fig. 2. Illustration for comparing the structures of systems developed on concepts of Web 2.0 and Web 3.0 in the model of Complete Decentralization by [9].

Jennings [9] states that a full decentralization model, as its name suggests, emphasizes that all components of your system must be decentralized. In Fig. 2 we can see a comparison between the centralized model (Web 2.0) and a decentralized model (Web 3.0), where the following situations can be noticed:

- Need to implement an open-source smart contract protocol on a Blockchain to form the backend of the system.
- The client layer is decentralized, meaning that all system software operates off-chain, being the access point to the smart contract protocol (which can vary from websites, front-end, or complex applications).
- Enabling governance layer through the DAO (Decentralized Autonomous Organization), through smart contracts that allow participants to make collective decisions about governance, operations, and leadership.
- Ensuring that user data is owned, retained, and controlled by the user and no longer centralized in servers and databases as seen in Web 2.0 systems.

To ensure that no individual or group can control a solution in the complete decentralization model, only through governance in the form of a DAO where he is responsible for taking control of the smart contract protocols of the company that developed the system [9].

Like the full decentralization model, open decentralization utilizes components such as decentralized Blockchain, digital assets, and governance through a DAO. However, its main difference lies in the shared smart contract layer in this model, allowing independent developers to build their off-chain applications to serve their customers, reusing the benefits and functionality offered by the contract already implemented on the Blockchain [9].

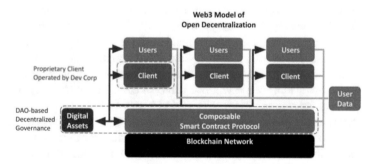

Fig. 3. Open decentralization model in Web 3.0 by [9].

In Fig. 3 we can see the difference from the open decentralization model, where developers build and operate different clients on the same smart contract protocols. In this scenario, the creation and evolution of clients/software would be encouraged in different ways, including the payment of premiums of digital assets (cryptocurrencies) based on performance metrics established by the DAO. It is expected that in more complex systems there may be a significant increase in the premiums offered [9].

Hanada et al. [14] in their work, state that longevity through decentralization is another advantage when we choose to develop a solution with an open model. According to the work, in 2016 the average age of a car in the United States was 11.6 years, a period that can outlast an app provider. Thus, the open decentralization model, with smart contracts on a public interface, any person or application can interact with it, thus keeping the main state and logic active, even in a situation where the original application provider ceases to exist.

The objective of the work by Hanada et al. [14] is a decentralized application that allows the automated purchase of fuels, where the smart contract is the main component. It is in the smart contract where the gasoline purchase protocol is defined and where the status and logic for completing the fuel purchase are stored.

In the project, a single smart contract can be used by several different vehicles and stations. It acts as a common interface and protocol for the purchase of fuel, removing the credit card company and the supplier as a third party because it uses a digital currency and the smart contracts themselves as a guarantee that both parties get the expected results (fueling and receipt). For this, the smart contract was developed in a way that clearly specifies the sequence of events to be executed, aiming to minimize the risk involved in the exchange [14]. Ex.: After the gasoline is dispensed into the vehicle's tank, it cannot be returned and therefore, to eliminate the risk of non-payment, the contract requires payment to occur before the vehicle is authorized to fill up (as already happens in the refueling vehicles in the United States). It is important to note that in the project by Hanada et al. [14]. The smart contract itself contains all the information needed to calculate the payment to be sent to the gas station and, the change to be returned to the vehicle.

In summary, the vehicle refueling process through the project by Hanada et al. [14] starts with the publication of a smart contract on the Blockchain that allows the purchase of fuel. To start, a service station connected to the platform publishes the prices charged for the different types of fuel offered and the minimum deposit amount. Next, a user previously adds funds (cryptocurrency) to the vehicle's wallet and, when he decides to refill, uses the vehicle's application to make a deposit at the gas station he wants to visit. The gas station, in turn, can then view this transaction on the Blockchain and identify which vehicles are authorized to fill up at its establishment. When the vehicle arrives at the pump, a short-range wireless protocol can be used to verify the identity of the vehicle and the station allowing fueling to begin. Once the filling is completed, the station sends a transaction indicating the amount and type of fuel dispensed, causing the smart contract to calculate the payment to the station, returning the deposit to the vehicle without subtracting the payment. Unlike our approach, the smart contract is used by several different vehicles and gas stations, storing the result within the Blockchain structure itself, which may cause, in the future, with an increase in the volume of data stored there, processing time and performance can be impacted, [14] has already signaled that smart contracts on Ethereum have a relatively low performance and that the dApps developed must be designed in order to allow delay in transactions from minutes to hours or, be prepared to increase lower transaction time *gasPrice* values thus encouraging miners to prioritize this transaction.

4 Platform Architecture for Charging Electric Vehicles

In our approach, the proposed architecture aims to provide a framework for the development of a platform for charging electric vehicles, supported by an application based on Web 3.0 where there is no intermediary agent, that is, there is no database that stores application state as little, a centralized web server where the backend logic resides. This approach is possible since applications are developed on a decentralized state machine, maintained on an anonymous node within the Blockchain in the form of a smart contract.

In the development of the architecture, some premises were elaborated for the choice of each one of the related components, they are:

- The backend of the platform will be developed in smart contracts deployed in a decentralized state machine, and its access and processing will be carried out through the Blockchain.
- Users will use a front-end as an access point to the system, and its communication with smart contracts must be performed using nodes provided by third-party services such as Infura or Alchemy.
- All write/post transactions on the Blockchain need to be "signed" using a private key. Metamask is a tool that stores a user's private keys and can be used as a simple browser extension when writing on the Blockchain.
- Smart contracts on the Blockchain must store only data essential for their execution, granular information and other volumes of data must be stored in IPFS, a distributed file system, having only their access hash stored within the transaction on the Blockchain.

In addition to the premises mentioned above, for the development of the architecture it is important to define which Blockchain protocol will be used in the back-end layer. Because decentralization and security use so much computing power from the network, Blockchain cannot keep up with the demand of new developments and response times, so the concept of different layers of Blockchain was devised. 1st layer blockchains act on the main network protocol such as Bitcoin and Ethereum, the applications implemented on it can verify, validate, and finalize the negotiations without depending on any other network, and the networks of this layer have their native cryptocurrencies that help to refund transaction fees. On the other hand, 2nd layer Blockchains are structures designed on top of an existing ecosystem, with the aim of increasing transaction execution speed and reducing scalability complexities.

Another important feature present in 2nd layer Blockchains is related to the fact that transaction processing takes place outside the main network (1st layer), that is, the speed of development of new applications is accelerated and, the prices charged for transactions are much higher. Lower when compared to the prices charged on 1st tier Blockchains (Ex.: An Ethereum network transaction – 1st tier – daily fluctuates between $50 and $125 while on Polygon – 2nd tier – the cost is approximately $0.05 per transaction).

For the construction of the architecture proposal, we chose to use the Polygon 2nd layer Blockchain, where we can highlight as relevant points [15] for this choice: Reduced value for transaction processing; High processing speed (TPS of 7000); High number of connectors (API) with wallet services and file storage systems; Greater volume of technical documentation available.

Based on the assumptions presented in this section, combined with the analyzes related to the choice of Blockchain to develop this approach, Fig. 4 presents an overview of the proposed architecture and components capable of supporting the platform and offering products and services to EV drivers. While waiting for their recharge at registered stations.

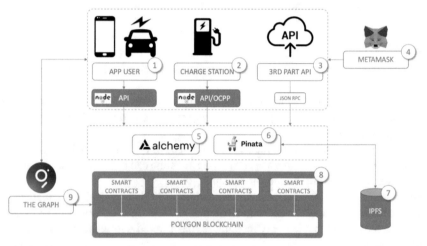

Fig. 4. Architectural proposal for electric vehicle recharging platform based on Web 3.0 concepts.

(1) *App User*: Front-end of the platform, allows communication between the user and the system. Developed with the Node.js programming language, it has available actions: Reserve charging station; Get charging station status; Share token with personal data to offer personalized services; View product and service offers.

(2) *Charge Station*: Represents the charging station for EV's, has communication with the platform, allowing to transact actions such as: Reserve the station for charging; Monitor station status; Authorization to start the recharge; Remote configuration of parameters for recharging and measuring values; The communication between the station and the platform is carried out through the OCPP protocol and an API in Node.js that acts as an intermediary in the connection with the Blockchain node service provider (Alchemy).

(3) *3rd Part API*: Third-party applications that can connect to the platform to interact with drivers. Due to the fact that the processing logic lies directly in the smart contract, third-party applications can access its functions reinforcing the open decentralized Web 3.0 model used in this approach.

(4) *MetaMask*: Wallet solution to store the private key needed to compose the signature when implementing smart contracts or performing writing operations on the Blockchain.

(5) *Alchemy*: Blockchain node service provider, where infrastructure maintenance and response time are managed by the provider through a compensation/incentive. It is through this API that the front-end or other applications can implement or interact with smart contracts on the Blockchain.

(6) *Pinata*: Service provider that generates a gateway between an application and IPFS, allowing the publication of new objects in the distributed file system.

(7) *IPFS*: Decentralized file storage system, aims to store the largest volume of data transacted by the platform, generating immutable hash that are stored on the Blockchain for later consumption.

(8) *Polygon Blockchain and Smart Contracts*: Backend layer of the architecture proposal. Smart contracts are implemented in Polygon, thus allowing interaction between the different actors and modules of the platform, in addition to having a rules engine capable of generating a personalized offer of products or services to the user.

(9) *The Graph*: Indexing protocol that allows querying information stored on networks such as Ethereum and Polygon. The component allows you to define which smart contracts to index as well as which events and calls it should listen for, allowing you to query data on the Blockchain with low latency.

Finally, the proposed architecture has as one of the main objectives, to enable the exchange of information by the data owner with the platform, in a safe way and ensuring control, transparency and domain over the shared data. Data access is performed within the smart contracts layer, thus ensuring that only data authorized by the owner is used and not collected as in centralized architectures.

5   Discussion

Our architecture comprises different components that aim to provide greater security and transparency about the data to the owner, as well as to obtain immutability benefits of a transaction when implemented within the Blockchain. This is possible due to smart contracts, which are the soul of this architecture, its development is carried out through the Solidity programming language and allows the development of business rules to a simple get to obtain the connection hash with the IPFS.

Preliminary results point to good performance results when we run on the Polygon testnet, implementation actions of smart contracts and interactions through functions and methods developed in them.

In our approach, we understand that it would be necessary to develop a smart contract called StationManager, responsible for managing the charging stations registered on the platform, as well as maintaining the address of the contract signed between an EV's and the charging station, this second smart contract is generated at the moment the station reservation was confirmed, that is, it is there that the details about the recharge and the personalized offer is available for access by both the user and the company that made the offer, thus guaranteeing its validity and authenticity.

6 Conclusion and Future Work

This work contributes with an architectural proposal for a charging platform for EV's that uses the advantages proposed by the Web 3.0 concept to provide personalized experiences for users while they wait for the charging process, making it possible to adapt it to other business scenarios such as for example, customer loyalty and shopping club.

The technology used in the architecture puts into practice concepts of ownership and ownership of data, seeking greater adherence to recent regulations aimed at preserving data privacy, in addition to bringing greater control and transparency in the use of this data by interested people and companies.

As future work, the architecture of the platform will be implemented in a way that allows evaluating its performance, providing empirical evidence of its feasibility and efficiency, allowing the analysis and optimization of resources at each interaction and the inclusion of new functionalities such as the possibility of the recharge payment is made through the platform itself, without the need for an intermediary for this action.

References

1. Adel Alabdulwahhab. F.: Web 3.0: the decentralized web blockchain networks and protocol innovation. In: 2018 1st International Conference on Computer Applications Information Security (ICCAIS), pp. 1–4 (2018). https://doi.org/10.1109/CAIS.2018.8441990
2. Open Charge Alliance: Prédios novos em SP serão obrigados a ter recarga para carros elétricos. https://www.openchargealliance.org/about-us/background/
3. Besaçon, L., Silva, C.F.D., Ghodous, P., PatrickGelas, J.: A blockchain ontology for DApps development. IEEE Access **10**, 49905–49933 (2022) https://doi.org/10.1109/ACCESS.2022.3173313
4. Chen, Y., Li, H., Li, K., Zhang. J.: An improved P2P file system scheme based on IPFS and Blockchain. In: 2017 IEEE International Conference on Big Data (BigData), pp. 2652–2657. https://doi.org/10.1109/BigData.2017.8258226
5. Advanced Micro Devices: Making Sense Out of Blockchain Tecnologies. Technical Report. Advanced Micro Devices (2019)
6. Egbue, O., Long, S., Samaranayake, V.A.: Mass deployment of sustainable transportation: evaluation of factors that influence electric vehicle adoption. Clean Technol. Environ. Policy **19**(7), 1927–1939 (2017). https://doi.org/10.1007/s10098-017-1375-4
7. Ali, M.S., et al.: Applications of Blockchains in the Internet of Things: a comprehensive survey. IEEE Commun. Surv. Tutor. **21**(2), 1676–1717 (2019)
8. Higueras-Castillo, E., Guillén, A., Herrera, L., Liébana-Cabanillas, F.: Adoption of electric vehicles: which factors are really important? Int. J. Sustain. Transp. **15**(10), 799–813 (2021)
9. Miles Jennings. 2022. Principles Models of Web3 Decentralization. Technical Report. A16z Crypto
10. Kaushal, R.K.; Kumar, N.; Panda, S.N.; Kukreja, V.: Immutable smart contracts on blockchain technology: Its benefits and barriers. In: 2021 9th International Conference on Reliability, Infocom Technologies and Optimization (Trends and Future Directions) (ICRITO), pp. 1–5 (2021)
11. Lazaroiu, C., Cristian Lazaroiu, G., Pagano, M., Roscia, M.: Smart agent to optimize recharge of electric vehicles (EVs) into smart cities. In: 2018 International Symposium on Power Electronics, Electrical Drives, Automation and Motion (SPEEDAM), pp. 437–442. https://doi.org/10.1109/SPEEDAM.2018.8445333

12. USA Department of Energy: History of the Eletric Car. https://www.energy.gov/articles/his
    tory-electric-car
13. Figueiredo, J.E.M., Lima, I.N.: Contratos inteligentes com ethereum. J. Innov. Sci. Res. Appl.
    1, 38–48 (2021)
14. Hanada, Y., Hsiao, L., Levis, P.: Smart contracts for machine-to-machine communication: pos-
    sibilities and Limitations. In: 2018 IEEE International Conference on Internet of Things and
    Intelligence System (IOTAIS), pp. 130–136 (2018). https://doi.org/10.1109/IOTAIS.2018.
    8600854
15. Verna, S.: Blockchain Layer 1 Vs. Layer 2: a detailed comparison. Tech-nical Report.
    Blockchain Council (2022)

Services and Operations of Electric Vehicle System by Virtual Power Plant in Rural Area

Yoshitaka Shibata[✉], Masahiro Ueda, and Akiko Ueda

Iwate Prefectural University, Sugo, Takizawa 152-89, Iwate, Japan
{shibata,ueda,aki}@iwate-pu.ac.jp

Abstract. In this paper, we introduce a new renewable based electric vehicle system in local areas. Various renewable energies including solar, wind power, small hydro power, geothermal power, etc. in the same areas are integrated and organized a virtual power plant (VPP). A part of the integrated energy with storage battery is used for electric power of autonomous vehicles as various service such as public transportation, personal vehicles for daily senior care for in the same area. Thus the generated renewable energies are locally used and consumed without any carbon emission. In this paper, a system configuration, operation and VPP are explained. As one of services of VPP, autonomous operation of electric vehicles is precisely shown.

1 Introduction

As progress of global warming due to mainly power generation and car driving using fossil fuels, CO2 emissions seriously get increasing and serious for environmental conditions on the earth. The recommendation that the zero emission of CO2 must be attained by the 2050 year has been agreed on the Paris Protocol. Thus the use of fossil fuel have to be reduced and replaced to other resource.

On the other hand, the renewable energy resources (RES) such as Photo Voltaic (PV), Wind Turbine (WT), Bio-Mass (BM), Hydraulic Power (HP), Geothermal (GT) Power Generations have being paying attentions as alternative power energy resources because they are clean, no carbine emission and reproducible. Many countries introduce those renewable resources instead of fossil fuels and atomic power generation.

The automobile industries also are producing electric vehicles based on based on battery power. However, the size of their energy are relatively small, not stabile of the energy supply distributed geologically on depending on the scale, and the stability of RES is not stable depending on the weather conditions. On the other hand, the scale and size of power station by RES are flexible from small to large and easily managed.

Therefore, by aggregating those RESs and controlling according to the power demand, more flexible usages as services can be realized as a Virtual Power Plant, called VPP.

In this paper, we discuss relatively small distrusted VPP and its application and operation of public transportation system small town using electric vehicle (EV) based on the locally produced RES as VPP.

L. Barolli (Ed.): AINA 2023, LNNS 655, pp. 370–375, 2023.
https://doi.org/10.1007/978-3-031-28694-0_35

The VPP is aggregated by PVs, WTs, BM power generation to supply the electric power to the residents and public uses. As application services for residents, autonomous EV service is provided. The EV runs on the road in the town and provides for senior residents to go clinic, shops, healthcare community center even though the weather conditions is not good such as rain, snow and icy conditions by quite low price. Thus, the VPP can provide very important functions for residents in even though the small town.

In the followings, the related works with VPP are introduced in section two. Then general topology of power system with VPP is discussed in the section there. After that, the application and operations of VPP power system for small town in snow country is considered in Sect. 4. Finally, the consideration and discussion of this research direction and future works are described in Sect. 5.

2 Related Works

There are a number of references with VPP. First, Awerbuch and Preston originally defined the concept of VPP [1] and established that individual participator can provide highly efficient power service to customers through virtually sharing of their private properties to improve individual utilization efficiency and avoid redundant construction using mathematical mode. In the reference [2], Zhou and Yang, et al. defined VPP as an energy Internet hub, in which relies on remote control technology and central optimization. On the other hand, Khodr and Halabi confined to a loose coalition between same type of energy resources, like microgrids (MGs) in [3], combined heat and power plants (CHPs) in [4] or active distribution networks (ADNs) in [5], and hybrid energy system in [6]. Through those researches in [7], the VPP is summarized as three key characteristics, Information gathering and processing, Geographical influence ignorance and Dynamic operation and optimization. As internal control of VPP, Mashhour [7, 8] established centralized control method to integrated DER. On the other hand, Yang [9] and Xin [10] established distributed control method of independent DES by communicating with the information among all of subsystems of DESs. This method could provide more flexible and expand as increase of distributed DES environment. In this paper, we apply distributed method because the various RES managed by different local energy organizations are aggregated as a VPP.

3 General Topology of Power System with VPP

General topology of VPP is defined in Fig. 1 and organized by various RES including PV (RES) such as Photo-Voltaic (PV), Wind Turbine (WT), Bio-Mass (BM), Hydraulic Power (HP), Geothermal (GT) Power Generations, Power Grid, Electricity Market, Independent System Operator (ISO), Supervisor, Utility User, Industries. Each RES with storages facilities generates electric power on different locations, and send to PG, and partially store their storage devices. The generated powers are aggregated and managed by Supervisor to suppliers, such as utility users and industries. Supervisor also bits to electricity market to sell to supply to other regions. The ISO in the region manages and control transmission line system independently from the RES to send power to utility user and, industries and other regions. Thus, the supervisor can manage the generated power in the same region and not only supply to the same region but sell to other region through power grid.

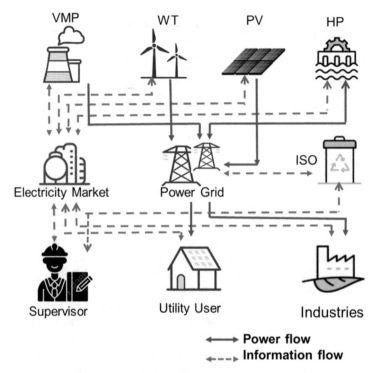

Fig. 1. General topology of power system with VPP

4   VPP System for Rural Area

The conventional power supply and transmission systems which are monopolized by major regional electric companies, in which are based on the fossil fuel. However, since the local areas in Japan have higher potential of RES, particularly northern parts of Japan have wide fields, wooden and volcanic mountains, strong wind areas at top mountains and coast lines. Those environments bring higher potentials of RESs, namely, PV power on wide fields, WT power by strong wind areas on both mountains and offshore, biomass power from wide wooden mountains, hydraulic power on the snowy mountains, and geothermal power from volcanic mountains. Therefore, VPP by RES can be easily constructed and locally operated for not only selling purpose but local consumption. By aggregating those RESs, local VPP company can be established and can manage and operate those power demand and supply to the residents and industrial company and offices. Those generated power also supply to the public transportation by electric vehicles and many other services. Thus precise and stable power demand supply plan based on the big data of precise weather and predicted total production, supply and demand of power can be realized. Even though the emergency such as disaster, typhoon, heavy rain and snow occurred, blackout can be avoid because various different RESs are aggregated and cooperated with back up line by major power generation company (Fig. 2).

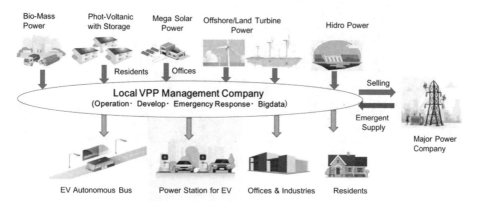

Fig. 2. Realization of VPP system in local area

5   Electric Vehicle Service System by VPP

There are a number of services by VPP system are considered in rural area. Since rural area have various RES, organization of VPP is relatively easy. The generated electric power can be provided as rural electricity thought the rural power grid lines and realize various services including resident. One of the typical and useful services by VPP for rural area is electric vehicles service system which includes electric charge of EV at resident house and offices, power stations for EV and the autonomous EV bus service [11, 12].

By supplying electric power to the autonomous EV bus service as public transportation, the senor peoples, students an residents who do not drivers licenses can easily go to hospitals, clinics and healthcare, schools, shops, office without driving. This autonomous EV bus service can also apply for delivering load as logistics.

Tourists can take the autonomous EV bus to go to tourist spots, historical places parking lots while using electric coupon and cashless payment.

On emergence response, evaluation activity for senor people and handicapped can be realized. Thus, electric EV service system can perform public transportation facility for rural area in local production in local consumption manner (Fig. 3).

Fig. 3. Autonomous EV bus service in local snow area

6 Conclusions

In this paper, we introduce a new renewable based electric vehicle system in local areas. Various renewable energies including solar, wind power, small hydro power, geothermal power, etc. in the same areas are integrated and organized a virtual power plant (VPP). As one of VPP usage, autonomous EV bus service as public transportation, personal vehicles for daily senior care for in the same area. Thus, the generated renewable energies are locally generated and consumed without any carbon emission. Now we are designing and constructing a prototype system for autonomous EV bus system and VPP using the renewable energy resources to evaluate the effects and functional and performance in one local snow area.

Acknowledgments. The research was supported by Japan Keiba Association, Grant Numbers 2021M-198 and Japan Science and Technology Agency (JST), Grant Numbers JPMUPF2003.

References

1. Awerbuch, S., Preston, A.: The Virtual Utility: Accounting, Technology and Competitive Aspects of the Emerging Industry. Springer, USA (1997)
2. Zhou, K., Yang, S., Shao, Z.: Energy internet: the business perspective. Appl. Energy **178**, 212–222 (2016)
3. Khodr, H.M., Halabi, N.E., García-Gracia, M.: Intelligent renewable microgrid scheduling controlled by a virtual power producer: a laboratory experience. Renew. Energy **48**, 269–275 (2012)
4. Zamani, A.G., Zakariazadeh, A., Jadid, S., et al.: Stochastic operational scheduling of distributed energy resources in a large scale virtual powerplant. Int. J. Electr. Power Energy Syst. **82**, 608–620 (2016)

5. Peik-Herfeh, M., Seifi, H., Sheikh-El-Eslami, M.K.: Two-stage approach for optimal dispatch of distributed energy resources in distribution networks considering virtual power plant concept. Int. Trans. Electr. Energy Syst. **24**(1), 43–63 (2014)
6. Abbassi, R., Chebbi, S.: Energy management strategy for a grid-connected wind-solar hybrid system with battery storage: policy for optimizing conventional energy generation. Int. Rev. Electr. Eng. **7**(2), 3979–3990 (2012)
7. Mashhour, E., Moghaddas-Tafreshi, S.M.: Bidding strategy of virtual powerplant for participating in energy and spinning reserve markets–part I: problem formulation. IEEE Trans. Power Syst. **26**(2), 949–956 (2011)
8. Mashhour, E., Moghaddas-Tafreshi, S.M.: Bidding strategy of virtual powerplant for participating in energy and spinning reserve markets – part II: numerical analysis. IEEE Trans. Power Syst. **26**(2), 957–964 (2011)
9. Shabanzadeh, M., Sheikh-El-Eslami, M.K., Haghifam, M.R.: A medium-term coalition-forming model of heterogeneous DERs for a commercial virtual power plant. Appl. Energy **169**, 663–681 (2016)
10. Giuntoli, M., Poli, D.: Optimized thermal and electrical scheduling of a largescale virtual power plant in the presence of energy storages. IEEE Trans. Smart Grid **4**(2), 942–955 (2013)
11. Shibata, Y.: Social experiment of realtime road state sensing and analysis for autonomous ev driving in snow country. In: The 10th International Conference on Emerging Internet, Data & Web Technologies (EIDWT2022), LNDECT 118, pp. 291–300, March 2022
12. Shibata, Y., Sakuraba, A., Arai, Y., Saito, Y.: Improved road state sensing system and its data analysis for snow country. In: Barolli, L., Hussain, F., Enokido, T. (eds.) Advanced Information Networking and Applications. AINA 2022. LNCS, vol. 449, pp. 321–329. Springer, Cham (2022). https://doi.org/10.1007/978-3-030-99584-3_28

A Triangulation Based Water Level Measuring System for a Water Reservoir Tank

Yuki Nagai[1], Tetsuya Oda[2]($\boxtimes$), Kyohei Toyoshima[1], Chihiro Yukawa[1], Kei Tabuchi[2], Tomoaki Matsui[2], and Leonard Barolli[3]

[1] Graduate School of Engineering, Okayama University of Science (OUS), 1-1 Ridaicho, Kita-ku, Okayama 700-0005, Japan
{t22jm23rv,t22jm24jd,t22jm19st}@ous.jp
[2] Department of Information and Computer Engineering, Okayama University of Science (OUS), 1-1 Ridaicho, Kita-ku, Okayama 700-0005, Japan
oda@ous.ac.jp, {t19j048tk,t19j077mt}@ous.jp
[3] Department of Information and Communication Engineering, Fukuoka Institute of Technology, 3-30-1 Wajiro-Higashi-ku, Fukuoka 811-0295, Japan
barolli@fit.ac.jp

Abstract. There are different water reservoir tanks such as septic tanks, agricultural water storage tanks and fire protection tanks. The water level of a water reservoir tank in outdoor environment changes depending on weather conditions. In the case of heavy rains, the water level rises leading to flooding. Therefore, it is necessary to monitor the water reservoir tank and predict the overflow in order to reduce the flood damages. However, it is difficult to install the water level gauge depending on the installation environment at the water reservoir tanks. Therefore, we consider to calculate the water level from a distant place by triangulation. In this paper, we propose a method for measuring the water level of a water reservoir tank based on triangulation using two cameras. We carried out some experiments and the experimental results shows that even if the water level in the water reservoir tank changes significantly, it is possible to measure the change. Thus, the proposed method is able to measure the water level in the water reservoir tank.

1 Introduction

There are different water reservoir tanks such as septic tanks, industrial water tanks, agricultural water tanks and fire protection tanks. In general, water reservoir tanks are installed outdoors and the water level depending on weather conditions. By monitoring the water level in the water reservoir tanks, we can predict the water overflowing and reduce flood damages.

In order to monitor the environment around a water reservoir, the authors proposed an intelligent sensor network [1–7] by developing and integrating sensing devices, intelligent systems and a mesh network [8–10]. The proposed

© The Author(s), under exclusive license to Springer Nature Switzerland AG 2023
L. Barolli (Ed.): AINA 2023, LNNS 655, pp. 376–383, 2023.
https://doi.org/10.1007/978-3-031-28694-0_36

intelligent sensor network aggregates sensing data from sensor nodes to sink nodes based on a mesh mechanism that can perform multi-hop communication, and analyzes and predicts sensing data based on deep learning.

A water level gauge is generally used to measure the water level, but it may be difficult to install a water gauge depending on the construction environment of the water reservoir tank. Even in locations where it is difficult to install a water level gauge, if the distance to the water surface can be measured from a remote location, the water level in the reservoir can be calculated based on the positional relationship with the observation point. Therefore, triangulation [11] is a good distance measurement method. Triangulation is a technique that measures the distance and angle of a polygon to determine its position and can be used to make measurements from distant locations.

In this paper, we propose a method for measuring the water level of a water reservoir using triangulation method with two cameras. We carried out some experiments and the experimental results shows that even if the water level in the water reservoir tank changes significantly, it is possible to measure the change. Thus, the proposed method is able to measure the water level in the water reservoir tank.

The paper is organized as follows. In Sect. 2, we present an overview of a water reservoir tank monitoring and predicting system. In Sect. 3, we discuss the evaluation results. Finally, in Sect. 4, we conclude the paper.

2 Proposed Method

Figure 1 shows the proposed method and Fig. 2 shows the image of triangulation. The proposed method consists of two camera that capture images of the water surface and a sink node that aggregates the images and calculates the water level. Each camera and sink node are connected by wireless communication. The communication method is based on a mesh mechanism and multi-hop communication is performed between each camera and sink node to aggregate images from each camera node to the sink node. The sink node sends a signal to each camera to take images simultaneously and align the times at which the cameras take images. Then, the parallax of two cameras is used to determine the distance by triangulation. The orientation and position of the cameras relative to the measurement point are adjusted before the measurement is taken so that the measurement point is projected onto the center of the captured image and the optical axes of the two cameras intersect at the measurement point. It is necessary the x-coordinate of the measurement point is the same with the midpoint between cameras.

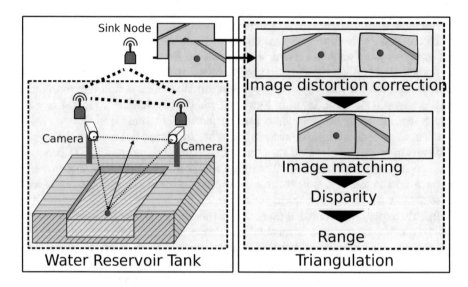

Fig. 1. Proposed method.

The water level measurement based on triangulation corrects the distortion of the acquired camera image and performs stereo matching [12–15]. This gives the disparity between two cameras.

In the proposed method, the measurement points and each camera draw an isosceles triangle. Therefore, the intersection point on the extended line of the coordinates with half the disparity is the distance to the water surface. The distance to the surface of the water becomes near when the water level increases and becomes far when the water level decreases.

The height from camera to water surface is obtained by Eq. (1) considering the distance L from the camera to the water surface and the angle θ between the optical axis of the camera and the vertical direction as the water level wh.

$$wh = L * \cos\theta \tag{1}$$

3   Evaluation Results

3.1   Experimental Environment

The experiment is conducted in a water reservoir tank installed at the Okayama University of Science, Japan as shown in Fig. 3. While, Fig. 4 shows the arrangement of camera, sink node and water level measurement points. The distance between cameras is 5.000 [m], the distance from the midpoint of two cameras to the measurement point is 5.630 [m] and the height from the water surface to the camera is 0.885 [m].

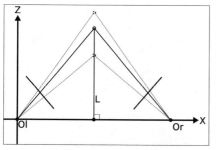
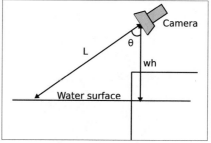

(a) Conceptual diagram of triangulation.   (b) Image of water level calculation method.

Fig. 2. Image of triangulation.

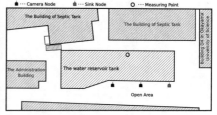

(a) Water reservoir tank at Okayama University of Science.

(b) Location of sink node, camera nodes and measuring point.

Fig. 3. Experimental environment.

Figure 4 shows the devices used in the experiment. Figure 4(a) shows the device of the camera node, where the camera used to get the image of the water surface is a C270n. A Raspberry Pi is used to control the camera. Figure 4(b) shows the device of the sink node. The sink node is composed of a Raspberry Pi and a battery. It just aggregates and stores images. The devices are also stored in a waterproof and dustproof case. Figure 5 shows surrounding environment of the camera node and the sink node.

3.2 Experimental Results

Figure 6 shows the visualization results of the camera. Figure 6(a) shows the left camera image and Fig. 6(b) shows the right camera image. Figure 6(c) and (d) are images obtained by processing distortion correction and parallelization for stereo matching. Figure 6(e) shows the distance visualization results. The center of the image is the measurement point. The darker is the color of the visualization result, the greater is the distance.

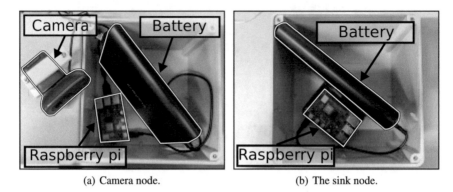

(a) Camera node.   (b) The sink node.

Fig. 4. Devices used for experiments.

(a) The camera node (Left).   (b) Camera node (Right).

(c) Sink node.

Fig. 5. Surrounding environment of camera node and sink node.

Figure 7 shows the measurement results of the water level. The sink node sends a signal to the camera node every second and acquires the image. The blue line in the figure is the change in height from the surface of the water to the camera. The larger is the value, the farther is the distance from the camera and the lower is the water level. The green line is the distance from the camera to the measurement point. We confirmed that the height from the camera to the surface of the

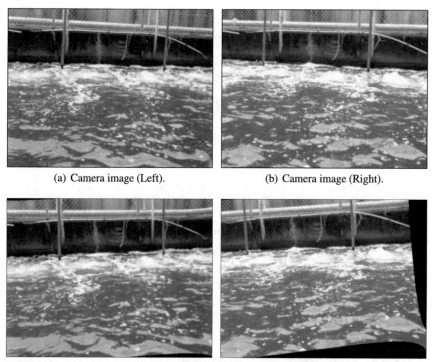

(a) Camera image (Left).   (b) Camera image (Right).

(c) Distortion correction and parallelization im- (d) Distortion correction and parallelization images (Left).   ages (Right)

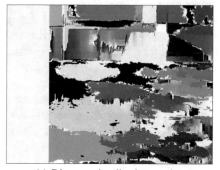

(e) Distance visualization results.

Fig. 6. Visualization results.

water changes from 1.052 [m] to 1.578 [m] and the difference is 0.53 [m]. Also, it can
be seen that the distance from the camera to the observation point changes from
2.525 [m] to 3.156 [m]. This is considered to be the increase and decrease of the
water level by the waves in the water reservoir tank. Therefore, even if the water
level in the water reservoir tank changes significantly, it is possible to measure the
change and the proposed method can measure the water level in the water reservoir
tank.

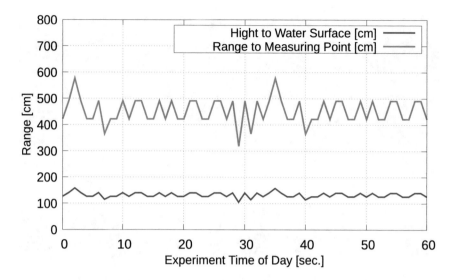

Fig. 7. Experimental results of measuring water level.

4   Conclusions

In this paper, we proposed an object recognition based measuring water level method for a water reservoir tank. For experiments, the proposed system was installed in a water reservoir tank at the Okayama University of Science, Japan and then we measured the water level. We carried out some experiments and the experimental results have shown that even if the water level in the water reservoir tank changes significantly, it is possible to measure the change. Thus, the proposed method is able to measure the water level in the water reservoir tank.

In the future, we would like to consider a different measurement method that takes into account the discrepancy between the camera imaging position and the measurement point. Also, we will develop an infrared camera for water level measurement at night.

Acknowledgement. This work was supported by JSPS KAKENHI Grant Number JP20K19793.

References

1. Lewi, T., et al.: Aerial sensing system for wildfire detection. In: Proceedings of the 18-th ACM Conference on Embedded Networked Sensor Systems, pp. 595–596 (2020)
2. Mulukutla, G., et al.: Deployment of a large-scale soil monitoring Geosensor network. SIGSPATIAL Special **7**(2), 3–13 (2015)

3. Gayathri, M., et al.: A low cost wireless sensor network for water quality monitoring in natural water bodies. In: The IEEE Global Humanitarian Technology Conference, pp. 1–8 (2017)

4. Gellhaar, M., et al.: Design and evaluation of underground wireless sensor networks for reforestation monitoring. In: Proceedings of the 41st International Conference on Embedded Wireless Systems and Networks, pp. 229–230 (2016)

5. Yu, A., et al.: Research of the factory sewage wireless monitoring system based on data fusion. In: Proceedings of the 3rd International Conference on Computer Science and Application Engineering, no. 65, pp. 1–6 (2019)

6. Suzuki, M., et al.: A high-density earthquake monitoring system using wireless sensor networks. In: Proceedings of the 5th International Conference on Embedded Networked Sensor Systems, pp. 373–374 (2007)

7. Oda, T., et al.: Design and implementation of a simulation system based on deep Q-network for mobile actor node control in wireless sensor and actor networks. In: Proceedings of the IEEE 31st International Conference on Advanced Information Networking and Applications Workshops, pp. 195–200 (2017)

8. Nagai, Y., Oda, T., et al.: A river monitoring and predicting system considering a wireless sensor fusion network and LSTM. In: Proceedings of the 10th International Conference on Emerging Internet, Data and Web Technologies, pp. 283–290. Okayama, Japan (2022)

9. Nagai, Y., et al.: A Wireless sensor network testbed for monitoring a water reservoir tank: experimental results of delay and temperature prediction by LSTM. In: Proceedings of the 25-th International Conference on Network-Based Information Systems, pp. 392–401 (2022)

10. Nagai, Y., et al.: A wireless sensor network testbed for monitoring a water reservoir tank: experimental results of delay. In: Proceedings of The 16th International Conference on Complex, Intelligent, and Software Intensive Systems, pp. 49–58 (2022)

11. Lee, D., et al.: A novel stereo camera system by a biprism. IEEE Trans. Robot. Autom. **16**(6), 528–541 (2000)

12. Hirschmuller, H., et al.: Evaluation of cost functions for stereo matching. In: The IEEE Conference on Computer Vision and Pattern Recognition, pp. 1–8 (2007)

13. Jones, G.R., et al.: Controlling perceived depth in stereoscopic images. Stereosc. Displays Virtual Real. Syst. VIII **4297**, 42–53 (2001)

14. Rodríguez, S., de la Prieta, F., Tapia, D.I., Corchado, J.M.: Agents and computer vision for processing stereoscopic images. In: Corchado, E., Graña Romay, M., Manhaes Savio, A. (eds.) HAIS 2010. LNCS (LNAI), vol. 6077, pp. 93–100. Springer, Heidelberg (2010). https://doi.org/10.1007/978-3-642-13803-4_12

15. Weng, J., et al.: Calibration of stereo cameras using a non-linear distortion model (CCD sensory). In: Proceedings, 10th International Conference on Pattern Recognition, pp. 246–253 (1990)

A System Architecture for Heterogeneous Time- Sensitive Networking Based on SDN

Hongrui Nie[✉]

School of Information and Commnunication Engineering,
Beijing University of Posts and Telecommunications, Beijing, China
nie_hy@bupt.edu.cn

Abstract. Time-Sensitive Networking (TSN) is a promising scheme to transfer real-time applications in future command and tactical systems. Combined with Software-Defined Networking (SDN) capacities, the network will be more flexible and controllable with global visibility. A combination of wireless networks and wired TSNs can be deployed to exploit the strengths of both, allowing for the widespread deployment of wireless sensors and actuators in charge of closed-loop command and control systems. This paper proposes a concept for heterogeneous TSN network architecture, with the help of OpenFlow protocol between SDN controllers and TSN switches, and presents detailed modules and their functionality. The proposed architecture enables integration with multiple wired and wireless systems, and the edge nodes access the backbone network through gateways. We have evaluated the proposed architecture on a test case via simulations in OMNet++. TT traffic can maintain its accurate transmission, and performance is not affected, moreover, the UDP traffic sent by the wireless end system is more stable in this framework.

1 Introduction

In the command and tactical system, with the development of network-centric battle theories, there is more reliance on sensor systems distributed over a wide operational area for information transfer, such as intelligence detection, advanced alert tracking, etc. Sensor information has grown sharply and become the core traffic in the network. Modern combat requires high-quality information to ensure the QoS requirements for the services [1]. Nowadays, Time-Sensitive Networking (TSN) task group has defined various standards to achieve deterministic and highly reliable transmission over Ethernet, while wireless networks (mobile devices (AR/VR applications, robotic communication) or portable devices (monitoring) are inseparable in many scenarios [2]. IEEE 802.1Qcc (Stream Reservation Protocol (SRP) Enhancements and Performance Improvements) [3] makes the configuration management more flexible and controllable during the runtime phase. Of the three user/network configuration models, the fully centralized

L. Barolli (Ed.): AINA 2023, LNNS 655, pp. 384–395, 2023.
https://doi.org/10.1007/978-3-031-28694-0_37

model has SDN features and supports a centralized user configuration (CUC) entity to discover end stations, retrieve end station features and user requirements, and configure TSN features in the end stations.

Compared with wired networks, wireless networks have higher latency and lower reliability due to uncertainty in transmission. Thus, it is necessary to further improve the deterministic performance of wireless networks by making them TSN-enabled, but this is beyond the scope of this paper. Importantly, it will be challenging to deploy a combination of wireless networks and wired TSNs that play different roles in different scenarios. A wired TSN network is expected to be the backbone network infrastructure and handle most of the traffic, whereas a wireless node enables communication over the network edge.

Owing to the superiority of SDN in management control aspect, there are some network architectures under centralized management in wired/wireless areas respectively. Boehm *et al.* [4] present a concept of a unified control-plane for the TSN-based fronthaul and the SDN-based core-network. The prototype and testbed is established to show that the proposed concept can establish paths using TSN- and SDN-devices. [5] present traffic management solution for large-scale distributed heterogeneous networks, seamlessly providing mobility management services. This work focuses on the collaboration between multiple controllers and the mobility management of end users, but the service deterministic guarantee mechanism is limited.

In this paper, we propose an SDN-enabled heterogeneous deterministic network architecture shown in Fig. 1. Essentially, a heterogeneous deterministic network must maintain the TSN features across the different communication domains. It allows the coexistence of traffic with different criticality requirements. There will be a domain translator, named gateway, to provide the interconnection between domains, including multiple types of access interfaces, conversion of data formats, intra-domain and inter-domain clock synchronization, etc. Thus, the heterogeneous TSN architecture consists of multiple wired/wireless interfaces, the gateway, the main TSN-enabled switch, and end systems. The detained functionality of each subsystem is shown in Sect. 3. In addition, we provide an evaluation of the proposed architecture for a test case based on the OMNet++ simulator.

The rest of this paper is organized as follows. First, Sect. 2 provides an overview of the related work of wired TSN and wireless TSN. Section 3 proposes a heterogeneous TSN network architecture design that can combine SDN, wired TSN, and wireless communications. Section 4 details the simulations used to test the performance of the proposed architecture. Finally, Sect. 5 summarizes our paper and outlines the future research directions.

2   Related Work

2.1   Wired TSN

There is a set of standards to provide deterministic communication for real-time traffic. The first important mechanism is time synchronization to allow all the network nodes to have a unique time reference. IEEE 802.1AS extends the Precision Time Protocol (PTP) with a specialized profile called generic Precision Time Protocol (gPTP) [6]. The IEEE 802.1Qbv [7] defines a time-awared shaper (TAS) to schedule time-triggered (TT) traffic and non-critical traffic controlled by Gate Control List (GCL). Based on the IEEE 802.1Q [8] standard encapsulated Virtual Local Area Network (VLAN) data format, the priority code point (PCP) field in the Ethernet header marks the priority of the traffic in a set of 3-bit fields as the IEEE 802.1p priority reference. The Ethernet frame can be classified into eight classes. Additionally, each queue is connected to a gate that can be either opened or closed. The state of gates is defined by GCL. There has been extensive research on scheduling and routing algorithms for IEEE 802.1Qbv network.

2.2   Wireless TSN

The 5th generation (5G) mobile communications technology is designed to support real-time applications based on ultra-reliable and low latency communications (URLLC). 5G is designed to optimize the design in terms of a protocol layer and architecture design to achieve low latency and high reliability, supporting different QoS Flows [9]. 3GPP Release 16 [10] defines the basic functionalities and architecture to support Time Sensitive Communications and integrate TSN and 5G. 5G System (5GS) is integrated within the TSN network as a logical TSN bridge. 3GPP introduces a TSN translator at the interconnection points between both networks. The device side and the network side act as TSN ingress and egress Ethernet ports in the logical 5GS TSN bridge. Recently, several studies have investigated 5G-TSN integrated networks. Martenvormfelde et al. [11] proposed a simulation model for the 5G-TSN integrated network and treated 5GS as a transparent bridge. Kehl et al. [12] implemented a 5G-TSN prototype designed for industrial mobile robotics use cases, evaluated and demonstrated that the 5G-TSN system is suitable for industrial use cases.

Another wireless communication technology WiFi has always been a natural extension of Ethernet networks, with less management and more complexity than 5G. WiFi-based TSN networks can prove to be feasible as well. 802.11 [13] already supports precise clock synchronization, and the last 802.11 amendment as well as 802.11ax [14] introduces mechanisms to enhance time-aware scheduling. Genc et al. [15] focused on improving the Quality of the Service of Wi-Fi and demonstrated that TSN and Wi-Fi technologies have the opportunity to be used in several industrial automation use cases. In addition, the wireless high-performance (WirelessHP) project designed and implemented a high-performance PHY layer for industrial applications, which can provide extremely low cycle time (<100 μs) [16].

3   Architecture Design

This chapter introduces a distributed heterogeneous TSN network architecture with multiple SDN controllers and explains the components and functionality in detail. It is separated into four planes: the end-terminal plane, the data plane, the control plane, and the application plane.

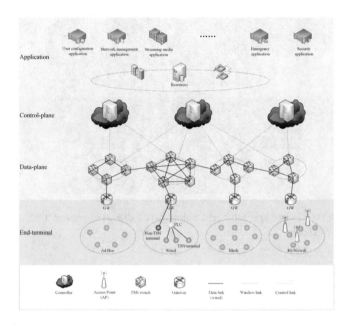

Fig. 1. An illustration of the proposed heterogeneous TSN network architecture based on SDN

3.1   Overall Architecture

A simplified depiction of a heterogeneous TSN network architecture based on SDN is shown in Fig. 1. The architecture supports multiple heterogeneous physical networks, such as wired networks, mobile ad hoc networks (MANET), mesh networks, 4G/5G, and Wi-Fi wireless networks. SDN controller is connected with one or more SDN-enabled TSN switches, which form the backbone network supporting access to heterogeneous end terminals through the gateway. The gateway is a super SDN-enable programmable TSN switch with wired and wireless communication capability. Each SDN controller uses IEEE 802.1 AS-Rev to maintain time synchronization among the SDN-enabled TSN switches, assuming that SDN controllers use the global time information to achieve time synchronization.

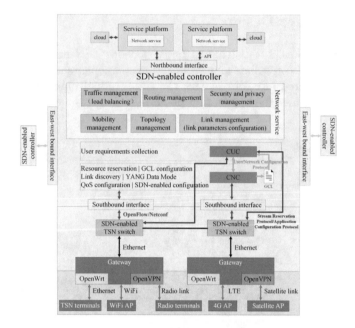

Fig. 2. An illustration of the proposed heterogeneous SDN-enabled TSN network functionality architecture

3.2   End-Terminal Plane

The end-terminal plane consists of TSN terminals, non-TSN terminals (wired terminals), and wireless terminals. Military communication systems can be divided into communication systems with infrastructure and tactical communication systems without infrastructure based on construction scale. Communication systems with infrastructure are usually used in military camps, military ports, and other communication systems that can rely on fixed military infrastructure. Tactical communication systems with complex network structures are based on wireless communications such as shortwave, ultra-short wave, and satellite. Small self-organizing networks with varied physical network forms will be connected to the backbone of the wide area coverage in different forms.

In our proposed architecture, they can be connected to the data-plane with different gateway selection algorithms, and they can take advantage of wired TSN networks with high bandwidth and high communication quality to deliver time-triggered critical flows (e.g., sensory data) to the controller, which calculates the execution action and delivers it to the actuators. The switch supporting multiple heterogeneous link access has high complexity and poor scalability performance. We propose a programmable SDN-enabled TSN switch supporting multidimensional software definable, including port type, accessible link type, supportable communication protocol, connection mode, etc. It can provide the ability to bridge across multiple heterogeneous networks and apply different wireless technologies, which can provide seamless mobile access services. For enabling wireless access capacities, the switches need to support the open-source operating system, e.g., OpenWrt to control wireless parameters and the Virtual Private Networking (VPN) technique connecting to the public server to achieve offsite networking. The gateway is a part of the TSN network, which must support part of TSN standards. It has to support time synchronization (IEEE 802.1 AS-Rev) and flow scheduling (IEEE 802.1 Qbv). Other standards are optional and gateways can be programmatically defined as needed.

3.3   Data-Plane

The data plane consists of SDN-enabled TSN switches which need to support the TSN bridges specification and SDN OpenFlow switch specification. All switches are precisely synchronized by IEEE 802.1 AS-Rev standard (Timing and Synchronization for Time-Sensitive Applications).

The TSN flow control process includes flow classification, flow shaping, flow scheduling and flow preemption. The detailed process is described as follows.

(1) The switch identifies and classifies data frames at the ingress based on identification (VTag), the TSN flow control standard provides the IEEE 802.1Q protocol to standardize the frame format and distinguish the QoS requirements of different applications through specific identification.
(2) the data frames enter the respective priority queues for queuing (TT flow queue, AVB flow queue and BE flow queue), i.e., 0–7 priority queue.
(3) The AVB streams that exceed the rate limit must be restricted after all flows are finished queuing. IEEE802.1Qav, and IEEE802.1 Qcr standards limit the maximum rate of burst flows by shaping the data flows and control the flows to be sent at a smoother rate, while TT and BE flows do not need to be shaped and go to the next operation.
(4) TSN switch performs orderly forwarding of flows according to different flow scheduling policies and algorithms, in addition, it performs preemption operations considering the special needs of high-priority flows. IEEE 802.1 Qbv and IEEE802.1 Qch standards provide time-aware shapers and gate control list (GCL) to achieve single flow scheduling, multiple flows scheduling, and packets orderly forwarding. IEEE 802.3 br and IEEE 802.1 Qbu standards provide flow preemption mechanisms for solving the flow priority inversion problem to ensure that high-priority flows can interrupt the scheduling of low-priority flows for timely forwarding.

For reliable traffic scheduling, IEEE 802.1 CB "Frame Replication and Elimination" supports redundant transmission through generating replicate frames and eliminating the duplicated frames. We adopt redundant transmission of frames and elimination of redundant frames on the data-plane to achieve compatibility of heterogeneous network communication. That is, the end terminal does not necessarily need to support the TSN protocol.

Furthermore, the switch contains Flow Table, Meter Table, and Group Table. The basis for data forwarding is the flow table. The controller has full knowledge of the entire network and can use custom intelligent routing algorithms to dynamically plan scheduling paths and distribute them to the switch via the OpenFlow protocol. Metering enables traffic monitoring at the ingress by counting and rate-limiting packets, which is a good extension of the flow control mechanism of TSN. Group tables can be used for link load balancing to improve the schedulability of the network.

3.4   Control-Plane

The control-plane provides a unified management for TSN networks and non-TSN networks, which has an overall view over its network topology, part of known cyclic flow and radio link parameters. It provides a fine-grained control into schedule traffic to the appropriate SDN-controlled heterogeneous wired and wireless network.

The IEEE 802.1Qcc standard is used for TSN networks to configure infrastructures and switches for plug-and-play capability. Concerning Fig. 2, the Centralized User Configuration (CUC) cooperates with end-terminals to manage QoS-aware flows. The control flow works in both directions, that is, end-terminals send configuration requirements for a QoS-aware flow to the CUC via some middlewares or TSN agents, e.g., OPC-UA, DDS, Zenoh, and the CUC can push configuration parameters towards end-terminals. The Centralized Network Configuration (CNC) uses remote management to discover the physical topology, fetch switch functionality, and configure TSN features in each switch. Path control and reservation (IEEE 802.1Qca) is used to collect network topology information in the TSN network. The common User/Network Interface (UNI) is between CUC and CNC to exchange Talker, Listener, and status configuration messages. A YANG Data Model defined in IEEE 802.1Qcp provides a data model language to express configuration data and status data for Network Configuration (NETCONF) Protocol, which provides standards for remote procedure calls (RPCs) and notification declarations operations to install, manage and delete configurations in the network devices, The CNC communicates with the network devices via NETCONF protocol.

Since TSN networks guarantee that packets are reachable within a low bounded delay, it is important to perform relatively optimized, strict path and time slot planning for all traffic at each node, including switches and end terminals. Different flow classes have different requirements with regard to latency and jitter. The control plane can manage complex traffic with strict planning of known traffic and handling of its burst conditions. The management of these traffic will be sent down to the TSN switches in the form of configuration information to update the GCL tables. Moreover, for the scheduling of unknown traffic, the control plane requires intelligent algorithms to predict the characteristics of network traffic and generate scheduling policies dynamically by end-terminal requirements to maximize network resource utilization and the schedulability of all traffic, to reduce conflicts between them and guarantee packet transmission without waiting. This requires complex computations that the SDN controller can achieve scheduling through an external high-performance infrastructure.

Inter-process communication between SDN controllers is achieved through a logically shared database. The authenticated SDN controller can store the request in the form of data items in another SDN controller database, enabling the sharing of control information. With the help of an east-west bound protocol, controllers in the control plane can negotiate the establishment of communication paths across heterogeneous networks for provisioning real-time end-to-end connection between multiple small networks with different types of end-terminals.

3.5   Application Plane

The application scenario of TSN technology is a complex multi-node network with limited bandwidth and where multiple traffic with different QoS requirements can be generated and transmitted in the network at the same time, i.e., mixed traffic can be mixed on the same physical medium. We can categorize the complex applications generated by different scenarios into the following three types of services.

a. *Time–triggered traffic* : Such type of applications is mainly generated by various types of monitoring devices and transmits data periodically in a fixed interval, and the data transmission bandwidth and latency requirements are known. The resource for planning this type of traffic can be reserved periodically on the switches.
b. *Event–driven traffic* : Such services are generated by the charging equipment and sensing devices, and such events are known and triggered by certain conditions. The data transmission bandwidth demand and the amount of data to be transmitted are not fixed, while they are with high priority, which requires prioritized and timely allocation of the required resources on the switches.
c. *Best Effort traffic* : Such services are mainly generated by auxiliary devices, and the data transmission bandwidth demand for such services is large and the latency requirement is not high.

4   Proof of Concept and Test Cases

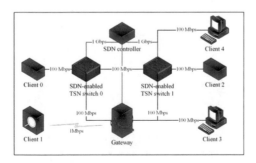

Fig. 3. Network topology of test case

For proof-of-concept, simulation and evaluation of the proposed model are made. The simulation environment used for our implementation is based on the discrete event simulator OMNeT++ (version 5.5.1) [17], the INET framework (version 3. x), CoRE4INET framework [18] and SDN4CoRe framework [19]. We implement the wireless access capability of the Ad Hoc mobile terminal for the gateway in this experiment, and this scenario serves as a verification of the networking capability of our proposed heterogeneous time-sensitive networking based on software-defined networking. The evaluation network topology is shown in Fig. 3, which consists of four nodes connected via two SDN-enabled TSN switches and a Gateway. These switches and gateway are connected to an SDN controller via two Ethernet switches which simulate time delay between the SDN controller and SDN-enabled TSN switches. The more detailed flow parameters are shown in Table 1.

Table 1. Flow parameters

Traffic	Parameter			
	Packet length	Period	Source	Destination
TSN traffic	100 B	1000 µs	Client[0].app[0]	client[2].app[0]
UDP traffic	1000 B	400–500 µs	client[1].udpApp[0], client[1].udpApp[1]	client[3].udpApp[0], client[4].udpApp[0]

Client 0 and Client 1 are the sources of the TSN traffic and UDP traffic. Client 2 is the receiver of TSN traffic from Client 0. Client 3 and 4 are the non-TSN end systems to receive UDP traffic from Client 1. We assume that client 1 and gateway are 30 m apart relative to each other in a physical location. Each wired link (black line) is 10 m long and has a speed of 100 Mbps. And wireless link (yellow lightning) has a transmission rate of 1 Mbps. The wireless propagation model includes small-scale Rician fading and path loss large-scale fading. Each control link (red line) has a data rate of 1 Gbps. Also, each switch has a forwarding delay of 3.5 µs. The gateway here is a wireless/wired switch

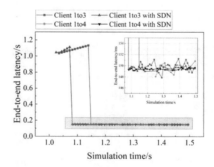

Fig. 4. Evaluation of UDP traffic end-to-end latency in test case.

with TSN capability, connected to a TSN switch, and can get the IP of the wireless NIC from the controller.

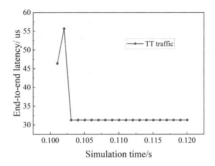

Fig. 5. Evaluation of TT traffic end-to-end latency in test case.

Figure 4 shows the end-to-end latency of all UDP traffic starting from the connection establishment phase. The latency of the network is compared under the heterogeneous TSN architecture and the original LAN. The gateway in the LAN will be replaced with an access point, and the SDN-enabled TSN switch will be a normal router. Before client 1 starts sending the frames, it resolves the traffic destination MAC address by ARP protocol.

In both cases, no ARP cache table exists in advance. In the LAN scenario, the client1 sends an ARP broadcast packet. Only the accessPoint has a wireless NIC, and the accessPoint forwards it to all hosts in the same LAN. In the heterogeneous TSN architecture, APR frames within the wired TSN network are forwarded by the SDN controller, and UDP frames are sent to the controller first to build the flow table and then update the switch configuration, thus introducing some additional latency. In the LAN scenario, the packet Latency is random. Since client 1's mobile state is not set, its delay fluctuates only within a certain range. In contrast, the priority of UDP packets is set to 0, and the delay of the network is more stable under the planning of GCL in TSN networks.

Figure 5 shows the end-to-end latency of TT traffic starting from the connection establishment phase lasting 0.101s. In the simulation start-up phase, all flow tables in the switch are empty. After the TCP connection between the switch and the controller is established, the switch starts processing data after the controller has acquired the switch characteristics. For packets that enter the switch without matching the flow table and do not know how to operate, the switch will encapsulate them in packet-in and forward them to the Controller for further decision. For this reason, an initial idle time of 0.1s is introduced in the network. When the network is stable, TT traffic can be guaranteed to be transmitted accurately without conflicts and jitter by the planning of GCL.

5   Conclusion

This paper presented our proposal of a heterogeneous TSN network architecture that combines SDN, TSN, and wireless communication capabilities. We pose that the proposed architecture described in this paper will help digitization in command and control systems. Simulation results show that our proposed heterogeneous architecture can support mixed traffic communication, which can be easily integrated with another existing network via the programming access interface of the gateway. In future work, the implementation and scheduling of combined wired/wireless networks should be optimized. The more detailed across-domain scenario should be investigated.

References

1. Lopes, R.R.F., Rettore, P.H., Eswarappa, S.M., Loevenich, J., Sevenich, P.: Performance analysis of proactive neighbor discovery in a heterogeneous tactical network. In: Proceedings of International Conference Military Communication Information System (ICMCIS), pp. 1–8, May 2021
2. IEEE 802.1 Time-Sensitive Networking Task Group. Time sensitive networking task group
3. IEEE Standard for Local and Metropolitan Area Networks-Bridges and Bridged Networks - Amendment 31: Stream Reservation Protocol (SRP) Enhancements and Performance Improvements. IEEE Std 802.1QCC-2018, pp. 1–208, October 2018
4. Boehm, M., Ohms, J., Kumar, M., Gebauer, O., Wermser, D.: Time-sensitive software-defined networking: a unified control- plane for TSN and SDN. In: Mobile Communication - Technologies and Applications, vol. 24, pp. 1–6. Osnabrueck, Germany, ITG-Symposium (2019)
5. Ishtaique ul Huque, T., Yego, K., Sioutis, C., Nobakht, M., Sitnikova, E., den Hartog, F.: A system architecture for time-sensitive heterogeneous wireless distributed software-defined networks. In: Military Communications and Information Systems Conference (MilCIS), Canberra, ACT, Australia 2019, pp. 1–6 (2019)
6. IEEE 802.1 Time and Synchronization for Time-Sensitive Applications in Bridged Local Area Networks (2020)
7. IEEE Standard for Local and Metropolitan Area Networks-Bridges and Bridged Networks-Amendment 25: Enhancements for Scheduled Traffic, pp. 1–57
8. IEEE. IEEE Std. 802.1Q, IEEE Standard for Local and Metropolitan Area Network-Bridges and Bridged Networks, IEEE Std. 802.1Q-2018 (Revision of IEEE Std. 802.1Q-2014[S] (2018)
9. 3GPP. Study on NR industrial internet of things (IoT)(R16)
10. Technical Specification Group Services and System Aspects; System architecture for the 5G System (5GS) Stage 2 (Release 16) 3GPP TS 23.501 V16.4.0, March 2020
11. Martenvormfelde, L., Neumann, A., Wisniewski, L., Jasperneite, J.: A simulation model for integrating 5G into time sensitive networking as a transparent bridge. In: 2020 25th IEEE International Conference on Emerging Technologies and Factory Automation (ETFA), vol. 1, pp. 1103–1106 (2020)

12. Kehl, P., Ansari, J., Jafari, M.H., Becker, P., Sachs, J., Koenig, N.: Prototype of 5G integrated with TSN for edge-controlled mobile robotics. Electronics **11**(11), 1666 (2022)
13. Ibrahim, M., et al.: Verification: accuracy evaluation of Wifi fine time measurements on an open platform. In: Proceedings of Annual International Conference on Mobile Computer Network, pp. 417–427 (2018)
14. IEEE Standard for Information Technology-Telecommunications and Information Exchange between Systems Local and Metropolitan Area Networks-Specific Requirements Part 11: Wireless LAN Medium Access Control (MAC) and Physical Layer (PHY) Specifications Amendment 1: Enhancements for High-Efficiency WLAN, IEEE Standard 802.11ax-2021 (Amendment to IEEE Standard 802.11-2020), pp. 1–767, 19 May 2021
15. Genc, E, Del Carpio, L.F.: Wi-Fi QoS enhancements for downlink operations in industrial automation using TSN. In: 2019 15th IEEE International Workshop on Factory Communication Systems (WFCS), pp. 1–6 (2019)
16. Luvisotto, M., Pang, Z., Dzung, D.: High-performance wireless networks for industrial control applications: new targets and feasibility. Proc. IEEE **107**(6), 1074–1093 (2019)
17. OMNeT++ Discrete Event Simulator. https://www.omnetpp.org. Accessed 25 Oct 2022
18. Hackel, T., Meyer, P., Korf, F.: A simulation model for software defined networking for communication over real-time ethernet. In: Proceedings of the 6th International OMNeT++ Community Summit (2019)
19. Häckel, T., Meyer, P., Korf, F., Schmidt, T.C.: SDN4CoRE: A Simulation Model for Software-Defined Networking for Communication over Real-Time Ethernet. In: Proceedings of the 6th International OMNeT++ Community Summit 2019, pp. 24–31, December 2019, EasyChair (2019)

A Parking System Based on Priority Scheme

Walter Balzano[✉], Antonio Lanuto, Erasmo Prosciutto,
Biagio Scotto di Covella, and Silvia Stranieri

University of Naples, Federico II, Naples, Italy
{wbalzano,silvia.stranieri}@unina.it,
{an.lanuto,e.prosciutto,b.scottodicovella}@studenti.unina.it

Abstract. The way the vehicles are placed in a parking space parti-
tioned into parking slots reachable through internal routes means that
the available space is not used at best of its capacity. In this paper, a
solution model is proposed, adopting a chequered parking layout, which
aims at optimizing the available surface and, consequently, increasing the
capacity in terms of available parking slots. To this aim, an isomorphism
with the game of fifteen is found and a solution is proposed.

Keywords: Parking · Queue · Chequered parking · The game of
fifteen

1 Introduction

With the considerable increasing of the number of cars on the roads, the problem
of finding a parking space has increased [7,8]. This not only affects the time
wasted by motorists in finding parking spaces, but also affects other aspects. It
has been seen that around 30% of urban traffic is also caused by cars looking
for parking or parked badly, thus affecting air pollution. But if we also consider
how the high demand for parking and the low availability of parking spaces has
led, especially in large cities, to an exponential increasing of parking prices. The
aim of this research is to find a way to optimise the available spaces in order to
have more and more parking slots. This would also solve the other side problems
explained above. Starting with a space at our disposal, the idea is to arrange the
cars side by side, in a column, with no movement routes but with a manoeuvring
area to move the cars. In particular, the cars will be arranged in column order,
so each car will have another car in front of it with a smaller remaining time.
This allows us having always the cars that will free the parking space before the
others near the exit. The manoeuvring area will serve us to ensure order within
the car park. The following sections are divided as follows: in Sect. 2 the related
works are described; in Sect. 3 an isomorphism with the fifteen puzzle game [6]
is presented; in Sect. 4 the adopted parking model and all its characteristics are
explained; in Sect. 5 the algorithm for managing chequered parking is described,
and an example is also presented; in Sect. 6 the final results of this research are
presented; finally in Sect. 7 the conclusions are explained.

L. Barolli (Ed.): AINA 2023, LNNS 655, pp. 396–405, 2023.
https://doi.org/10.1007/978-3-031-28694-0_38

2   Related Works

Smart parking, due to the growth of urban population and traffic congestion, is a crucial issue to address both in research and economically. Advances in technology allow drivers to more easily find parking spots through smart parking services. However, implementing a smart parking system is a complex and multidisciplinary process that requires extensive surveying and inspection. In [12], authors provide a smart parking guidance algorithm that supports drivers to find the most appropriate parking facility considering real-time status of parking facilities in a city. In [14], they present a feasible method to do parking planning, by transforming the problem into a kind of linear assignment problem and by taking vehicles as jobs and parking spaces as agents. They take distances between vehicles and parking spaces as costs for agents doing jobs, then they design an algorithm for this particular assignment problem and solve the parking planning problem. In [13], instead, authors provide a fingerprinting based solution for smart parking in indoor settings.

In this work we found an isomorphism with existing scenarios, in particular the fifteen game, similarly to what authors of [1] did by finding a solution to the parking problem by reducing it to the ant colony optimization problem.

The fifteen puzzle game has been largely studied in literature, and several solutions have been proposed [5,10,11], but it has never been applied to a parking scenario.

As far as we know, this is the first work designing a new parking environment with the aim of optimizing the available surface of the car park.

3   Fifteen Puzzle Isomorphism

In order to best describe the problem and the solution to be adopted, it was decided to resort to an isomorphism with a very famous puzzle which is *the game of fifteen*. The aim of the game is to reorder a matrix divided into four rows and four columns, which contains 15 tiles numbered from 1 to 15, using only one empty place in it. As can be seen, the game of 15 is based on the use of a matrix that can be represented by a very trivial chessboard $N * N$, and already here we begin to see a similarity with the proposed problem. In fact, the objective is to place the cars in a chessboard car park in such a way as to order them, from the bottom to the exit of the car park, in order of how long the car will remain in the car park. To do this, as in the game of fifteen, it is necessary to have a space where the parking spaces will be delineated, and an area that will represent the empty space of the game of fifteen. In this way, this area can be used as a temporary car park to facilitate the movement of cars within the car park itself.

4   Parking Model

It has been seen how *game of fifteen* represents a true isomorphism for checkerboard parking lots. Starting from the characteristics of the game, the goal is to demonstrate how, by using a buffer zone or cache zone, one can optimize the arrangement of cars in a checkerboard parking lot. To best clarify the basic idea of this paper, we will analyze the operation of the system in an ideal case. Starting from a parking lot having $n * n$ parking spaces, with a maneuvering area of n size and a checkerboard layout, we imagine that the parking lot is initially empty. When cars arrive at the parking lot, they tell the system how long they intend to stay inside it. The system, having taken this information, processes a priority queue in which at the top we will have the cars with a longer remaining time(Tr) than those following. Once this queue is processed, an arrangement of cars within the parking lot will be schematized, where at the bottom we will have the cars with a higher Tr. Importantly, it is not important to have a total sorting of the parking lot, but it is enough to have a sorting by column. In fact, in this way, the cars in front of everything, thus those at the top of each column of the matrix with which the parking lot is represented, will never have any obstacles to exit. Inside the parking lot we will have another queue that represents the actual presence of the cars in the parking lot, and again there is a temporal ordering of the cars, but in this case at the top we will have the cars with Tr lower than the others, and which will therefore exit the parking lot before the others The use of the buffer zone is necessary in case a car arrives with such a Tr that it cannot be parked in any free space, as it would not respect the ordering we set out to have. This situation makes it necessary to move cars so that the latter arrives in the right place in relation to its Tr. Using the buffer zone, the cars that will have to make room for the newcomer will be moved to this zone, so that the correct place for the newly arrived car will be freed up. Having done this, the previously moved cars will be repositioned to the correct location without any problems. Note how using this Tr, each car always has a car in front of it that will start before it. Of course, in a real case it may happen that a user wants to pick up his car before the Tr expires, and even in this case the buffer zone can be made use of so that the car can be returned to the customer without any problems. So far we have talked about buffer zone and cache zone referring to the same parking area, and we have used the two terms interchangeably. Actually a difference is there even if it is the same parking area. The different designation given to this zone, refers to the use that is made of it. Specifically, when we speak of a buffer zone, we are referring to a temporary use of this zone to facilitate the movement of cars in the parking lot. When, on the other hand, we speak of a cache zone, we are referring to a different use. In fact, this zone could be used to put cars that are about to leave and free up spaces for new cars. Obviously, in this case, it must be a use that is not prolonged, since the cars stopped in this zone do not allow to move within the parking lot. We can therefore say that the proposed parking model can be divided into two main areas: (i) **Buffer/Cache Parking Area** - Part of the parking area used for moving cars or for temporary parking of exiting cars; (ii) **Effective Parking Area** - The main parking area, where cars will park in their assigned position according to the remaining time reported.

In addition to these two areas, one can consider a third parking access area, called **waiting area**, dedicated to the arrival of cars. It is here that the user, will communicate to the system the time he will stay and his car data (license plate) and then leave the car to those who will have to place it inside the parking lot.

5   Chequered Parking Algorithm

In this section, the he Chequered Parking Algorithm (CPA) is explained in its main steps and through a running example. The CPA algorithm can be divided into four working steps. It is taken for granted that the algorithm knows, from the first working stage, basic information such as the number of total parking spaces, those that are usable as parking spaces, and those that are reserved for the maneuvering zone or buffer zone.

- **Phase 1, Initialization**: the system asks the user for information about the incoming cars, such as the license plate number and the time he intends to stay(Tr) the car in the parking lot. The number of incoming cars accepted, of course, is equal to the number of vacancies, which at this stage will be equal to the number of total spaces in the parking lot. Once this information is acquired, the algorithm creates a queue of incoming cars. Then this queue is sorted according to the Tr of each car, giving higher priority to cars with a higher Tr. Each car is saved in the form of a list of type [x,y] with x representing the car identifier and y the remaining time.
- **Phase 2, Parking Arrangement**: the system processes this priority queue and arranges the cars in the parking lot, according to a sorting by column. The cars will then be arranged so that each car has ahead of it either no car, or another car but with a lower Tr than it.

Having completed these first two steps, we enter the heart of the algorithm. In particular, we can say that the next steps represent dynamic parking management.

- **Phase 3, New Car Insertion**: this phase deals with the management of incoming new cars, implementing the same process as described above but taking into account the cars already in the parking lot. The algorithm, then, will manage the queue of incoming cars and look for the free place to place the incoming car. Of course, the vacancy may not coincide with the column sorting we are using. Starting then from the column of the free place found, the algorithm will check how many cars in that column need to be moved before the car can be placed, in order to comply with the sorting used. The algorithm then looks for the vacancy that allows the least number of cars to be moved.
- **Phase 4, Parking Rearrangement**: the algorithm now knows which column to work on and through a series of car moves, using the buffer zone as a maneuvering space, it manages to reposition them to the correct location while respecting a sorting by column.

5.1   Running Example

In order to better understand how it works, example images of how each algorithm presented above works are shown on the following pages.

Figure 1 shows an example of how the CPA algorithm works, specifically phases 1 and 2 are presented. Before going on to explain the tables, let us give some brief information. Specifically, the matrix represents a possible car park, within which, after trivial calculations, the two areas, presented above, were created with their respective parking spaces. In particular, the area *Buffer Parking Area* has at least as many spaces as there are in a column (in the event that there are columns with different sizes, the one with the most parking spaces is considered); empty spaces are marked *[0,0]*. Also to the left of the matrix is the initial queueEntry and the queueEntrySort, which is the queue we will be working on to arrange the cars in the car park. Finally, each time a car is placed in the car park, it is removed from the queueEntry and placed in the queueParking. For convenience, not all the steps are explained, but they are quite clear. In Table1 in Fig. 1, the matrix representing the car park is initialised at [0,0], so that it represents the empty car park. Then in table2 the queueEntry is filled and subsequently sorted so that it can be processed. In table3 the cars in the queueEntry are placed in their respective parking spaces, respecting the required column sorting. Finally in table4 we enter the heart of the algorithm with the arrival of new cars to park, but this time with the parking space not empty.

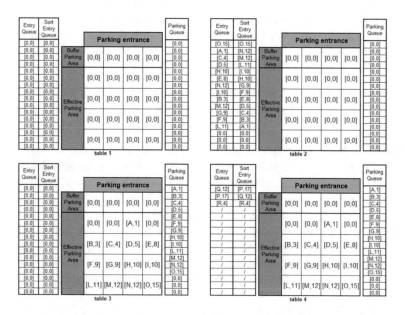

Fig. 1. CPA algorithm: example Phase 1 and 2

In Figs. 2 the CPA algorithm works in steps 3 and 4 presented above. In particular, we notice how the algorithm, once it has taken the incoming car, calculates the most efficient place, i.e. one that allows the least number of cars to be moved, where to park it. The free place, therefore, does not necessarily represent the place where the car should be parked, but represents the column where that place is and where we are going to move it. It is important to note that the place in the cache at the working column must always remain free to allow the car to be entered. As explained for Fig. 1, here too the cars are deleted from the queueEntry and inserted in the queueParking, always respecting an order of priority, in the case of the queueParking at the top we will always have the car with the lowest Tr. The insertion of the cars ['Q',12] and ['R',4] is not presented because it is superfluous, but would follow the same logic as the example in Fig. 2.

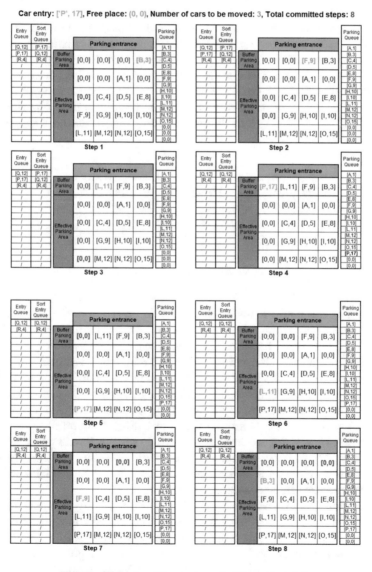

Fig. 2. CPA algorithm: example Phase 3 and 4

6   Evaluation

In this section, we discuss the size of the Buffer that is needed for a fair compromise between space and time waste. We also provide a comparison of our solution with existing ones.

6.1   Buffer Size Evaluation

We now verify how the size for the buffer should be chosen. In particular, we start with an n*n matrix, and show that using the buffer area will only take k steps. Vice versa, we use a smaller and smaller buffer zone to show that with the latter we will employ at least k + 1 steps. In this example we make use of a 3*3 matrix, with a buffer zone of size 3. We only demonstrate the worst case which is that the car park has only one free space and the incoming car must be placed in the last row of the matrix. Let us start by verifying how many steps it will take to sort this matrix n*n with the use of a buffer zone of size n. As one can see in Fig. 3, in order to place the car with Tr = 10 in the right place, thus respecting the sorting by column, we need to move two cars and thus take a total of **6 steps**.

Car entry: [10], Free place: (0, 0), Number of cars to be moved: 2, Total committed steps: 6

Starting Matrix			Step1			Step2			Step3			Step4			Step5			Step6		
0	0	0	0	0	6	0	8	6	10	8	6	0	8	6	0	0	6	0	0	0
0	1	2	0	1	2	0	1	2	0	1	2	0	1	2	0	1	2	6	1	2
6	3	4	0	3	4	0	3	4	0	3	4	0	3	4	8	3	4	8	3	4
8	7	5	8	7	5	0	7	5	0	7	5	10	7	5	10	7	5	10	7	5

Fig. 3. BPA: example with CPA

Now let us check how many steps it will take to get from the *Starting Matrix* to the *Matrix Step6* using a buffer zone with fewer available parking places, in particular we first use a buffer zone of one place, then one of two places. Notice that, in a real scenario, in order for a car to be inserted in a column, the place corresponding to it in the buffer zone must remain available. It is important to note that in the counting of steps, neighbouring empty boxes can be considered as a single box. Furthermore, to explain these two cases, we use the logic of the game of 15 to move the cars around the board (Figs. 4, 5 and 6).

In the figure, we can see how using a BPA with only one parking space, a solution has not yet been reached after six steps. We can therefore already see how the BPA with only one parking space is less efficient than a BPA with three parking spaces used in the example in Fig. 3.

Also in this figure we can see how, using a BPA with only two parking spaces, after five steps we still have not arrived at a solution. We have stopped at step five because it is trivial to see how the step does not lead to the hoped-for solution, and it would also be very complex to show all possible combinations in the figure. We can, therefore, see how the BPA with two parking spaces is less efficient than a BPA with three parking spaces used in the example in figure three. These two simple examples prove that the buffer parking area must be at least as large as a column of the matrix.

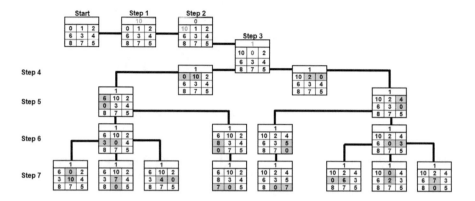

Fig. 4. BPA: example case 1

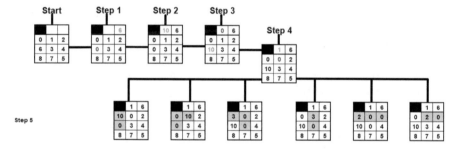

Fig. 5. BPA: example case 2

6.2   Comparison with Existing Algorithms

In this section we will show, with the help of an example, how chequerboard car parks, provided with a manoeuvring area, are an excellent solution not only for optimising movement within the car park itself, but also increase the capacity of available parking spaces and reduce wasted space. Let us start by constructing a chequerboard car park that has four parking spaces in each row and four in each column. With the algorithm used so far, we will then need an additional row with four parking spaces to create the Buffer Parking Area.

Under current regulations, a parking space must have an area of at least 2.5*5.0 m (12.5 m^2). The chequered parking space to be designed must therefore have one side 10 m long and another 25 m long. As one can see in the figure, the car park has **16** parking spaces for the *Effective Parking Area*, and 4 parking spaces for the *Buffer Parking Area*. The total area occupied by the chequered

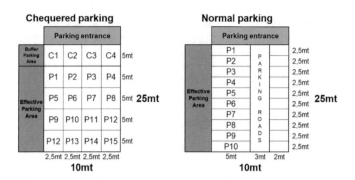

Fig. 6. Differences between two types of parking

parking area drawn above is $250\,\mathrm{m}^2$ ($25 * 10\,\mathrm{m}$). With these dimensions we will see how a normal car park, with even internal ways to move around, makes almost 50% fewer spaces available.

In fact, if we were to arrange the parking spaces in a single row on the long side, in order to comply with the regulations on the minimum dimensions of parking spaces and internal streets, we would not have more than 10 parking spaces available. Moreover, it can be seen that, in the subdivision of the space, a space of $2\,\mathrm{m} * 25\,\mathrm{m}$ would remain, which would not allow the creation of new parking spaces and would therefore remain unused.

With this simple example we have shown how chequered parking spaces are a valid solution both for optimising space but also for having a higher actual parking capacity than normal.

7   Conclusions

In this work, an algorithm for smart vehicle allocation into parking slot is provided. Precisely, it is designed through the fifteen puzzle game. Indeed, the parking surface is designed as a chequered with some parking slots are reserved to move vehicle from one side to another (buffer).

Experimental results prove the effectiveness of this approach in terms of optimization of the parking surface. Indeed, it is compared with other algorithms relying on the standard parking surface division and the occupied space is seriously increased.

As hints for future development, we plan to study the parking problem from the perspective of a strategic reasoning approach through multi-agent systems [9], by possibly investigating learning techniques [2–4].

References

1. Agizza, M., Balzano, W., Stranieri, S.: An improved ant colony optimization based parking algorithm with graph coloring. In: Barolli, L., Hussain, F., Enokido, T. (eds.) AINA 2022. LNNS, vol. 451, pp. 82–94. Springer, Cham (2022). https://doi.org/10.1007/978-3-030-99619-2_8

2. Amato, F., Coppolino, L., Cozzolino, G., Mazzeo, G., Moscato, F., Nardone, R.: Enhancing random forest classification with NLP in DAMEH: a system for data management in eHealth domain. Neurocomputing **444**, 79–91 (2021)

3. Amato, F., Coppolino, L., Mercaldo, F., Moscato, F., Nardone, R., Santone, A.: Can-bus attack detection with deep learning. IEEE Trans. Intell. Transp. Syst. **22**(8), 5081–5090 (2021)

4. Amato, F., Cozzolino, G., Moscato, F., Moscato, V., Xhafa, F.: A model for verification and validation of law compliance of smart contracts in IoT environment. IEEE Trans. Industr. Inf. **17**(11), 7752–7759 (2021)

5. Hasan, D.O., Aladdin, A.M., Talabani, H.S., Rashid, T.A., Mirjalili, S.: The fifteen puzzle-a new approach through hybridizing three heuristics methods. Computers **12**(1), 11 (2023)

6. Hollist, J.T.: The fifteen puzzle. Math. Teacher **72**(8), 603–607 (1979)

7. Idris, M.Y.I., Leng, Y.Y., Tamil, E.M., Noor, N.M., Razak, Z., et al.: Car park system: a review of smart parking system and its technology. Inf. Technol. J. **8**(2), 101–113 (2009)

8. Lin, T., Rivano, H., Le Mouël, F.: A survey of smart parking solutions. IEEE Trans. Intell. Transp. Syst. **18**(12), 3229–3253 (2017)

9. Malvone, V., Stranieri, S.: Towards a model checking tool for strategy logic with simple goals. In: ICTCS, pp. 311–316 (2021)

10. Morris, B., Raymer, A.: Mixing time of the fifteen puzzle. Electron. J. Probab. **22**, 1–29 (2017)

11. Setyobudhi, C.T.: Comparison of A* algorithm and greedy best search in searching fifteen puzzle solution. Int. J. Comput. Inf. Technol. (2279-0764) **11**(3) (2022)

12. Shin, J.-H., Jun, H.-B.: A study on smart parking guidance algorithm. Transp. Res. Part C Emerg. Technol. **44**, 299–317 (2014)

13. Stranieri, S.: An indoor smart parking algorithm based on fingerprinting. Future Internet **14**(6), 185 (2022)

14. Zhao, X., Zhao, K., Hai, F.: An algorithm of parking planning for smart parking system. In: Proceeding of the 11th World Congress on Intelligent Control and Automation, pp. 4965–4969. IEEE (2014)

DTAG: A Dynamic Threshold-Based Anti-packet Generation Method for Vehicular DTN

Shota Uchimura[1], Masaya Azuma[1], Makoto Ikeda[2(✉)] ⓘ,
and Leonard Barolli[2] ⓘ

[1] Graduate School of Engineering, Fukuoka Institute of Technology (FIT),
3-30-1 Wajiro-Higashi, Higashi-Ku, Fukuoka 811-0295, Japan
{mgm21102,mgm21101}@bene.fit.ac.jp
[2] Department of Information and Communication Engineering,
Fukuoka Institute of Technology (FIT), 3-30-1 Wajiro-Higashi, Higashi-Ku,
Fukuoka 811-0295, Japan
makoto.ikd@acm.org, barolli@fit.ac.jp

Abstract. In this paper, we propose a Dynamic Threshold-based Anti-packet Generation (DTAG) method, which considers replication progress of the adjacent nodes. We considered Epidemic and Spray and Wait (SpW) protocols and combined with the proposed DTAG method and conventional anti-packet. Thus, we implemented four scenarios by simulations. From the simulation results, we found that the combination of the proposed DTAG method with Epidemic and SpW protocols reduces the overhead compared with combination of anti-packet method with Epidemic and SpW protocols. Also, the combination of SpW protocol with the DTAG method can reduce the storage usage.

Keywords: DTAG · DTN · Epidemic · SpW · Recovery scheme

1 Introduction

Delay-/Disruption-/Disconnection-Tolerant Networking (DTN) has been getting a lot of attention for communication in space, on land and underwater with a large delay and frequent disconnection [8–10,14,21]. Storage management is typically difficult because of the overhead involved in the usual DTN technique. The architecture and bundle specification are opened in [6,16] and some implementations are still under development [3,5]. Many DTN protocols have been proposed around the world [2,7,11,12,18,22–24]. Epidemic [13,20], Spray and Wait (SpW) [17] and MaxProp [4] are typical DTN protocols.

In [19], we have proposed an Enhanced Adaptive Anti-packet Recovery (EAAR) for Vehicular DTN. The EAAR is classified as enhanced recovery techniques for Epidemic routing, but it used a threshold to generate anti-packet.

ⓒ The Author(s), under exclusive license to Springer Nature Switzerland AG 2023
L. Barolli (Ed.): AINA 2023, LNNS 655, pp. 406–414, 2023.
https://doi.org/10.1007/978-3-031-28694-0_39

In this paper, we propose a Dynamic Threshold-based Anti-packet Generation (DTAG) method, which considers replication progress of the adjacent nodes. For evaluation, we considered Epidemic and SpW protocols with the proposed DTAG method and conventional anti-packet. From the simulation results, we found that the combination of the proposed DTAG method with Epidemic and SpW protocols reduces the overhead compared with combination of anti-packet method with Epidemic and SpW protocols. Also, the combination of SpW protocol with the DTAG method can reduce the storage usage.

The rest of the paper is organized as follows. In Sect. 2, we provide the overview of Epidemic and SpW protocols. In Sect. 3, we describe the DTAG method for Vehicular DTN. In Sect. 4, we present the simulation environment and the evaluation results. Finally, we conclude the paper in Sect. 5.

2   Overview of Epidemic and SpW Protocols

In Epidemic protocol a group of messages is sent with two separate control messages across the whole network and each node occasionally distributes a Summary Vector (SV). The SV records the bundle messages that have been stored for each node and is verified by the nodes that have received data against their SV. If the node detects an unknown bundle message in the received SV, the node generates a REQUEST.

Because nodes copy bundle messages to neighbor nodes within their communication range, the consumption of network resources and storage space becomes a major challenge in Epidemic protocol. Thus, the bundle messages are constantly copied and stored, so they remain in storage even after the endpoint receives them. Also, some copied bundle messages are deleted too late by recovery methods such as anti-packet.

The SpW protocol is able to optimize network resources because it places a strict limit on the total number of copied bundle messages. There are two stages at operation, which are regarded as the spraying stage and the waiting stage, respectively.

Traditional anti-packets are broadcasted by the endpoints and provide a list of bundles received by the endpoints. After the bundle messages are verified, nodes can delete them from their storage.ă Then, the nodes distribute the anti-packet to other nodes, but the anti-packet may have a harmful impact on network resources.

3   Dynamic Threshold-based Anti-packet Generation Method

In Fig. 1, we show a flowchart of the proposed DTAG method for Vehicular DTN. We present the procedure regarding each received message, such as SV, REQUEST, MESSAGE and anti-packet. In DTAG method, anti-packets are generated by intermediate nodes, before the bundle message reach the destination. In this approach, the node counts the number of received SVs to calculate the number of the same messages and new messages, respectively. There is no predetermined threshold in the DTAG procedure.

If the Eq. (1) is true, the DTAG method generate anti-packet. Otherwise, if Eq. (1) is false, the node stores the message in their storage.

$$\frac{\text{Count New Messages}}{\text{Count All Messages}} \geq \frac{\text{Count Same Messages}}{\text{Count All Messages}}, (\text{Count All Messages} \neq 0). \tag{1}$$

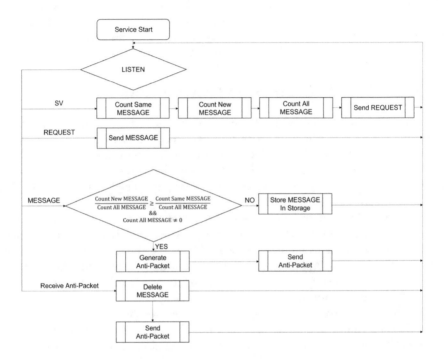

Fig. 1. Flowchart of DTAG method.

4   Evaluation of DTAG Method

We evaluate the proposed DTAG method using three metrics: the mean delivery ratio, mean delay, mean overhead and mean storage usage. The mean values are calculated from 10 different seeds and can consider various patterns of vehicle movement. For evaluation, we implemented the DTAG method on the Scenargie [15] and combined with the Epidemic and SpW protocols.

4.1   Simulation Environment

A grid road model is used from 50 to 150 vehicles per square kilometer. The simulation parameters are shown in Table 1. Vehicles continue to move on the roads based on the random way-point mobility model. We consider the following four scenarios.

- Case-A: Epidemic protocol with DTAG method.
- Case-B: Epidemic protocol with conventional anti-packet.
- Case-C: SpW protocol with DTAG method.
- Case-D: SpW protocol with conventional anti-packet.

The originator sends a bundle message to the adjacent nodes, which continue to forward the message to other nodes until the message reaches the destination. We consider 600 s for conducting the simulations. In simulations, we consider a radio frequency of 5.9 GHz and apply the ITU-R P.1411 [1] propagation model.

Table 1. Simulation parameters.

Items	Setup values
Simulation duration	600 [seconds]
Area dimensions	$1,000$ [meters] $\times$ $1,000$ [meters]
Number of vehicles	50–150 [vehicles/km^2]
Nodes: minimum and maximum velocity	8.333–16.666 [m/s]
Message: start and end time	10–400 [seconds]
Message: size	500 [bytes]
Message: generation interval	10 [seconds]
SV: generation interval	1 [second]
DTN protocol	Epidemic and SpW
DTN: recovery method	Conventional anti-packet and DTAG method
SpW: maximum number of copies	50
SpW: binary mode	True

4.2   Evaluation Results

We present the simulation results of mean delivery ratio in Fig. 2. In the Case-A, the delivery ratio is slightly decreased. However, it is above 90% regardless of the vehicle density. In the Case-C, the delivery ratio significantly decreases with increasing vehicle densities. This is because the proposed method deletes messages by anti-packet when the conditions are true before the messages reach the destination. Also, the SpW updates the number of copies to half for every replication of a message. Therefore, when the number of vehicles is increased, there are more frequent replications and the number of possible copies of SpW decreases.

We present the simulation results of the mean delay in Fig. 3. For Case-A and Case-B, we observed that the difference is small and the mean delay is shorter than Case-C and Case-D.

We present the simulation results of the mean overhead in Fig. 4. We found that the mean overhead of Case-A and Case-B is higher than Case-C and Case-D. The proposed DTAG method (Case-A and Case-C) reduces the overhead for both protocols, compared with Case-B and Case-D, respectively.

We present the simulation results of mean storage usage for 75 and 150 vehicle/km^2 in Fig. 5. Comparing Case-A with Case-B, the storage usage of Case-A is increased compared with Case-B regardless of the vehicle densities. Also in Case-A, the storage usage is not zero even after 400 s. This means that the messages are deleted independently from each vehicle when the condition is true. While for Case-C and Case-D, we found that in Case-C the storage usage is reduced significantly.

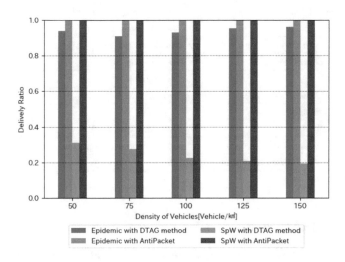

Fig. 2. Mean delivery ratio for different cases.

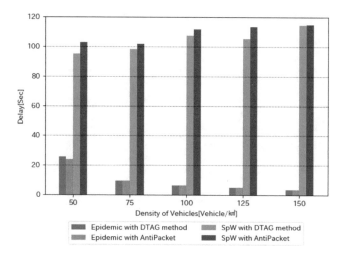

Fig. 3. Mean delay for different cases.

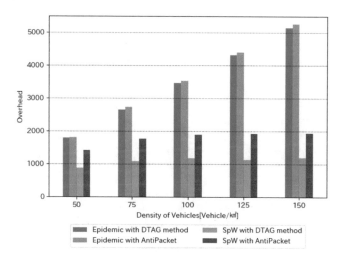

Fig. 4. Mean overhead for different cases.

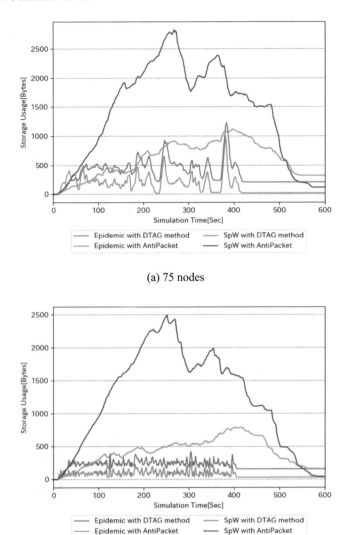

(a) 75 nodes

(b) 150 nodes

Fig. 5. Storage usage for different vehicles.

5   Conclusions

In this paper, we proposed DTAG method for Vehicular DTN. Then, we combined the proposed method and conventional anti-packet method with Epidemic and SpW protocols and implemented four scenarios. We evaluated by simulations the implemented scenarios. From the simulation results, we found that the combination of the proposed DTAG method with Epidemic and SpW protocols reduces the overhead compared with combination of anti-packet method with Epidemic and SpW protocols. Also, the combination of SpW protocol with the DTAG method can reduce the storage usage.

In future work, we would like to consider the other methods, real road map and other parameters.

References

1. Recommendation ITU-R P.1411-11: propagation data and prediction methods for the planning of short-range outdoor radiocommunication systems and radio local area networks in the frequency range 300 MHz to 100 GHz. ITU, September 2019. https://www.itu.int/rec/R-REC-P.1411-11-202109-I/en
2. Barroca, C., Grilo, A., Pereira, P.R.: Improving message delivery in UAV-based delay tolerant networks. In: Proceedings of the 16th International Conference on Intelligent Transportation Systems Telecommunications (ITST-2018), pp. 1–7, October 2018. https://doi.org/10.1109/ITST.2018.8566956
3. Baumgärtner, L., Höchst, J., Meuser, T.: B-DTN7: browser-based disruption-tolerant networking via bundle protocol 7. In: Proceedings of the International Conference on Information and Communication Technologies for Disaster Management (ICT-DM-2019), pp. 1–8, December 2019. https://doi.org/10.1109/ICT-DM47966.2019.9032944
4. Burgess, J., Gallagher, B., Jensen, D., Levine, B.N.: MaxProp: routing for vehicle-based disruption-tolerant networks. In: Proceedings of the 25th IEEE International Conference on Computer Communications (IEEE INFOCOM-2006), pp. 1–11, April 2006. https://doi.org/10.1109/INFOCOM.2006.228
5. Burleigh, S., Fall, K., E. Birrane, I.: Bundle protocol version 7. IETF RFC 9171 (Standards Track), January 2022
6. Cerf, V., et al.: Delay-tolerant networking architecture. IETF RFC 4838 (Informational), April 2007
7. Chuah, M.C., Ma, W.B.: Integrated buffer and route management in a DTN with message ferry. In: Proceedings of the IEEE Military Communications Conference (MILCOM-2006), pp. 1–7, October 2006. https://doi.org/10.1109/MILCOM.2006.302288
8. Davarian, F., et al.: Improving small satellite communications and tracking in deep space - a review of the existing systems and technologies with recommendations for improvement. Part ii: small satellite navigation, proximity links, and communications link science. IEEE Aerosp. Electron. Syst. Mag. $35(7)$, 26–40 (2020). https://doi.org/10.1109/MAES.2020.2975260
9. Fall, K.: A delay-tolerant network architecture for challenged internets. In: Proceedings of the International Conference on Applications, Technologies, Architectures, and Protocols for Computer Communications, pp. 27–34, August 2003. https://doi.org/10.1145/863955.863960
10. Fraire, J.A., Feldmann, M., Burleigh, S.C.: Benefits and challenges of cross-linked ring road satellite networks: a case study. In: Proceedings of the IEEE International Conference on Communications (ICC-2017), pp. 1–7, May 2017. https://doi.org/10.1109/ICC.2017.7996778
11. Henkel, D., Brown, T.X.: Delay-tolerant communication using mobile robotic helper nodes. In: Proceedings of the 6th International Symposium on Modeling and Optimization in Mobile, Ad Hoc, and Wireless Networks and Workshops 2008, pp. 657–666, April 2008. https://doi.org/10.1109/WIOPT.2008.4586155

12. Iranmanesh, S., Raad, R., Raheel, M.S., Tubbal, F., Jan, T.: Novel DTN mobility-driven routing in autonomous drone logistics networks. IEEE Access **8**, 13661–13673 (2020). https://doi.org/10.1109/ACCESS.2019.2959275

13. Ramanathan, R., Hansen, R., Basu, P., Hain, R.R., Krishnan, R.: Prioritized epidemic routing for opportunistic networks. In: Proceedings of the 1st International MobiSys Workshop on Mobile Opportunistic Networking (MobiOpp 2007), pp. 62–66, June 2007. https://doi.org/10.1145/1247694.1247707

14. Rüsch, S., Schürmann, D., Kapitza, R., Wolf, L.: Forward secure delay-tolerant networking. In: Proceedings of the 12th Workshop on Challenged Networks (CHANTS-2017), pp. 7–12, October 2017. https://doi.org/10.1145/3124087.3124094

15. Scenargie: Space-time engineering, LLC. http://www.spacetime-eng.com/

16. Scott, K., Burleigh, S.: Bundle protocol specification. IETF RFC 5050 (Experimental), November 2007

17. Spyropoulos, T., Psounis, K., Raghavendra, C.S.: Spray and Wait: an efficient routing scheme for intermittently connected mobile networks. In: Proceedings of the ACM SIGCOMM Workshop on Delay-Tolerant Networking 2005 (WDTN 2005), pp. 252–259, August 2005. https://doi.org/10.1145/1080139.1080143

18. Sugihara, K., Hayashibara, N.: Message delivery of Nomadic Lévy walk based message ferry routing in delay tolerant networks. In: Barolli, L., Hussain, F., Enokido, T. (eds.) AINA 2022. LNNS, vol. 449, pp. 259–270. Springer, Cham (2022). https://doi.org/10.1007/978-3-030-99584-3_23

19. Uchimura, S., Azuma, M., Ikeda, M., Barolli, L.: An enhanced adaptive anti-packet recovery method forÂ inter-vehicle communications. In: Proceedings of the International Conference on Network-Based Information Systems (NBiS-2022), pp. 374–383 (2022). https://doi.org/10.1007/978-3-031-14314-4_38

20. Vahdat, A., Becker, D.: Epidemic routing for partially-connected ad hoc networks. Duke University, Technical report (2000)

21. Wyatt, J., Burleigh, S., Jones, R., Torgerson, L., Wissler, S.: Disruption tolerant networking flight validation experiment on NASA's EPOXI mission. In: Proceedings of the 1st International Conference on Advances in Satellite and Space Communications (SPACOMM-2009), pp. 187–196, July 2009. https://doi.org/10.1109/SPACOMM.2009.39

22. Yasmeen, F., Huda, N., Yamada, S., Borcea, C.: Ferry access points and sticky transfers: improving communication in ferry-assisted DTNs. In: Proceedings of the IEEE International Symposium on a World of Wireless, Mobile and Multimedia Networks (WoWMoM-2012), pp. 1–7, June 2012. https://doi.org/10.1109/WoWMoM.2012.6263746

23. Zhao, W., Ammar, M., Zegura, E.: Controlling the mobility of multiple data transport ferries in a delay-tolerant network. In: Proceedings IEEE 24th Annual Joint Conference of the IEEE Computer and Communications Societies, vol. 2, pp. 1407–1418, March 2005. https://doi.org/10.1109/INFCOM.2005.1498365

24. Zhao, W., Ammar, M.: Message ferrying: proactive routing in highly-partitioned wireless ad hoc networks. In: The Ninth IEEE Workshop on Future Trends of Distributed Computing Systems, FTDCS 2003, Proceedings, pp. 308–314, May 2003. https://doi.org/10.1109/FTDCS.2003.1204352

Optimal and Suboptimal Routing Protocols for WSN

Rahil Bensaid[1,2,3(✉)], Adel Ben Mnaouer[2], and Hatem Boujemaa[1]

[1] UCAR COSIM Lab, Higher School of Communications of Tunis, Ariana, Tunisia
{rahil.bensaid,boujemaa.hatem}@supcom.tn
[2] Faculty of Engineering and Architecture, Canadian University Dubai, Dubai, UAE
adel@cud.ac.ae
[3] ENIG National Engineering School of Gabes, Gabes, Tunisia

Abstract. In this paper, we propose one hop, optimal and suboptimal routing protocols for Wireless Sensor Networks (WSNs). The WSN is organized in L hops and N branches between the source S and the destination D. One hop routing activates the best relay in each hop. Optimal routing selects the best path among all N^L paths. Suboptimal routing decomposes the network in K subnetworks and activates the best path in each subnetworks. We derive the outage probability of the three studied routing protocols and compare them to simulation results.

1 Introduction

Wireless Sensor Networks (WSNs) still gaining attention since their intensive deployment in growing Internet of Things (IoT) applications. Many challenges face the effective deployment of a WSN such as reliability and longevity. Routing protocols and data aggregation techniques are considered as key solutions to improve the network lifetime. In literature, different routing protocols have been suggested for WSN to minimize the outage probability [1–10] and provide good quality of service. A modified data aggregation technique has been suggested in [1] for secure routing techniques in WSN. Deep learning techniques have been used to optimizing the routing protocol in [2]. In [3], authors proposed the use of super capacitors instead of batteries as energy source for wireless nodes. The implemented technique improved the network lifetime and efficiently balanced the energy load among clusters. Improved Whale Algorithm (WA) has been used in [4] to improve the routing protocol in WSN. A survey on energy efficient routing protocols for WSNs and possible optimization techniques has been presented in [5]. Routing protocol with energy harvesting has been studied in [6]. Authors in [7], proposed a weighted cluster routing protocol which defines a redundant cluster head to ensure the end-to-end connectivity in a crisis case scenario. An hybrid routing technique, which takes benefit from cluster-based and chain-based routing protocols, was used in [8] to extend the network lifetime. In [9], a modified Low Energy Adaptive Clustering Hierarchy (LEACH) algorithm has been

© The Author(s), under exclusive license to Springer Nature Switzerland AG 2023
L. Barolli (Ed.): AINA 2023, LNNS 655, pp. 415–426, 2023.
https://doi.org/10.1007/978-3-031-28694-0_40

suggested to enhance the energy efficiency. A bootstrapped Particle Swar Optimization (PSO) clustering has been suggested in [10] to obtain an efficient WSN
routing protocol.

In this paper, we aim at decreasing the outage probability of the end-to-end
link between the sensor node and the base station and so increasing the energy
efficiency of the network. Outage probability is considered among important
performance metrics to assess and improve the system reliability. A high outage
probability will affect the rate of erroneous packets and will result on a lower
energy efficiency since the node consumed energy at the transmission process,
yet the receiver didn't successfully retrieve the transmitted packet. In this work,
we consider a multi-hop and multi-branch network architecture between the
source and the destination. We first, propose one hop routing that maximizes
the Signal to Noise Ratio (SNR) at each hop. We also suggest optimal routing
that selects the best end-to-end (e2e) path between the source and destination.
The best path is chosen among all paths so that the e2e SNR is maximized. In
order to reduce the complexity and obtain a low outage probability, we suggest
a suboptimal routing protocol that decomposes the network in K subnetworks.
The best e2e path is chosen in each subnetwork. We derive the outage probability
of the three suggest routing protocols and compare it to simulation results.

The paper contains seven sections. Section 2 describes the system model.
Sections 3, 4 and 5 suggest one hop routing, optimal and suboptimal routing protocols. The outage probabilities of the three protocols are also derived. Section 6
gives the theoretical and simulation results. Section 7 concludes the paper.

2   System Model

We consider a wireless network where sensor nodes are uniformly distributed, as
shown in Fig. 1. Sensor nodes can cooperate with each other using the amplify
and forward (AF) protocol [11] to transmit data to the base station (BS), located
at the edge of the network. All sensor nodes in the neighborhood can serve as
potential relay nodes.

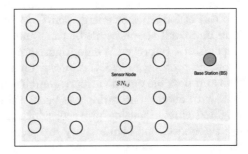

Fig. 1. Wireless sensor network architecture

To transmit a data packet to the BS, a $L - hop$ chain of sensor nodes is constructed between the source sensor node (S) and the BS by selecting in each hop a relay node from N branches, as illustrated in Fig. 2. We assume that all sensor nodes in the same branch operate in orthogonal transmission times. So only one path over possible paths is active in the same time. Our main focus here is to assess the improvement of reliability through the selected transmission path by the evaluation of the end-to-end outage probability.

For simplicity, we denote the transmitting source node by S, the destination or the base station node by D, sensor nodes in between the source and the destination are called relaying nodes $R_{i,j}$ where i, $i \in \{1,...(L-1)\}$, is hop number and $j, i \in \{1,...N\}$, is the branch number.

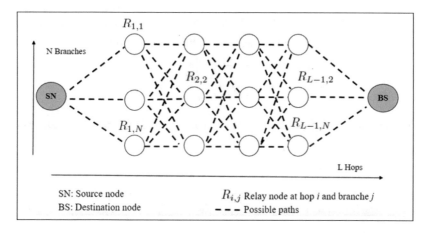

Fig. 2. System Model showing relaying nodes between a source node S and the destination D

As for the channel model, we assume a Rayleigh channel. The channel coefficient $h_{i,j}$ of the link between two sensor nodes i and j is modeled as: $h_{i,j} = d_{i,j}^{-\alpha}$ where $d_{i,j}$ is the normalized distance between i and j and α is the path loss exponent. The normalized distance is calculated as $d_{i,j} = \frac{d}{d_0}$, where d is the effective distance and d_0 is a reference distance

The received signal at the selected relay node R_1, at the first hop, from the source node is given by

$$Y_{S,R_1} = \sqrt{P_S}h_{S,R_1}X_S + n_{S,R_1} \qquad (1)$$

where P_S is the transmitting power, h_{S,R_1} is the Rayleigh fading coefficient of the link, X_S is the transmitted signal and n_{S,R_1} is an additive Gaussian noise with variance N_0.

Since sensor nodes are cooperating using AF relaying scheme, the received signal at the selected relay node R_i at hop i, from the selected relay node at hop $i-1$, R_{i-1} is given by

$$Y_{R_{i-1},R_i} = h_{R_{i-1},R_i} G_{i-1} Y_{R_{i-2},R_{i-1}} + n_{R_{i-1},R_i} \qquad (2)$$

where G_{i-1} is the amplification gain used by the relay R_{i-1}.

$$G_{i-1} = \sqrt{\frac{P_{R_{i-1}}}{(P_{R_i}\|h_{R_{i-1},R_i}\|^2 + N_0)}} \qquad (3)$$

The received signal at the BS node from the selected relay R_{L-1} at hop L is given by

$$Y_{R_{L-1},D} = h_{R_{L-1},D} G_{L-1} Y_{R_{L-2},R_{L-1}} + n_{R_{L-1},D} \qquad (4)$$

where G_{L-1} is the amplification gain used by the relay R_{L-1}.

$$G_{L-1} = \sqrt{\frac{P_{R_{i-1}}}{(P_{R_i}\|h_{R_{i-1},R_i}\|^2 + N_0)}} \qquad (5)$$

The received instantaneous SNR for each link between two sensor nodes i and j is given by

$$\gamma_{i,j} = \frac{P_i\|h_{i,j}\|^2}{N_0} \qquad (6)$$

For Rayleigh channels, the probability density function (PDF) and the cumulative density function (CDF) are given by

$$f_{\gamma_{i,j}}(x) = 1 - e^{-\frac{x}{\overline{\gamma_{i,j}}}} \qquad (7)$$

$$F_{\gamma_{i,j}}(x) = \frac{1}{\overline{\gamma_{i,j}}} e^{-\frac{x}{\overline{\gamma_{i,j}}}} \qquad (8)$$

where $\overline{\gamma_{i,j}}$ is the expectation of $\gamma_{i,j}$.

3   One Hop Routing

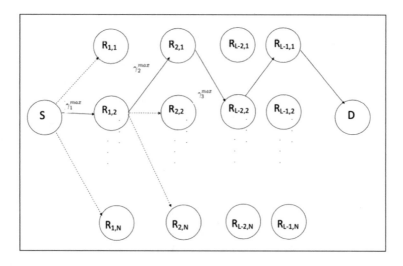

Fig. 3. One hop routing protocol

As shown in Fig. 3, one hop routing consists to choose the relay with largest SNR at the i-th hop:

$$\gamma_i^{max} = max(\gamma_{i,1}, ..., \gamma_{i,N}), \tag{9}$$

where $1 \leq i \leq L$, L is the number of hops and N is the number of branches which is also the number of available relays at each hop.

The end-to-end (e2e) SNR of Amplify and Forward relaying is tightly upper bounded by

$$\gamma_{e2e} < \gamma_{e2e}^{max} = min(\gamma_1^{max}, ..., \gamma_L^{max}) \tag{10}$$

The outage probability is equal to

$$P(\gamma_{e2e} < T) > P(\gamma_{e2e}^{max} < T) = 1 - \prod_{i=1}^{L}[1 - F_{\gamma_i^{max}}(T)] \tag{11}$$

where T is the SNR threshold.

The Cumulative Distribution Function (CDF) of the maximum SNR at i-th hop is computed as

$$F_{\gamma_i^{max}}(T) = \prod_{j=1}^{N} F_{\gamma_{i,j} < x} = \prod_{j=1}^{N}[1 - exp(-\frac{T}{\overline{\gamma}_{i,j}})] \tag{12}$$

where $\overline{\gamma}_{i,j}$ is the average SNR over i-th hop and j-th branch.

4   Optimal Routing

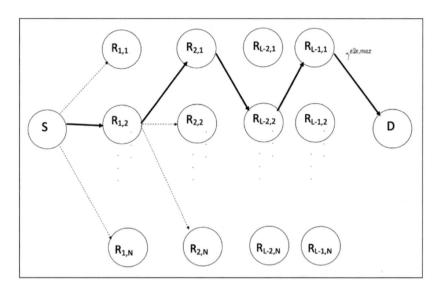

Fig. 4. Optimal routing protocol

As shown in Fig. 4, optimal routing consists to search the path with highest SNR between the source and destination. The SNR of optimal routing is the maximum over all available paths, i.e. N^L paths:

$$\gamma^{optimal} = max_{1 \leq p \leq N^L}(\gamma_p^{e2e}) \tag{13}$$

where γ_p^{e2e} is the e2e SNR of p-th path.
    We deduce:

$$P_{\gamma^{optimal}}(x) = P(max_{1 \leq p \leq N^{L-1}}(\gamma_p^{e2e}) \leq x) = \prod_{p=1}^{N^{L-1}} P_{\gamma_p^{e2e}}(x) \tag{14}$$

where $P_{\gamma_p^{e2e}}(x)$ is the CDF of end-to-end SNR of p-th path written as

$$\gamma_p^{e2e} < min(\gamma_1^p, \gamma_2^p, ..., \gamma_L^p) \tag{15}$$

where γ_i^p is the SNR of i-th hop for the p-th path.
    We have

$$P_{\gamma_p^{e2e}}(x) > 1 - \prod_{i=1}^{L}[1 - P_{\gamma_i^p}(x)] \tag{16}$$

5   Suboptimal Routing

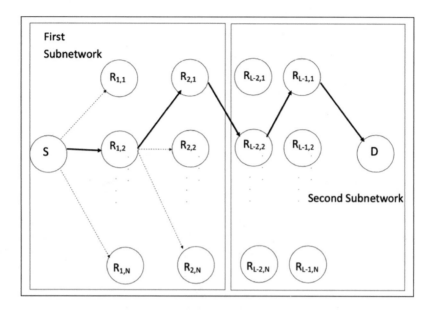

Fig. 5. Suboptimal routing protocol

As shown in Fig. 5, suboptimal routing decomposes the network in K subnetworks then the best path is activated in each subnetwork. In order to not have an outage, we should not have an outage in all K subnetworks.

$$P_{\gamma^{suboptimal}}(x) = 1 - \prod_{k=1}^{K}[1 - P_{\gamma^{k,optimal}}(x)] \qquad (17)$$

where $\gamma^{k,opitmal}$ is the e2e SNR in the k-th subnetwork expressed as (15).

6   Theoretical and Simulation Results

We did some simulations where there are $L = 4$ hops and $N = 2$ branches in each hop as illustrated in Fig. 6.

The normalized distance between consecutive node is $d_{normalized} = 1.5 = \frac{d}{d_0}$. The path loss exponent α is three and the average power of channel coefficient between consecutive node is $\frac{1}{d^{\alpha}_{normalized}}$.

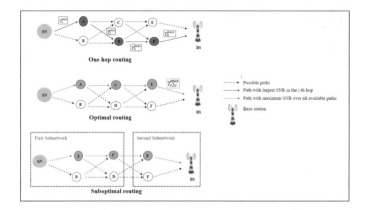

Fig. 6. System with $L = 4$ hops and $N = 2$ branches

Figures 7, 8 and 9 show the outage probability of one hop routing, optimal routing and suboptimal routing protocols for Rayleigh channels. We observe that the simulation results are in accordance with theoretical outage probability obtained fo $T = 0$ dB.

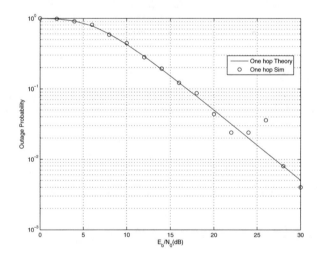

Fig. 7. Outage probability of one hop routing protocol

Figure 10 compares the outage probability of the three studied routing protocols. Optimal routing offers 13 dB and 1.5 dB gain versus one hop routing and suboptimal routing.

Figures 11, 12 and 13 show the outage probability of one hop routing, optimal and suboptimal routing for $T = 0, 10$ dB. As threshold T increases as the outage probability increases.

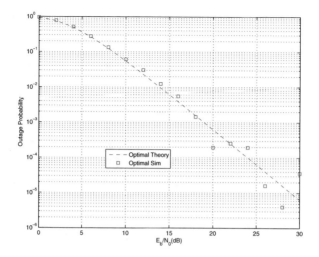

Fig. 8. Outage probability of optimal routing

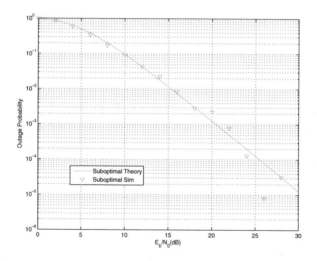

Fig. 9. Outage probability of suboptimal routing

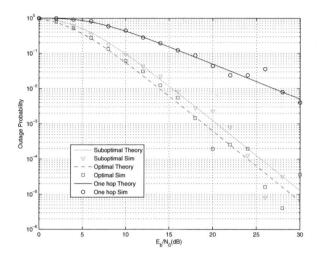

Fig. 10. Performance comparison of routing protocols

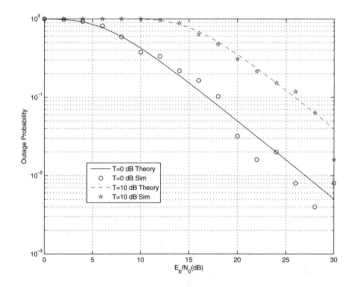

Fig. 11. Outage probability of one hop routing for $T = 0, 10$ dB

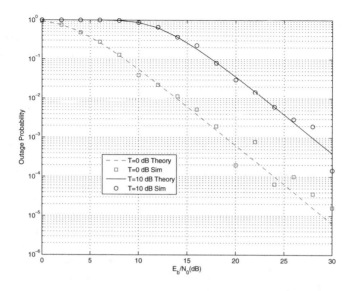

Fig. 12. Outage probability of optimal routing for $T = 0, 10$ dB

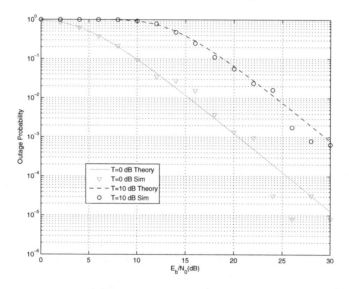

Fig. 13. Outage probability of suboptimal routing for $T = 0, 10$ dB

7 Conclusion and Perspectives

In this paper, we suggested three routing protocols for WSN. One hop routing operates hop per hop to activate the relay with largest Signal to Noise Ratio (SNR). Optimal routing activates the best end-to-end path. Suboptimal routing decomposes the network in K subnetwork where the best path is selected in each subnetwork. We have shown that optimal routing offers 13 dB and 1.5 dB gains versus one hop and suboptimal routing. As a perspective, we can study WSN with radio frequency energy harvesting and study the same routing protocol as well as studying their complexities.

References

1. Pralhad, K.-S.B., Singla, C.R., Patil, S.B.: Design of an energy efficient data aggregation method for secured routing protocol in WSN. In: 2022 2nd Asian Conference on Innovation in Technology (ASIANCON) (2022)
2. Ahmed, A.J., et al.: An effectual secure cryptography scheme for multipath routing in a WSN-based IoT environment. In: 2022 5th International Conference on Engineering Technology and its Applications (IICETA) (2022)
3. Charaan, R.M.D., Therasa, P.R., Jeya, B.L.: Solar aware routing with super capacitors to balance energy in unequal clusters for WSN. In: 2022 3rd International Conference on Electronics and Sustainable Communication Systems (ICESC) (2022)
4. Zhang, K., He, W., Liu, L., Gao, N.: A WSN clustering routing protocol based on improved Whale algorithm. In: 2022 4th International Conference on Natural Language Processing (ICNLP) (2022)
5. Ahlawat, B., Sangwan, A.: Energy efficient routing protocols for WSN in IOT: a survey. In: 2022 International Conference on Machine Learning, Big Data, Cloud and Parallel Computing (COM-IT-CON) (2022)
6. Xin, W., Cuiran, L., Jianli, X.: Research on clustering routing protocol for energy-harvesting WSN. In: 2022 IEEE 2nd International Conference on Electronic Technology, Communication and Information (ICETCI) (2022)
7. Khayat, G., Mavromoustakis, C.X., Mastorakis, G., Batalla, J.M., Pallis, E.: Weighted cluster routing protocol with redundant cluster head for damaged WSN. In: 2022 IEEE 7th International Energy Conference (ENERGYCON) (2022)
8. Bensaid, R., Boujemaa, H.: A combined cluster-chain based routing protocol for lifetime improvement in WSN. In: 2022 International Wireless Communications and Mobile Computing (IWCMC) (2022)
9. Goyal, P.K., Gupta, N.K.: Design of the modified D-LEACH based routing protocols for the dense deployment based WSN. In: 2022 Second International Conference on Advances in Electrical, Computing, Communication and Sustainable Technologies (ICAECT) (2022)
10. Obad, A.T., Ilyas, M.: Efficient WSN routing using bootstapped PSO clustering. In: 2022 International Congress on Human-Computer Interaction, Optimization and Robotic Applications (HORA) (2022)
11. Laneman, J.N., Tse, D.N., Wornell, G.W.: Cooperative diversity in wireless networks: efficient protocols and outage behavior. IEEE Trans. Inf. Theory **50**, 3062–3080 (2004)

CL-DECCM-SA: A Cluster-based Delaunay Edge and Simulated Annealing Approach for Optimization of Mesh Routers Placement in WMNs

Aoto Hirata[1], Yuki Nagai[1], Kyohei Toyoshima[1], Chihiro Yukawa[1], Tetsuya Oda[2(✉)], and Leonard Barolli[3]

[1] Graduate School of Engineering, Okayama University of Science (OUS), Okayama, 1-1 Ridaicho, Kita-ku, Okayama 700–0005, Japan
{t21jm02zr,t22jm23rv,t22jm24jd,t22jm19st}@ous.jp

[2] Department of Information and Computer Engineering, Okayama University of Science (OUS), 1-1 Ridaicho, Kita-ku, Okayama 700-0005, Japan
oda@ous.ac.jp

[3] Department of Information and Communication Engineering, Fukuoka Institute of Technology, 3-30-1, Wajiro-Higashi, Higashi-Ku, Fukuoka 811-0295, Japan
barolli@fit.ac.jp

Abstract. Wireless Mesh Networks (WMNs) enable large and stable networks to be built at low cost through wireless communication among multiple mesh routers. There are many methods, algorithms and systems to optimize the placement of mesh routers. In our previous work, we proposed Coverage Construction Method (CCM), CCM-based Hill Climbing (HC), and CCM-based Simulated Annealing (SA) systems for the mesh router placement problem. We also proposed a hybrid system by combining Delaunay Edge and CCM-based SA (DECCM-SA) in order to increase the probability of finding distant mesh clients. In this paper, we propose a Clustering-DECCM-SA (CL-DECCM-SA) as an improvement of the DECCM-SA system. The proposed CL-DECCM-SA approach clusters the mesh clients and performs Delaunay triangulation on the average point of each cluster. After that the derived Delaunay edges are approximated to integer values and used as the range of possible mesh router locations for the CCM to generate the initial solution. We evaluated the proposed approach by simulations. For the simulations, we consider the evacuation areas in Okayama City, Japan, which is the target to be covered by mesh routers. The simulation results show that the proposed approach is able to place mesh routers near mesh clients that are clustered far away from each other and can cover all mesh clients.

1 Introduction

Wireless Mesh Networks (WMNs) enable large and stable networks to be built at low cost through wireless communication among multiple routers. Mesh router placement affects the cost and size of wireless coverage area. There are many

L. Barolli (Ed.): AINA 2023, LNNS 655, pp. 427–434, 2023.
https://doi.org/10.1007/978-3-031-28694-0_41

research works to optimize the placement of mesh routers. In our previous work [1–12], we proposed and evaluated different meta-heuristic algorithms such as Genetic Algorithms (GA) [13], Hill Climbing (HC) [14], Simulated Annealing (SA) [15], Tabu Search (TS) [16] and Particle Swarm Optimization (PSO) [17] for mesh router placement optimization. We also proposed a Coverage Construction Method (CCM) [18], CCM-based Hill Climbing (CCM-HC) [19] and CCM-based Simulated Annealing (CCM-SA) approaches. The CCM can rapidly create a set of mesh routers that maximize the wireless coverage of the mesh routers. The CCM-HC covered many mesh clients generated with normal and uniform distributions, as well as many of mesh clients in the two-island model [20]. We also performed comparative studies by changing the parameters [21,22]. The CCM-SA [23] had a better performance than CCM-HC and was able to cover many mesh clients. We also proposed the Delaunay edge and CCM-based SA (DECCM-SA) approach [24], which considered more realistic scenarios.

In this paper, we propose a Clustering-DECCM-SA (CL-DECCM-SA) as an improvement of the DECCM-SA system. The proposed CL-DECCM-SA approach clusters the mesh clients and performs Delaunay triangulation on the average point of each cluster. After that the derived Delaunay edges are approximated to integer values and used as the range of possible mesh router locations for the CCM to generate the initial solution. We evaluated the proposed approach by simulations. As evaluation metrics, we consider the Size of Giant Component (SGC) and the Number of Covered Mesh Clients (NCMC). For the simulations, we consider the evacuation areas in Okayama City, Japan, which is the target to be covered by mesh routers. The simulation results show that the proposed approach is able to place mesh routers near mesh clients that are clustered far away from each other and can cover all mesh clients.

The structure of the paper is as follows. In Sect. 2, we give a short description of mesh router placement problem. In Sect. 3, we present the proposed approach. In Sect. 4, we discuss the simulation results. Finally, in Sect. 5, we conclude the paper and give future research directions.

2   Mesh Router Placement Problem

The mesh router placement problem is a bi-objective optimization problem to maximize network connectivity and coverage of mesh clients in a WMN. The network connectivity of the mesh routers is measured by SGC, which is derived from the mesh router connectivity graph and the coverage of mesh clients is NCMC, which is the number of mesh clients that fall within the communication range of at least one mesh router. We use SGC and NCMC as optimization metrics for the mesh router placement problem and make the following considerations.

- There is an area $Width \times Height$ which is the considered area for mesh router placement.
- Each mesh router has its own radio communication range.
- Each mesh client can be located arbitrarily in the considered area.

3 Proposed Clustering-DECCM-SA Approach

In this section, we describe Clustering-Delaunay edge and CCM-based SA (CL-DECCM-SA) approach. In our previous work, we proposed some algorithms for mesh routers placement in WMNs. but the objective was to cover randomly generated mesh clients based on a normal or uniform distribution by mesh clients. We also proposed DECCM-SA for mesh router placement optimization, which considered more realistic scenarios. The DECCM-SA can perform Delaunay-triangulation on mesh clients, set the coordinates of the derived Delaunay edges as possible mesh client placement area in the CCM and applies CCM-SA. The DECCM-SA uses Delaunay edges to place mesh routers and can place mesh routers close to mesh clients even when NCMC cannot be updated.

The previous DECCM-SA used a GIS application for deciding Delaunay edge and image processing for approximation of Delaunay edges to integer values. However, this approach can be used only for simulations based on GIS information.

Our proposed Clustering-DECCM-SA approach performs clustering of mesh clients by applying Delaunay triangulation and approximating Delaunay edges to integer values using Bresenham's line algorithm [26]. The clustering is performed to separate the mesh clients in several groups. There are different clustering methods, and the suitable clustering method depends on the distribution of data. The proposed approach uses DBSCAN [27], which is suitable for separating groups of mesh clients. The DBSCAN is a data density-compliant clustering method that assigns nearby mesh clients to the same cluster. We consider two parameters: $min\_samples$ and ϵ. DBSCAN randomly selects a mesh client and assigns a cluster label to the selected mesh client if there are at least $min\_samples$ of other mesh clients within an ϵ radius circle of that mesh router. Then, if a mesh client with a cluster label assigned exists within the epsilon radius circle, the same cluster label as the mesh client is assigned to the selected mesh client. These operations are repeated to assign cluster labels to all mesh clients.

We determine the average coordinates of mesh clients belonging to each cluster and perform a Delaunay-triangulation on those coordinates. However, the resulting Delaunay edge coordinates are not integer values. Therefore, we use Bresenham's line algorithm to approximate the coordinates of the Delaunay edge to integer values. The Bresenham's line algorithm can approximate a line segment by placing consecutive points between the coordinates of starting and end

(a) Original map image.   (b) Evacuation area.

Fig. 1. Visualization of placement area.

points. The coordinates of starting point and end point are used as coordinates
of both ends of each Delaunay edge to apply Bresenham's line algorithm.

Table 1. Parameters and values.

Parameters	Values
Width	260 [$unit$]
Hight	180 [$unit$]
Number of Mesh Routers	256
Radius of radio communication range of mesh routers	4 [$unit$]
Number of mesh clients	3089
Number of loops (CCM)	2000 [$times$]
Number of loops (SA)	10000 [$times$]
T_{max}	100 [$unit$]
T_{min}	1 [$unit$]
α	10
min_samples	1
ϵ	3

Table 2. Simulation results.

Approaches	Best SGC	Average SGC	Best NCMC	Average NCMC [%]
CL-DECCM	256	256	1745	55.864
CL-DECCM-SA	256	256	3089	99.833

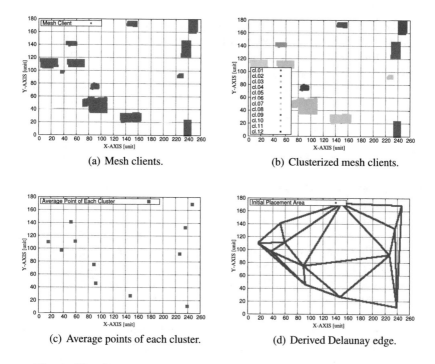

(a) Mesh clients.

(b) Clusterized mesh clients.

(c) Average points of each cluster.

(d) Derived Delaunay edge.

Fig. 2. Visualization results of clustering and Delaunay triangulation.

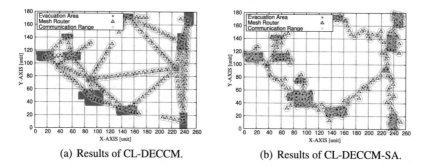

(a) Results of CL-DECCM.

(b) Results of CL-DECCM-SA.

Fig. 3. Visualization results of mesh router placement optimization.

4   Simulation Results

4.1   Simulation Settings

In this section, we present simulation results. The simulation parameters are shown in Table 1. The placement area is around Okayama Station in Okayama City, Okayama Prefecture, Japan. The buildings are used as evacuation areas and are considered as mesh clients.

For displaying the geographic information, we use QGIS, which is a GIS application. We also used shapefiles of buildings from OpenStreetMap and evacuation site information from open data published by Okayama City [25]. The original map image is shown in Fig. 1(a). The red dots indicate the position of evacuation areas. In Fig. 1(b), the buildings designated as evacuation sites are shown painted in red. These red buildings are considered as mesh clients for simulations.

4.2   Simulation Results

The simulation results of 100 runs of each approach are shown in Table 2. In all simulations, the SGC is maximized for each approach and the CL-DECCM-SA covered most of the mesh clients.

In Fig. 2 are shown the visualization results of clustering and Delaunay triangulation. In Figs. 2(a) and 2(b), we show the mesh clients and clusterized mesh clients. From Fig. 2(b), it can be seen that mesh clients are grouped according to the distance. Figures 2(c) and 2(d) shows the average point of each cluster and derived Delaunay edge.

In Fig. 3, we show the visualization results of mesh router placement optimization. Figure 3(a) shows the placement of mesh routers by CL-DECCM and Fig. 3(b) shows the placement of mesh routers optimized by CL-DECCM-based SA, which covers all mesh clients.

5   Conclusions

In this paper, we proposed a Clustering-DECCM-SA (CL-DECCM-SA) approach in order to improve our previous DECCM-SA system. The proposed CL-DECCM-SA approach clusters the mesh clients and performs Delaunay triangulation on the average point of each cluster. After that the derived Delaunay edges are approximated to integer values and used as the range of possible mesh router locations for the CCM to generate the initial solution.

We evaluated the proposed approach by simulations. For the simulations, we considered the evacuation areas in Okayama City, Japan, which is the target to be covered by mesh routers. The simulation results have shown that the proposed approach is able to place mesh routers near mesh clients that are clustered far away from each other and can cover all mesh clients.

In the future work, we will compare the performance of the proposed approach with other algorithms, methods and systems.

Acknowledgement. This work was supported by JSPS KAKENHI Grant Number JP20K19793.

References

1. Oda, T., et al.: Evaluation of WMN-GA for different mutation operators. Int. J. Space Based Situat. Comput. **2**(3), 149–157 (2012)
2. Oda, T., et al.: Performance evaluation of WMN-GA for different mutation and crossover rates considering number of covered users parameter. Mob. Inf. Syst. **8**(1), 1–16 (2012)
3. Oda, T., et al.: WMN-GA: a simulation system for WMNS and its evaluation considering selection operators. J. Ambient. Intell. Humaniz. Comput. **4**(3), 323–330 (2013)
4. Oda, T., et al.: Node placement in WMNs using WMN-GA system considering uniform and normal distribution of mesh clients. In: Proceedings of the IEEE 8-th International Conference on Complex, Intelligent and Software Intensive Systems (IEEE CISIS-2014), pp. 120–127 (2014)
5. Oda, T., et al.: A GA-based simulation system for WMNs: performance analysis for different WMN architectures considering TCP. In: Proceedings of the IEEE 9-th International Conference on Broadband and Wireless Computing, Communication and Applications (IEEE BWCCA-2014), pp. 120–126 (2014)
6. Oda, T., et al.: Effects of population size for location-aware node placement in WMNs: evaluation by a genetic algorithm-based approach. Pers. Ubiquit. Comput. **18**(2), 261–269 (2014)
7. Ikeda, M., et al.: Analysis of WMN-GA simulation results: WMN performance considering stationary and mobile scenarios. In: Proceedings of the 28-th IEEE International Conference on Advanced Information Networking and Applications (IEEE AINA-2014), pp. 337–342 (2014)
8. Oda, T., et al.: Analysis of mesh router placement in wireless mesh networks using Friedman test. In: Proceedings of the IEEE 28-th International Conference on Advanced Information Networking and Applications (IEEE AINA-2014), pp. 289–296 (2014)
9. Oda, T., et al.: Effect of different grid shapes in wireless mesh network-genetic algorithm system. Int. J. Web Grid Serv. **10**(4), 371–395 (2014)
10. Oda, T., et al.: Analysis of mesh router placement in wireless mesh networks using Friedman test considering different meta-heuristics. Int. J. Commun. Netw. Distrib. Syst. **15**(1), 84–106 (2015)
11. Oda, T., et al.: A genetic algorithm-based system for wireless mesh networks: analysis of system data considering different routing protocols and architectures. Soft. Comput. **20**(7), 2627–2640 (2016)
12. Sakamoto, S., et al.: Performance evaluation of intelligent hybrid systems for node placement in wireless mesh networks: a comparison study of WMN-PSOHC and WMN-PSOSA. In: Proceedings of the 11-th International Conference on Innovative Mobile and Internet Services in Ubiquitous Computing (IMIS-2017), pp. 16–26 (2017)
13. Holland, J.H.: Genetic algorithms. Sci. Am. **267**(1), 66–73 (1992)
14. Skalak, D.B.: Prototype and feature selection by sampling and random mutation hill climbing algorithms. In: Proceedings of the 11-th International Conference on Machine Learning (ICML-1994), pp. 293–301 (1994)
15. Kirkpatrick, S., et al.: Optimization by simulated annealing. Science **220**(4598), 671–680 (1983)
16. Glover, F.: Tabu search: a tutorial. Interfaces **20**(4), 74–94 (1990)

17. Kennedy, J., Eberhart, R.: Particle swarm optimization. In: Proceedings of the IEEE International Conference on Neural Networks (ICNN-1995), pp. 1942–1948 (1995)
18. Hirata, A., et al.: Approach of a solution construction method for mesh router placement optimization problem. In: Proceedings of the IEEE 9-th Global Conference on Consumer Electronics (IEEE GCCE-2020), pp. 467–468 (2020)
19. Hirata, A., et al.: A coverage construction method based hill climbing approach for mesh router placement optimization. In: Proceedings of the 15-th International Conference on Broadband and Wireless Computing, Communication and Applications (BWCCA-2020), pp. 355–364 (2020)
20. Hirata, A., et al.: Simulation results of CCM based HC for mesh router placement optimization considering two islands model of mesh clients distributions. In: Proceedings of the 9-th International Conference on Emerging Internet, Data & Web Technologies (EIDWT-2021), pp. 180–188 (2021)
21. Hirata, A., et al.: A coverage construction and hill climbing approach for mesh router placement optimization: simulation results for different number of mesh routers and instances considering normal distribution of mesh clients. In: Proceedings of the 15-th International Conference on Complex, Intelligent and Software Intensive Systems (CISIS-2021), pp. 161–171 (2021)
22. Hirata, A., et al.: A CCM-based HC system for mesh router placement optimization: a comparison study for different instances considering normal and uniform distributions of mesh clients. In: Proceedings of the 24-th International Conference on Network-Based Information Systems (NBiS-2021), pp. 329–340 (2021)
23. Hirata, A., et al.: A simulation system for mesh router placement in WMNs considering coverage construction method and simulated annealing. In: Proceedings of the 16-th International Conference on Broadband and Wireless Computing, Communication and Applications (BWCCA-2021), pp. 78–87 (2021)
24. Hirata, A., et al.: A Delaunay edge and CCM-based SA approach for mesh router placement optimization in WMN: a case study for evacuation area in Okayama City. In: Proceedings of the 10-th International Conference on Emerging Internet, Data & Web Technologies (EIDWT-2022), pp. 346–356 (2022)
25. Integrated GIS for all of Okayama Prefecture. http://www.gis.pref.okayama.jp/pref-okayama/OpenData, 16 November 2021
26. Bresenham, J.E.: Algorithm for computer control of a digital plotter. IBM Syst. J. 4(1), 25–30 (1965)
27. Ester, M., et al.: A density-based algorithm for discovering clusters in large spatial databases with noise. In: Proceedings of the 2-nd International Conference on Knowledge Discovery and Data Mining (KDD-1996), pp. 226–231 (1996)

An Analytical Queuing Model Based on SDN for IoT Traffic in 5G

Aliyu Lawal Aliyu[ID] and Jim Diockou[(✉)]

Leeds Trinity University, Leeds, UK
{a.aliyu,j.diockou}@leedstrinity.ac.uk

Abstract. The latest mobile and wireless communication technology 5G will revolutionise the way we communicate and interact in the digital world. 5G is expected to have a large-scale impact on society, industries and the digital economy. The technology will unleash an ecosystem that enables Ultra-Reliable Low Latency Communication (URLLC) and massive Machine-Type Communication (mMTC), this will heavily benefit IoT devices. However, despite the lucrative advantages offered by 5G, the network infrastructure and operations will come with huge financial cost making capital expenditure (CAPEX) and operational expenditure (OPEX) an issue. With the advent of Software Defined Networking (SDN) and Network Function Virtualisation (NFV), most of the financial burden can be reduced through virtualisation of the access network infrastructure (eNodeB, gNodeB), these access networks send traffic from ubiquitous IoT devices to IP network switches. Considering the massive machine-type traffic and the need for URLLC, we need an efficient queuing model that can cater for the network packets in transit. This paper proposes an analytical Markovian queuing model based on M/M/C/∞/∞ to offer efficient and scalable traffic engineering for the massive traffic that transit via the 5G access networks to SDN architecture. The SDN controller and NFV will be used to implement the Markovian queuing model and to intelligently route the traffic efficiently that comes from the various 5G access networks to their final destination and egress point through the use of virtual switches.

Keywords: SDN · 5G · IoT · mMTC · URLLC · Queue

1 Introduction

The next generation mobile network 5G is set to be a major player in the growth of IoT applications [1]. 5G is expected to provide network connectivity to ubiquitous devices ranging from smartphones, smart-wears, home appliances, cars, bio-chips and miscellaneous devices that require a network connection to receive, send and process vital data [2]. Even though 5G is at its early stage of adoption, mobile wireless technology offers increased throughput, better reliability and connectivity with high capacity to allow more connected devices [3].

The evolving 5G technology is expected to provide the interface of connectivity for future IoT applications. However, this will come with a cost and a change of architecture

Supported by organisation.

where new radio access technology is introduced, with antenna improvements and the use of high frequencies. As of 2020, there are about 20.41 billion IoT devices connected via Machine-to-Machine (M2M) and these numbers are expected to rise exponentially [4]. With the prospects of 5G providing an enabling platform for ubiquitous IoT communications, this will bring improvement in various business sectors that require the services of IoT to support mission-critical applications [5].

Because of the diverse service requirements from miscellaneous IoT devices, it will be costly for the 5G architecture to provide such on-demand requirements without high capital expenditure, because the infrastructures that make up the 5G architecture are expensive [6]. However, there is a big opportunity in terms of service delivery, flexibility, scalability and on-demand service provisioning that can be integrated with the 5G architecture to meet the demands of divergent IoT devices and business application requirements [7]. This opportunity can be realised by leveraging two main technologies which are Software Defined Networking (SDN) and Network Function Virtualisation (NFV) [8].

The harmonisation of SDN and NFV in 5G architecture will provide an opportunity to build flexible and on-demand network function services [9]. One crucial advantage of using NFV is the ability to dynamically scale and react to traffic load changes. NFV can instantiate and allocate more resources based on traffic surge and can dynamically scale down the resources of the traffic load [10].

This paper leverages the SDN and NFV technology to introduce an efficient traffic management scheme based on Markovian M/M/C/∞/∞ model to handle massive IoT traffic in 5G networks. The remainder of the paper is structured in to several sections. Section 2 provides background on 5G architecture for IoT, Sect. 3 gives an overview of SDN and NFV for IoT. Section 4 presents the problem statement and Sect. 5 presents the proposed model for handling the massive traffic in 5G network. Section 6 provides insight into how the model is implemented and Sect. 7 presents the concluding remarks and future work.

2   Revolutionised 5G Architecture that Supports IoT Communication

The 5G architecture has been revolutionised to accept traffic from different Radio Access Networks (RANs) ranging from Wi-Fi, GSM, GPRS/EDGE, UMTS, LTE, LTE-advanced, WiMAX, CDMA2000 and the new 5G-RAN [3,6]. The reason for this architectural design is backward compatibility to allow massive Machine Type Communication (MTC) from miscellaneous IoT devices and User Equipment (UE) sending traffic across different Radio Access Networks (RANs). With this architecture a Wi-Fi client connected under any of the Wi-Fi standards a/b/g/n or a cellular device connected via LTE-RAN can establish a network connection via 5G infrastructure [1].

The 5G architecture consists of the previous generation RANs, the new RAN aggregator, IP network, the nanocore and network elements. For the purpose of broad outreach and connectivity, a new radio interface is introduced in the architecture known as the 5G New Radio (5G-NR) [11].

Figure 1a shows the 5G architecture with IP network support.

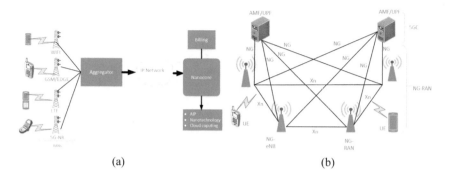

Fig. 1. (a) 5G Architecture with previous RANS and IP network (b) 5G Interfaces

2.1 5G New Radio

The RAN for 5G is backward compatible and will be a mix of technologies, nodes and varying frequencies [12]. And this results in one of the biggest challenges for the deployment of 5G. The components of the 5G architecture are expensive, most of these components are hardware based that implement network and service function with little agility and scalability. By virtue of this problem to provision network services at a cheaper and more efficient manner, virtualisation of most of the services is needed, this leads to the integration of network function virtualisation and software defined networks in the 5G Architecture. Figure 1b shows the different 5G-NR interfaces.

3 SDN and NFV for IoT Based Applications in 5G

Software Defined Network (SDN) and Network Function Virtualisation (NFV) are two enabling technologies that promise to cost-effectively provide the scale, agility and versatility necessary for IoT services and applications in 5G ecosystem. With the application of these two technologies (SDN and NFV), a more efficient and fine-grained network management of IoT applications and services can be achieved.

NFV allows the running and provisioning of complex network function on top of a virtualised infrastructure that is hosted on a central cloud or edge cloud. The NFV helps in building flexible and on demand network services, which is highly needed in 5G. The acquisition and provisioning of 5G network equipment are expensive, with NFV, these costs can be alleviated because complex network functions and critical core elements can be virtualised.

The ability of NFV to divide a physical network substrate into distinct virtual networks with resources that can be managed by a centralised entity called the controller makes it appealing to the ecosystem of 5G. The controller provides the ability to manage the virtualised resources in a scalable, flexible and agile manner. This controller is integral part of the SDN architecture. That is why NFV and SDN are important to 5G.

The introduction of SDN and NFV bring significant contributions and benefits to the 5G core network. The crucial ones are dynamic control of mobile traffic and redirection to appropriate gateways based on network requirements and conditions. Figure 2 presents the 5G base architecture for SDN and NFV.

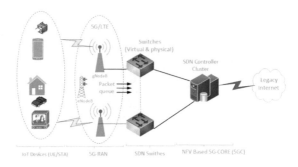

Fig. 2. SDN Architecture for 5G

4 Problem Statement

The exponential growth in the number of resource constrained Machine-Type communication devices brings a critical challenge of satisfying the different communication requirements in dynamic and ultra-dense wireless environments. Among the different use case scenarios, the upcoming 5G cellular network is expected to support Ultra-Reliable and Low Latency Communication (URRLc), massive Machine Type Communications (mMTC) and enhanced Mobile Broadband (eMBB), the mMTC presents the unique challenge of supporting large amount of MTC devices.

The challenge at hand is how the ubiquitous packet traffic in millions from the RAN get served by the SDN switch that have finite computing and storage (memory and buffer) resources so that the network policies and internet access can be granted for any request. The traffic surge becomes overwhelming and slow the rate at which packet request from the nodes get services.

The nature of the packets arrival to the SDN switch is random within a specified time frame. These packet arrival by default are handled in a first in first out paradigm and these raised concerns on mission critical application and time sensitive traffic. This problem impacts business process and disrupt vital services that are crucial to business goals. In essence the real interpretation of this is the lost of valuable time and return on investment on the business front. Next Sect. 5 presents the method and approach that solve the issue of mMTC congestion and offer enhanced traffic management.

5 Proposed Model for Handling Massive Traffic Between 5G and SDN Architecture

The massive MTC traffic from the nodes pass through the access network (gNb, ng-eNB) to the SDN switch. To have a global view of how these packets transactions are handled at the side of the SDN switch, an analytical model is required to evaluate the *average queue length, average waiting time* and *the loss probability as a result of buffer overflow.*

This is needed to expedite packet processing, the time it takes for a packet to be treated at the switch buffer before it is forwarded to any egress port that is known as the service rate. The analytical model helps determine the service with tolerable waiting time and provide guaranteed packet loss probability. This provides flexible control of the queue so that differential services to user packets can be applied.

Based on the nature of packet arrival and departure in the network the most suitable analytical model that is used to optimise traffic engineering is the Queuing Model. The queuing model has proven to be effective in several 5G and SDN related traffic engineering models [13].

The presented problem at hand is traffic from the 5G core as it ingress the SDN switch as seen in Fig. 2. There is one SDN switch that contains multiple instances of virtualised switches using the (hypervisor) and one controller in charge of the network activities and as such the multiple server queuing model will be used.

The server is the transmission point and represent the instructions installed on the switch on how received packets can be handled. The packet arrival and the annotated service time are random. In addition, the service time takes in cognisant of the length of the packet and the buffer provides the available contention room for waiting.

An interaction with a single server is presented to show what happens at every server when packets arrive. The packets first arrive to an empty server, if there are multiple empty servers then the packets choose a random server.

5.1 Queuing Model Representation

The analytical queue modelling shows the evolution over time of the number of events (packets) in the system (waiting or being served). Queues are represented with the Kendall's notation (A/B /C):(D /E /F) based on the Queuing theory model.

- **A :** This represents the arrival process of packets, and the analytical model of a random variable arriving in a specified time is best evaluated as a Markovian Poisson Process (M); let λ be the arrival rate (the average number of packets arriving in unit time). The inter arrival time is the time between the arrival of two separate packets at random. And following the Poisson model the inter arrival time is exponentially distributed with mean $(1/\lambda)$. Figure 3a shows packet arrival at different time T_i.
- **B :** The packet departure process, the inter departure time (service time) is exponentially distributed with average service time of $(1/\mu)$, where μ is the service rate and it is Markovian.
- **C :** Indicates the number of parallel servers in the system.
- **D :** This represents the service discipline of the queue. For example, First in First out (FIFO), Last in First Out (LIFO), Priority and Service in Random order (SRO).
- **E :** This stands for the maximum number of packets in the queue. In any analytical model if K is not specified then it is assumed to infinity.
- **F :** Calling source, in this instance the 5G RAN traffic is our source of all events.

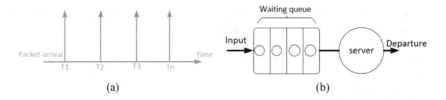

Fig. 3. (a) Packet arrival (b) Queue representation

Based on the Kendall's notation of A/B/C/D/E/F, the Poisson arrival (A) and the departure process (B) are Markovian (M). The representation of our model is M/M/C/D/E/F where C represents the number of the servers. Figure 3b shows a simple representation of a queuing system.

5.2   Packet Arrival from 5G

The packet arrival from the 5G core to the SDN switch follows a Poisson Distribution model with parameter λt. The arrival process of each packet is random and independent. Let X be the number of arrivals in unit time [0, t] then the arrival can be modelled as:

$$P(X = k) = (\lambda t)^k \cdot e^{-\lambda t}/k!, \quad k = 0, 1, 2, 3, ...n$$

The arrival rate (λ) is the average number of arrival per time unit. It is measured in seconds, in this context (packet/second). However, if the inter arrival time of packets is known at $t + \tau$ interval, where τ is random arrival then the arrival rate can be evaluated as follows:

$$P(\tau > x) = P(\text{no arrival occurs in}(t, t + x))$$
$$= P(X = 0 \text{ in } (t, t + x))$$
$$= e^{-\lambda t}, \tau > 0, x \geq 0$$

The inter arrival rate for (t, t + x) will be (1 − P(X = 0 in (t, t + x)). And let $F_\tau x = (1 − P(X = 0 \text{ in } (t, t + x))$.

$$F_\tau x = 1 - P(\tau > x)$$
$$= 1 - e^{-\lambda t}, x \geq 0$$

The average inter-arrival rate is represented as (1/λ). Figure 4a represents the behaviour of random packet arrival at different time epoch and Fig. 4b shows the inter-arrival (U) sequence where $U_i = T_{i+1} - T_i$ between packet i and (i +1). Figure 4c represent the default packet arrival curve. There are situation where the arrival is continuous under steady condition as seen in Fig. 4d.

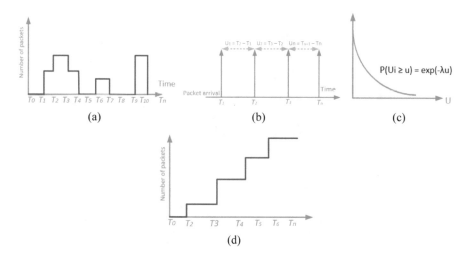

Fig. 4. (a) Inter-arrival graph (b) Inter-arrival flow (c) Packets arrival and service (d) Continuous packets arrival

5.3   Packet Service Process and Departure from 5G to SDN Architecture

To estimate the average waiting time for the arriving packets to be served by the SDN switch. To evaluate this, the switch takes a random length of time T_s to serve or process each arriving packet. The processing times are independent for every unit packet due to varying length of packet and the applied differential service. T_s has exponential distribution based on queuing model with parameter μ. Where μ represents the average number of packets being served per unit time. Therefore, the packet departure process F_{T_s} can be evaluated as follows:

$$F_{T_s(t)} = \mu e^{-\mu t}, t \geq 0$$

The service duration denoted as S_i of packet i is the time during which the server is processing a packet. The service time in SDN is a process in the controller known as the *flow_mod* operation. It is assumed that service time are independent and identically distributed. They are characterised by their probability distribution. The service rate (μ) is the average number of packets served per time unit. The mean service duration is ($1/\mu$) in seconds.

Figure 5a shows the processing time in SDN from the input (*packet_in*) to the service duration (*flow_mod*) and finally to the exit (*packet_out*). The arrival follows the Poisson while the departure follows the exponential model.

Just like packet arrival time, the assumption around service time is the same, thus, they are independent and identically distributed. They are characterised by their probability distribution as seen in Fig. 5b.

For the processing at the server side which is the controller. The average performance of the system which is also known as the utilisation factor is represented as $\rho = \lambda/\mu$.

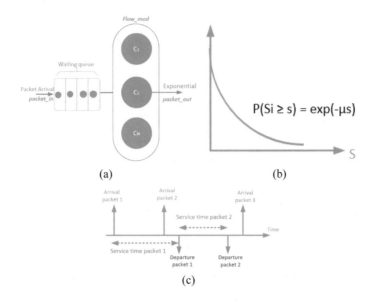

Fig. 5. (a) Service duration in SDN architecture (b) Behaviour of exponential service time (c) Arrival and service time processing

5.4 Buffer Capacity

There are series of steps that should be evaluated before the system can accurately predict the average number of packets the buffer in the switches can handle. The assumption here is that all switches have the same buffer size. These steps are:

- Average number of packets in the system N.
- The average time a packet spent in the system before being served T.
- Finally, the buffer capacity (W), waiting time for packets.

The service time and inter-arrival rate play a major role in the number of packets in transit a buffer can hold before it starts dropping packets. Figure 5c shows the two relationships of arrival and servicing differentiation.

6   Implementation in SDN and NFV Environment

There are frameworks that implement and integrate 5G architectural functions into SDN and NFV environment. Our work is a complimentary effort. However, this is how the proposed efficient massive traffic handling model is realised in SDN environment. The components in the 5G core which includes UPF and AMF are implemented as a virtualised network function in the control layer of SDN architecture. The data plane functions are performed by the gNBs and the ng-eNBs. For backward compatibility the RAN accommodates LTE-A.

The control plane comprises of the SDN controller, in this case ONOS controller with firmware version 1.14 running inside Mininet [14] emulation environment and is hosted on a Virtual Machine (VM) Ubuntu 16.04 64-bit with processor

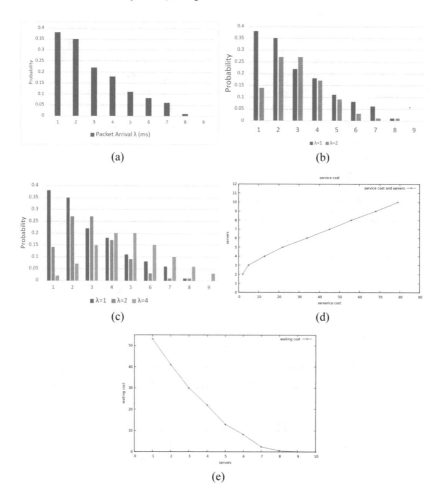

Fig. 6. (a) PDF of average arrival for packets in hundred thousand (b) PDF of average millionth packet arrival when $\lambda = 1$ and $\lambda = 2$ (c) PDF of average millionth packet arrival when $\lambda = 1$, $\lambda = 2$ and $\lambda = 4$. (d) Service cost for packets (e) Waiting cost for packets.

Intel(R)Core(TM) i5-4200M, CPU processing speed of 2.50 GHz and installed RAM capacity of 16.0 GB. The controller manages the applications that manage the data plane request from the nodes and provide access to data network (Internet). The proposed model is developed and deployed as an application, the application compiled as part of the controller and have access to various controller classes, methods and data. The solution is integrated as part of the ONOS controller instance in order to demonstrate the feasibility of the proposed model. Due to the complexity of 5G networks, the topology used in modelling the network is an SDN torus with a fanout of 20 is *sudo mn –topo torus,20,20*

The testing and evaluation are presented as seen in Fig. 6a thousands of packet arriving where the Probability Distribution Function (PDF) of is spread over time. We can see how the queue processing jumps up at the beginning of the packet processing and then it gradually comes down in a near linear manner. At long run, the PDF keeps reducing. Figure 6b shows a two line arrival of the packets, the behaviour of the PDF with additional line of incoming ingress, the processing (mu) time starts averagely then it peaked, from there a gradual descent is seen as more packets arrive. In Fig. 6c we have a four line distribution where packets follow the Poisson behaviour and arrive independently, the PDF shows that processing rate is efficient at the start and then it peaked at the middle then normalised at the end.

Service cost is the invested capital in expanding or adding more infrastructure to handle in-coming packets. The higher you spend the lower the waiting cost of packets because now we have more processing channels. This can be seen in Fig. 6d. The waiting cost, is the lost of time and value incurred when a packet is in transit. Figure 6e shows a downward curve of waiting cost the more the number of servers increases.

7   Summary and Conclusion

The proposed analytical Markovian queuing model that implements an efficient method for handling mMTC in 5G ecosystem and enabling URLLC has been presented and evaluated. The benefits that will be derived out of these are enormous.

The proposed model can improve and alleviate the burden of overwhelming traffic arrival, departure and handling in the 5G architecture. The proposed model leverage SDN and NFV technology to provide an efficient solution to handle massive traffic in 5G architecture. The benefit and opportunities from business perspective can be itemised as follows:

– Reliable Communication: There will be guaranteed and reliable communication between nodes at the data plane and the network applications in the 5G core that provide resources to them. These include delay sensitive applications like real time traffic and mission critical applications. Packet loss due to congestion and contention is drastically reduced with efficient method of handling massive traffic for end to end communication.
– Cost Saving: In terms of infrastructure costs, 5G components and equipment are prohibitively expensive. This can contain fast deployment by Telco-provider and stakeholders in adopting solutions provided by the 5G technology. However in our case we consider SDN and NFV where all the functions of the physical components are virtualised and implemented as a software solution which provides huge capital expenditure saving.
– Unified Architecture: The proposed model is implemented in software with all the network functions virtualised. This abstracts most of the disparity in the data plane where different access networks exist. The proposed model advocates for a soft RAN where functions are not vendor dependent but can evolve as business requirement changes. This will make nodes oblivious of the access network that connects them to the 5G network.
– Fast Provisioning: Business needs and requirements change with time requiring new methods to manage mission critical applications. With the unified architecture, new applications especially for IoT networks can be implemented in a timely manner.

For future work and further development, this work will be improved by running the model on a test-bed of SDN infrastructure and fine-tuning the Markovian queuing model using different use cases in handling and processing network traffic.

References

1. Marwat, S., et al.: Method for handling massive IoT traffic in 5G networks. Sensors **18**(11), 3966 (2018)
2. Rahimi, H., Zibaeenejad, A., Safavi, A.A.: A novel IoT architecture based on 5G-IoT and next generation technologies. In: IEEE 9th Annual Information Technology, Electronics and Mobile Communication Conference (IEMCON), pp. 81–88. IEEE (2018)
3. Kekki, S., et al.: MEC in 5G networks. ETSI White Paper, vol. 28, pp. 1–28 (2018)
4. Alsirhani, A., et al.: Securing low-power blockchain-enabled IoT devices against energy depletion attack. ACM Trans. Internet Technol. (2022, accepted). https://doi.org/10.1145/3511903
5. Hsieh, H.-C., Chen, J.-L., Benslimane, A.: 5G virtualized multi-access edge computing platform for IoT applications. J. Netw. Comput. Appl. **115**, 94–102 (2018)
6. Li, S., Da Xu, L., Zhao, S.: 5G internet of things: a survey. J. Ind. Inf. Integr. **10**, 1–9 (2018)
7. Kazim, M., Liu, L., Zhu, S.Y.: A framework for orchestrating secure and dynamic access of IoT services in multi-cloud environments. IEEE Access **6**, 58 619–58 633 (2018)
8. Xie, J., et al.: A survey of machine learning techniques applied to software defined networking (SDN): research issues and challenges. IEEE Commun. Surv. Tutorials **21**(1), 393–430 (2018)
9. Yousaf, F.Z., Bredel, M., Schaller, S., Schneider, F.: NFV and SDN-key technology enablers for 5G networks. IEEE J. Sel. Areas Commun. **35**(11), 2468–2478 (2017)
10. Ma, L., Wen, X., Wang, L., Lu, Z., Knopp, R.: An SDN/NFV based framework for management and deployment of service based 5G core network. China Commun. **15**(10), 86–98 (2018)
11. Alay, Ö., et al.: End to end 5G measurements with MONROE: challenges and opportunities. In: IEEE 4th International Forum on Research and Technology for Society and Industry (RTSI), pp. 1–6. IEEE (2018)
12. Enescu, M.: 5G New Radio: A Beam-based Air Interface. Wiley, New York (2020)
13. Rahouti, M., Xiong, K., Xin, Y., Ghani, N.: A priority-based queueing mechanism in software-defined networking environments. In: 2021 IEEE 18th Annual Consumer Communications & Networking Conference (CCNC), pp. 1–2. IEEE Press (2021). https://doi.org/10.1109/CCNC49032.2021.9369614
14. Aliyu, A.L., Aneiba, A., Patwary, M., Bull, P.: A trust management framework for software defined network (SDN) controller and network applications. Comput. Netw. **181**, 107421 (2020)

An Expert Survey for the Evaluation of 5G Adoption in Bangladesh

Md. Zahirul Islam[1(✉)], Md. Abdur Rahim[2], Md. Salahuddin[3], Syed Md. Galib[1], and Rahamatullah Khondoker[4]

[1] Jashore University of Science and Technology, Jashore, Bangladesh
zahirete@student.just.edu.bd, galib.cse@just.edu.bd
[2] Banglalink Digital Communication Ltd., Dhaka, Bangladesh
abrahim@banglalink.net
[3] Bangladesh High Commission Canberra, Canberra, Australia
labour.canberra@mofa.gov.bd
[4] Business Informatics, THM University of Applied Sciences, Friedberg, Germany
rahamatullah.khondokar@mnd.thm.de

Abstract. The main objective of this study is to find the challenges and opportunities for the adoption of 5G technology for the development and growth of the business environment in Bangladesh. An online survey based on an opinion poll (questionnaire) has been selected as the research method due to the short turnaround time, quick delivery, and simple return. Due to the low response rate, some of the responses were collected physically at the workplaces of the poll participants. The summary of the total 34 responses from Mobile network operator (MNO), National Telecommunication Transmission Network (NTTN), International Internet Gateway (IIG), and Internet Service Provider (ISP) is as follows: 17 of the respondents agreed that 5G technology should be implemented over a period of up to 3 years. With regard to the availability and interest in the implementation of 5G technology, the result shows three key hurdles for 5G in Bangladesh: first, service price is unreasonably high; second, the network coverage area is not adequate; and third, a shortage of 5G-enabled devices. The next step will be to expand this research work by identifying potentially workable options from a technical perspective such as the use of virtual network functions (VNF), software defined networking (SDN), adaptive network slicing for network resource management, cyber security in the context of next generation of mobile networks 5G and beyond for Bangladesh.

1 Introduction

The fifth generation of cellular mobile communication is known as 5G, which is succeeded by the earlier generations including 2G, 3G, and 4G networks. 5G enables a new kind of network that is designed to connect virtually everyone and everything together including machines, objects, and devices. 5G can be significantly faster than 4G, delivering up to 20 Gigabits-per-second (Gbps) peak data rates and 100+ Megabits-per-second (Mbps) average data rates. The current 5G networks will operate in a high-frequency range of the wireless spectrum, ranging from 30 GHz to 300 GHz [1]. The

L. Barolli (Ed.): AINA 2023, LNNS 655, pp. 446–457, 2023.
https://doi.org/10.1007/978-3-031-28694-0_43

rate at which data is transferred from one device to another is extremely fast, however, it does not travel as far as the lower-frequency waves utilized in 4G networks. As 5G uses a high-frequency wave, it has trouble getting around obstacles such as walls, buildings, and other structures.

Multiple new techniques, such as new frequency bands with millimeter-wave (mmWave) and optical spectra, are introduced in 5G networks, boosting spectrum practice and management, and it can mix permitted and abandoned bands. Designers are already searching for design goals for the next phase, with 5G hitting its limits in 2030. As a result, researchers around the world are beginning to consider 6G in the next ten years. On 3G and 4G networks, we are still having problems getting complete coverage. Bangladesh is the latest country to get 5G. It has also taken steps to launch the 5G network and work has started to expand the network in a phased manner as per the need rather than blanket activation of the fifth generation network. The existing 5G design is cutting-edge where the network components and terminals will have to be fully or partially modified to meet the requirements of 5G. In terms of overall technical performance, 5G aims to achieve the following: 10 to 100 times higher data usage rate, obtaining 1 Gbit/s to 20 Gbit/s for dense urban clusters; 1000 times more mobile data per area, reaching traffic density values of $10\,Tb/s/km^2$; 1000 times more connected devices, reaching a density of 1 M $terminals/km^2$; 10 years of battery life for large M2M devices (sensors, actuators, smart meters, etc.). Similarly, by employing new technology, service organizations can easily include value-added services.

To determine the challenges of deploying 5G technology in Bangladesh, we must first comprehend the current state of the telecommunications industry. There are several issues that must be addressed such as the environment (the place where the 5G network will operate), users (business and end users of 5G technology), handsets (5G capable), technical (5G implementation challenges), and organizational (spectrum allocation) issues. To find out the congenial environment for the aforementioned issue, a survey was conducted to collect and analyze opinions from mobile network operators (MNOs) and the national telecommunication transmission network (NTTN). According to our knowledge, this is the first survey of its sort to engage mobile network operators for their opinions. Determining the implementation gap should be possible through analysis of the data acquired from the survey.

2   Benefits of 5G Network: Bangladesh Perspective

According to PwC, the impact of using 5G network will be USD 1.339 trillion by the 2030 in the global economy. Although the lion share will be enjoyed by the developed countries especially the United States, China, Japan, but its impact on the economies of developing countries will also very significant. Economy of Bangladesh also be impacted hugely by taking proper policy and procedure of using 5G networks in timely manner. The government of Bangladesh has taken various steps, including increasing the use of Internet in the country, to build a Digital Bangladesh. Using 5G networks, its implementation can be more accurate and at a specific time. 5G powered healthcare, smart utilities, consumer and media applications, industrial manufacturing, financial service application will be key economic driving indicators for any countries.

Healthcare: During the COVID-19 pandemics, the rapid rise of tele-medicine provides a glimpse of the future of healthcare sector, but remote care is just one area in which 5G can enable both cost savings and better health outcomes. As Broad band is not covered the rural area mostly, where 5G can play significant role and saves lives and times. There are several app-based medical service provider platforms in Bangladesh, such as Prava, Doctor Koi and Doctime.

E-Commerce: E-commerce is now playing an important role in the economy of the country. According to the e-Commerce Association of Bangladesh (e-CAB), official reports, and industry insiders, online sales rose about 70% in 2020 from the previous year, and market size of the industry stood at nearly 2 billion USD as of August that year which crossed about 2.32 billion USD in 2021. E-commerce industry will be rapidly increased by using 5G networks.

Mobile and Online Banking: Mobile banking in Bangladesh continues to grow fast. As per Bangladesh Bank data, the number of transactions through mobile financial service surged to Tk 7,70,166.7 crore in 2021 from Tk 5,61,395.8 crore in the previous year. By using 5G technology financial institution will be driven their business with smooth and timely manner.

5G will generate significant employment throughout the economy. This represents a mix of temporary, part-time and full-time jobs across all industries. So, for achieving its vision 2040 Bangladesh should be implemented 5G technology with timely manner.

3   Related Work

In addition to a revolution in network architecture that enables the creation of flexible, cost-effective networks, 5G technology marks the evolution of radio access networks to satisfy future data transfer requirements [2]. There will always be enough capacity available for whatever type of data transport required thanks to the 5G network's unlimited network capacity, or boundless internet perception [3]. The prospect and standard of 5G technology are well known by the researchers and engineers, however, it is useful to refresh the concept of this technology. The fundamental aspects of 5G business potential is discussed by Soós, Gábor, et al. in [4]. Oughton et al. [5] identified six main factors for 5G techno-economic assessment which leads the recommendation for the design and standardization for next generation 6G wireless technologies. With its attributes of high bandwidth, low latency, and widespread connectivity, 5G, a new generation of mobile communication technology, is crucial in many sectors of smart manufacturing [6]. Radio network evaluation of 5G technology and several applications in the context of Internet of Things (IoT), Heterogeneous Networks (HetNet), Big Data, Healthcare etc. is discussed in the survey performed by Al-Namari et al. [1]. Chen et al. [7], illustrated the overview of 5G network architecture with new radio interface, use case for 5G communication by the standards development organizations (SDO) and spectrum bands for the deployment of the various use cases is illustrated. A 4×4 multiple-input, multiple-output (MIMO) antenna is proposed for 5G New Radio (NR) mobile handset with a dimension of 10.2 mm length and 2.8 mm width by Chen et al. [7]. The prototype is designed for operating 3300MHz-3600MHz and 4800MHz-5000MHz bands. Triple band dual antenna based MIMO system for 5G mobile handset is proposed to reduce the coupling between different access ports of the antenna to reduce the correlation between received signal by de Mingo Sanz et al. [8]. A new power amplifier Integrated Circuit (IC) for 5G new radio handset is designed using InGaP/GaAs which can operate in 2.8–3.8 GHz [9]. mmWave communications, backhaul technology, technology maturity, energy consumption, and business aspects including business models, ecosystem maturity, coordination of industry verticals and regulatory aspects including spectrum management and fragmentation are the biggest challenges in the implementation of 5G mentioned by Taheribakhsh et al. [10]. The 5G New Radio (NR) technology, distributed massive MIMO, sub-millimeter wave and Tera-hertz spectrum bring challenges both in terms of implementation as well as deployment mentioned by Gustavsson et al. [11]. For the allocation of spectrum frequencies in 5G regulation, it is recommended to use 3.5 GHz spectrum frequency for the implementation of 5G technology in Indonesia which is mentioned by Sastrawidjaja et al. [12].

4   Salient Features of 5G

As the successor of 4G, 5G technology is the fifth generation of broadband cellular networks. Different use cases will benefit from 5G technology, which will also fulfill the high expectations of users in new application areas. Prior to the final development of 5G, International Mobile Telecommunication (IMT) establishes the standards and outlines three usage scenarios illustrated in Fig. 1:

- **eMBB** (Enhanced Mobile Broadband, also called Extreme Mobile Broadband). A use-case scenario of eMBB is to download a 3D full-length movie which requires **high speed**.
- **mMTC** (Massive Machine Type Communications). Use-case scenarios of mMTC are IoT based smart home, smart industry and smart city which will require massive Machine-to-Machine (M2M) communication with **high-reliability & low latency**.
- **uRLLC** (Ultra-Reliable and Low Latency Communications). Use-case scenarios of uRLLC are self driving smart car, mission critical operation, and playing game in cloud would require **both high speed and low latency**.

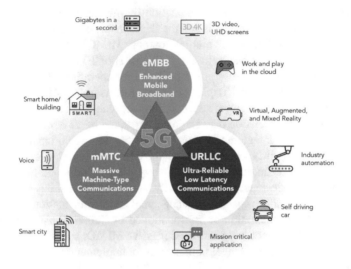

Fig. 1. Features of 5G [17]

5 Overall 5G Architecture

The overall architecture envisioned in this paper is shown in Fig. 2. The architecture is built on top of the 5G platform, which is composed of fundamental elements that adhere to the most recent 3GPP version. We add some modules that offer the functionalities that are needed for the various vertical use cases that are not offered inside the underlying platform on top of the baseline components. To set up and control the underlying functions, these modules use industry-standard interfaces.

On top of the 5G baseline components, the architecture envisages the following two modules:

- In accordance with the vertical requirements, the "MANO (Management and Orchestration) & Control" module is in charge of orchestrating and controlling the various network functions (NFs) and resources [13].
- The "Artificial Intelligence (AI)" module uses AI approaches to govern, control, and configure the network generally. This includes setting up the NFs and MANO algorithms.

The so-called service layer, which is the top layer in the suggested design, gives vertical customers and other network slice tenants a user-friendly interface for requesting and managing network services. The user requests coming from the verticals are mapped by this service layer to primitives in the interfaces with the underlying modules.

The broadcast is a crucial part of the suggested architecture in addition to the aforementioned features. In fact, the vertical use cases involving the simultaneous distribution of a certain piece of content to numerous people necessitate the broadcast. This is

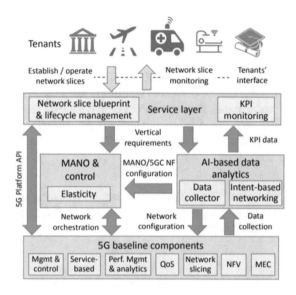

Fig. 2. Overall 5G Architecture for Industries [18]

the case, for instance, with applications like immersive ultra-HD movies or IoT device software upgrades. In order to achieve this, support for broadcast and multicast technologies offering point-to-multipoint (PTM) transmissions is required because point-to-point (PTP) unicast transmissions would be incredibly ineffective. The proposed architecture uses the discoveries of the 5GPPP 5G-Xcast project [14] to provide this feature, implementing PTM through the use of cutting-edge algorithms in the 5G Core and the 5G Radio Access Network (RAN); specifically (i) two new NFs are added into the 5G Core, the 5G-Xcast User Plane Network Function (XUF) and the 5G-Xcast Control Plane Network Function (XCF), responsible for the delivery of multimedia content and for the control of the active broadcast sessions, and (ii) a 5G NR Broadcast solution is offered in the RAN relying on the idea of RAN Multicast Area (RMA) to broadcast a given service to a group of cells.

6   Methodology

On the basis of the MNO and NTTN's technological reviews, the survey questionnaire was created. Due to its advantages, as mentioned by Banchs, A et al. [15], online surveying was chosen as the best approach because of the short turnaround time and need for quick delivery and simple return.

We disperse the leading operators in Bangladesh among a significant number of MNO and NTTN via the Internet. It was possible to incorporate several question forms, organise the data in a complex way using filters, collect data straight into a database, and it allowed for data quality assurance. Additionally, confidentiality might be readily protected. Each item on the list of inquiries connected to the ecology of 5G, as described above, was compiled. In addition, a list of general inquiries was created to learn more about our participants.

Pilot Study: Some surveys were distributed as softcopies during the pre-testing phase to a small sample of the technical personnel who are typically working for MNO. They were requested to read the questionnaire carefully, focusing on the question's general language, consistency, understandability, and redundancy. Additionally, the understandability of each individual question was assessed. The questionnaire was improved in light of the feedback received.

Equipment and Software: The revised questionnaire was then entered into Google Form, a web-based form developed by Google that offers all features needed to plan, carry out, and assess online surveys.

Reliability of the Survey: The questionnaire was prepared for publication after a month of diligent preparation. Only technical personnel employed by MNOs and NTTN received the questions.

Motivational Quality: Following the initial poor return rate (8 responses from the participants), a number of actions such as phone call, SMS, email and physical meeting with the technical persons were done. They received a few emails i.e. we are expecting the valuable feedback from you in respect as representative of your prestigious organization for taking part in the technical survey to find the technological gap for implementing 5G technology in Bangladesh, informing them of the survey's deadline. Additionally, the responses were physically gathered at their place of employment such as for Grameenphone at GPHOUSE, Bashundhara, Banglalink at Tigers' Den Gulshan 1, Robi, Airtel and Teletalk at Gulshan 1.

7   Survey Questions

Mobile Phone Operators (MNO)s are trying to implement 5G communication technology in Bangladesh. This survey is conducted for finding out the technological gap for implementing the 5G communication technology in Bangladesh. In this survey, most of the responses came from the technical persons of the MNO's. The questionnaire was highlighted based on the application on 5G handset in 5G technology, technical challenges or gap for implementing 5G technology, a possible spectrum which will be used for 5G and finally the expected year of launching 5G technology in Bangladesh.

First Version for Testing and Getting Feedback: At the very outset, a panel discussion over online is conducted with the core technical persons of different MNOs in order to find out the implementation gap o technology in Bangladesh. In the discussion several topics come out as limitations for implementing 5G. These topics were then classified as policy level limitation, technical limitation, and business related limitation. Then, it was decided to develop a survey questionnaire to find out the actual scenario from the different MNOs in the context of Bangladesh. As the 5G technology operates on different frequency spectrum, traditional mobile handset will not operate properly and new antenna based handset is required [17]. In the discussion, there are several technical topics come out as shortcomings for implementation of 5G in the context of Bangladesh. Comparing to technical topics, the non-technical topics specially policy related issues popup. Service price and coverage are the vital points for implementing the 5G. The discussion concluded with the debate of possible operating spectrum and expected year to launch 5G in Bangladesh.

Final Version: After preparing the draft survey questionnaire from the discussion, the final question form was prepared. In the final question, several modifications have been made after the discussion with the technical person from MNOs. In the first question, we included National Telecommunication Transmission Network (NTTN), International Internet Gateway (IIG) and Internet Service Provider (ISP) in the survey as they require some contribution for implementing 5G. A free text field is added with each question so that participant can give their opinion also. The second and third question remain same. In fourth question, several issues related with 5G handset in the context of Bangladesh has been added here. Policy level issue, NTTN issue and also revenue issues are included in the fifth question because these issues are directly related with the implementation factor. Finally other two questions are remain same for the survey.

8   Results

The results collected from Google Forms are analyzed in detail as follows. As it is seen in the last section, "Please mention your work area," was the first query. The question is a single select multiple choice question. In total, thirty four participants answered this question. thirty out of the thirty four respondents use a mobile network operator (MNO), along with two NTTN, one IIG, and one ISP participant.

Mobile network operator (MNO) wise responses are illustrated participated in Fig. 3. The next question was to retrieve the working area of the participants, "If you are working in MNO, please mention which company are you working with?". The question is a single select multiple choice question. A variety of domains were responded by the participants. Majority of the participants form the Banglalink 23.5% (n = 8). Other domains are Grameenphone (5), Robi (6), Airtel (4), Teletalk (7) and others (4) as illustrated in Fig. 3.

National Telecommunication Transmission Network (NTTN) wise responses are illustrated for participants in this section. The next question was to receive the working are of the participants, "If you are working with NTTN, please mention which NTTN are you working with?". Only single response was allowed for this question. A variety of domains were responded by the participants. Majority of the participants do not work with NTTN 88.2% (n=30). Other domains are Summit Communication Limited (2), Bangladesh Telecommunication Company Limited (1) and IIG (1).

The next question was related with the application of 5G handset in 5G technology stated, "What do you think about the role of 5G handset in 5G technology?". Open and Multiple responses were allowed in this question. A variety of domains were responded by the participants. 76.5% of the participants gave their opinion with the 5G enabled

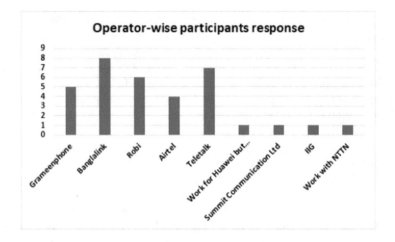

Fig. 3. Operator wise participants response

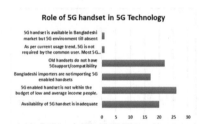

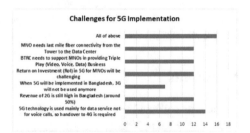

Fig. 4. Role of 5G handset in 5G technology   **Fig. 5.** Challenges for 5G implementation

handset is not within the budget of low and average income people (n = 26). 64.7% thought that Old handsets do not have 5G support/compatibility (n = 22). Then 58.8% of the participants assume that availability of 5G handset is inadequate (n = 20). 50% of the participants thought Bangladeshi importers are not importing 5G enabled handsets (n = 17). Other domains are, As per current usage trend, 5G is not required by the common user. Most 5G scenarios will require big businesses using the technology for their ease (1) and 5G handset is available in Bangladeshi market but 5G environment till absent (1) are illustrated in Fig. 4.

The next question was describes about the challenges for 5G implementation stated, "What do you think the reason for not implementing 5G?". Open and Multiple responses were allowed in this question. A variety of domains were responded by the participants. 47.1% of the participants gave their opinion with all of the options which are provided with the question (n = 16). 41.2% thought that 5G technology is used mainly for data service not for voice calls, so handover to 4G is required (n = 14). Then 38.2% of the participants assume that MNO needs last mile fiber connectivity from the Tower to the Data Center (n = 13).35.3% of the participants thought both Return on

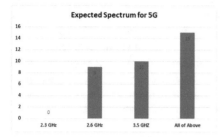

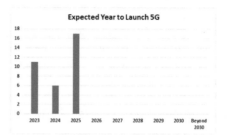

Fig. 6. Expected Spectrum for 5G   **Fig. 7.** Expected Year to Launch 5G

Investment (RoI) in 5G for MNOs will be challenging and Bangladesh Telecommunication Regularity Commission (BTRC) needs to support MNOs in providing Triple Play (Video, Voice, Data) Business (n = 12). 20.6% assume that When 5G will be implemented in Bangladesh, 3G will not be used anymore (n = 7) are illustrated in Fig. 5. However, 2G voice will be used for longtime.

The next question was related with the possible spectrum which will be used for 5G technology in Bangladesh and the question stated, "Which spectrum is expected for 5G in Bangladesh?". Only single response was allowed for this question. A variety of domains were responded by the participants. 44.1% of the participants gave their opinion with all of the options (2.3 GHz, 2.6 GHz and 3.5 GHz) which are provided with the question (n = 15). 29.4% thought that 3.5 GHz spectrum will be used for 5G technology (n = 10) whereas 26.5% participants assumes that 2.6 GHz spectrum will be used for 5G technology (n = 9). The overall scenario is illustrated in Fig. 6.

The next question was related with the launching of 5G technology in Bangladesh and the question stated, "What is the predicted year for the launch of 5G in Bangladesh?". Only single response was allowed for this question. A variety of domains were responded by the participants. 50% of the participants gave their opinion that 5G technology will launch in Bangladesh by 2025 (n = 17). 32.4% of the participants assumes that 5G technology will launch in Bangladesh by 2023 (n = 11) whereas 17.6% participants assumes that 5G technology will launch by 2024 in Bangladesh (n = 6). The overall scenario is illustrated in Fig. 7.

9   Discussion and Future Work

To build a digital Bangladesh, the role of 5G is vital. However, the implementation of 5G is challenging due to the necessity of cooperation with the regulation agency which requires time and effort. A survey is conducted to find out the open challenges for the implementation of 5G in Bangladesh. In this survey, 88.2% of the participants are mobile network operators. For the role of 5G handset in 5G technology, 76.5% participants assume that 5G enabled handset is not within the budget of low and average-income people. 47.1% participants assume that all the reasons which are mentioned in the survey cause the implementation gap in Bangladesh. The possible spectrum for 5G technology will be applicable in Bangladesh is 2.3 GHz, 2.6 GHz and 3.5 GHz, which are assumed by the 44.1% of the participants. Finally, the survey concluded with the assumption of 50% participants that 5G technology will be launched in Bangladesh by 2025. In future, this survey will be conducted based on technical aspects and surveys with the engagement of more participants. In addition, technological perspectives for the next generation of mobile networks including 5G and beyond will be addressed: usage of virtual network functions (VNF), software defined networking (SDN), adaptive network slicing for network resource management and cyber security.

Acknowledgement. The authors are grateful for a Doctor of Philosophy (Ph.D.) Research Fellowship from the ICT Division of the Ministry of Posts, Telecommunications, and Information Technology of the Government of the People's Republic of Bangladesh. The authors also acknowledge the partial funding from Jashore University of Science and Technology for this research. The authors would like to express their gratitude to the technical persons of various Mobile Network Operators for their assistance and contributions.

References

1. Al-Namari, M.A., Mansoor, A.M., Idris, M.Y.I.: A brief survey on 5G wireless mobile network. Int. J. Adv. Comput. Sci. Appl. **8**(11) (2017)
2. Guevara, L., Auat Cheein, F.: The role of 5G technologies: challenges in smart cities and intelligent transportation systems. Sustainability **12**(16), 6469 (2020)
3. Oughton, E.J., Frias, Z., van der Gaast, S., van der Berg, R.: Assessing the capacity, coverage and cost of 5G infrastructure strategies: analysis of the Netherlands. Telematics Inform. **37**, 50–69 (2019)
4. Soós, G., Ficzere, D., Seres, T., Veress, S., Németh, I.: Business opportunitied evaluation of non-public 5G cellular networks-a survey. Infocommun. J. **12**(3), 31–38 (2020)
5. Oughton, E.J., Lehr, W.: Surveying average-incomenomic research to inform the evaluation of 6G wireless technologies. IEEE Access **10**, 25237–25257 (2022)
6. Cheng, J., Yang, Y., Zou, X., Zuo, Y.: 5G in manufacturing: a literature review and future research. Int. J. Adv. Manuf. Technol. 1–23 (2022)
7. Chen, I.F., Peng, C.M., Liu, H.A., Chen, Y.M.: 4 × 4 MIMO antenna design for 5G NR FR1 mobile handset applications. In: 2020 International Workshop on Electromagnetics: Applications and Student Innovation Competition (iWEM), pp. 1–2. IEEE, August 2020
8. de Mingo Sanz, J., Ceballos, P.L.C., Ducar, P.G., Bardaji, A.V., Cerezo, J.E.G.: Triple-band dual-antenna decoupling system for mobile handset. In: 2022 16th European Conference on Antennas and Propagation (EuCAP), pp. 1–5. IEEE, March 2022
9. Oh, H., et al.: 2.8–3.8 GHz broadband InGaP/GaAs HBT Doherty power amplifier IC for 5G new radio handset. In: 2022 IEEE/MTT-S International Microwave Symposium-IMS 2022, pp. 849–852. IEEE (2022)
10. Taheribakhsh, M., Jafari, A., Peiro, M.M., Kazemifard, N.: 5G implementation: major issues and challenges. In: 2020 25th International Computer Conference, Computer Society of Iran (CSICC), pp. 1–5. IEEE, January 2020
11. Gustavsson, U., et al.: Implementation challenges and opportunities in beyond-5G and 6G communication. IEEE J. Microwaves **1**(1), 86–100 (2021)
12. Sastrawidjaja, L., Suryanegara, M.: Regulation challenges of 5G spectrum deployment at 3.5 GHz: the framework for Indonesia. In: 2018 Electrical Power, Electronics, Communications, Controls and Informatics Seminar (EECCIS), pp. 213–217. IEEE, October 2018

13. Erunkulu, O.O., Zungeru, A.M., Lebekwe, C.K., Mosalaosi, M., Chuma, J.M.: 5G mobile communication applications: a survey and comparison of use cases. IEEE Access **9**, 97251–97295 (2021)
14. Bulakci, Ö., et al.: Overall 5G-MoNArch architecture and implications for resource elasticity (2018)
15. Banchs, A., Gutierrez-Estevez, D.M., Fuentes, M., Boldi, M., Provvedi, S.: A 5G mobile network architecture to support vertical industries. IEEE Commun. Mag. **57**(12), 38–44 (2019)
16. Jamsen, J., Corley, K.: E-survey methodology. In Handbook of Research on Electronic Surveys and Measurements, pp. 1–8. IGI Global (2007)
17. Zhang, J., Zhang, S., Pedersen, G.F.: Frequency reconfigurable endfire vertical polarized array for 5G handset applications. In: 2020 14th European Conference on Antennas and Propagation (EuCAP), pp. 1–3. IEEE, March 2020
18. https://software.org/wp-content/uploads/softwareorg5Gsoftware.pdf . Accessed 15 Oct 2022

Optical Advanced Hybrid Phase Shift Approach for RF Beamforming and 5G Wideband Radar

Yosra Bouchoucha[✉], Dorsaf Omri, and Taoufik Aguili

SysCom Laboratory, National Engineering School of Tunis (ENIT),
University of Tunis El Manar, Tunis, Tunisia
yosra.bouchoucha@etudiant-enit.utm.tn,
taoufik.aguili@enit.utm.tn

Abstract. Massive multiple-input multiple-output (MIMO) systems combined with Phased shifted antenna array or smart antenna technologies are expected to play a key role in next generation wireless communication systems and for wideband radar applications. In this paper, a phase shift-based beamforming network for microwave applications is discussed. The main purpose of this comparative study is to discuss the state of-the-art research about the most advantageous types of Phased shifted antenna array techniques. In addition, it aims at highlighting the importance of these techniques in massive MIMO systems for mitigating against numerous technical implementation issues particularly the problem of beamsquint over wide bandwidths. Finally, a hybrid structure of optical phase shifter is proposed which is based on the combination of True Time Delay (TTD) and Phase Shift (PS) techniques that can provide the highest performance in terms of throughput as it reduces intra- and inter-cell interference.

1 Introduction

The continuously growing requirement for higher data rates in wireless communications drives modern applications into the mm-wave frequency domain. In various microwave and RF systems, phase shifters (PSs) are crucial components, as they are needed for monitoring the relative phase between the elements of a phased array antenna in radars or in steerable communications links, or for managing the signal phase in microwave devices, in electronic warfare. With the capability of optical phase shifts mapping into radio frequency phase shifts, microwave photonics (MWP) have been suggested since several years as a promising alternative to conquer the restrictions of electronic circuits for PS implementation.

These mm-wave applications would profit from the advantages of using advanced phased-array technologies. The latter, in fact, offers fast electronic beam steering, multi beam operation, adaptive pattern shaping, as well as multiple-input, multiple-output (MIMO) capabilities [1–3].

However, the traditional directly radiating phased-array electronics solutions have major drawbacks. These limitations are far too expensive and consume a great deal of power due to the low efficiency of state-of-the-art mm-wave integrated circuits. Added to this, 5G requires beam forming with high precision beam steering, which is not possible due to several issues in RF components. As a result, we can use photonics.

In the last two decades, the worlds of microwave and of photonics met in the technological field known nowadays as Microwave Photonics [1]. This favorable hybridization between the two disciplines increased the potential of both, on one hand providing microwave systems with desired enhanced features like remarkable broad bandwidth (BW), increased signal purity at higher frequencies and higher flexibility. On the other hand, stimulating a relentless research in photonics to attain the required performance. The recent striking advances in photonic integrated circuits (PICs) fabrication and packaging processes [1, 2] provided photonic devices with fundamental features like small footprint, low power consumption, and better mechanical ruggedness, making them competitive with their classical, electronic counterparts.

One of the fields that has particularly taken advantage of this cross-fertilization is beamforming (BF). BF is the ability of a phased-array antenna (PAA) to change the pointing direction of its radiation pattern without any mechanical movement [2]. This feature will be highly desirable for next future microwave systems, since it will enable large adaptability and flexibility, which will be crucial in the domain of 5G communications, as well as in remote sensing. BF enables wireless access networks with higher throughput and optimized power consumption, along with a more efficient use of the frequency spectrum [2]. Moreover, in the field of remote sensing, BF will allow for a widespread use of compact radar systems with spatial scanning capabilities. However, regardless of the considered application, pointing accuracy, low losses, and reduced power consumption will be necessary [2].

In this perspective, photonics can provide promising solutions to meet these requirements. In fact, the requisites of increased bandwidth and carrier frequency of signals are very demanding for present-day digital RF phase shifters (PSs), which exhibit important insertion loss and power consumption, together with non-negligible phase errors. Optical phase shifter instead presents a good linearity, and can be easily employed together with the almost-lossless optical distribution of signals for antenna remoting.

In general, beamforming (BF) can be implemented following two possible approaches: true-time delay (TTD) or phase shift (PS). TTD techniques present the advantage of totally avoiding the beam squint effect, since a real time delay is introduced between the signals feeding the PAA elements. However, PS allows for easier implementation, although still frequency-dependent and therefore not immune to distortions in wideband communications [3].

In fact, two main undesired effects can be caused by the PS: signal impairments and beam squint. Impairments are due to the non-constant group delay over the signal BW, as in a dispersive channel, which is caused by the frequency-constant PS. However, transmission protocols normally compensate for this kind of this activity has been impairments. Beam squint, on the other hand, due to the same cause, is related to the superposition of the signal replicas radiated by the PAA elements, and cannot be mitigated. Nonetheless, if the fractional BW (FBW) of the transmitted signal is not too wide, and the number of array elements (NAE) is not too high, the effect of squint can be tolerable. As an example, Fig. 1 reports the array gain of a PAA, which is calculated as the integral over the whole signal frequency band (normalized to the signal BW) of the frequency-dependent array factor, such a FBW implies the transmission of signals with a BW of hundreds of MHz or some GHz. In addition, broad operational BW and small size antennas will be needed in many cases as, e.g., in automotive radars for obstacles detection and tracking for the next-future advanced driver assistance systems (ADAS) [4].

These solutions exhibit high loss, low power handling or limited BW. Moreover, they often do not allow for large scanning angles [4]. On the other hand, analog or digital electronic phase shifters often introduce high losses and signal distortions, and work on limited frequency ranges, which represent an important limitation to systems that would benefit from operating frequency flexibility [5].

A solution to overcome these limitations is offered by photonics, either with TTD, or PS, with high expectations put on photonic integrated circuits (PICs) [6]. In fact, PICs allow performing stable coherent operations with ultra-compact footprint and low operating power. The TTD and PS based beamforming systems have their limitations namely large size, higher cost, beam squint and limited bandwidth tunability. For the mmWave communication, an optical hybrid phase shifter approach is proposed here.

This paper is structured as follows: In Sect. 2, brief and Basic concepts of BF are given. Then, Sect. 3 reports simulated results of Phased array pattern using different methods phased shift (PS) and true time delay (TTD) equation and its effect on Beamsquint (Bs) issues, whereas in Sect. 4, a comparison between the performance of the photonics based and some commercial RF phase shifters are shown, confirming the suitability of the photonic phase shifters solution for 5G access networks and radar systems. Finally, before the Conclusion, in Sect. 6, there is an optical hybrid phase shifter, its block chain is proposed as a good alternative to profit from the advantages of the two techniques and to mitigate the beam squint effect.

2 Concepts of Beamforming

2.1 Equations of Time-Delay Beamsteering

Phased array antennas (PAAs) play important role in modern radar and wireless communication systems. An antenna can transmit radiation in a particular direction only. The direction of main lobe or beam is fixed in space. The beam of antenna can be steered by forming an array of antenna elements. The antennas can be arranged along an axis (one dimensional) or plane (two dimensionals). The beam emitting from an array can be steered in space by phase delaying signal at each antenna element. As shown in Fig. 1, beamforming for a linear array of four elements array spaced by a distance d, can steer a beam b through an angle ϴ by providing a phase delay of $\Delta \varphi$ at element 2 with respect to the first element.

The required phase shift can be calculated as follows:

$$\Delta_\varphi = 2\pi d \, \sin \frac{\theta}{\lambda}. \tag{1}$$

From Fig. 1, it can be observed that the required phase delay $\Delta \varphi$ at each element is a strong function of wavelength. As most of the radar systems as well as communication

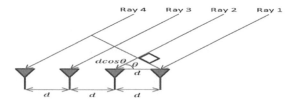

Fig. 1. Scenario of beamforming for a linear array of four elements

ones operate over a certain bandwidth, the variation of phase delay with frequency results in multiple antenna beams steering at different angles. This phenomenon is known as beam squint. This latter occurs due to constant phase delay to each spectral component. In order to avoid squint, a progressive wavelength dependent phase delay is needed.

Optical waveguides/fibers can provide a phase delay based on the applied optical frequency. This technique is used to provide phase delay to antenna elements and it is commonly known as true time delay (TTD).

True-time delay beamforming based on photonic technologies has been extensively researched in the last few years. As most of the radar systems as well as communication systems operate over a certain bandwidth, the variation of phase delay with frequency results in multiple antenna beams steering at different angles. This phenomenon is known as beam squint. The latter occurs due to constant phase delay to each spectral component. In order to avoid squint, a progressive wavelength dependent phase delay is needed.

2.2   Electronic Versus-Optical Phase Shifter

BF can be realized following two possible approaches, namely true-time delay (TTD) or phase shift (PS). In this paper, we discuss a photonics-assisted BFN for microwave and mm-wave signals, whose phase shifters have been realized as PICs in Silicon-On-Insulator (SOI) technology [1].

However, electronically implemented TTD may be based either on switched delay lines using micro-electromechanical systems (MEMS), or on monolithic microwave integrated circuits (MMICs). The main limitations of these solutions lie in their relatively limited BW and operation frequency range, or low power handling, often with the additional constraint of relatively narrow scanning angles [1–3]. In particular, they do not guarantee the same performance, e.g. in terms of delay/phase shift minimum step or maximum range over their operating bandwidth. Then, when they show excellent performance, it is at the expense of flexibility [5].

In particular, besides being immune to electromagnetic interferences, photonics potentially offer the additional advantages of broadband operation at large RF carrier values, high phase accuracy with small amplitude variations, and fast response time. All of these are key features for several emerging applications, including those of the coming 5G systems. Amongst the different proposed techniques, solutions based on photonic integrated circuits (PICs) are distinguished as the most promising approaches for practical MWP-based PS implementation. This is due to numerous factors: increased stability, drastically minimized size and weight, and the possibility of low operating power. By resorting to different technological platforms, several PICs have been realized in the last years to implement microwave PS operations. The architecture employment of an optical beamformer will be investigated. For it has many advantages over electronic beamformers, such as a light weight, low loss, and large bandwidth, but many research needs to be done to control the delays with this device.

Knowledge about optics, signals, electronics, but also optimization and control, is required to make this device operate with high performance. In the next-future mobile and wireless communications scenario, BFNs are expected to drive PAAs with potentially several elements, their size will shrink due to the employed high frequency. Therefore, antenna pointing accuracy, low losses, reduced power consumption, and small size are crucial characteristics in future BFNs. In this perspective, photonics can provide promising solutions to meet these requirements, potentially reducing also the cost of BFN elements, thanks to photonic integration [6].

2.3 Phase Shift versus. True-Time Delay

There are two basic methods for implementing the delay lines for TTD: optical and electronic. Therefore, Optical methods modulate the RF signal onto an optical carrier and use long fibers to delay the signal. Whereas, electronic methods use traditional microstrip lines or coax cable to delay the signal [5–7].

Take into account a uniformly spaced linear array with element spacing d as appeared in Fig. 1, the wave must travel an additional distance d sin θ to arrive at each successive element [3]. Array theory tends to The realization of tunable true-time delays based on a fiber-optic prism consisting of an array of dispersive delay lines was demonstrated. The delay could be converted by the signal into a phase shift at a given frequency [3]:

$$\Delta_t = \frac{d \, \sin \theta}{c} \tag{2}$$

$$\Delta_\varphi = 2 \pi f \Delta_t = \frac{2 \pi \sin \theta}{\lambda} \tag{3}$$

Skolnik shows in [3] that the array factor of a uniformly spaced linear array

$$G_a(\theta, \lambda) = \frac{\sin^2(\frac{N\Delta_\varphi}{2})}{N^2 \sin^2(\frac{\Delta_\varphi}{2})} \tag{4}$$

$$= \frac{\sin^2(N\pi(\frac{d}{\lambda}) \sin \theta)}{N^2 \sin^2(\pi(\frac{d}{\lambda}) \sin \theta)} \tag{5}$$

The array can be steered by applying a phase shift such as $\varphi = 0$ in terms of interest. The required phase shift can be applied using two phase shifters that produce constants. phase shift, or time delays, which produce a frequency dependent phase shift. Each method will be explored and the trade-offs explained [3].

The conventional procedure of steering a phased array is with phase shifters. Since the phase shift is fixed λ becomes a fixed λ_0 and φ_0 is defined as [3]:

$$\Delta_{\varphi 0} = 2 = \frac{2 \pi d \sin \theta_o}{\lambda_o} \tag{6}$$

The following expression for the array factor of an array steered with phase shifters is defined as [3]:

$$G_a(\theta, \lambda) = \frac{\sin^2(N\pi d(\frac{\sin\theta}{\lambda} - \frac{\sin\theta_o}{\lambda_o}))}{N^2\sin^2(\pi d(\frac{\sin\theta}{\lambda} - \frac{\sin\theta_o}{\lambda_o}))}. \tag{7}$$

A plot of the array factor for three different frequencies is given in Fig. 2. Note how the beam position changes with frequency when steered using the phase shifter method. The formula for the array pattern when steered with time-delay mode is:

$$G_a(\theta, \lambda) = \frac{\sin^2(N\pi d(\frac{\sin\theta}{\lambda} - \frac{\sin\theta_o}{\lambda_o}))}{N^2\sin^2(\pi d(\frac{\sin\theta}{\lambda} - \frac{\sin\theta_o}{\lambda_o}))}. \tag{8}$$

A plot of the array factor for the same three frequencies as above is given in Fig. 2. Note how the beams are now all pointing at 20∘ and it is simply the beam width that varies with frequency.

The Beamforming uses multiple antennas to generate/receive electromagnetic wave with controllable beam pattern. As shown in Fig. 1, to make specific slanted RF signals, assuming the example of 20°, the feeding signals of each antenna have their own designated signal delay which is proportional to antenna space and slanted angle [3].

As mentioned earlier, the TTD and PS based beamforming systems have their limitations namely large size, higher cost, beam squint and limited bandwidth tunability. As for the mmWave communication, a hybrid approach is proposed here. The suggested system benefits from the best of both worlds i.e. TTD and PS based systems [3].

As we know from the previous section, TTD approach provides a frequency dependent phase shift with no beam squint. On the other hand, the minimum resolution of phase shift is dependent linearly on the number of paths/waveguides. For obtaining a phase resolution of 1, 360°' paths are required. This approach is costly as it requires a larger area on the Photonic Integrated Circuit (PIC). Phase shift based approach implemented using ring resonators and PN junctions can provide a higher degree of phase resolution but as the phase shift is independent of wavelength, it creates beam squint for large bandwidth signals (as required by 5G). Thus, we propose a hybrid approach which benefits from TTD as well as PS.

A useful computation is to figure out how much deviation from the nominal frequency the system can tolerate before the beam is pointed away from the target. To do this, we first develop an equation for the beam squint as a function of frequency. The squinted beam peak occurs at Angle θ_P when $\frac{\theta_P}{\lambda} = \sin\frac{\theta_0}{\lambda_0}$. The beam squint can be defined as the difference between the actual peak and the desired peak [3]:

$$\theta_{BS} = \theta_p - \theta_o = \sin^{-1}(\frac{f_o}{f}\sin\theta_o) - \theta_o \tag{9}$$

Skolnik [3] provides an approximate equation for the 3 dB beamwidth of an array, $\theta_{3dB} = \frac{102}{N}$, ,, where N is the number of elements in the array. Setting the beam squint is (9) equal to the 3_{dB} beamwidth and solving for frequency leads to.

$$f = \frac{f_o\sin\theta_o}{\sin(\theta_o \pm \frac{102}{N})}. \tag{10}$$

This equation will tell you at what frequencies (above and below f0) your beam will have moved off the target by the 3 dB beamwidth. A similar equation is derived in [2] by requiring that the beam squint should be less than the beamwidth. The result is a limit on the bandwidth of the system (3):

$$B \ll \frac{c}{L \sin \theta_o},$$ (11)

3   Simulated Results

Active electronically –scanned arrays (AESAs) are becoming popular for use in radar as well as other RF systems. They allow control of amplitude and phase of each element which enables fine manipulation of beam direction and shape. It can be changed much more rapidly than mechanically steered array [4]. This present problem of beam squint for (AESAs) that has traditionally been steered with the phase shift because of the instantaneous bandwidth waveforms and narrow bandwidths, this beam squint can be enough to steer off the target resulting in a greatly reduced return [4]. This simulation derives the expression describing beam squint and shows how the array factor changes when using phase shifters versus time-delay using Eqs. (7) and (8).

3.1   Expression for the Array Factor of an Array Steered with Phase Shifters and TTD

Plots of the array factor for three different frequencies using Eqs. (7) and (8) are given in Figs. 2 and 3 and plotted with Matlab, which is entitled "Beam Pattern Steered with Phase shifters as a function of Frequency». It notes how the beam position changes with frequency when steered using the phase shift method [3].

To plot this equation, we assumed these parameters:

Frequency Range: 9–11 GHz, $Theta_0 = 20°$ (Steering angle $= 20°$), $f_0 = 10$ GHz, d $= 0.5*\lambda$ for 12 GHz, N $= 64$ elements.

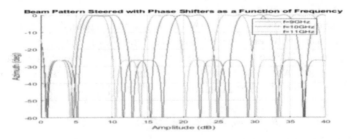

Fig. 2. Beam pattern with phase shifters as a function of frequency based on PS with (7)

A plot of the array factor for the same three frequencies as described above note how the beams are now all pointing at 20° and it is simply the beamwidth that varies with frequency with Matlab.

Obviously, both plots show the benefits of true time delay beam steering, wide instantaneous bandwidths that vary with frequency and can be accommodated without beam squint.

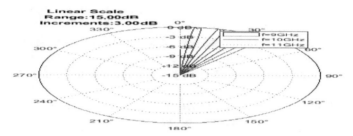

Fig. 3. Polar Plot of Array factor steered with Phase Shift (PS)

The benefits of true time-delay were shown, along with some of the equations to describe beam squint. Rules of thumb relating instantaneous bandwidth to allowable beam squint were also presented here to show the benefits of true time-delay beam steering.

The plots in Fig. 3 clearly show the benefits of true time delay beamsteering: wide instantaneous bandwidths can be accommodated without beamsquint.

Since resolution improves with wider bandwidth, imaging radar and SAR tend to use wide bandwidths to improve image quality. For example, [2] states that extremely fine-resolution multi-mode SAR systems may use as much as 2 GHz of bandwidth to obtain a better resolution.

2 GHz of bandwidth is what is shown in Fig. 4. If phase shifters are used, the beam would actually steer between three adjacent resolution cells during the pulse, blurring the image. Such applications would obtain a great benefit from TTD beamsteering.

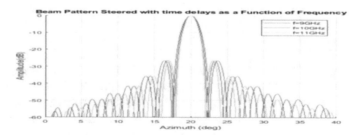

Fig. 4. Beam Pattern with Phase Shifters as a Function of Frequency based on TTD with (8)

3.2 Expression of Important Parameters

First parameter θBS:

A useful computation is to figure out how much derivation from the nominal frequency, the system can tolerate before the beam is pointed away from the target. Through the developed equation for the beam squint as a function of frequency. The beam squinted occurs at θBS as shown in (9).

Second parameter:
This equation will tell you at what frequencies (above and below f0) your beam will have moved off the target by the 3 dB beamwidth.

Skolnik [1] provides an approximate equation for the 3 dB beamwidth of an array, $\theta_{3dB} = \frac{102}{N}$, where N is the number of elements in the array. Setting the beam squint in (8) is -equal to the 3 dB beamwidth and solving for frequency leads to Eq. (10).

Third parameter:
A similar equation is derived in [3] by requiring that the beam squint should be much less than the beamwidth; the result is a limit on the bandwidth of the system Bas explained previously in (11).

Note that this equation does not depend on wavelength or frequency, just on the length of the array. Equations can be used to determine whether exotic techniques like stepped chirp waveforms or intra-pulse beam steering are needed for a given system.

Finally, it is worth noting that beam squint has not always been a negative thing. In fact, some of the earliest it is the change in the beam direction as a function of frequency, therefore, arrays used this property to steer the beam in what are called frequency-scan arrays [3]. Frequency-scan arrays choose the beam direction by changing the frequency of the transmitter. This is relatively easy to implement, but results in some problems. Since frequency is used to steer the beam, it cannot be used for other tasks such as improved resolution or moving target detection with Doppler. As a result, frequency-scan arrays are not used much in modern systems [3].

3.2.1   Beamsquint and Instantaneous Bandwidth (IBW)

Beamsquint Effect
Antennas using phased-array technique have widespread range of applications for occurrence in radar, imaging and communication systems [1, 3, 4]. The design of phased array systems becomes thought-provoking, particularly when we require a wide band of operation. A significant phenomenon that can limit bandwidth in phased array antenna systems is beam squinting [1For example the altering of the beam direction as a function of the operating frequency. Beam squinting simply means that an antenna pattern points to θ0 + Δθ at frequency f0 + Δf rather than θ0, which was the pointing direction at frequency f0, it is the alter in the beam direction as a function of the operation frequency, polarization, or orientation Thus, there is an associated pattern loss at the commanded scan angle. However, it can limit the bandwidth in phased array antenna systems and may affect uniform linear arrays (ULA) characteristics. In fact, as bandwidth increases, beam squint increases too [4].

Instantaneous Bandwidth (IBW)
When describing instantaneous bandwidth (IBW) it is advantageous to first think of it from a phase shifter perspective. For an Electronically Scanned Array ESA employing phase delay, the phase shifters are set at each element to scan the beam. The phase shifters have the characteristic of constant phase vs. frequency. The Instantaneous Bandwidth

(IBW) is the range of frequencies over which the loss is acceptable and is 2 * Δf, typically, the IBW specified is the 3 or 4 dB IBW [8]. From the plots of beam squint for an Electronically Scanned Array illustrated in Fig. 5, we can observe that the array suffers more from loss at the steered scan angle of 20°.

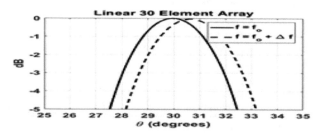

Fig. 5. Beam squint using phase-shifter steering for ESAs with the solid line pattern represents operation at the tune frequency f0 and the dashed line represents beam squint at f0 + Δf

4 A Comparison Between RF and Photonic Phase Shifter

In this section, we introduce some state-of-the-art, off-the shelf commercial devices [1]. Table 1 reports the comparison between the photonic integrated OPS and classical RF phase shifters, available in the market from different companies. Fundamental parameters like number of elements, BW, PS range, PS resolution, PS error, gain, maximum power consumption, OP1, IIP3, switching time are considered [1].

As shown in Table 1, the proposed OPS exhibits beter performance than the average of the RF devices, especially in terms of PS range, error, power consumption, and switching time. The power loss, calculated as 28 dB RF-equivalent loss, but can be possibly reduced to 20 dB, by reducing the coupling loss [1].

The five reported RF phase shifters work in the range 5–18 or 18–40 GHz i.e. the possibility of having the same performance at any RF frequency. Some RF phase shifters, on the other hand, change their specifications in different portions of the operating band. The switching time is a very important parameter in communication as well as in remote sensing. As reported in the table (Table 1), the OPS switching time is much lower than the RF counterparts [9].

5 Proposed Solutions

From the above discussion we can deduce two important facts, regardless of bandwidth the PAAs can be used with the small number of antenna array. Added to this, large antenna array need TTD circuits that have large total delay amount. In this context, we suggest hybrid beamforming (PAAs and TTDs) architecture that will cover wide bandwidth and large array size as illustrated by Fig. 4. Firstly, large antenna array is divided into sub-groups of antennas which have almost beam squint free even using PAAs structure. Secondly, phase shifters are manipulated in order to control subgroup beam direction.

Table 1. Comparison between the proposed optical phased Shift OPS and RF commercial phase shifters [2]

	Kratos 7928A	Analog Devices HMC247	RF-Lambda RFPSHT	MiteqDSP-0618-360-5-5.6	Photonic PS
BW	6–18	5–18	6–18	6–18	10–16
PS range [°]	0–360	0–400	0–360	0–360	62–68 62–68 0–475
PS resolution [°]	1.4	Continuous	5.625	5.6	Continuous
PS error[°]	±15	n/a	±15	3	<2
Max Power Consumption[W] Switching Time [ns] Analog/Digital	500 Digital (8 bits)	20 Analog	100 Digital (6bits)	20 Digital (6bits)	<1 Analog

The proposed hybrid structure provides not only high resolution (resolution of PAAs and resolution of TTD) but also large total delay (due to analog TTD). The additional PLLs and analog TTDs are little burdensome but hybrid architecture has many advantages considering the beam squint free and design difficulties in mm-wave TTD.

We suggest that using phase shifter at RF frequency and using analog TTD will be a good choice. With this combination, large total delay with fine resolution is achieved without beam squint effect. We also design analog TTD circuits at IF. As mentioned earlier, the TTD and PS based beamforming systems have their limitations namely large size, higher cost, beam squint and limited bandwidth tenability for the mm Wave communication.

As we know from previous section, TTD approach provides a frequency dependent phase shift with no beam squint. On the other hand, the minimum resolution of phase shift is dependent linearly on number of paths/waveguides provides a higher degree of phase resolution but as the phase shift is independent of wavelength, it creates beam squint for large bandwidth signals (as required by 5G). Thus, we propose a hybrid approach which benefits from the best of the both worlds i.e. TTD and PS based systems.

The schematic structure and the operation principle of the proposed MWP-PS are illustrated in Fig. 4. The core element of the architecture, highlighted in the dashed box of the figure, comprises an optical Hybrid phase shifter (OHPS), a Laser source (LS), Modulator (MZM), and a photodiode (PD). At the input of the circuit, the microwave signal to be phase-shifted, which is considered to show a given power spectral density around the carrier frequency fRF, drives a single-sideband electro-optic modulator (SSB-EOM) to generate a sideband centered at the optical frequency vsb, spaced by fRF from the optical carrier at frequency vc provided by a laser source (LS). The generated full-carrier SSB modulated optical signal is then fed into the MWP-HPS. The operation of the scheme of the MWP-HPS relies on optical domain (Fig. 6).

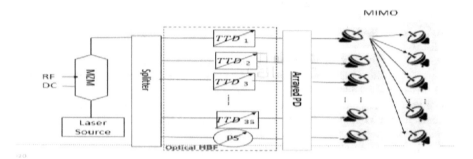

Fig. 6. Block Chain of Advanced Beamforming Phase shifter

6 Conclusion

In this paper, hybrid beamforming architecture was proposed using the analog TTD and Phase shift circuit for wide bandwidth and large array antenna system which will be highlighted in future. In the next years, it is predicted that PAAs with efficient BF capabilities will be, on one hand, massively employed in 5G wireless access networks. On the other hand, radars, that are becoming commonly used systems, particularly in the automotive field, can benefit from beamforming to sweep directive beams, improving obstacles detection. Moreover, photonics is likely to be more and more employed in many RF systems, in the domain of communications as well as radars.

With the latest advances in PICs, photonics may disclose the possibility of implementing BFNs fundamental blocks achieving better performance than BFNs realized with electronic solutions in terms of precision, BW, signal integrity, frequency agility. In addition to these performances, the photonic BFNS enabling PS-based as well as TTD-based approaches, at the same time guarantee a small footprint. A thorough comparison with state-of-the-art, RF commercial devices is also reported, showing that the proposed PIC-based BFN is a suitable candidate for the practical implementation of rapidly reconfiguring beamsteering in PAAs for future 5G systems with high efficiency and adaptability, as well as for compact high-resolution radars.

References

1. Reza, M., et al.: Multi-static multi-band synthetic aperture radar (SAR) constellation based on integrated photonic circuits. Electronics **11**(24), 4151 (2022)
2. Giovanni, S., Claudio, P., Bilal, H., et al.: High-performance beamforming network based on Si-photonics phase shifters for wideband communications and radar applications. IEEE J. Sel. Top. Quant. Electron. **26**, 1 (2020)

3. Yosra, B., Salem, H., et al.: Beamforming using linear antenna arrays for photonic applications. In: 2018 International Conference on Internet of Things, Embedded Systems and Communications (IINTEC), pp. 59–63. IEEE (2018)
4. Matt, L.: True time-delay beamsteering for radar. In: 2012 IEEE National Aerospace and Electronics Conference (NAECON), pp. 246–249. IEEE (2012)
5. Seyed Kasra, G., Eric-AM, K., Bram, N., et al.: Phased-array antenna beam squinting related to frequency dependency of delay circuits. In: 2011 41st European Microwave Conference (2011)
6. Porzi, C., et al.: Photonic integrated microwave phase shifter up to the mm-wave band with fast response time in silicon-on-insulator technology. J. Lightwave Technol. **36**(19), 4494–4500 (2018). https://doi.org/10.1109/JLT.2018.2846288
7. Giovanni, S., Claudio, P., Vito, S., et al.: Design and characterization of a photonic integrated circuit for beam forming in 5G wireless networks. In: Giovanni, S., Claudio, P., Fabio, F., et al. International Topical Meeting on Microwave Photonics (MWP), pp. 1–4. IEEE, 2017 (2017)
8. Serafino, G., et al.: Photonics-assisted beamforming for 5G communications. IEEE Photon. Technol. Lett. **30**(21), 1826–1829 (2018). https://doi.org/10.1109/LPT.2018.2874468
9. Lee, S.-S., Oh, Y.-H., Shin, S.-Y.: Photonic microwave true-time delay based on a tapered fiber Bragg grating with resistive coating. IEEE Photon. **16**, 2335–2337 (2004)

Quo Vadis, Web Authentication? – An Empirical Analysis of Login Methods on the Internet

Andreas Grüner[✉], Alexander Mühle, Nils Rümmler, Adnan Kadric, and Christoph Meinel

Hasso Plattner Institute (HPI), University of Potsdam, 14482 Potsdam, Germany
{andreas.gruener,alexander.muehle,tatiana.gayvoronskaya,
christoph.meinel}@hpi.uni-potsdam.de

Abstract. Identity management technology (IdM) is fundamental to an application's security. Over time, distinct identity management schemes have been developed with their specific advantages and disadvantages. Furthermore, the novel Self-Sovereign Identity (SSI) paradigm emerged along the invention of blockchain technology. However, the distribution of applied concepts on the Internet is an open research field. In particular, any use of the emergent SSI solutions remain undetermined. To close this gap, we research the login method distribution based on the most popular domains identified by the Majestic project. In a period of 3 months, we analyzed the Internet for the IdM model distribution. Thereby, we identified an average use of 27.3% centralized, 65.5% isolated and 0% SSI model. In the centralized model, we identified Facebook with an average of 16.2% as the primary Single Sign-On (SSO) provider.

1 Introduction

IdM technology provides functions to securely personalize services and recognize users. Along the increasing number of online services, adversaries challenge the security of IdM implementations. In particular, identity theft due to leaked passwords [1] has been a growing threat on the Internet. The user experiences IdM as a registration process to provide personal information and to create an account upon first visit of the service. Subsequently, the user accesses its account by providing a secret credential. As an alternative, clients may authenticate with an existing account at an Identity Provider (IdP), e.g. a social login like Facebook. Thus, the user is redirected from the service to an IdP. After successful authentication, the IdP forwards the user and information about her/ him to the service.

These authentication options represent various IdM paradigms that have evolved over the last decades. Starting with the isolated model, a user's registration is specific to a dedicated Service Provider (SP). As a result, the number of accounts increases along to the enrolled services. Thus, the user must protect several authentication credentials. With the establishment of the centralized scheme, the IdP becomes an independent entity that serves several SPs.

L. Barolli (Ed.): AINA 2023, LNNS 655, pp. 471–479, 2023.
https://doi.org/10.1007/978-3-031-28694-0_45

Therefore, the user requires only a single registration to access various SPs. As shortcoming, the centralized IdP holds a powerful position with vast trust dependencies towards the user and the SP [2]. A popular centralized IdP may deny service to a user or SP. Thus, these entities cannot interact. Moreover, the IdP is a primary target for stealing a vast amount of identity data. In identity federations, interconnected IdPs are used across trust boundaries to accept users of foreign organizations. This network establishes a circle of trust between several IdPs and SPs.

Moreover, Allen [3] proposed the SSI paradigm to bring the user back in control of its identity. In this setting, the traditional actors changed to the issuer (from IdP or attribute provider), the verifier (from the SP) and the identity holder (the user). The issuer originates attributes of an identity holder. These attributes are called Verifiable Credentials (VC) [4]. In contrast, the verifier receives a VC and confirms its validity. The blockchain provides a viable implementation option for this concept although other approaches are possible. In this context, the IdP is implemented on the blockchain in a decentralized manner. Thus, the IdP is ressolved as a Trusted Third Party (TTP) and trust requirements shift in favour of the user [2]. In particular, the IdP cannot exert illegitimate control about any identity.

The theoretical advantages and disadvantages of the IdM models have been researched. However, the adoption of these models on the Internet might contravene their advantages. Furthermore, the popular SSI paradigm might not have proven its suitability yet in a productive environment. In this paper, we analyze empirically the login method distribution on popular websites on the Internet to derive the prevailing IdM models. Additionally, we examine the actually integrated centralized IdPs to determine their popularity. To achieve this, we deploy the Majestic most popular domains [5] as a foundation. Starting form this set, a crawler identifies implemented login methods and integrated IdPs. We examine the domains in three monthly iterations to determine shifts in the model adoption. Thus, we investigate the following research questions.

- Which are the used IdM models on the Internet?
- What are the most used SSO provider?

The remainder of the paper is organized as follows. In Sect. 2, we present related research work. Subsequently, we depict in Sect. 3 the methodology for data retrieval and analysis. In Sect. 4, we evaluate the gathered data to identify the login methods. Finally, we discuss our results in Sect. 5 and conclude our findings in Sect. 6.

2   Related Work

Ghasemisharif et al. [6] conducted an empirical analysis to uncover security flaws on websites using SSO techniques. Thereby, the researchers investigate the Alexa's top 1 million [7] websites and focussed on the attack vector of account hijacking. As a result, 6.30% of the domains provide SSO authentication. Additionally, the authors propose an extension to the OpenID Connect [8] specification, called SSO Off, to simultaneously revoke access to all relations of an hijacked IdP account. Moreover, Zhou and Evans [9] concentrate in their research on vulnerabilities by using SSO with the Facebook IdP. They studied 20.000 popular websites and found that 8,3% integrated Facebook. The authors built the SSOScan application to conduct a full registration process by using a Facebook test account. SSOScan examines security flaws in the IdM integration. Additionally, Cho and Kim [10] conveyed a user study with 364 participants to investigate the usage of SSO providers for applications. As a result, there is a dependency of using a social IdP in privacy-sensitive domains due to data protection concerns.

In contrast, our research focus in general on the distribution of different IdM models on the Internet and the popularity of various IdPs. Our aim is also to identify if any SSI solution is already productively used on the Internet.

3   Data Collection and Processing

In this section, we describe the selection of the data set (Sect. 3.1), the structure of the web crawler (Sect. 3.2) and specifics of the process for identifying the authentication methods and IdPs (Sect. 3.3.)

3.1   Data Set Selection

To determine the distribution of login mechanisms across the Internet, our analysis requires a representative data set as foundation. The Majestic [5] data set comprises popular websites ranked based on referring subnets. The data set is actively maintained and updated daily. Furthermore, the Cisco Umbrella [11] project provides a collection of prevalent domains on the Internet. The ranked domains are based on about 620 billion domain requests that are generated users globally. The data set is updated daily and considers any domain requests besides web traffic. As a disadvantage, technical sub domains are included that do not refer to a website with user authentication. Further free of charge available collections are Amazon Alexa's top 1 million list [7] and the OpenPageRank project [12]. However, the data set of Amazon Alexa is deprecated. The OpenPageRank project's collection is only updated every three months. Therefore, we select the Majestic Million data set for our analysis.

3.2   Crawler Setup

We developed a crawler for automatically processing websites and extracting the required information. Our crawler is implemented in Python and executes on a virtual Linux machine. Furthermore, we apply the Selenium [13] automation library in connection with the chrome driver to parse the examined websites. The results are stored in a PosgreSQL database. The crawler runs in the data center of the Hasso Plattner Institute in Germany. Additionally, we consider only the English version of the crawled websites.

3.3   Data Processing

The data processing for each domain follows an analogous pattern. First the crawler determines the domain's validity (Sect. 3.3.1). Subsequently, the agent searches the login page (Sect. 3.3.2) and finally, determines the login options (Sect. 3.3.3).

3.3.1   Validate Domain

The crawler investigates for each domain if a website is accessible via the HTTP or HTTPS protocol. In case the crawler can open the web site of the domain, it can proceed with the next step. Otherwise, the analysis stops for the domain at this point.

3.3.2   Login Link Search

After the domain's web site is loaded completely, the crawler searches the link to the login page. In general, a web presence offers all login options combined on a single page for the user. Thereby, the crawler examines the links, link tags and surrounded text on the main page to identify the most probable link to the login page. Regular expression supports the crawler for the link identification. Figure 1 lists an extract of the regular expressions. Per our analysis, the web sites login link predominantly apply these words.

Being the first operation on the loaded web site, a challenge arises with GDPR cookie consent banners. If the banner blocks access to the content of the website, the consent dialog must be detected and subsequently accepted. The crawler searches the amount of links and determines if they do not exceed a threshold. If this is the case, a cookie consent banner is detected. Afterwards, we accept the banner by searching for an accept consent button.

```
(.*?)login(.*?)
(.*?)log-in(.*?)
(.*?)logon(.*?)
(.*?)log-on(.*?)
(.*?)signin(.*?)
(.*?)sign-in(.*?)
(.*?)account(.*?)
([^a-zA-Z]*?)auth([^a-zA-Z]*?)
([^a-zA-Z]*?)authorize([^a-zA-Z]*?)
([^a-zA-Z]*?)authentication([^a-zA-Z]*?)
([^a-zA-Z]*?)authenticator([^a-zA-Z]*?)
```

Fig. 1. Extract of login link search regular expressions

3.3.3 Login Options Search

After the identification of the login page, the crawler evaluates the site to determine available authentication options. Thereby, password authentication, SSO authentication by referring to another IdP, and SSI-based authentication are differentiated. To distinguish these options, the agent selects all links and matches them against SSO regular expressions. Figure 2a shows these indicators. Subsequently, all form fields and buttons are parsed to identify password-based authentication. Figure 2b depicts an extract of the search terms. Finally, the crawler analyzes QR codes to identify any SSI solutions.

```
(.*?)login(.*?)              (.*?)email(.*?)
(.*?)log-in(.*?)             (.*?)phone(.*?)
(.*?)log_in(.*?)             (.*?)number(.*?)
(.*?)oidc(.*?)               (.*?)username(.*?)
(.*?)saml(.*?)               (.*?)password(.*?)
(.*?)oauth(.*?)              (.*?)passphrase(.*?)
(.*?)identity(.*?)           (.*?)e-mail(.*?)
```

(a) SSO indicators   (b) Password indicators

Fig. 2. Extract of login option search regular expressions

3.4 Frequency and Time Period

The crawler runs once per month in June, July and August 2022. The average execution time per run is 15 days. At the start of the run, the current popular domain list from Majestic is retrieved. The crawler evaluates per run the most popular 100.000 domains.

4 Web Authentication Distribution on the Web

Figure 3 shows general results of the three crawler runs in June, July and August. As basis serve the 100.000 most popular domains of the day the crawler commence its search. On average 11.6% of the domains are inaccessible. The average is based on the 11.8% in June, 11.9% in July and 11.1% in August of non available domains. Reasons that a domain is not accessible encompasses connection time-outs, connection resets and Domain Name System (DNS) resolution errors. Furthermore, the crawler could not determine a link to the login page for on average of 49.7% of the domains. The average is based on 50.6% in June, 49.8% in July and 48.8% in August. As a conclusion, the website has no protected area that requires authentication. In addition to that, the crawler could not identify login options for an average 6.2% of the domains by analyzing the login page. This outcome can either be based on a falsely identified login page or the inability of the crawler to recognize certain login option patterns.

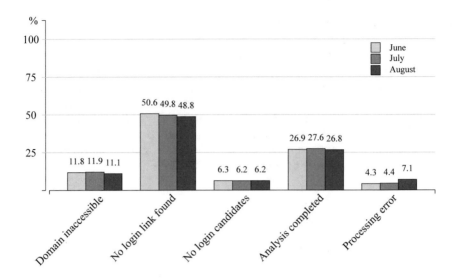

Fig. 3. Overview general analysis results

Moreover, for an average of 27.1% of the domains, the crawler completely finished the analysis. These domains enable the analysis of further details concerning the IdM model use. Finally, about 5.3% of the domains evaluation ended with a processing error of the crawler. This partial set of the domains does not contribute to further insights in the IdM model distribution. In general, the percentages for the three runs in June, July and August are close to each other. Therefore, one time errors seem not influence the results.

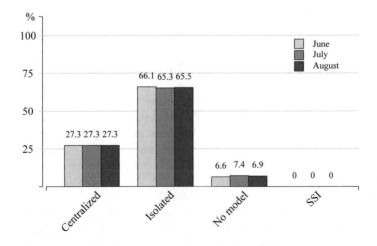

Fig. 4. Detected IdM models for completely analyzed domains

The category of completely analyzed domains encompasses the richest information. In this category, the crawler could identify the login page and login candidates. As a result, conclusions of the applied IdM models are possible. On average 65.6% of the completely analyzed domains offer an isolated login model. Furthermore, about 27.3% of the domains provide a centralized authentication method. In addition to that, we could not identity the usage of any SSI authentication method at any domain. The crawler identifies the isolated model by the use of username and password authentication scheme. The centralized model is determined by the usage of any link to an SSO provider. This outcome answers our first research question about the IdM model distribution and the prevalence of the SSI paradigm. Furthermore, no relevant change of the IdM model distribution during the analyzed three months could be identified. An overview of these results is shown in Fig. 4.

Analyzing the centralized IdM models, the most popular SSO providers are of particular interest. Table 1 presents the most prevalent SSO providers of the examined domains. Facebook and Google are the primarily used SSO providers with a large distance to their followers. Facebook is leading with an average of 16.2%. Moreover, Google is placed second with an average of 15.6% of the fully investigated domains. The third located RTBMarket seems to be a falsly identified SSO provider as it is the only non well-known provider. The subsequently following SSO providers have only a single digit percentage use of the analyzed domains. Twitter lists with an average of 3.2% and then, for instance, Apple is ranked with an average of 3.6%. This clearly shows a dominant usage of Facebook resp. Google and answers our second research question.

Table 1. Top 10 SSO provider in the centralized model

	June	July	August	Average
Facebook	16.3%	16.5%	15.8%	16.2%
Google	15.5%	16.0%	15.4%	15.6%
RTBMarket	4.6%	4.7%	4.3%	4.5%
Twitter	4.2%	1.1 %	4.2%	3.2%
Apple	3.6%	3.7%	3.6%	3.6%
LinkedIn	3.4%	3.3%	3.5%	3.4%
Live	2.1%	2.1%	1.9%	2.0%
Microsoft	1.3%	1.3%	1.4%	1.3%
GitHub	0.9%	0.9%	0.9%	0.9%
Pinterest	0.7%	0.7%	0.7%	0.7%

Evaluating closer the isolated model, Table 2 depicts the main phrases that indicate this scheme. On the very top, the phrase "email" and "username" with an average of 26.9% resp. 20.2% are listed. If these terms are written on the login page for input fields or as button, the web page most likely have a separate registration form.

Table 2. Top 5 phrases to indicate the isolated model

	June	July	August	Average
Email	26.9%	27.2%	26.6%	26.9%
Username	20.4%	20.3%	19.8%	20.2%
Password	10.8%	10.8%	10.6%	10.7%
Enter your username	1.4%	1.3%	1.3%	1.3%
Email address	1.2%	0.9%	0.8%	0.9%

5   Discussion

Using a web crawler, we were able to evaluate a set of most popular domains of the Internet for the applied IdM models. However, there are certain challenges to conduct the evaluation. First, a decision for a data set, that reflects the most popular domains, is required. This determines the quality of the underlying data. Subsequently, the crawler faces enormous variants of web site implementations. Thus, specific implementations might not be captured by the crawler during the analysis.

6   Conclusion

Web applications on the Internet provide protected areas that require users to authenticate. To realize the authentication different IdM models have emerged over time. Lately, the SSI scheme originated to bring the user back in control of its identity. We analyzed the most popular domains on the Internet to evaluate the IdM model distribution. As a result, 27.3% of the completely examined domains apply the centralized model, 65.5% use the isolated scheme and 0% implement the SSI paradigm. Additionally, the primarily integrated SSO providers are Facebook and Google. Further providers solely follow at a long distance. Thus, the SSI model's advantages cannot be seen yet on the Internet.

References

1. Hunt, T.: PWNed passwords, version 5 (2019). https://www.troyhunt.com/pwned-passwords-version-5/. Accessed 05 Jan 2022

2. Grüner, A., Mühle, A., Gayvoronskaya, T., Meinel, C.: A comparative analysis of trust requirements in decentralized identity management. In: Proceedings of the 33rd International Conference on Advanced Information Networking and Applications (AINA), pp. 200–213 (2019)

3. Allen, C.: The path to self-sovereign identity (2016). http://www.lifewithalacrity. com/previous/2016/04/the-path-to-self-sovereign-identity.html. Accessed 04 Mar 2022

4. Sporny, M., Longley, D., Chadwick, D.: Verifiable credentials data model v1.1. (2022). https://www.w3.org/TR/vc-data-model/. Accessed 29 July 2022

5. majestic. The majestic million. https://majestic.com/reports/majestic-million. Accessed 02 Apr 2022

6. Ghasemisharif, M., Ramesh, A., Kanich, S.C.C., Polakis, J.: O single sign-off, where art thou? An empirical analysis of single sign-on account hijacking and session management on the web. In: Proceedings of USENIX Security 2018 (2018)

7. Amazon. Alexa top sites. https://aws.amazon.com/de/alexa-top-sites/. Accessed 02 Apr 2022

8. OpenID Foundation. Openid connect core 1.0. https://openid.net/specs/openid-connect-core-1_0.html. Accessed 04 Mar 2022

9. Zhou, Y., Evans, D.: SSOScan: automated testing of web applications for single Sign-On vulnerabilities. In: 23rd USENIX Security Symposium (USENIX Security 2014), pp. 495–510 (2014)

10. Cho, E., Kim, J., Sundar, S.S.: Will you log into tinder using your Facebook account? Adoption of single sign-on for privacy-sensitive apps. In: Extended Abstracts of the CHI Conference on Human Factors in Computing Systems 2020, pp. 1–7 (2020)

11. Hubbard, D.: Cisco umbrella 1 million (2021). https://umbrella.cisco.com/blog/ cisco-umbrella-1-million. Accessed 02 Apr 2022

12. majestic. The open page rank. https://www.domcop.com/openpagerank/ frequently-asked-questions. Accessed 02 Apr 2022

13. Selenium. Selenium. https://www.selenium.dev. Accessed 02 Apr 2022

Device Tracking Threats in 5G Network

Maksim Iavich[1], Giorgi Akhalaia[2(✉)], and Razvan Bocu[3]

[1] Scientific Cyber Security Association, Cyber Security Direction at Caucasus University,
Tbilisi, Georgia
miavich@cu.edu.ge
[2] Georgian Technical University, Tbilisi, Georgia
akhalaia.g@gtu.ge
[3] Transylvania University of Brasov, Brasov, Romania
razvan@bocu.ro

Abstract. Development of IoT, mobile devices and AI have triggered 5G network deployment. The new standard of telecom communication will exceed the limitations of the existing network and create new ecosystem with incorporated different industries, including emergency and critical services, security, healthcare. A new standard, design, solution always triggers new threats. When we are talking about mobile devices, generally they are used for everyday activities by end users. Therefor the risk of PII, sensitive data leakage is significantly increased. Our research was oriented on assessing end user privacy related to LBS threats in 5G networks. During the research, different experimental works were done to compare which method of locating device is less noisy for determining end-user location without their additional permissions. According to the results of our experimental study, cell-towers for device locating is more effective than using standard approach - GNSS method. By theoretical studies and practical analysis of 5G telecom standard, study shows that high-band from operating spectrum, can be used to determine device location using only with one cell tower, instead of standard approach – by the information from minimum three visible cell-towers. The only thing which can limit the process is switch function, which is responsible for smooth roaming between towers. We have intercepted the switching process and forced device to stay connected on mmWave tower, without measuring the signal strength. Our research objectives were: how 5G network architecture effect on UE location privacy? Which operating spectrum of 5G network is more vulnerable and to assess the scale of this vulnerability.

1 Introduction

World leading manufacturers, operating in tech market, are struggling to make their products more portable, mobile to expand their use case. Extensive development of portable devices like smartphones, IoT, microcomputers (raspberry pi, Arduino) approve their importance and strengthen their role in everyday life. Developing mobile devices acts as catalyst for the improvement of telecom standard. Hence the engineers have started working on 5th generation network.

New telecom standard of communication, 5th generation network, will be more diverse by incorporating different industries into one huge network. Therefore, this arises cyber threats against the network. The goal of our research is to reassess location-based vulnerabilities (in terms of UE - User Equipment) in 5G ecosystem. By satisfying three KPIs 5G standard: Enhanced mobile broadband, massive machine type communications and ultra-reliable low latency communication new standard will exceed the limitation of telecom communication and will start new era in mobile communication. [1] Hence, network engineers have to make some software and technical changes to achieve requirements provided by 3GPP. Research was done on these modifications, new solutions to assess how software/technical improvement effects on security level of devices in terms of location tracking. Our study objectives were:

- How 5G network architecture effect on UE location privacy?
- Which of the three operating spectrum is more vulnerable?
- What is the scale of this vulnerability?

2   Literature Overview

Ideas overviewed and developed in this paper are fulfilled with the latest scientific articles. Technical documents, official reports, standard analysis and overviews from world leading companies, mobile network operators were considered. Experts have discussed major improvement, technical and design changes of 5G network, their research results, and the various threats. For the experimental works done during the research, we have used our self-written tools and open-source projects, available on internet. Our research was done on basis of theoretical analysis and was approved with practical experiments. Majority of the similar research are oriented on standard method of device tracking – triangulation or trilateration. How to improve method efficiency. Authors are discussing the signal strength parameter and how it can be used to calculate approximate location of the device. Some of them approve of the idea that it can be done with high accuracy but some of them think that system is so complex, it cannot be exactly determined by which factor was lessen the strength of signal: by distance or by limitation factors. The main difference with other articles and our study is that we are locating devices by guessing the frequencies on which the device is operating, and intercepting device attach request to cell-towers. Manipulating with frequencies has been successfully done by other researchers too. However, working with operating spectrum and its usage in tracking device has not been found in scientific or practical journals yet.

The idea of manipulating with radio frequencies is fulfilled with the studies about how people are fighting with RF pollution. How engineers are trying to minimize the RF exposure indoors by using different building materials. Processing the information provided with their research approves that high frequencies can be used for tracking the device. Which was approved by our experimental work.

3   Techniques for Locating Devices

There are well-known methods for locating, tracking devices. Tracking means to determine device's precise location and it's variations in a specific time period. Concept of resolving device location for different methods is same: we need a reference system, by which UE calculates its coordinates related to it (reference system). Generally, GNSS satellites or cell-towers are used for reference systems. By processing data and measuring signals tracked from GNSS satellites or cell-towers UE calculates it's coordinates. Usually, arrival time, frequencies, signal strength and angle are used to solve device location. When we use mobile cell-towers for the calculation, level of accuracy depends on telecom service providers, how accurate do they configure cell-towers. In some regions, for emergency services like 9–1-1 or 1–1-2 government regulates and requires certain accuracy of cell-towers. [2].

Mobile positioning is used every day: for checking in some places (for social media); for navigation; emergency/critical service, for the marketing purposes like advertising and so on. However, sometimes hackers or illegally motivated people are trying to track devices, end-users without prior permissions. During the study, we analyzed GNSS techniques and cell-tower method of locating/tracking devices.

The most accurate and precise method for calculating device's location (on the earth) is Geodetic technique – GNSS (Global Navigation Satellite System). GNSS satellites send signal from the space, which is received by GPS enabled devices. By processing signal data (Fig. 1) devices solve their real location. Mm level accuracy and high precision can be achieved only by scientific, geodetic level GNSS systems. Because they support multiple type signal, by which they can overcome ionospheric affect and various type noises. User friendly devices is compatible (usually) only with L1 band, so they provide up to 3–5 m accuracy. But, it is enough to track the devices. However, GNSS method also has some limitations. For our case study, the most interesting is that it could not be used indoor. Devices must have direct view with satellites and the satellites should have good geometric positions. In some articles, this method is called Global Positioning System - GPS. GPSs are satellites operated by Unite States. As they were the first satellite providers, GPS is used instead of GNSS.

Assisted-GPS (A-GPS) is a method where cell-towers are used for locating device. That is the best solution for in-door use, like in building or for underground use. But it has a lower accuracy than standard GPS method. (Fig. 2).

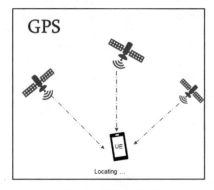
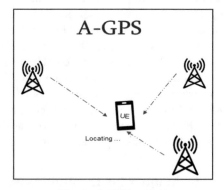

Fig. 1. GPS method   **Fig. 2.** A-GPS method

GSM network operators try to setup a good network coverage by installing cell-towers with specific geometry. By knowing their x and y coordinates, the device can calculate its location(need minimum three visible cell-towers coordinates). There are two well-known techniques to use: triangulation and trilateration. In triangulation two lines from the cell-towers to UE and 3rd line between cell-towers are used. This will create the triangle (Fig. 4). Sides of cell-towers and Alfa, Beta angles are known. While, in case of trilateration, distances are recalculated from each tower and common area represents estimated coordinate of device. (Fig. 3) In some papers trilateration is called - distance measuring techniques. [2].

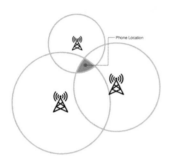

Fig. 3. Trilateration method   **Fig. 4.** Triangulation method

Figure 3 describes trilateration method. Green circles represent possible locations of device (maximum coverage area of the tower) for a specific radius of tower. To solve the equation, we need to compute the intersection of the circles.

For simplifying the calculation process, 2D model is used (x, y):

1. Equations per circle:

$$(x - x_1)^2 + (y - y_1)^2 = r_1^2$$

$$(x - x_2)^2 + (y - y_2)^2 = r_2^2$$

$$(x - x_3)^2 + (y - y_3)^2 = r_3^2$$

2. Open the parentheses for each equations:

$$x^2 - 2x_1x + x_1^2 + y^2 - 2y_1y + y_1^2 = r_1^2$$

$$x^2 - 2x_2x + x_2^2 + y^2 - 2y_2y + y_2^2 = r_2^2$$

$$x^2 - 2x_3x + x_3^2 + y^2 - 2y_3y + y_3^2 = r_3^2$$

3. Rewrite the system with A,B,C,D,E,F

$$A_x + B_y = C$$

$$D_x + E_y = F$$

4. Solution for the system is:

$$x = \frac{CE - FB}{EA - BD}$$

$$y = \frac{CD - AF}{BD - AE}$$

This is a simplified 2D version of the trilateration [3].

4 Experimental Work

4.1 Obtain GNSS Data from Devices

We have simulated various cases to find the easiest way of device tracking. Market is full with the software for stealing data from device. For our study, we used storm-braker (Fig. 5) [4].

Fig. 5. Experimental work

Which is open-source and runs on Linux. Storm-Braker generates malicious link for the victim. After opening the link, hacker gets device lat/long coordinates. However, there it has limitations: the method is too noise, as it requires enabled GPS module and the permission from user. When the tool tried to steal GNSS data, end-users were alerted several times: "software/Link/Webpage wants to use device's location" (Fig. 6). It should be mentioned that activating GPS module is not mandatory for devices, so rest of users disable GNSS module to optimize battery usage.

Second important note is that, usually device does not track GNSS signals in background. Hence, when we need to track them using satellites, we have to enable GPS module and startup measurements process on device. Because of the security aspects, operating systems (like Android, Windows, iOS) automatically draws sign of "location", alerting user that GNSS measurement has been started. Therefor this method is very noisy. Also, because of the GNSS method limitations, if the victim is in building or at any other location, where they do not have "open sky", hacker cannot locate using satellites. Hence, according to the research this method is not the best solution for device tracking without user permission.

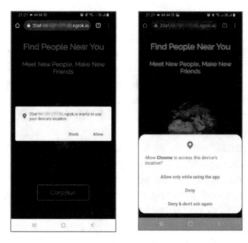

Fig. 6. Experimental work (GNSS)

4.2 Collect Cell-Towers Using Smartphones

Cell-towers spread their information, including Lat/Long coordinates, IDs by defaults. We used this information to collect and map all available towers in the city.

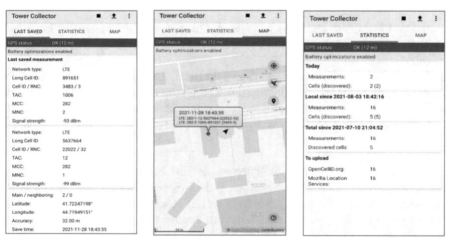

Fig. 7. Collecting cell tower information

During the experimental work we used "Tower Collector", which is available on "Play Store" market. Figure 7 describes a specific tower. Network type shows that this is LTE network tower. RNC (Radio Network Controller) is responsible for managing and controlling connected NODE BS. It should be mentioned that encryption is done at this level. LAC/TAC (for UMTS/LTE networks) is the unique identifier for current location area. Mobile Network Code (MNC) identifies network operator. Mobile Country

Code (MCC) identifies the country. Cell-ID represents unique ID for identification Base Transceiver Station (BTS) or sector for the specific LAC [5].

The most arguable factor is the signal strength parameter. There are various articles about its value in solving device location. However, it is too complicated: as the signal can be affected by numerous things. So, low signal strength does not always represent long distance from towers, it might be caused by any influence that affects radio waves, for example: buildings. Figure 14 shows the results (device location and towers) on map. Figure 15 presents statistical information.

As detailed information is provided by telecom towers, attackers can map whole network and try to track the devices. The only data to collect from victims is visible towers. The most important note is that the device connected to telecom network constantly searches for cell-towers and switches to the tower with the strongest signal. Hence, catching that information from device, attacker can calculate approximate coordinates of the device. The process of tracking cell-towers runs in background on devices, so hackers do not need to start any additional software or any hardware module (which is required for GNSS method) for that reason.

4.3 Locating Device Using MmWave (High-Band)

5G Network standard requires some major changes to satisfy the requirements of 3 GPP. For our research area, the most interesting is operating spectrum. Which is divided into 3, following categories:

1. Below 1 GHz, also known as Low-Band. This range is less affected by buildings. Therefor it is used in urban areas. But, also has the bandwidth limitation: up to 100 Mbps.
2. From 1 GHz to 6 GHz, also known as Mid-Band. This level has better bandwidth, its about up to 1 Gbps. But, the signal from this range is more affected by buildings.
3. From 6 GHz to 100 GHz (High-Band / mmWave). Maintain the best bandwidth in 5G Network - 10Gbps. On the other hand, this range is critically affected by buildings. [1]

According to our research, limitation of the High-Band can be used for locating devices by only on tower instead of three visible towers. Merging the two factors: 1. High band is highly affected by buildings and 2. Device always scan cell-towers to find the one with the strongest signal, we can assume that device has to be close to the mmWave tower to connect it. Otherwise, it will connect to the tower from different categories. So, if we steal the information from device when it is connected or is in the range of mmWave tower, we can calculate estimated location. Therefore, we do not need to get information about three visible towers or get the data of a minimum of three visible GNSS satellites.

Knowing that devices will always connect to the tower with the strongest signal, and mmWave technology is highly affected by building, we assume that in most cases devices will not be connected to high-band, most vulnerable category. So, we have simulated the cases when attacker controls device Switch Function, which is responsible for smooth roaming between towers. We manually made devices to stay connected on mmWave towers. We have created a simulated map of cell-towers. When the device could not connect to the high-band tower, it means that it was outside the coverage area. In real life (in countries with 5G network service) this means that devices are not in populated areas. However, with a little bit more processing, known that device is not in high-band tower coverage area, also gives the possible locations. For the experimental work, we have used Raspberry PIs (Fig. 8), with GPS module and 5G antennas, Kali OS and simulation tools/software (Table 1) .

Table 1. Details of experimental work

Device	Quantity	Usage
Raspberry Pi with LTE and GPS Modules	10	5-For Base Station 3 - For Fake Base Station 2 - For User Equipment
Smartphone with GPS support	5	For User Equipment
Laptop	2	Manage and Monitor Experimental Work
Results		
Algorithm Type	**Success/Fail**	**Comment**
GPS (Catch data from UE)	Success	Success with noise if GPS module was enabled. User interaction was needed. As they were alerted by the system
A-GPS (Catch data from UE)	Success	10/10
MITM by Fake towers	Success	10/10
Stealing Frequency Info	Success	8/10

Fig. 8. Raspberry Pis, with GPS and 5G Modules

An important point related to location-based services in 5G Network is that new standard is not safe from MITM type of attack. [6] So, the chance of fake towers is real. As it was explained in previous section, during the emergency telecom towers are used to the determine location of UE. Results, without exception, depend on the input data quality. A crucial detail in our equation is the Lat/Long coordinates of towers. So, when the network suffers with fake towers (MITM), which sends falsificated coordinates, the calculated location for device will not be accurate.

By the experimental activities, we have compared our new approach of tracking devices with existing one.

- We have analyzed High-Band, mmWave frequencies and its limitation by the building, building materials and other objects.
- We sniffed the attach request, sent by user device. Where all necessary parameters of the device and cell-towers are provided
- We have forced the device to always connect on High-Band frequencies in 5G networks
- By knowing locations of the towers, also which cell-tower is serving target device and limitations of high-frequency radio waves, we have tracked user by only one tower.

5   Conclusion

Development of 5G Network will play a crucial role for existing services and for future progress. Securely and successfully implemented networks will clear up existing limitations and will create new possibilities. Considering the scale of 5G target market, interest of hackers will totally arise. Therefore, working on security protocols, policies, and design of 5G network should be on the highest priority. Our theoretical and practical analysis approved location-based vulnerabilities in 5G networks. The study was oriented to assess device location identification in 5G networks. Hence, after the theoretical analysis, we have conducted experimental works.

According to our results, the most precise method of locating device – GNSS technique, is the noisiest, as the operating systems alert users when a third party tries to steal location, GPS data. Also, it is not effective when the GPS module is disabled. The second experiment showed that, stealing A-GPS information from the device is less noisy than previous one, but needs information from minimum three visible towers for solving well-known equation: trilateration. The most interesting experiment was about using the limitation of 3rd category of operating spectrum in 5G network, to calculate device location. By which we located user equipment only with one tower, instead of three towers. However, this attempt might be interrupted by Switch Function of the device, which tries to find the tower with the strongest signal and connect to it. So, we tried to manually control the switching function and leave the device on third operating spectrum, despite the signal strength. Our simulated experiments were successful and approved the location-based vulnerabilities in 5G networks, that should be solved before large-scale deployment.

Acknowledgment. The work was conducted as a part of PHDF-21–088 financed by Shota Rustaveli National Science Foundation of Georgia.

References

1. Huawei Technologies CO., LTD. in "5G Network Architecture – A high Level Perspective (2016)
2. Asad Hussain, S., Ahmed, S., Emran, M.: Positioning a mobile subscriber in a cellular network system based on signal strength. IAENG Int. J. Comput. Sci. **34**, 2 (2007). https://www.res earchgate.net/publication/26492533
3. Cell Phone Trilateration Algorithm: Online Journal "Computer Science (2019). https://www.101computing.net/cell-phone-trilateration-algorithm/. Accessed 10 Dec 2021
4. Ultrasecurity, "Strom-Breaked" (Software Package). https://github.com/ultrasecurity/Storm-Breaker. Accessed 8 Dec 2021
5. Johhny: How to find the Cell Id location with MCC, MNC, LAC and CellID (CID) (2015). https://cellidfinder.com/articles/how-to-find-cellid-location-with-mcc-mnc-lac-i-cellid-cid
6. Iavich, M., Akhalaia, G., Gnatyuk, S.: Method of improving the security of 5G network architecture concept for energy and other sectors of the critical infrastructure. In: Zaporozhets, A. (ed.) Systems, Decision and Control in Energy III. SSDC, vol. 399, pp. 237–246. Springer, Cham (2022). https://doi.org/10.1007/978-3-030-87675-3_14
7. Maheshwari, M.K., Agiwal, M., Saxena, N., Abhishek, R.: Flexible beamforming in 5G wireless for internet of things. IETE Techn. Rev. **36**(1), 3–16 (2017). https://doi.org/10.1080/02564602.2017.1381048
8. Ivezic, M., Ivezic, L.: 5G Security & Privacy Challenges. In: 5G Security Personal Blog (2019). https://5g.security/cyber-kinetic/5g-security-privacy-challenges/
9. Shaik, R.B., Park, S., Selfert, J.P.: New vulnerabilities in 4G and 5G cellular access network protocols: exposing device capabilities. In: WiSec 2019: Proceedings of the 12th Conference on Security and Privacy in Wireless and Mobile Networks (2019). https://doi.org/10.1145/3317549, ISBN: 9781450367264
10. Purdy, A.: Why 5G Can Be More Secure Than 4G. In: Forbes online Journal (2019). https://www.forbes.com/sites/forbestechcouncil/2019/09/23/why-5g-can-be-more-secure-than-4g/?sh=2ffcdf1657b2
11. Qualcomm Technologies inc.: What is 5G. https://www.qualcomm.com/5g/what-is-5g
12. SK Telecom: 5G architecture design and implementation guideline (2015)
13. Hanif, M.: 5G Phones Will Drain Your Battery Faster Than You Think (2020). https://www.rumblerum.com/5g-phones-drain-battery-life/
14. Samsung in online report: Samsung Phone Battery Drains Quickly on 5G Service. https://www.samsung.com/us/support/troubleshooting/TSG01201462
15. Yusof, R., Khairuddin, U., Khalid, M.: A new mutation operation for faster convergence in genetic algorithm feature selection. Int. J. Innov. Comput. Inf. Control **18**(10), 7363–7380 (2012)
16. The EU Space Programme. https://www.euspa.europa.eu/european-space/eu-space-pro gramme. Accessed 10 Dec 2021
17. Gagnidze, A., Iavich, M., Iashvili, G.: Novel version of Merkle cryptosystem. Bull. Georgian Natl. Acad. Sci. **11**(4), 28–33 (2017)
18. Akhalaia, G., Iavich, M., Gnatyuk, S.: Location-based threats for user equipment in 5G network. In: Hu, Z., Dychka, I., Petoukhov, S., He, M. (eds.) Advances in Computer Science for Engineering and Education, pp. 117–130. Springer International Publishing, Cham (2022). https://doi.org/10.1007/978-3-031-04812-8_11
19. Iashvili, G., Avkurova, Z., Iavich, M., Bauyrzhan, M., Gagnidze, A., Gnatyuk, S.: Content-based machine learning approach for hardware vulnerabilities identification system. In: Hu, Z., Petoukhov, S., Dychka, I., He, M. (eds.) Advances in Computer Science for Engineering and Education IV, pp. 117–126. Springer International Publishing, Cham (2021). https://doi.org/10.1007/978-3-030-80472-5_10

Trusted and only Trusted.
That is the Access!
Improving Access Control Allowing only Trusted Execution Environment Applications

Dalton C. G. Valadares[1,2(✉)], Álvaro Sobrinho[3], Newton C. Will[4],
Kyller C. Gorgônio[1], and Angelo Perkusich[1]

[1] UFCG/VIRTUS RDI Center, Campina Grande, PB, Brazil
dalton.valadares@embedded.ufcg.edu.br, kyller@computacao.ufcg.edu.br,
perkusic@dee.ufcg.edu.br
[2] Federal Institute of Pernambuco, Caruaru, Brazil
[3] Federal University of the Agreste of Pernambuco, Garanhuns, Brazil
alvaro.alvares@ufape.edu.br
[4] Federal University of Technology - Paraná, Dois Vizinhos, Brazil
will@utfpr.edu.br

Abstract. Security concerns should always be considered when deploying distributed systems that deal with sensitive data. Generally, the software components responsible for storing these sensitive data are protected, having access control systems to allow or deny external requests. A Policy Enforcement Point (PEP) Proxy is one of these systems which allows or denies access to protected data by checking if the requester is authorized and has permission to access. Despite these two validations about the requester (authorization and data access permission), the traditional PEP Proxy does not guarantee anything more about the requester which will process the data. This work proposes an improvement to the PEP Proxy protection in a way that it can also verify if the requester runs on a Trusted Execution Environment (TEE) application. A TEE is responsible for trusted computing, processing data in a protected region of memory, which is tamper-resistant and isolated from external resources, and keeping code and data protected even if the operating system is hacked. The Trusted PEP Proxy (TruPP) performs the remote attestation (RA) process to guarantee that the requester runs on a TEE. We created a Coloured Petri Net (CPN) model to help validate our proposal by checking some security properties.

1 Introduction

The advances in the Internet of Things (IoT) technologies have attracted the attention of industries, researchers/academies, and different kinds of customers. Since IoT devices are increasingly smarter, they are frequently called smart objects [8,12]. Despite this, these smart objects have specific constraints regarding battery, memory, and processing capabilities. Thus, a requirement for many

© The Author(s), under exclusive license to Springer Nature Switzerland AG 2023
L. Barolli (Ed.): AINA 2023, LNNS 655, pp. 490–503, 2023.
https://doi.org/10.1007/978-3-031-28694-0_47

applications is the use of a more robust device, generally called as gateway, with more processing and memory power and fewer constraints.

According to the local where this gateway is placed/deployed and the pre-processing is performed, we have the concepts of edge and fog computing, which are alternatives to cloud computing. When the gateway is located together with the IoT devices, with local data processing, this scenario is named edge computing [23]. When IoT applications demand even more processing power, with increasing restricted time requirements, we can deploy specific servers locally to provide the needed resources for gateways and IoT devices instead of sending data directly to the cloud. This scenario is the basis of fog computing, in which IoT devices collect data from the environment and send them to a fog server. This server preprocesses these data before sending them to a cloud server, for instance [9,24]. Fog computing can reduce the latency when certain situations require fast responses in the field, and it can also decrease costs with cloud processing and communication.

Regardless of edge or fog computing, IoT applications often have to deal with sensitive data, like Personally Identifiable Information (PII), such as medical, financial, or educational information. These kinds of information can include name, fingerprint, telephone number, and social security number, which requires special care regarding overall data security. A particular type of PII is named Protected Health Information (PHI), which is related to health information from patients in healthcare organizations. This information has become an attractive target for criminals since the PHI records do not include only health data but also valuable personal information (e.g., address or social security number).

The interest in these data is even greater because they do not use to change, unlike credit card and bank account numbers, which can be canceled. In addition, while credit card numbers can be sold for US$1 on the black market, and PII can be sold for US$10 up to US$20, medical information such as PHI can reach between US$20 and US$50, being the most valuable kind of information [14]. The number of PHI breaches reported by the US federal government in 2016 reached the health information of more than 15 million people, with the biggest case happening in Arizona through a network server hacking attack, which involved more than 3.7 million patients' health information [18]. Also, in the same year, a successful ransomware attack forced a hospital in Los Angeles to pay US$17000 in order to get its resources and network released.

Another known scenario that demands data protection, although it does not deal with PII, is a smart metering application that is responsible for periodically measuring the energy consumption of a place. Suppose an adversary gets access to a data set containing energy consumption measurements from a specific period (e.g., a month). In that case, this adversary can use a Non-Intrusive Appliance Loading Monitoring (NIALM) technique to estimate what appliances were being used at a specific time and, by doing this, estimate how many people were in that place and what these people were doing [15,24].

Security mechanisms should be adopted to avoid this kind of problem, considering, for example, data and communication channel protection. Encryption techniques can be used to protect data, and the communication channel can be secured through TLS (Transport Layer Security). Besides, access to sensitive data must be controlled, allowing only authenticated and authorized parties. A common way to control data access is deploying a Policy Enforcement Point (PEP) Proxy before the data service that needs to be secured. A PEP Proxy is responsible for intercepting requests to a protected service and checking if this request comes from an authenticated and authorized party. The intercepted request usually contains a token that should be validated with an Identity Management (IdM) system to check if the requester was authenticated and is authorized. In a positive case, the requester gets access to the data service; otherwise, the access is denied, and the data remain safe [24].

Although data can be accessed only by authenticated and authorized requesters, there is no reason to trust their machines where data are processed since there is the risk that an adversary can get access to their operating systems with high-level privileges and thus get access to sensitive data. We established two research questions as guidance to deal with this concern and achieve a solution:

1. How to improve security by decreasing the distrust in the data consumers/requesters?
2. How to improve access control in a way that only trusted parties can access sensitive data?

In this sense, Trusted Execution Environments (TEE) have become frequently adopted to increase the security of data and systems. A TEE is enabled due to an extension of the processor instructions set, which allows the creation of shielded memory spaces isolated from the rest of the system. The two TEE technologies more common in the market and the scientific communities are ARM TrustZone and Intel Software Guard Extensions (SGX). The protected memory regions are commonly called "secure world" (TrustZone) or "enclaves" (SGX), and even an adversary with high-level privileges should fail when trying to access these specific memory regions. Attestation protocols allow third parties to validate that an application is running on a valid TEE. These protocols provide specific information about the secure world/enclave, verifying its authenticity by enabling third parties to check and validate this information [13,27].

In this work, considering both research questions, we propose an improvement to the PEP Proxy in a way that it can also perform the attestation process. Thus, after authentication and authorization checking, the attestation process is initialized to verify if the requester application, which is interested in sensitive data, runs on a TEE and, therefore, is considered a trusted entity. We created a Coloured Petri Net (CPN) model representing the communication flow with a PEP Proxy that also performs the attestation process besides authentication and authorization. With the CPN model, we can validate security properties. Thus, the main contribution of this solution is that we improve the trust in data consumers/requesters by allowing sensitive data distribution only to trusted entities, which means that these entities were previously authenticated, authorized, and attested.

The remainder of this paper is organized as follows: we present the background in Sect. 2; we describe the fundamentals of our proposal, a trusted access control, in Sect. 3; we demonstrate the CPN model for our proposal in Sect. 4; in Sect. 5, we present the related works, and we finish this paper presenting some final considerations in Sect. 6.

2 Background

This Section briefly describes the essential concepts we use, such as Policy Enforcement Point (PEP) and Trusted Execution Environment (TEE). Besides, we also describe the Coloured Petri Nets as we used this formalism to model our proposal and verify security behaviors.

2.1 Policy Enforcement Point (PEP) Proxy

A PEP Proxy works as a gatekeeper controlling the access to a protected service or resource, verifying and validating the requesters to allow or deny their accesses. Generally, a PEP acts together with some policy-based access management system, which can be a Policy Decision Point (PDP) or an Identity and Access Management (IAM) system.

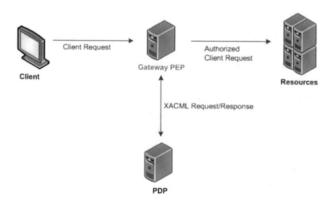

Fig. 1. Architecture with a Policy Enforcement Point [17].

Figure 1 shows a basic architecture using a PEP. When the PEP proxy/gateway receives a request, it checks the access policies with the PDP or IAM system to determine whether the requester has permission to access the protected resources. Since the PEP enforces the access policies application, it forwards the request to the protected service or sends an unauthorized message to the requester, respectively, whether the access was allowed or denied. An IAM is also known as IdM (Identity Management system).

2.2   Trusted Execution Environment and Intel SGX

Despite the many definitions found in the literature, a TEE can be defined as a processing environment, tamper resistant, that runs on separated kernel space [19,25,26]. Besides guaranteeing an isolated execution, TEE also provides secure storage and an attestation mechanism that proves its trustworthiness for third parties [22,30].

Software Guard Extensions (SGX) is the TEE implementation from Intel. SGX is a set of instructions and mechanisms to provide data confidentiality and integrity. Sensitive data and code are placed in a container called *enclave*, which is a cipher memory area protected by the CPU against external code, even the BIOS, kernel, or hypervisor code [30].

SGX has a reduced Trusted Computing Base (TCB) involving a piece of hardware and software. The developer chooses which software parts will be placed inside the enclave and that data are protected even if the operating system or hypervisor is compromised. The communication between trusted and untrusted parts of the application is performed by the edge routines, which are defined by the developer [7].

Each enclave has a self-signed certificate by the enclave author, which contains information that allows the SGX to detect if any part of the enclave was tampered with and allows an enclave to prove that it was correctly loaded and can be trusted. The Enclave Measurement computation is a unique 256 bits hash code, which identifies the code and initial data to be placed within the enclave, the order and position in which they are to be placed, and the security properties of those pages. A change in any of these variables will result in a different measure. [7].

To enable two or more enclaves to share data securely, the SGX also supports Remote Attestation (RA), which is the process of proving that software was properly initialized on a TEE platform. In the case of Intel SGX, RA is the mechanism by which a third party confirms that a particular software entity is running on an Intel SGX-enabled platform and protected within an enclave before provisioning that software with secrets and protected data.

2.3   Coloured Petri Nets

CPN is a graphical and formal language for modeling and validating concurrent systems [11]. CPN extends the classical Petri nets with a functional programming language (i.e., CPN ML language). A CPN model includes elements such as places (graphically represented by an ellipse), transitions (graphically represented by a rectangle), arcs, color sets, variables, and tokens. Modelers can only connect places to transitions and transitions to places using arcs (never places to places and transitions to transitions). The distribution of tokens among all places represents one marking of a CPN model. The state space of a model contains the set of all reachable markings from the initial one. Readers can use the book of Jensen and Kristensen [10] for more technical details on the syntax and semantics of CPN.

The CPN Tools software provides features to develop and formally analyze CPN models. For instance, CPN Tools enables automatic state space analysis using standard and user-defined query functions based on the CPN ML language. Besides, using a specific library (i.e., ASK-CTL) to conduct model checking [5], modelers can verify and ensure that a CPN model satisfies temporal modal logic formulas.

3   Trusted PEP Proxy

As stated before, although a PEP Proxy can be deployed to protect a sensitive data service, nothing can be inferred about the security of the requester machine, which processes the sensitive data. This Section explains our proposal to improve the protection provided by a PEP Proxy, allowing access to protected service only to TEE applications, which securely process sensitive data.

3.1   Application Scenario

A smart metering application is a scenario considered as the base case for this work. In this case, residences have smart meters responsible for measuring energy consumption and sending this information to be processed by the energy supply company. This application can be divided into producers and consumers: producers are the smart meters that generate energy consumption data, and consumers are the applications responsible for processing these data.

As seen before, this kind of application is susceptible to NIALM attacks, which hit the residents' privacy once their behavior can be inferred by someone that analyzes the residences' energy consumption in a specific period.

3.2   Threat Model

As the case base application contains three main components (producers, data storage service, and consumers), the attack surface considers data in transit between any of these components, data in rest inside the data storage service, and data in processing, mainly in the consumers. This work's focus is the improvement of the PEP Proxy. Thus we consider the data are encrypted in the producers and decrypted in the consumers, with the cryptographic keys exchange process out of scope (not described here). With this deliberation, the attack surface for this work considers only the consumers since the main worry is regarding the data to be processed once consumers can be a target of attacks, and attackers can get access to the sensitive data.

3.3   Design Principles

As a standard PEP Proxy works validating just if a specific requester is authorized to access a specific service, we decided to improve in a way the data owner could decrease the concern regarding possible attacks in the requester machine during data processing. For this, the application requires that the data service be accessed only by requesters running TEE applications, ensuring a better security level for data during processing.

Thus, the PEP Proxy should also validate if the requester runs a TEE application, which will perform secure data processing. To accomplish this desired improvement, the PEP Proxy should execute the remote attestation (RA) process to check if the requester runs a valid TEE application. To validate our idea, we created a Coloured Petri Net (CPN) model representing all three main components (i.e., producer, protected data service, and consumer) and the communication flow between them. The CPN model also considers the authentication, authorization, and attestation process at a high level. With this model, we can verify security properties by simulating scenarios where consumers run and do not run with TEE, for instance.

3.4   Trusted PEP Proxy (TRUPP)

Since the improved PEP Proxy verifies if the requester runs on a TEE, allowing only trusted applications to get access to the protected data service, we call it Trusted PEP Proxy (TRUPP). Figure 2 shows an architecture with the deployment of the TRUPP.

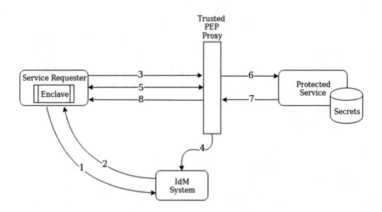

Fig. 2. Architecture with the Trusted PEP Proxy

The communication flow between the components seen in Fig. 2, which is enumerated, is described below:

1. The service requester, which is supposed to run a TEE application (inside an Intel SGX enclave, for instance), authenticates with the Identity Management (IdM) System, informing valid credentials;
2. After the authentication process, validating the service requester credentials, the IdM System returns a valid OAuth2 token, which will be used later to check authorization (access permissions);

3. The service requester sends a request to the protected service, which is intercepted by the TruPP, informing the OAuth2 token received from the IdM System after the authentication process;
4. The TruPP verifies the OAuth2 token with the IdM System, checking if the service requester has the right permissions to access the protected service;
5. Once the service requester has access permissions to the protected service, the TruPP starts the RA process to verify if the service requester runs a TEE application;
6. After the RA process is finished, if it succeeded, the request reaches the protected service, which will process the return;
7. The protected service returns the requested data to the TruPP;
8. The TruPP forwards the requested data received from the protected service.

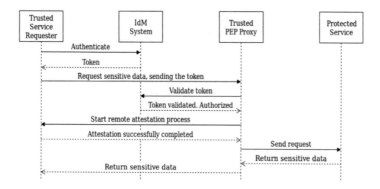

Fig. 3. Message sequence diagram

The main difference between a standard PEP Proxy and the improved TruPP is step 5, which is the execution of the RA process, checking for a TEE application in the service requester. We can see the message sequence diagram of this communication flow in Fig. 3. As explained, the trusted service requester authenticates in the IdM System and receives a valid OAuth2 token. Then, it sends a sensitive data request to the protected service, which the TruPP intercepts. The TruPP then validates the OAuth2 token with the IdM System and starts the remote attestation process. Once the trusted service requester is successfully attested, the TruPP sends the request to the protected service and receives back the sensitive data requested. Finally, the TruPP forwards the sensitive data to the trusted service requester, which will securely perform the processing (inside an enclave, i.e., a trusted application).

4   Formal Model and Analysis

We used the CPN formal modeling language to represent and validate the behaviors of our solution. Figure 4 presents the main module of our hierarchical CPN model. The model comprises four sub-modules: *Service Requester with*

TEE Protection, Trusted PEP Proxy, IdM System, and *Protected Service.* The marking of the *Entity* place represents that there is one valid entity to request sensitive data and one fake entity that should not be able to consume sensitive data.

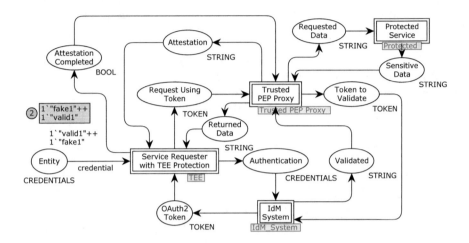

Fig. 4. Main module of the hierarchical CPN model.

Figure 5 presents the sub-module related to the substitution transition *IdM System.* The IdM system maintains valid identities to enable the validation of credentials. Besides, the system validates the token received from the trusted PEP proxy. The other sub-modules are detailed representations of the solution presented in the previous section (e.g., see Fig. 3). We omitted the remaining sub-modules due to space constraints.

To increase confidence in our solution, we formally verified the model using the state-space analysis tools available from the CPN Tools software. We considered the initial marking of the model that represents one valid entity able to request sensitive data and a list of valid identities registered in the IdM system. Figure 6 shows two examples of CPN ML functions used to analyze if the model presents desired behaviors, protecting the consumption of sensitive data. CPN ML is a functional programming language defined based on the ML standard and embedded in the CPN Tools software. We defined the CML ML function of Fig. 6a to verify if the consumption of sensitive data is denied when a valid entity sends a request without the existence of a valid TEE (i.e., the entity failed the attestation process). The standard CPN ML function *PreAllNodes* conducts an exhaustive search throughout the state space, verifies the predicate (represented using a new CPN ML function defined by a modeler) for each state, and returns the states satisfying the predicate. We also defined the CPN ML function of Fig. 6b to analyze the remaining states in which a valid entity requests and consumes sensitive data after passing the attestation process (i.e., the entity can consume sensitive data).

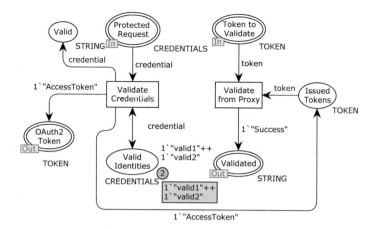

Fig. 5. Sub-module of the hierarchical CPN model, representing the IdM system.

```
fun failedOrReadyForAttestation n =
  let
    val valid = Mark.IdM_System'Valid 1 n
    val result = Mark.Trusted_PEP_Proxy'Result_Att 1 n
    val received = Mark.TEE'Received 1 n
  in
    (valid == 1`"valid1") andalso (result == empty)
    andalso (received == empty)
  end;

val invalid = PredAllNodes (fn n =>
                  failedOrReadyForAttestation n)

                    (a)
```

```
fun canConsume n =
  let
    val valid = Mark.IdM_System'Valid 1 n
    val result = Mark.Trusted_PEP_Proxy'Result_Att 1 n
    val received = Mark.TEE'Received 1 n
  in
    (result == 1`true) andalso (valid == 1`"valid1")
  end;

val invalid = PredAllNodes (fn n => canConsume n)

                    (b)
```

Fig. 6. Examples of CPN ML functions for verification of desired behaviors

After running both functions, we observed that all states are covered because, from the 13 existing states, nine represent that the entity is ready for attestation or failed the attestation. Besides, five states represent the entity that passed the attestation and is prepared to consume (or have already consumed) sensitive data. Although we only consider a unique, valid entity in our initial marking of the model (which leads to only 13 states), it is enough to formally analyze the expected behaviors for protecting sensitive data. Adding more valid or fake entities would only increase the number of states in the state space without negatively impacting our previous analysis. Our model also guarantees that only a valid entity can request sensitive data for consumption.

5 Related Work

In this section, we present some works that apply Policy Enforcement Point (PEP) Proxy and Trusted Execution Environment (TEE) to improve the security of sensitive data applications.

Bruno et al. [4] used a PEP Proxy written in Python to improve access control in IoT applications that use the MQTT protocol. PEP Proxy acts between IoT nodes and the MQTT broker, intercepting the messages sent from the nodes to the broker and from the broker to the nodes. A Policy Decision Point (PDP) checks all the requests. The authors performed an experimental evaluation and concluded that their solution presents an overhead of 100ms, considered a viable performance. Achillas et al. [1] used the FIWARE Wilma PEP Proxy to control access in a fleet management system responsible for optimizing task and time management in agricultural material handling operations. Valadares et al. [24] also applied the Wilma PEP Proxy to control access of data producers and consumers to protected services. Only authorized producers and consumers have access to a data broker and a key distribution system in their solution.

Shaghaghi et al. [21] proposed PEP as a Service (PEPS), aiming to protect applications in a Software Defined Network (SDN) domain. The authors implemented a prototype and reported an analysis regarding the challenges and performance of the PEPS, suggesting improvements for future work. Al-Hasnawi et al. [2] presented a Policy Enforcement Fog Module (PEFM), which aims to protect data in fog-based IoT applications. The PEFM employs a PEP and is positioned between fog servers and IoT sensors, controlling data access through the enforcement of privacy policies. They evaluated the PEFM in different scenarios (local and remote) through simulations, and the results showed acceptable performance overhead. Shahraki et al. [20] proposed a decentralized multi-authority attribute-based access control (DMA-ABAC) model to protect healthcare data. The DMA-ABAC uses a PEP too, and enables authorities to control their security settings directly. The proposal aims to resist against replay and third-party storage attacks and also attribute collusion, allowing users to access healthcare data according to security regulations.

Condé et al. [6] used the Intel SGX to implement a novel file protection scheme, protecting authentication credentials in a common authentication framework (PAM). According to the authors, tests with the implemented proposal presented acceptable performance overhead. Will et al. [29] propose the use of TEE to enforce the security of IPC mechanisms, creating a trusted communication channel between the applications and ensuring the confidentiality and integrity of data exchanged between them. Valadares et al. [24] proposed using TEE to establish a trusted architecture for IoT data dissemination with protection, the Trusted IoT Architecture (TIoTA). In the TIoTA, two components run on TEEs: a key distribution system and the data consumer. They conducted tests with a TIoTA implementation that uses Intel SGX, resulting in a low latency overhead (3ms on average). A privacy-preserving IoT data aggregation scheme that ensures data and user privacy in the processing of heterogeneous data and in the presence of heterogeneous devices is proposed by [28].

Nguyen et al. [16] proposed a distributed, scalable, fault-tolerant, and trusted logger for IoT device data using Intel SGX. This proposal, called LogSafe, achieved data confidentiality, integrity, and availability, providing tamper detection and protecting against replay, injection, and eavesdropping attacks.

Through experiments, LogSafe presented high scalability and a high data transmission rate. Ayoade et al. [3] presented a decentralized system for data management in IoT applications using Intel SGX and blockchain, enforcing access control with smart contracts and storing the raw data in an SGX application, while the data hashes are stored in the blockchain. According to performed experiments with an implementation of the proposal, the authors conclude that it has an acceptable efficiency.

6   Conclusion

Access control mechanisms are present in many systems to allow access only to authorized parties/entities. A PEP Proxy is often used to intercept requests and verify if the requesters have permission to access the protected data/service by checking tokens, policies, or other attributes. Generally, a PEP Proxy validates if the requester is authenticated and has the authorization to access what is requested.

In this work, we proposed an improvement to PEP Proxies: adding the remote attestation protocol to verify if the requester runs in a TEE, guaranteeing that data will be processed more securely, warranted by the TEE capabilities and protections. This way, only authenticated and authorized requesters running in a TEE can access the protected data/services.

To validate our proposal, we modeled a Coloured Petri Net representing the architecture with the main entities: a trusted service requester, an IdM System, a Trusted PEP Proxy, and a protected service. The CPN model considered the communication flow among the entities and the main operations (authentication, authorization, and attestation). With this model, we verified that the proposal behaves as desired, keeping the security properties and improving the data security provided by the PEP Proxy.

For future work, we suggest implementing a PEP Proxy in a TEE application, ensuring that it processes the requests with more security, and also adding timestamps to attested requesters, with an expiration time, to identify when the PEP should perform a new attestation. Besides, we plan to perform model checking of security properties with the CPN model and a few experiments to measure the time and CPU overhead of the PEP improvements.

References

1. Achillas, C., Bochtis, D., Aidonis, D., Marinoudi, V., Folinas, D.: Voice-driven fleet management system for agricultural operations. Inf. Process. Agric. **6**(4), 471–478 (2019)
2. Al-Hasnawi, A., Carr, S.M., Gupta, A.: Fog-based local and remote policy enforcement for preserving data privacy in the internet of things. Internet of Things **7**, 100069 (2019)
3. Ayoade, G., Karande, V., Khan, L., Hamlen, K.: Decentralized IoT data management using blockchain and trusted execution environment. In: International Conference on Information Reuse and Integration. IEEE, Salt Lake City, UT, USA (2018)

4. Bruno, E., Gallier, R., Gabillon, A.: Enforcing access controls in IoT networks. In: International Conference on Future Data and Security Engineering. Springer, Nha Trang City, Vietnam (2019)
5. Clarke, E.M., Henzinger, T.A., Veith, H., Bloem, R.: Handbook of Model Checking, 1st edn. Springer, Cham (2018). https://doi.org/10.1007/978-3-319-10575-8
6. Condé, R.C.R., Maziero, C.A., Will, N.C.: Using Intel SGX to protect authentication credentials in an untrusted operating system. In: Symposium on Computers and Communications. IEEE, Natal, RN, Brazil (2018)
7. Costan, V., Devadas, S.: Intel SGX explained. IACR Cryptology ePrint Archive 2016 (2016)
8. Henze, M., Hermerschmidt, L., Kerpen, D., Häußling, R., Rumpe, B., Wehrle, K.: A comprehensive approach to privacy in the cloud-based internet of things. Future Gener. Comput. Syst. **56**, 701–718 (2016)
9. Hu, P., Dhelim, S., Ning, H., Qiu, T.: Survey on fog computing: architecture, key technologies, applications and open issues. J. Netw. Comput. Appl. **98**, 27–42 (2017)
10. Jensen, K., Kristensen, L.M.: Coloured Petri Nets, 1st edn. Springer, Heidelberg (2009). https://doi.org/10.1007/b95112
11. Jensen, K., Kristensen, L.M.: Colored petri nets: a graphical language for formal modeling and validation of concurrent systems. Commun. ACM **58**(6), 61–70 (2015)
12. Kortuem, G., Kawsar, F., Sundramoorthy, V., Fitton, D.: Smart objects as building blocks for the internet of things. IEEE Internet Comput. **14**(1), 44–51 (2010)
13. Küçük, K.A., Paverd, A., Martin, A., Asokan, N., Simpson, A., Ankele, R.: Exploring the use of Intel SGX for secure many-party applications. In: Workshop on System Software for Trusted Execution. ACM, Trento, Italy (2016)
14. Lee, K.: Despite risks, healthcare IT professionals stick with mobile (2018). https://searchhealthit.techtarget.com/feature/Despite-risks-healthcare-IT-professionals-stick-with-mobile
15. McLaughlin, S., McDaniel, P., Aiello, W.: Protecting consumer privacy from electric load monitoring. In: Conference on Computer and Communications Security. ACM, Chicago, IL, USA (2011)
16. Nguyen, H., Ivanov, R., Phan, L.T.X., Sokolsky, O., Weimer, J., Lee, I.: LogSafe: secure and scalable data logger for IoT devices. In: International Conference on Internet-of-Things Design and Implementation. IEEE, Orlando, FL, USA (2018)
17. Oracle: XACML Policy Enforcement Point (2019). https://docs.oracle.com/cd/E27515_01/common/tutorials/authz_xacml_pep.html
18. Rouse, M., Sutner, S.: PHI breach (protected health information breach) (2018). https://searchhealthit.techtarget.com/definition/PHI-breach-protected-health-information-breach
19. Sabt, M., Achemlal, M., Bouabdallah, A.: Trusted execution environment: what it is, and what it is not. In: International Conference on Trust, Security and Privacy in Computing and Communications. IEEE, Helsinki, Finland (2015)
20. Salehi Shahraki, A., Rudolph, C., Grobler, M.: A dynamic access control policy model for sharing of healthcare data in multiple domains. In: International Conference on Trust, Security and Privacy in Computing and Communications. IEEE, Rotorua, New Zealand (2019)
21. Shaghaghi, A., Kaafar, M.A., Scott-Hayward, S., Kanhere, S.S., Jha, S.: Towards policy enforcement point as a service (PEPS). In: Conference on Network Function Virtualization and Software Defined Networks. IEEE, Palo Alto, CA, USA (2016)

22. Shepherd, C., et al.: Secure and trusted execution: past, present, and future - a critical review in the context of the internet of things and cyber-physical systems. In: International Conference on Trust, Security and Privacy in Computing and Communications. IEEE, Tianjin, China (2016)
23. Shi, W., Cao, J., Zhang, Q., Li, Y., Xu, L.: Edge computing: vision and challenges. IEEE Internet Things J. **3**(5), 637–646 (2016)
24. Valadares, D.C.G., da Silva, M.S.L., Brito, A.M.E., Salvador, E.M.: Achieving data dissemination with security using FIWARE and intel software guard extensions (SGX). In: Symposium on Computers and Communications. IEEE, Natal, RN, Brazil (2018)
25. Valadares, D.C.G., Will, N.C., Caminha, J., Perkusich, M.B., Perkusich, A., Gorgônio, K.C.: Systematic literature review on the use of trusted execution environments to protect cloud/fog-based internet of things applications. IEEE Access **9**, 80953–80969 (2021)
26. Valadares, D.C.G., Will, N.C., Spohn, M.A., de Souza Santos, D.F., Perkusich, A., Gorgônio, K.C.: Confidential computing in cloud/fog-based internet of things scenarios. Internet of Things. **19**, 100543 (2022)
27. Wang, J., Hong, Z., Zhang, Y., Jin, Y.: Enabling security-enhanced attestation with Intel SGX for remote terminal and IoT. IEEE Trans. Comput. Aided Des. Integr. Circuits Syst. **37**(1), 88–96 (2018)
28. Will, N.C.: A privacy-preserving data aggregation scheme for fog/cloud-enhanced IoT applications using a trusted execution environment. In: International Systems Conference on IEEE, Montreal, QC, Canada (2022)
29. Will, N.C., Heinrich, T., Viescinski, A.B., Maziero, C.A.: Trusted inter-process communication using hardware enclaves. In: International Systems Conference on IEEE, Vancouver, BC, Canada (2021)
30. Will, N.C., Valadares, D.C.G., de Souza Santos, D.F., Perkusich, A.: Intel software guard extensions in internet of things scenarios: a systematic mapping study. In: International Conference on Future Internet of Things and Cloud. IEEE, Rome, Italy (2021)

A Survey of Intrusion Detection-Based Trust Management Approaches in IoT Networks

Meriem Soula[1]([✉])(iD), Bacem Mbarek[2](iD), Aref Meddeb[3], and Tomáš Pitner[2](iD)

[1] Tunisia Polytechnic School, University of Carthage, BP 743, Al Marsa 2078, Tunisia
meriemsoula7@gmail.com
[2] Faculty of Informatics, Masaryk University, Brno, Czech Republic
bacem.mbarek@mail.muni.cz, tomp@fi.muni.cz
[3] National Engineering School of Sousse, University of Sousse, Sousse, Tunisia
aref.meddeb@eniso.u-sousse.tn

Abstract. Trust is a primary concern, particularly for the Internet of Things (IoT) applications, where constrained devices should be able to effectively detect attacks based on their trust models. The main purpose of this survey is to summarize outstanding research activities on trust-based Intrusion Detection Systems (IDS) in IoT networks. We first summarize the state-of-the-art of trust approaches in IoT according to the trust assessment method, i.e., behavior analysis and recommendation systems. We then discuss major external and internal threats and enumerate the most likely and potentially dangerous attacks on IoT networks to stimulate research efforts in designing and developing effective and efficient IDS systems. Further, we present the state of the art and new insights into trust-based collaborative IDS systems.

Keywords: IoT · Intrusion detection · Security · Trust

1 Introduction

The Internet of Things (IoT) plays an important role in the majority of our daily lives. The huge number of IoT devices that are interconnected to the internet may raise concerns about private information protection [1]. Therefore, IoT application security comes with a set of serious challenges, like how to protect the privacy of its users ? How to detect attacks over the network ? and What is the most viable Intrusion Detection System (IDS) in IoT network ?. An IDS would detect adversaries that are injecting malicious activity into the IoT devices [1]. There are several IDS approaches that have been developed to find intruders who have crossed the border of the IoT network. However, many up-to-date IDS techniques are not effective against some attacks [3–5]. To mitigate the risks of attacks, different IDS techniques use a variety of approaches such as Artificial Intelligence (AI), Machine Learning (ML), fuzzy logic, and trust approaches. Trust is one of the most effective approaches to detect intrusions

© The Author(s), under exclusive license to Springer Nature Switzerland AG 2023
L. Barolli (Ed.): AINA 2023, LNNS 655, pp. 504–517, 2023.
https://doi.org/10.1007/978-3-031-28694-0_48

in a network, as it enables the evaluation of the trustworthiness of a connected device [6]. By combining trust approaches and IDS Systems, various proposed solutions have been proposed in the literature to improve the detection of attacks in IoT networks [7]. Moreover, many validation strategies have been proposed in the literature to validate the performance measurement of an IDS approach [8]. However, combining Trust and IDS may not cover all security requirements. In fact, the current IDS systems are not sufficient due to the myriad of new types of cybersecurity threats and attack vectors in IoT environments. Hence, a fundamental concern is how to design an appropriate IDS system for IoT, which can be a combination of different trust-based IDS approaches.

The purpose of this survey is to stimulate research efforts to develop Trust based IDS approaches for IoT Networks. We present the most popular trust methods and IDS models and discuss their combinations. In Sect. 2, we present the definition of Trust approaches in IoT, and we describe the state-of-the-art solutions that have been proposed to use the these trust approaches in IoT. Section 3 presents a comprehensive review of IDS systems in IoT. In Sect. 4, we discuss Trust based IDS approaches presented in the literature. In Sect. 5, we conclude the paper by giving recommendations on how to design and develop appropriate Trust-based IDS approaches.

2   Trust in IoT

This section outlines the various definitions of Trust in IoT systems and highlights the most outstanding trust approaches proposed by researchers. Trust can be sought as an abstruse concept used in various contexts and is influenced by many factors such as belief, reliability, availability, or ability. Although its importance is widely accepted, there is no consensus on the definition of trust. Therefore, trust should be considered as one key element to the success of IoT [5]. Over the past few years, several researchers proposed trust models for IoT networks that have been widely used to detect untrusted devices by analyzing their behavior and make appropriate decisions, such as isolating these untrusted devices [9].

2.1   Trust Models in IoT

This subsection summarizes the main components used to classify trust models and to evaluate the device's trustworthiness by estimating the value of trust. In [6], the authors introduced a method for classifying trust models in IoT. The classification method uses several key characteristics including *Trust Source*, *Trust Metric*, *Trust Algorithm*, *Trust Architecture*, and *Trust Propagation*. Below, we define each one of these concepts. Table 1 summarizes advantages and disadvantages of different trust models.

The first categorical level is called a *Trust Source*. Trust source is divided into three types: (a) Direct Trust; (b) Indirect Trust; (c) Hybrid Trust. Each IoT device is able to evaluate the direct trust based on its own observation. For example, in [10], the authors proposed a direct trust based on direct observations, where each node collects its own information such as energy consumption, delay, and data transmission. The indirect trust is calculated based on information given by the neighboring nodes. In [11], an indirect trust model based on the observations of neighboring nodes was proposed. Hybrid trust includes a combination of direct and indirect trust approaches [12]. The second categorical level is called *Trust Metric*. This category segregates the trust in two types: Quality of Service (QoS) performances and social interactions, wherein QoS includes a set of parameters such as the Packet Delivery Ratio (PDR), the delay, the error rate; whereas social interaction depends on communications, interactions, friendship, and the community of interest between mutual devices. In [13], the authors proposed a similarity-based trust model, where the trust computational model uses parameters like community of interest, cowork relationship, and friendship similarity in an IoT environment. Another trust-based Software Defined Network (SDN) model is presented in [14], where a clustering method based on QoS and Quality of experience (QoE) parameters is presented.

The third category level is referred to as *Trust Algorithm*. The main algorithms and approaches used to evaluate trust implement Bayesian Analysis, Fuzzy Logic, Weighted Sum, and Machine Learning. **Bayesian Analysis** is a statistical paradigm used to analyze data based on estimating parameters theorem, known as the "Bayes' theorem" [15]. It calculates the value of trust according to the probabilistic distribution and the occurrence of events [4]. This value is presented as a random variable ranging from 0 to 1 [5] to classify the misbehavior of a node. In [16], the authors proposed a Lightweight Trust Management model based on Bayesian analysis and Entropy (LTMBE) for wireless Sensor Networks (WSN). The Bayesian principle is used to compute the direct trust value for the current information and the history information to keep balance between trust accuracy and node's energy and memory usage. Correspondingly, the Entropy Theory is used for distributing weights to different trust values instead of random weights. **The Fuzzy Logic approach** is used to deal with approximate logic rather than exact logic by computing the fuzzy value and presenting it in the range 0 to 1 [3]. This allows the use of a linguistic variable (i.e., trusted, untrusted, high, low) managed by specific membership functions. The fuzzy logic principle is adopted for the computation and analysis of trust among devices in order to make a suitable decision and block the malicious nodes. A Fuzzy Trust-based Aggregator (FTBA) algorithm is proposed in [3] as a node election technique for WSNs. The FTBA method enhances the decision-making function to authenticate a node in order to act as a trusted aggregator through four factors, namely, physical, remoteness, reputation, and aggregation intelligence. Based on the above factors, Fuzzy logic is used to decide the trusted node that qualifies to be an aggregator.

Weighted Sum is a method that aims to aggregate the score of each IoT device depending on its experiences or rating [4]. This sum is exploited to evaluate the credibility of the node interacting in the transaction, wherein the greater weight represents the trustworthy node in the network [5]. Therefore, the weighted sum can be obtained by self-observation trust (direct trust) or recommendation trust (indirect trust). In [13], the authors addressed the issue of trust by proposing Similarity-based Trust Computational Model for Social Internet of Things, which separates trust in two types: direct and indirect trust and took advantage of the dynamic weighted sum method to aggregate the similarity based on the interactions between entities. The proposed model managed the trustworthiness by extracting the features to designate the node's trust using three social trust *similarity* metrics: Community of Interest, Friendship, and Co-work. In the process of decision-making, the object is denoted as trustworthy or untrustworthy using a boundary value obtained by aggregating the different trust similarity metrics to unify a final trust score. **Machine learning (ML)** algorithms learn autonomously to perform a task or make predictions from data and improve their performance over time. These algorithms can be classified into supervised learning, unsupervised learning, semi-supervised learning, and reinforced learning [17]. ML is applied in an algorithm that calculates the trust value in an IoT environment to enhance the identification of trusted and untrusted nodes [6]. In [18], the authors present a computational trust algorithm for IoT services that employs unsupervised learning to predict malicious attacks and analyzes the trustworthiness depending on two techniques: 1) k-means to identify the clusters or labels i.e., trustworthy or untrustworthy using Support Vector Machine (SVM) as a classification method and 2) an optimum threshold level to separate trustworthy from untrustworthy interactions.

The fourth Category level is called Trust Architecture. This category includes Centralized and Decentralized architecture, considering the way information is propagated in the network between devices [5]. The centralized trust requires a trusted central authority that provides an efficient data control, makes the distribution of trusted information easier, and maintains the trust computation. In [19], the authors proposed a centralized trust management mechanism (ShareTrust) to ensure the secure sharing of resources and identify compromised or malicious nodes. The ShareTrust architecture contains three layers: resource provider, resource seeker, and centralized layer. In the Decentralized/Distributed trust approach, each node in the network independently computes the value of trust to prevent malicious nodes without preconditioning the centralized node [19]. The crucial advantage of such architecture is that compromised node cannot affect the degree of trust. The *RobustTrust* system presented [20] is a cross-domain trust management system based on distributed trust dissemination, which makes the device suitable to evaluate the trust locally, depending on its own. The fifth Category level is called Trust propagation. Trust propagation describes the process of how the node disseminates its degree of trust with other nodes. The trust propagation can be controlled at the node-level or cluster-level. The node-level is a propagation model that aims to deploy independently the degree of trust to other nodes, without the necessity of a cluster head. In [21], the authors suggest a trust management for service-oriented IoT architecture based on the node-level approach referred to as "Adaptive IoT Trust". The proposed approach uses distributed collaborative filtering considering the similarity rating of friendships, social, and community of interest as filters to select trust feedback. In a cluster-level model, the propagation of trust values is performed by a cluster head. A study [22] developed a lightweight trust computing mechanism designed with a cloud platform. The architecture of the proposed mechanism is composed of three layers: network layer, broker layer, and device layer.

Table 1. Pros and cons of trust models in IoT

Ref.	Year	Category	Benefits	Limitations
[10]	2019	Trust source: Direct Trust	Fast node compromise detection and revocation	False positive risk, Trust update is difficult and expensive
[11]	2015	Trust source: Indirect Trust	Detects hello flood, jamming and selective forwarding attack	Overhead due to communication
[12]	2021	Trust source: Hybrid Trust	Balance the benefits of direct and indirect trust	Trust is a complicated, multidimensional
[13]	2020	Trust Metric: Social Interactions	Dynamic trust score based on a dynamic weighted sum which combines direct and indirect trust	An automatic and dynamic threshold selection is not considered
[14]	2020	Trust Metric: Quality of Service	Energy efficiency, reliable communication, lower latency	Centralised Software Defined Networking (SDN), controller, centralised risk management process
[16]	2015	Trust Algorithm: Bayesian Analysis	Reduces energy consumption, prevents on-off attacks and bad-mouthing attacks	Estimation can be wrong in some cases, Bayes Theorem is extremely intensive computationally
[3]	2016	Trust Algorithm: Fuzzy Logic approach	Balances the energy consumption by using a Bio-inspired Energy Efficient Cluster (BEE-C) protocol, Increases nodes lifetime	Higher failure rate at the initial iterations
[13]	2020	Trust Algorithm: Weighted Sum	Combines direct and indirect trust, Provides an efficient computation technique	An effective trust update mechanism is absent

(*continued*)

Table 1. (*continued*)

Ref.	Year	Category	Benefits	Limitations
[18]	2018	Trust Algorithm: Machine learning	True positive rate, High performance against untrustworthy objects	High computation time when having a large database, the drawback of using predefined features. A feature selection algorithm is not used
[21]	2016	Trust propagation: node-level model	Human intervention errors because of the used of distributed collaborating filtering to select trust feedback from owners	Trust coming from owners is not guaranteed and not trustful
[22]	2018	Trust propagation: cluster-level model	IoT edge computing cluster to facilitate low-overhead trust computing and communication algorithms	It requires more storage capacity, Edge computing is not suitable for a huge amount of data
[19]	2022	Trust architecture: centralised trust	Distinguish the positive and the negative characteristics of shared information, Two-way trust evaluation, Requires less computation resources, Reduces the consumption of energy by using Event-based trust evaluation	Risk of compromised central authority, then there is no substitute gadget to control the trust of IoT devices
[20]	2019	Trust architecture: distributed trust	An effective protection against bad-mouthing, good-mouthing and on-off attacks, Supports scalability	Ignores the movement of nodes, Inability to adapt to changes in dynamic environments, High energy consumption

3   Intrusion Detection Systems

Since security of pervasive IoT systems is critical, it is important to identify IoT threats and specify existing Intrusion Detection Systems (IDS) [1]. In this section, we briefly define IDS and enumerate the various IDS approaches in IoT environments.

An Intrusion Detection System (IDS) is a system (software or hardware) that detects unusual or suspicious activity [1]. In general, IDS tools are passive and do not interfere with traffic. They only detect and inform about potential threats. In contrast, Intrusion Prevention Systems (IPS) take the necessary actions to stop or prevent malicious activities. While IPS systems are of crucial importance in IoT security, they are outside the scope of this article. In IoT networks, IDS techniques monitor the network to discover malicious activities [2]. IDS enables real-time detection of possible threats in IoT networks by using specific algorithms and different technologies such as Network Scanning, Data Analytics, Network Monitoring, Attack Signature Matching, Abnormal Behavior, Blockchain, Machine Learning, Fuzzy Logic, etc.

3.1   Relevant IDS Approaches in IoT

In this subsection, we enumerate the most recent IDS methods that can be applied in IoT environments. IDS approaches may be categorized according to the following attributes: *Placement of IDS*, *Detection approach*, *Security threat* and *Validation strategy*. Table 2 summarizes advantages and disadvantages of different IDS approaches in IoT.

   Placement of IDS can be Centralized, Distributed, or Hybrid. In [23], the authors proposed a centralized IDS for IoT to intercept and capture packets that pass through the border router. In [24], authors proposed a distributed IDS approach based on Deep Learning in IoT-Fog networks. The authors used a data training near the IoT devices. In [2], a hybrid IDS approach was proposed. The approach is based on Deep Neural Networks (DNN) and K-Nearest Neighbor Methods. **Detection approach** can be Signature-based detection, Anomaly-based detection, Specification-based detection, Hybrid Approaches. In [25], the authors propose a distributed approach based on combining Blockchains with signaturebased IDS. Blockchain uses a distributed ledger that enables recording transactions across various connected IoT devices. These latter use an IDS-based blockchain to share required information with each other in order to enhance the detection capability. **Security threat** can be a Conventional attack, Routing attack, Man-in-the-Middle (MitM) attack, or Denial of Service (DoS) attack. In [26], the authors discuss various IDS techniques and classify them based on threat models. They classified IDS into two categories, Host Intrusion Detection System (HIDS) and Network-based Intrusion Detection system (NIDS). HIDS aims to monitor the activities of packets and the behavior of the edge devices. NIDS, on the other hand, detects malicious traffic in the network, such as root or DoD attacks and monitors the networking devices such as routers and switches. **Validation strategy** can be Hypothetical, Empirical, Simulation-based, Theoretical, or None [27]. Validation strategy is the process to validate the functionality of software. Performance measurement is an automated approach that takes into consideration several metrics such as Detection Rate (DR) and False Alarm Rate (FAR) [8].

Table 2. Pros and cons of IDS models in IoT

Ref.	year	Category	Benefits	Limitations
[23]	2009	Placement of IDS: Centralized	Detection of Botnet attack (launch spam-mail, key-logging, and DDoS attacks) on 6LoWPAN with lower false positive rates and higher detecting rates	The detection module is installed on the 6LoWPAN Gateway, then the risk that the 6LoWPAN Gateway gets compromised
[24]	2018	Placement of IDS: Distributed	Offloading storage and computational overheads of models by using a distributed deep learning-based IoT/Fog network attack detection system	The distribution point "Master node" is the weak point of the solution, because centralised master node could be compromised
[2]	2020	Placement of IDS: Hybrid	The DNN-kNN method works with low overhead in terms of memory and processing costs. The DNN-kNN is able to achieve greater precision by using DNN and kNN algorithm	The DNN-kNN method is not capable of detecting routing attacks. Moreover, KNN has a small error rate
[25]	2019	Detection approach	The proposed collaborative Blockchained signature-based IDS has a high processing speed for known attacks and secure and easy updates of trust based on Blockchain	Signature-based IDS can deal only with known attacks, and not able to detect advanced insider attacks
[26]	2021	Security threat	Classifications of threats in the layers of the network by using Various machine learning solutions	The proposed threat model is not able to show the attacks on different layers
[8]	2015	Validation strategy	KDD Dataset Attributes (four classes which are Basic, Content, Traffic and Host) can help enhance the suitability of data set to achieve maximum Detection Rate (DR) with minimum False Alarm Rate (FAR)	Huge number of redundant record and doesn't contain the appropriate features

4 Trust Based IDS Approaches

Combining trust-based and IDS-based approaches seems to be a very efficient way to improve the protection of IoT networks and devices. The combination of the two approaches allows faster and easier detection of attacks. Trust-based intrusion detection approaches can be classified into five categories: Trust based on neural networks and machine learning, trust based on fuzzy reputation or Bayesian analysis, trust based on recommendation systems, trust based on node/network behavior, and trust based on navigation in distributed systems. In Table 3, we summarize the different trust-based IDS approaches in IoT networks.

Table 3. Overview: Intrusion detection-based trust management approaches

Ref	year	Contributions	Limitations
Trust-based on neural Network / Machine Learning			
[7]	2022	Training based on Recurrent Neural network (RNN), issues a trust degree to each node	Training can be difficult in terms of: decision, time, inputs.
[28]	2022	Deep learning based, KDD features, and trust factors	Centralized IDS
Trust based on Bayesian / fuzzy reputation			
[15]	2019	Fuzzy clustering, fuzzy Naive Bayes, direct trust, indirect trust, recent trust, BDE-based trust	Complexity and difficulty of learning/analysing information from different sources
[29]	2011	Fuzzy theory based trust and reputation model	The fuzzy rationale isn't always exact
Trust based on recommendation systems			
[11]	2015	Trust based on the observations of neighboring Nodes	The main drawback is overhead due to communication
[10]	2019	Direct trust: energy effective data transmission	Information can be unreliable
Trust based on nodes/networks behavior detection			
[30]	2017	Multi-agent system, node behavior	Trust between agents may not be a sufficient guarantee
[31]	2019	According to the node behaviors, a trust score will be attributed to the examined node	Optimal threshold is not determined
Trust-based navigation in distributed systems			
[32]	2020	Blockchain based trust management scheme	High Processing Time
[33]	2019	IDS nodes use Blockchain to gain knowledge from each other by sharing information	Communication overhead, High Processing Time, possibly fake nodes

A trust-based IDS approach, named AutoTrust, was proposed in [7]. The authors propose a Recurrent Neural Network (RNN)-based autonomic trust management approach. The RNN model uses a predefined trust parameter process as an input. In [28], the authors propose a trust model using Deep Learning of suspicious activities. In [15], the authors propose an IDS method based on fuzzy clustering. They implement fuzzy naive Bayes approaches to cluster the nodes based on various trust factors such as direct, indirect, and recent trust, as well as Bayesian Dirichlet likelihood-Equivalence score (BDE) Trust. In [29], the authors proposed a trust and reputation model for WSN based on fuzzy logic theory. The model uses the reputation evaluation of all interactions between nodes to detect malicious activities. In [11], the authors propose a trust-based IDS system that calculates the trust of each node by using observation and recommendations given by the neighboring nodes. In [10], a direct trust model was introduced to detect malicious nodes in Wireless Body Area Network (WBAN). The proposed model uses the energy-effective data transmission as a trust factor to detect malignant nodes. In [30], a multi-agent trust-based intrusion detection scheme was proposed that establishes a multi-agent model in both the cluster heads and the sensor nodes. The Agents judge the malicious activities based on the calculated and attributed trust score of each node. Further, in [31], a trust-based intrusion detection mechanism was proposed to select the clusters and to detect malicious activity in the network by monitoring the node behavior. In [32], the authors propose a Blockchain-based trust management scheme for IoT, which allows nodes to detect propagated malicious packets on the chain. In [33], a trust-chain based on Blockchain has been proposed to distribute information between nodes and to secure communication between nodes through the trust chain

4.1   Generic Approach to Select Trust-based IDS

In this section, we propose a generic approach to select the most suitable Trust-based IDS, i.e., the one that achieve a high probability of successful attack detection and with the lowest probability of false detection. In fact, most IDS systems cannot identify all injected attacks. For this reason, potential solutions could be based on combining various IDS approaches. The trust-based IDS approaches can be divided into three distinct phases: Network monitoring based on collective trusting decision, Anomaly detection based on combining several IDS approaches, and the validation of IDS approaches. For each phase, a combination of several protocols can improve the efficiency of intrusion detection.

Network Monitoring Based on Collective Trusting Decision: In [34], the authors proposed a collaborative intrusion detection networks using collaboration between heterogeneous trusting agents. In [35], a combining of BCT (Blockchain technology) and MAS (Multi-agent system) has been proposed to detect an anomaly and distribute alert-related information through the Blockchain network. In [36], authors proposed an IDS system based on a MAC system, BCT, and deep learning. The agents will communicate through the

Blockchain network and all transaction will be recorded on Blockchain. The distributed platform helps agents to trust each other and to collaborate for detection the anomaly.

Anomaly Detection Based on Combining Several IDS Approaches: To improve the detection rate, authors in [37] propose a combination between genetic fuzzy systems and pairwise learning. In [38], the authors proposed an approach based on combining multiple learning techniques for effective and efficient intrusion detection. The proposed approach is based on extracting data from cluster centers and nearest neighbors to provide higher classifier training and testing of attack detection.

Validation of IDS Approaches: A feature selection for IDS is an important task for effective intrusion detection. Therefore, feature selection is required for intrusion detection in IoT that builds the models in minimum time and achieves higher performance. Several IoT dataset for feature selection and intrusions has been proposed to test an IDS approach. Most of feature selection dataset includes UDP, TCP, and HTTP-based DoS and DDoS attacks. For example, in [39], the authors proposed an effective anomaly detection by using a subset of features, namely Information Gain (IG) and Gain Ratio (GR). The proposed a feature selection has been tested with BoT-IoT and KDD Cup 1999 dataset.

5   Conclusion

In this survey, we presented a literature review of Intrusion Detection Systems, trust management systems, and trust based IDS approaches for IoT environments. We presented a classification of IDS, Trust, and Trust-based IDS approaches. We have also proposed generic approach to select Trust-based approaches, where we explained that combination of different approaches is a potential solution.

Research dealing with Trust-based IDS approaches in IoT is still in its infancy. Nevertheless, combinations of various trust-based IDS approaches seems to be the most viable way to get the best of existing solutions.

As future work directions, we envision to propose a holistic architecture for Trust-based IDS that takes advantage of all existing solutions. Major issues include:

- How to increase accuracy of intrusion detection without additional overhead?
- How to use Machine and Deep learning in constrained IoT environments?
- How to use blockchain to reinforce trust?
- How to implement IDS in dynamic fog/edge-based IoT networks?
- How to use fuzzy logic to increase trust?
- What can be defined as an attack in an IoT network to identify attack signature?
- How to determine if a node is compromised or simply depleted its energy?

References

1. Sherasiya, T., Upadhyay, H., Patel, H.B.: A survey: intrusion detection system for internet of things. Int. J. Comput. Sci. Eng. (IJCSE) **5**(2), 91–99 (2016)
2. de Souza, C.A., Westphall, C.B., et al.: Hybrid approach to intrusion detection in fog-based IoT environments. Comput. Netw. **180**, 107417 (2020)
3. Rajesh, G., Raajini, X.M., Vinayagasundaram, B.: Fuzzy trust-based aggregator sensor node election in internet of things. Int. J. Internet Protocol Technol. **9**(2/3), 151–160 (2016)
4. Guo, J., Chen, R., Tsai, J.J.: A survey of trust computation models for service management in internet of things systems. Comput. Commun. **97**, 1–14 (2017)
5. Ahmed, A.I.A., Ab hamid, S.H., Gani, A., Khan, M.K.: Trust and reputation for Internet of Things: fundamentals, taxonomy, and open Research Challenges. J. Netw. Comput. Appl. **145**, 102409 (2019)
6. Najib, W., Sulistyo, S., et al.: Survey on trust calculation methods in Internet of Things. Proc. Comput. Sci. **161**, 1300–1307 (2019)
7. Awan, K.A., et al.: AutoTrust: a privacy-enhanced trust-based intrusion detection approach for internet of smart things. Future Gener. Comput. Syst. **137**, 288–301 (2022)
8. Aggarwal, P., Sharma, S.K.: Analysis of KDD dataset attributes-class wise for intrusion detection. Proc. Comput. Sci. **57**, 842–851 (2015)
9. Meddeb, A.: Internet of things standards: who stands out from the crowd? IEEE Commun. Mag. **54**(7), 40–47 (2016). https://doi.org/10.1109/MCOM.2016.7514162
10. Anguraj, D.K., Smys, S.: Trust-based intrusion detection and clustering approach for wireless body area networks. Wirel. Person. Commun. **104**(1), 1–20 (2019)
11. Sajjad, S.M., Bouk, S.H., Yousaf, M.: Neighbor node trust based intrusion detection system for WSN. Proc. Comput. Sci. **63**, 183–188 (2015)
12. Narang, N., Kar, S.: A hybrid trust management framework for a multi-service social IoT network. Comput. Commun. **171**, 61–79 (2021)
13. Subhash, S., Adnan, M., Jitander, P., Quan, S.: A time-aware similarity-based trust computational model for social internet of things (2020). https://doi.org/10.1109/GLOBECOM42002.2020.9322540
14. Kalkan, K.: SUTSEC: SDN utilized trust based secure clustering in IoT. Comput. Netw. **178**, 107328 (2020)
15. Veeraiah, N., Krishna, B.T.: Trust-aware FuzzyClus-Fuzzy NB: intrusion detection scheme based on fuzzy clustering and Bayesian rule. Wirel. Netw. **25**(7), 4021–4035 (2019)
16. Shenyun, C., Renjian, F., Xuan, L., Xiao, W.: A lightweight trust management based on Bayesian and entropy for wireless sensor networks. Securi. Commun. Netw. **8**, 168–175 (2015). https://doi.org/10.1002/sec.969
17. Wang, J., Yan, Z., Wang, H., Li, T., Pedrycz, W.: A survey on trust models in heterogeneous networks. IEEE Commun. Surv. Tutor. (2022). https://doi.org/10.1109/COMST.2022.3192978
18. Jayasinghe, U., Lee, G., Myoung, T.-W., Shi, Q..: Machine learning based trust computational model for IoT services. IEEE Trans. Sustainab. Comput. **4**, 39–52 (2018). https://doi.org/10.1109/TSUSC.2018.2839623
19. Din, I.U., Awan Kamran, A., Almogren, A., Byung-Seo, K.: ShareTrust: centralized trust management mechanism for trustworthy resource sharing in industrial Internet of Things. Comput. Elect. Eng. **100**, 108013 (2022). https://doi.org/10.1016/j.compeleceng.2022.108013

20. Awan, K.A., Din, I.U., Almogren, A., Guizani, M., Altameem, A., Jadoon, S.U.: RobustTrust - a pro-privacy robust distributed trust management mechanism for internet of things. IEEE Access. **7**, 62095–62106 (2019). https://doi.org/10.1109/ACCESS.2019.2916340

21. Chen, I.-R., Guo, J., Bao, F.: Trust management for SOA-based IoT and its application to service composition. IEEE Trans. Serv. Comput. **9**(3), 482–495 (2016). https://doi.org/10.1109/TSC.2014.2365797

22. Yuan, J., Li, X.: A reliable and lightweight trust computing mechanism for IoT edge devices based on multi-source feedback information fusion. IEEE Access. **6**, 23626–23638 (2018)

23. Cho, E.J., Kim, J.H., Hong, C.S.: Attack model and detection scheme for botnet on 6LoWPAN. In: Asia-Pacific Network Operations and management symposium, pp. 515–518 (2009)

24. Diro, A.A., Chilamkurti, N.: Distributed attack detection scheme using deep learning approach for Internet of Things. Future Gener. Comput. Syst. **82**, 761–768 (2018)

25. Li, W., Tug, S., Meng, W., Wang, Y.: Designing collaborative blockchained signature-based intrusion detection in IoT environments. Future Gener. Comput. Syst. **96**, 481–489 (2019)

26. Kapil, D., Mehra, N., Gupta, A., Maurya, S., Sharma, A.: Network security: threat model, attacks, and IDS using machine learning. In: International Conference on Artificial Intelligence and Smart Systems (ICAIS), pp. 203–208 (2021)

27. Nweke, L.O.: A survey of specification-based intrusion detection techniques for cyber-physical systems. Int. J. Adv. Comput. Sci. Appl. **12**(5), 1–9 (2021)

28. Bhor, H.N., Kalla, M.: TRUST-based features for detecting the intruders in the Internet of Things network using deep learning. Comput. Intell. **38**(2), 438–462 (2022)

29. Chen, D., Chang, G., Sun, D., Li, J., Jia, J., Wang, X.: TRM-IoT: a trust management model based on fuzzy reputation for internet of things. Comput. Sci. Inf. Syst. **8**(4), 1207–1228 (2011)

30. Jin, X., Liang, J., Tong, W., Lu, L., Li, Z.: Multi-agent trust-based intrusion detection scheme for wireless sensor networks. Comput. Elect. Eng. **59**, 262–273 (2017)

31. Dang, N., Liu, X., Yu, J., Zhang, X.: TIDS: trust intrusion detection system based on double cluster heads for WSNs. In: International Conference on Wireless Algorithms, Systems, and Applications, pp. 67–83 (2019)

32. Meng, W., Li, W., Yang, L.T., Li, P.: Enhancing challenge-based collaborative intrusion detection networks against insider attacks using blockchain. Int. J. Inf. Secur. **19**(3), 279–290 (2020)

33. Kolokotronis, N., Brotsis, S., Germanos, G., Vassilakis, C., Shiaeles, S.: On blockchain architectures for trust-based collaborative intrusion detection. In: IEEE world Congress on Services (SERVICES), vol. 2642, pp. 21–28 (2019)

34. Grill, M., Bartos, K., et al.: Trust-based classifier combination for network anomaly detection. In: International Workshop on Cooperative Information Agents, pp. 116–130 (2008)

35. Mbarek, B., Ge, M., Pitner, T.: An adaptive anti-jamming system in HyperLedger-based wireless sensor networks. Wirel. Netw. **28**(2), 691–703 (2022)

36. Liang, C., Shanmugam, B., et al.: Intrusion detection system for the internet of things based on blockchain and multi-agent systems. Electronics. **9**(7), 11–20 (2020)

37. Elhag, S., Fernández, A., Bawakid, A., Alshomrani, S., Herrera, F.: On the combination of genetic fuzzy systems and pairwise learning for improving detection rates on intrusion detection systems. Expert Syst. Appl. **42**(1), 193–202 (2015)
38. Lin, W.-C., Ke, S.-W., Tsai, C.-F.: CANN: an intrusion detection system based on combining cluster centers and nearest neighbors. Knowl. Based Syst. **78**, 13–21 (2015)
39. Nimbalkar, P., Kshirsagar, D.: Feature selection for intrusion detection system in Internet-of-Things (IoT). ICT Express. **7**(2), 177–181 (2021)

Context-Aware Security in the Internet of Things: A Review

Everton de Matos[1(✉)], Eduardo Viegas[1], Ramão Tiburski[2],
and Fabiano Hessel[3]

[1] Technology Innovation Institute, Abu Dhabi, UAE
{everton.dematos,eduardo.viegas}@tii.ae
[2] Federal Institute of Education Science and Technology of Santa Catarina, São
Lourenço do Oeste, Brazil
ramao.tiburski@ifsc.edu.br
[3] Pontifical Catholic University of Rio Grande do Sul (PUCRS), Porto Alegre, Brazil
fabiano.hessel@pucrs.br

Abstract. Security and privacy are hot topics when considering the
Internet of Things (IoT) application scenarios. By dealing with sensitive and sometimes personal data, IoT application environments need
mechanisms to protect against different threats. The traditional security mechanisms are usually static and were not designed considering the
dynamism imposed by IoT environments. Those environments could have
mobile and dynamic entities that can change their status at deployment
time, needing novel security mechanisms to cope with their requirements.
Thus, a flexible approach to security provision is imperative. Context-
Aware Security (CAS) provides dynamic security for IoT environments
by being aware of the context. CAS solutions can adapt the security
service (e.g., authentication, authorization, access control, and privacy-
preserving) provision based on the context of the environment. This work
reviews the concepts around CAS and presents an extensive review of
existing solutions employing CAS in their architecture. Moreover, we
define a taxonomy for CAS based on the context-awareness area.

1 Introduction

The Internet of Things (IoT) has received significant attention in both academia
and industry. By integrating mobile networking and information processing into
a variety of devices and everyday objects, IoT has expanded the realm of information and communication technology [6]. The global IoT Security Product
market is estimated to grow from USD 12 Bn in 2017 to USD 48 Bn by 2027
[8], leading us to believe that efforts to provide novel security solutions for IoT
environments will be needed by the market.

The traditional security mechanism usually provides static and non-aware
security services. Considering the high dynamism of IoT scenarios, Context-
Aware Security (CAS) appears as a suitable mechanism to provide dynamic
security to IoT environments, as many recent works have been proposed in this

L. Barolli (Ed.): AINA 2023, LNNS 655, pp. 518–531, 2023.
https://doi.org/10.1007/978-3-031-28694-0_49

regard [18,21]. CAS uses context information to provide security. Context information can be considered any high-level (i.e., human-readable) information that characterizes an entity [19]. As the context of an entity (i.e., device, user, network) may change on the fly in IoT environments, it is crucial to care about those changes when providing security services.

The present work aims to shed light on CAS solutions for IoT environments. We present definitions of CAS, considering the four main security services that it can provide: (i) authentication, (ii) authorization, (iii) access control, and (iv) privacy-preserving. We also introduce a novel taxonomy based on the well-established context-awareness area to classify the CAS solutions. Moreover, we review recent works that employ CAS in their architecture. The review considers the possible security services provided by the works and also the requirements to provide CAS.

Recent works have extensively discussed and reviewed the available solutions in the context-aware domain, but just a few discuss the challenges in the CAS area [19,21]. Grimm et al. [11] present a survey on CAS for vehicles and fleets through a detailed analysis of the context information relevant to future vehicle security. Sylla et al. [26] conduct a survey of the CAS solutions that have been proposed for smart city IoT applications. In light of this, the present work shows its novelty by providing an extensible review of CAS characteristics and requirements, introducing a novel taxonomy for CAS, and analyzing a significant amount of CAS solutions considering different application areas.

The remainder of the paper is structured as follows: Sect. 2 presents an overview of CAS concepts. Section 3 presents a taxonomy on CAS. Section 4 presents the requirements of CAS solutions. Section 5 introduces available solutions for CAS. Section 6 presents a discussion on CAS solutions. Finally, Sect. 7 concludes the paper.

2   Context-Aware Security

It was believed that security requirements were static due to the fact that security decisions do not alter based on the situation, nor do they take into consideration any changes in the surrounding environment [2]. On the other hand, using context information in determining security decisions is a vital aspect of addressing certain security issues [19,29]. The CAS does not remove the need for traditional security mechanisms. It adds a layer of security and privacy, focusing on the dynamism of such IoT environments.

The definition of Context-Aware Security (CAS) is characterized by a combination of information gathered from both the user's and the application's surroundings, which is critical to the security infrastructure of both the user and the application, as stated in the works of Most'efaoui and Br'ezillon [7,20]. Additionally, CAS refers to a security approach that takes into account a set of information (context) when making security decisions. For instance, during communication intrusion detection, the security system may switch to a stronger authentication method. In contrast, context-unaware security mechanisms may fall short in the ever-changing and diverse environment of the Internet of Things. Context information can be used to reconfigure security mechanisms and adjust security parameters. The contextual information can be integrated into various security mechanisms such as authentication, access control, encryption, etc. [13].

While the notion of context awareness has been well spread through the scientific community [1], at present, there is a scarcity of security and privacy-enhancing mechanisms that consider dynamic context factors for the Internet of Things. [19,24]. For the implementation of CAS in IoT environments, four main security services can be provided: (i) authentication, (ii) authorization, (iii) access control, and (iv) privacy-preserving. The next items present an overview of each security service [2,13,29].

Authentication: Conventional authentication procedures entail a considerable amount of user involvement, such as manual login, logout, and file permission processes. Currently, passwords are the most prevalent mode of authentication. Nevertheless, passwords pose major security risks as they are frequently simple to guess, reused, forgotten, shared, and open to social engineering attacks. Furthermore, widely accepted authentication technologies such as facial recognition, iris scanning, and biometric techniques can be employed. In addition to these technologies, the integration of contextual information enhances the authentication process by providing an additional layer of security.

Authorization and Access Control: This refers to granting access if the user's or device's credentials match pre-existing credentials and denying access if there is a mismatch. Traditionally, this kind of system operates by following a set of fixed rules and ignoring other important factors, such as the user or device context, when making access decisions. In the dynamic environment of the Internet of Things, incorporating flexible security policies based on contextual information has the potential to enhance the accuracy of security decisions.

Privacy-Preserving: Information that reflects individuals' daily activities, such as travel routes and buying habits, is often seen as private by many users. It is, therefore, not surprising that privacy preservation is a key requirement for IoT applications. For instance, users may hesitate to share their current location with a Location-based Service (LBS) server due to privacy concerns. Context information can be leveraged to determine when to keep user information confidential.

3   Taxonomy of Context-Aware Security in IoT

We have developed a taxonomy of CAS in IoT considering the extensive published research on context-awareness, [13,19,21], and given the unique characteristics of highly heterogeneous IoT environments that may vary in terms of processing power for its devices, storage space for data, network and connectivity conditions, and various users and applications per domain. The present taxonomy is derived from a previous work on context-aware domain [19]. The taxonomy outlines the key features of CAS solutions and their potential deployment variations, as shown in Fig. 1. It is organized into three sections: (i) Context Modeling, which deals with context management, (ii) Key Architectural Components, which focuses on architectural characteristics, and (iii) Applicability, which covers the different ways in which CAS can be provided.

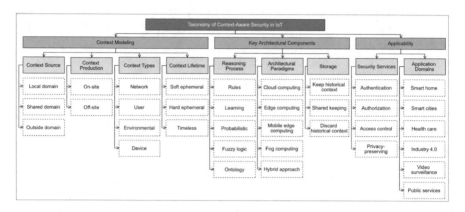

Fig. 1. A taxonomy that represents Context-Aware Security (CAS) in the Internet of Things (IoT).

3.1 Context Modeling

Context Source: To deliver security services, CAS solutions require context information which can be sourced from different domains such as the *local domain, shared domain,* and *outside domain.* The domain refers to the physical location where the solution is deployed (e.g., healthcare, smart city, industry 4.0). With *local domain,* the CAS solution has access only to the context information from its deployed domain. In the case of *shared domain,* context can be obtained from the same domain as the solution but from a different deployment site (e.g., two instances of the same healthcare solution). Lastly, *outside domain* refers to context obtained from a domain different from the deployed one.

Context Production: The context information for the CAS solution can be produced in two ways: *on-site* or *off-site.* When the CAS solution itself is responsible for acquiring raw data from IoT entities and transforming it into context information, it is called *on-site* context production. On the other hand, if a third-party software is responsible for producing context information, it is referred to as *off-site* context production.

Context Types: The type of context is determined by the characteristics of the source from which it was obtained. The various types of context information are: *network* (e.g., bandwidth, congestion, faulty nodes), *user* (e.g., location, activities, routes, preferences), *environmental* (e.g., weather, crowding), and *device* (e.g., battery life, potential errors). A CAS solution may have context information of one or multiple context types.

Context Lifetime: Context information can have a limited lifespan, especially in fast-changing IoT environments where data can quickly become outdated. Context Lifetime can be classified into three categories: *soft ephemeral*, *hard ephemeral*, and *timeless*. *Soft ephemeral* context information is valid for a limited period, while *hard ephemeral* context information may change in each interaction. *Timeless* context information remains relevant over time. The lifespan of context information depends on the specific deployment scenario and must be established by the CAS solution.

3.2 Key Architectural Components

Reasoning Process: The Reasoning Process in a CAS solution plays a crucial role in converting raw data into context information and utilizing that information to provide security services. Several reasoning techniques are commonly used, including *rules, learning, probabilistic, fuzzy logic*, and *ontology*. Among these techniques, *Rules* stand out for their simplicity and lightweight nature, making them a suitable option for IoT environments with limited resources. These rules are constructed using IF-THEN-ELSE conditions. On the other hand, *Learning* techniques such as Bayesian Networks and Decision Trees need a large data set to achieve accurate results. *Probabilistic* reasoning also requires a substantial amount of data, but it only reasons using past acquired data or context. *Fuzzy logic* offers a more intuitive representation of the environment, but it can be prone to errors since it is defined manually. On the other hand, the use of *ontologies* provides the ability to perform complex reasoning and representations. However, input data must be modeled in a compatible format, such as Web Ontology Language (OWL) or Resource Description Framework (RDF), which can result in low performance and a higher demand for computational resources than is typically available in resource-constrained IoT environments.

Architectural Paradigms: The architecture of a CAS solution can vary depending on the specific requirements, such as the availability of resources, storage space, network conditions, and processing power. These architectural paradigms can be divided into five categories: *cloud computing, edge computing, mobile edge computing, fog computing*, and a *hybrid approach*. *Cloud computing* is focused on performing crucial processing tasks in the cloud, and utilizing the cloud for data storage if required. On the other hand, both *edge computing* and *mobile edge computing* aim to process critical tasks directly on the data source devices at the edge of the network. These paradigms reduce network latency and are often more scalable. The main difference between the edge and mobile edge approaches is the mobility of the entities in mobile edge, which can change locations frequently. The *fog computing* paradigm extends cloud computing to an intermediate layer that is located closer to data source devices [12]. Lastly, *hybrid approaches* are solutions that integrate multiple architectural paradigms.

Storage: CAS solutions can handle context information in three ways, i.e. *keep historical context, shared keeping*, or *discard historical context*. *Keep historical* *context* involves storing all the context information internally, used to provide security services. This information can also be used for probabilistic processing in conjunction with CAS reasoning. *Shared keeping* involves accessing previous context information stored in a different instance, distinct from the one that performs the CAS process. The context is kept in a shared manner. On the other hand, some solutions may *discard historical context* by only using the context information during the time of execution and then discarding it.

3.3 Applicability

Security Services: The CAS solutions can provide various Security Services in the security and privacy field, including *authentication, authorization, access control*, and *privacy-preserving*. These services are provided by utilizing the Reasoning Process to analyze the context information. The number of services offered by a particular solution will vary depending on the deployment environment and the reasoning method used. For instance, the Context-Aware Role Based Access Control (CARBAC) scheme [14] proposed by Hosseinzadeh et al. combines traditional Role-based Access Control (RBAC) with current context information to control access for users.

Application Domains: A number of different application domains can benefit from using CAS solutions. Examples include *smart homes, smart cities, health care, industry 4.0, video surveillance*, and *public services*. In a smart city, for example, privacy-preserving can be achieved by allowing individuals to share their personal context information with the city infrastructure only in specific contexts (e.g., when they are not at home). Additionally, in an industry 4.0 environment, context-based access control policies can be defined to limit employee access to certain areas based on the current context information. These are just a few examples of how Security Services can be applied in different domains.

4 Key Requirements of Context-Aware Security in IoT

The taxonomy discussed in Sect. 3 highlighted the key features of CAS solutions in the IoT and their potential applications. Based on this understanding, the current section outlines the essential features required for the seamless delivery of security services utilizing context information [15, 21].

Context Acquisition: The acquisition process is comprised of three key aspects outlined in the taxonomy (as seen in Fig. 1): Context Source, Context Production, and Context Types. It is imperative that the CAS solution has a reliable source for acquiring context information. This information can come from a variety of sources, such as network, user, environmental, or device. For instance, a CAS solution deployed in a smart city environment can gather context information from a traffic light through a publish/subscribe method and from an ambulance through a web service simultaneously.

Context Processing: This component can also be referred to as reasoning or security services inference. It is closely tied to the Reasoning Process, as seen in Fig. 1. The purpose of context processing is to make secure decisions using context information as input. The amount of context information needed to reach a secure decision can vary based on the reasoning technique and the application domain. The design of context processing must be adaptable to the IoT environment in which it is implemented, taking into account resource limitations to ensure a seamless and effective CAS solution.

Interoperability: The format, data type, size, and representation of context information can vary, making interoperability a key requirement for CAS solutions in IoT. The compatibility of different context information must be ensured so that it can be used as input for the provision of security services. However, the absence of a standard for representing context information presents a challenge in implementing this requirement [21].

Privacy: The handling of sensitive information is a crucial aspect in IoT systems, including CAS solutions. For instance, in a hospital setting, a CAS solution would be handling the health information of patients, which is highly sensitive. To ensure the protection of such information, security measures must be in place to prevent theft or to tamper with the context information, whether it is stored or transmitted over a network. This requires the use of secure communication protocols to maintain data integrity and privacy, as well as authentication and access control methods for protecting stored information and utilizing cryptography/anonymization.

Reliability: Ensuring a high level of dependability is crucial for the effective functioning of CAS solutions in IoT environments. Given the wide range of applications that these solutions can be used in, it is important to establish trust among users and improve the overall security of the system. Reliability plays a key role in fostering usage and improving the overall security of the solution.

5   Projects Considering Context-Aware Security in IoT

For many years, researchers have been studying the concept of CAS, dating back to the era of pervasive computing [13,15]. During this time, important definitions were established that have continued to shape the field. Initially, the technological limitations restricted the applicability of CAS solutions to a specific set of use cases. However, with the widespread adoption of IoT applications, the field has become more relevant and applicable. Currently, various efforts are underway to provide CAS solutions that are tailored to the unique characteristics of IoT environments.

Table 1 evaluates CAS systems based on the security services they offer, including (i) authentication, (ii) authorization, (iii) access control, and (iv) privacy-preserving. A dash (—) is used throughout all columns to indicate that the feature is either absent or not specified in the available literature. A check mark (✓) indicates that the evaluated work implements the feature from a certain perspective. We choose the solutions by searching for the terms *Context-Aware Security* and *context-aware + privacy + secure + authentication + authorization + access control* on academic repositories (i.e., Google Scholar, IEEE Xplore, ACM Digital Library, Springer). We focused on solutions developed in recent years, and deployed for embedded systems applications.

Table 1. Overview of Context-Aware Security Solutions by Security Service Provided.

Solutions	Ref	Year	Authentication	Authorization	Access Control	Privacy-Preserving	Scope
SocIoTal	[24]	2015	✓	✓	✓	✓	IoT
Rachid et al.	[23]	2015	—	—	—	✓	IoT
Gansel et al.	[9]	2015	✓	—	✓	—	Automotive
SVM-CASE	[17]	2015	✓	—	—	—	VANET
CAS RBAC	[29]	2016	—	✓	✓	—	User
CARBAC	[14]	2016	✓	✓	✓	✓	WSN
ContexIoT	[16]	2017	—	✓	✓	—	IoT
CAPP	[30]	2017	—	—	—	✓	Smartphones
CRBAC	[4]	2018	✓	✓	✓	✓	Healthcare
CSIP	[3]	2018	✓	—	✓	—	Industrial IoT
Aegis	[25]	2019	—	✓	✓	—	Smart Home
Gheisari et al.	[10]	2019	—	—	—	✓	Smart City
Psarra et al.	[22]	2020	—	✓	✓	—	Healthcare
SETUCOM	[27]	2021	✓	✓	✓	✓	IoT
STAC	[5]	2021	✓	✓	✓	✓	IoT

In **SocIoTal**, the authors describe a framework that was developed as part of the EU FP7 SocIoTal project [24]. The framework consists of two key components: the Group Manager and the Context Manager. The Group Manager handles the sharing of data among a group of devices, while the Context Manager is responsible for providing CAS services through a Complex Event Processing (CEP) approach.

Rachid et al. proposed a context-aware architecture for preserving privacy in IoT [23] that can offer an ontology-based service for privacy processing. **Gansel et al.** presented a context-aware access control model for automotive scenarios [9]. The model grants permissions to access certain display areas based on the car's current context. The SVM-based Context-Aware Security Framework (**SVM-CASE**) [17] uses Support Vector Machine (SVM) learning to automatically differentiate between well-behaved and misbehaving nodes in VANETs by analyzing vehicle context information.

Trnka et al. present an RBAC extension incorporating context awareness [29]. Users are granted security levels based on context. **CARBAC** is a Context-Aware Role-Based Access Control scheme [14]. It employs ontological techniques and OWL and is implemented through CLIPS business rules. **ContexIoT** is a smartphone app providing contextual integrity through context-based permission system for IoT [16]. Developed using Samsung SmartThings, it analyzes context data flow.

The context-aware privacy-preserving algorithm (**CAPP**) algorithm is designed to decide how to share smartphone user context [30]. The authors of **CRBAC** propose a role-based access control sensitive to context [4]. The CRBAC model uses context conditions and roles/attributes to establish critical-situation policies. Context-sensitive seamless identity provisioning (**CSIP**) [3] is a mutual authentication framework for the Industrial Internet of Things (IIoT) that builds profiles of inhabitants (people, devices, services, systems, sensors, 5G smartphones) using their activities and usage patterns to create an identity proxy to verify during interactions.

Aegis is a framework that identifies malicious activity in Smart Home Systems (SHS) [25]. It gathers information from smart home sensors and devices to comprehend the user's activity context, allowing it to detect harmful behavior and alert users. **Gheisari et al.** equip IoT-based smart cities with a Software Defined Networking (SDN) approach and implement a privacy-protecting method for managing the data flow of IoT device packets [10]. The SDN categorizes all IoT device data based on context.

Psarra et al. [22] present a CAS model that can serve as background knowledge for creating and enforcing access control rules for electronic health records (EHR) using a combination of the Attribute Based Access Control (ABAC) and Attribute Based Encryption (ABE) models. **SETUCOM** [27] secures context information exchange by using a hybrid encryption system adapted to IoT devices and manages trust through artificial intelligence techniques such as Bayesian networks and fuzzy logic. **STAC** [5] uses ontologies to provide CAS of data regarding patient context, such as treatment history, test description, current location, or cause of the disease.

All of the analyzed solutions in the field of Context-Aware Security (CAS) have different approaches to ensuring security, including differences in their architecture and focus. Some systems aim to protect the entire infrastructure, while others have a specific security goal. These CAS solutions utilize context information to adapt to dynamic situations, taking into account factors such as location and status.

6   Summary of Context-Aware Security Solutions

A compilation of the most recent CAS solutions for the Internet of Things can be found in Table 2. The analysis is based on the CAS requirements outlined in Sect. 4. The table indicates the features of each solution with a dash (—) symbol representing missing or unmentioned features in the related publications. The focus is on the characteristics of the solutions in addressing the requirements rather than the specific technologies used. For instance, a solution can be considered as fulfilling the Privacy requirement if it includes a method to conceal user information in specific contexts, not just by utilizing a secure communication protocol such as DTLS or wolfSSL.

Table 2. Overview of Context-Aware Security Solutions by Requirements.

Solutions	Ref	Context acquisition	Context processing	Interoperability	Privacy	Reliability
SocIoTal	[24]	Gets context from internal or external device's sensors	Uses key-value pairs and markups for modeling and Complex Event Processing for reasoning	Translates raw context data into a proprietary common format	A repository of privacy rules is used to define privacy preferences of users	It has a component that allows smart objects to obtain data from other entities in a reliable way
Rachid et al.	[23]	Gets context trough a sensing layer from a WSN	Uses an ontology-based reasoning technique	An ontology provides a shared language for the context across the system components	Uses an ontology that describes the privacy of users	—
Gansel et al.	[9]	Obtains information regarding the car's condition, the surrounding environment, and the user	Uses rules for a dynamical access control	—	An application needs a permission that is limited to specific contexts	It uses a microkernel based hypervisor
SVM-CASE	[17]	Gets context from network and nodes/devices	Uses the Support Vector Machine (SVM) algorithm to classify nodes	Provides interoperability between different automotive nodes	—	Verifies if a node becomes misbehaving using context information
CAS RBAC	[29]	Uses both real-time and historical users context (e.g., location, time)	Uses rules to determine access control levels	It is interoperable with solutions using Role-based Access Control (RBAC)	—	It has mechanisms to verify the validity of context information
CARBAC	[14]	Gets context from users devices (e.g., smartphone)	Uses OWL and CLIPS rules to perform reasoning	By using ontologies, it is able to deal with generic data	Provides data privacy by security rules	—
ContexIoT	[16]	Collects context from the smartphone in installation and runtime	Uses rules to compare contexts in a key-value pair	It uses the Samsung SmartThings platform to prototype the solution	—	It keeps a runtime logging for the context events
CAPP	[30]	It acts as a middleware for the smartphone and can get context directly from smartphones sensors	It uses Markov Chain for both model and reasoning on context to provide privacy	It can be used by different smartphones	It protects user's privacy context from untrusted smartphone apps	It performs check methods before release the context of a user
CRBAC	[4]	Gets context from previously connected devices	Uses a rule-based processing for access control	—	Replaces the original data identities with unique privacy labels	It records the history of every data exchange between the entities
CSIP	[3]	Gets the context from sensors/smartphones of medical domain	Uses data mining techniques to extract patterns for future reasoning by comparison	Synchronizes data blocks with a cloud-based side	Uses a secure session-key for access to private data	Stores the processed data for learning purposes
Aegis	[25]	Collects the state of smart home devices and sensors (active or inactive) autonomously	A Markov Chain-based machine learning model is used to detect malicious activities	It has a data array format for parsing the collected context	It assigns an anonymous ID for each user to ensure privacy	—
Gheisari et al.	[10]	It has a Software-Defined Networking (SDN) for getting the device's data	It uses rules to determine the sensitivity level of each data	Different devices can participate a defined SDN	It defines that sensitive data not be disclosed unintentionally	It splits sensitive data and sends split parts through a secure route
Psarra et al.	[22]	Gets plaintext from medical records and attribute values from the environment	Uses a rule-based approach with ontologies for vocabulary	An ontology acts as a common vocabulary for the healthcare domain	—	—
SETUCOM	[27]	Acquisition through MQTT protocol. Subscription is also available	Uses fuzzy logic for the reasoning process	Context modeler module to help in interoperability	Modules to deal with devices trust and network security	Device trust module responsible for the reliability of the context information
STAC	[5]	Collected data is transferred between various devices to the knowledge base	Semantic rules applicable on OWL ontologies	—	Provides data privacy by security rules	—

Works in the realm of CAS, such as [16,17], and [25], employ context information from both users and the environment to detect unusual network or system activity via anomaly detection. This approach, when implemented effectively, can significantly improve the security of IoT environments by using historical data and recognizing unusual behavior. However, anomaly detection is just one of the many security provisions that can be achieved through a CAS solution and can serve as a trigger for various security services.

All analyzed works fulfilled the requirements of *Context acquisition* and *Context processing*, which are essential for CAS solutions. The majority of the works acquired context information from the user's mobile device [3,14,16]. However, Rachid et al. [23] proposed a distinct approach by acquiring context from a wireless sensor network through a sensing layer. Gheisari et al. [10] have designed a Software-Defined Networking (SDN) system for acquiring context. For context processing, the most common approach is to use rules [5,9,10,14,16,22,29]. Rules are simple, easy to use, and efficient, making them a popular choice in the context-aware field [21]. The trend in the field is the use of ontologies and machine learning techniques [14,23,27] to improve interoperability.

The requirement of *Interoperability* involves the ability of the solution to handle different entities (e.g., data, devices). Some analyzed works only provide interoperability with similar entities [16,17,30]. SocIoTal [24] and Aegis [25] work towards providing a standard data format for easier interoperability after parsing. Rachid et al. [23] and CARBAC [14] employ a similar approach, utilizing an ontology to establish a common vocabulary between entities.

The requirements of *Privacy* and *Reliability* are not widely achieved by the set of analyzed works. For *Privacy* preservation, a number of the reviewed works utilize either rule-based systems or condition managers to control the level of privacy for context information [14,23,24]. CRBAC [4] implements privacy processing by replacing certain elements of context information with unique data, preserving privacy. Aegis [25] does not store user data from smart home devices, reducing privacy risks compared to previous solutions. The analyzed works employ various approaches to ensure the *Reliability* requirement. Many of the works in the field operate with redundancy processing in order to reduce the likelihood of failures [3,10,16].

The significance of researching, examining, and enhancing Context-Aware Security solutions for the growth and establishment of the Internet of Things (IoT) cannot be overstated. The analyzed solutions implement Context-Aware Security in various ways in their respective applications. However, the most frequent challenges faced by Context-Aware Security solutions are in regards to data protection, particularly the context information. It is crucial to adhere to recent legal regulations such as the General Data Protection Regulation (GDPR) [28] to guarantee that privacy-sensitive data is not disclosed. Additionally, most Context-Aware Security solutions overlook the high diversity of IoT environments by failing to provide a comprehensive, interoperable mechanism for context information.

7   Conclusion

The field of Context-Aware Security in IoT is rapidly growing and gaining attention, both in academia and industry. With the increasing number of IoT devices in everyday use, it is crucial to have dynamic security solutions that can adapt to the changing context and provide the necessary protection. In this review, we aimed to provide a comprehensive overview of Context-Aware Security in IoT by addressing the key requirements and proposing a novel taxonomy. Additionally, we presented a detailed analysis of various existing Context-Aware Security solutions, showcasing their strengths and weaknesses. We also discussed the challenges and open questions surrounding these solutions and offered potential avenues for improvement.

Through this review, we hope to provide valuable insights and guidance for researchers, practitioners, and developers working in the field of Context-Aware Security in IoT. By highlighting the most important aspects of Context-Aware

Security and presenting a comprehensive overview of existing solutions, we aim to facilitate the development of new, more effective and efficient Context-Aware Security solutions in IoT.

References

1. Abowd, G.D., Dey, A.K., Brown, P.J., Davies, N., Smith, M., Steggles, P.: Towards a better understanding of context and context-awareness. In: Gellersen, H.-W. (ed.) HUC 1999. LNCS, vol. 1707, pp. 304–307. Springer, Heidelberg (1999). https://doi.org/10.1007/3-540-48157-5_29
2. Al-Muhtadi, J., Ranganathan, A., Campbell, R., Mickunas, M.D.: Cerberus: a context-aware security scheme for smart spaces. In: Proceedings of the 1st IEEE International Conference on Pervasive Computing and Communications, pp. 489–496, Machr 2003. https://doi.org/10.1109/PERCOM.2003.1192774
3. Al-Turjman, F., Alturjman, S.: Context-sensitive access in industrial internet of things (IIoT) healthcare applications. IEEE Trans. Industr. Inf. 14(6), 2736–2744 (2018). https://doi.org/10.1109/TII.2018.2808190
4. Alagar, V., Alsaig, A., Ormandjiva, O., Wan, K.: Context-based security and privacy for healthcare IoT. In: Proceedings of the 2nd IEEE International Conference on Smart Internet of Things, pp. 122–128, August 2018. https://doi.org/10.1109/SmartIoT.2018.00-14
5. Nasir, A.: An ontology based approach for context-aware security in the internet of things (IoT). Int. J. Wirel. Microwave Technol. 11(1), 28–46 (2021). https://doi.org/10.5815/ijwmt.2021.01.04
6. Atzori, L., Iera, A., Morabito, G.: The internet of things: a survey. Comput. Netw. 54(15), 2787–2805 (2010). https://doi.org/10.1016/j.comnet.2010.05.010
7. Brezillon, P., Mostefaoui, G.K.: Context-based security policies: a new modeling approach. In: Proceedings of the 2nd IEEE Annual Conference on Pervasive Computing and Communications Workshops, pp. 154–158, March 2004. https://doi.org/10.1109/PERCOMW.2004.1276923
8. Future Market Insights: Global Internet of Things (IoT) Security Product Market Overview (2017). https://www.futuremarketinsights.com/reports/internet-of-things-security-products-market
9. Gansel, S., Schnitzer, S., Gilbeau-Hammoud, A., Friesen, V., Dürr, F., Rothermel, K., Maihöfer, C., Krämer, U.: Context-aware access control in novel automotive HMI systems. In: Jajodia, S., Mazumdar, C. (eds.) ICISS 2015. LNCS, vol. 9478, pp. 118–138. Springer, Cham (2015). https://doi.org/10.1007/978-3-319-26961-0_8
10. Gheisari, M., Wang, G., Khan, W.Z., Fernández-Campusano, C.: A context-aware privacy-preserving method for IoT-based smart city using software defined networking. Comput. Secur. 87, 101470 (2019). https://doi.org/10.1016/j.cose.2019.02.006, http://www.sciencedirect.com/science/article/pii/S0167404818313336
11. Grimm, D., Stang, M., Sax, E.: Context-aware security for vehicles and fleets: a survey. IEEE Access 9, 101809–101846 (2021). https://doi.org/10.1109/ACCESS.2021.3097146
12. Gupta et al., H.: iFogSim: a toolkit for modeling and simulation of resource management techniques in the Internet of Things, Edge and Fog computing environments. Softw. Pract. Exp. 47(9), 1275–1296 (2017). https://doi.org/10.1002/spe.2509

13. Habib, K., Leister, W.: Context-aware authentication for the internet of things. In: Proceedings of the 11th International Conference on Autonomic and Autonomous Systems, p. 6 (2015)

14. Hosseinzadeh, S., Virtanen, S., Díaz-Rodríguez, N., Lilius, J.: a semantic security framework and context-aware role-based access control ontology for smart spaces. In: Proceedings of the 1st International Workshop on Semantic Big Data, pp. 8:1–8:6. SBD 2016, ACM, New York, NY, USA (2016). https://doi.org/10.1145/2928294.2928300

15. Hu et al., J.: A dynamic, context-aware security infrastructure for distributed healthcare applications. In: Proceedings of the 1st Workshop on Pervasive Privacy Security, Privacy, and Trust, pp. 1–8. Citeseer (2004)

16. Jia, Y.J., et al.: ContexIoT: towards providing contextual integrity to appified IoT platforms. In: Proceedings of the 21st Network and Distributed System Security Symposium, pp. 1–15 (2017)

17. Li, W., Joshi, A., Finin, T.: SVM-CASE: an SVM-based context aware security framework for vehicular Ad-Hoc networks. In: Proceedings of the 82nd IEEE Vehicular Technology Conference, pp. 1–5, September 2015. https://doi.org/10.1109/VTCFall.2015.7391162

18. de Matos, E., Tiburski, R.T., Amaral, L.A., Hessel, F.: Providing context-aware security for IoT environments through context sharing feature. In: Proceedings of the 17th IEEE International Conference On Trust, Security And Privacy In Computing And Communications, pp. 1711–1715, August 2018. https://doi.org/10.1109/TrustCom/BigDataSE.2018.00257

19. de Matos, E., et al.: Context information sharing for the Internet of Things: a survey. Comput. Netw. **166**, 1–19 (2020). https://doi.org/10.1016/j.comnet.2019.106988, http://www.sciencedirect.com/science/article/pii/S1389128619310400

20. Mostefaoui, G.K., Brezillon, P.: Modeling context-based security policies with contextual graphs. In: Proceedings of the IEEE Annual Conference on Pervasive Computing and Communications Workshops, pp. 28–32, March 2004. https://doi.org/10.1109/PERCOMW.2004.1276900

21. Perera, C., Zaslavsky, A., Christen, P., Georgakopoulos, D.: Context aware computing for the internet of things: a survey. IEEE Commun. Surv. Tutor. **16**(1), 414–454 (2014). https://doi.org/10.1109/SURV.2013.042313.00197

22. Psarra, E., Verginadis, Y., Patiniotakis, I., Apostolou, D., Mentzas, G.: A context-aware security model for a combination of attribute-based access control and attribute-based encryption in the healthcare domain. In: Barolli, L., Amato, F., Moscato, F., Enokido, T., Takizawa, M. (eds.) Web, Artificial Intelligence and Network Applications, pp. 1133–1142. Springer International Publishing, Cham (2020). https://doi.org/10.1007/978-3-030-44038-1_104

23. Rachid, S., Challal, Y., Nadjia, B.: Internet of things context-aware privacy architecture. In: Proceedings of the 12th IEEE/ACS International Conference of Computer Systems and Applications, pp. 1–2, November 2015. https://doi.org/10.1109/AICCSA.2015.7507247

24. Ramos, J.L.H., Bernabe, J.B., Skarmeta, A.F.: Managing context information for adaptive security in IoT environments. In: Proceedings of the 29th IEEE International Conference on Advanced Information Networking and Applications Workshops, pp. 676–681, March 2015. https://doi.org/10.1109/WAINA.2015.55

25. Sikder, A.K., Babun, L., Aksu, H., Uluagac, A.S.: Aegis: a context-aware security framework for smart home systems. In: Proceedings of the 35th Annual Computer Security Applications Conference, pp. 28–41. ACSAC 2019, Association for Computing Machinery, New York, NY, USA (2019). https://doi.org/10.1145/3359789.3359840

26. Sylla, T., Chalouf, M.A., Krief, F., Samaké, K.: Context-aware security in the internet of things: a survey. Int. J. Autonom. Adapt. Commun. Syst. **14**(3), 231–263 (2021). https://doi.org/10.1504/IJAACS.2021.117808, https://www.inderscienceonline.com/doi/abs/10.1504/IJAACS.2021.117808

27. Sylla, T., Chalouf, M.A., Krief, F., Samaké, K.: Setucom: secure and trustworthy context management for context-aware security and privacy in the internet of things. Secur. Commun. Netw. **2021**, 6632747 (2021). https://doi.org/10.1155/2021/6632747

28. Tikkinen-Piri, C., Rohunen, A., Markkula, J.: EU general data protection regulation: changes and implications for personal data collecting companies. Comput. Law Secur. Rev. **34**(1), 134–153 (2018). https://doi.org/10.1016/j.clsr.2017.05.015, http://www.sciencedirect.com/science/article/pii/S0267364917301966

29. Trnka, M., Cerny, T.: On security level usage in context-aware role-based access control. In: Proceedings of the 31st Annual ACM Symposium on Applied Computing. pp. 1192–1195. SAC 2016, ACM, New York, NY, USA (2016). https://doi.org/10.1145/2851613.2851664

30. Zhang, L., Li, Y., Wang, L., Lu, J., Li, P., Wang, X.: An efficient context-aware privacy preserving approach for smartphones. Secur. Commun. Netw. **2017**, 1–11 (2017). https://doi.org/10.1155/2017/4842694

Cybersecurity Attacks and Vulnerabilities During COVID-19

Sharmin Akter Mim[1]([✉]), Roksana Rahman[2], Md. Rashid Al Asif[3],
Khondokar Fida Hasan[4], and Rahamatullah Khondoker[5]

[1] Department of CSE, Bangladesh University of Engineering and Technology,
Dhaka, Bangladesh
1017052070@grad.cse.buet.ac.bd
[2] IDLC Finance Limited, Dhaka, Bangladesh
rroksana@idlc.com
[3] Department of CSE, University of Barishal, Barishal, Bangladesh
mraasif@bu.ac.bd
[4] Queensland University of Technology, Brisbane, Australia
fida.hasan@qut.edu.au
[5] Department of Business Informatics, THM University of Applied Sciences,
Friedberg, Germany
rahamatullah.khondoker@mnd.thm.de

Abstract. As a result of quick transformation to digitalization for providing the employees teleworking/home office services with the capabilities to access company resources from outside the company over Internet using remote desktop and virtual private network (VPN) applications and the increase in digital activity during COVID-19 such as the usage of audio/video conferencing applications, many businesses have been victims of cyber attacks. This paper investigates whether there was an increase in the frequency of cyber attacks during COVID-19. It also identifies the motivations for such attacks in light of software/hardware/system vulnerabilities. Following this research, we also categorize vulnerabilities and develop a taxonomy. Such a taxonomy helped to identify the type of attacks on their frequency and their impact. To do that, we developed a research methodology to collect attack and vulnerability information from the selected databases. Using relevant key words, we developed the taxonomy that led us to create insightful information to answer the research questions that are thoroughly analyzed and presented accordingly. This work also recommended a list of mitigation measures that can be considered in the future to prepare the industry for a similar pandemic including establishing and maintaining a Information Security Management System (ISMS) by following relevant standards (ISO/SAE 2700x, BSI-Standards 200-x, SMEs: CISIS12®).

1   Introduction

The coronavirus, also known as SARS-CoV-2 or COVID-19 and first identified in December 2019, which transmits from one person to another through close interactions, is the cause of the COVID-19, which the World Health Organization designated a global pandemic on March 11, 2020. For minimizing the impact of COVID-19 by reducing close contacts, many organizations allowed their employees to stay at home. Those organizations that had good IT infrastructure allowed their employees to work from home. However, to do that, some organizations needed to react quickly to adapt their IT infrastructure (for example, the deployment of servers to handle the data traffic, virtual meeting infrastructures, etc.), which is called fast digitization. Due to this fast digitization, cybersecurity defense mechanisms were not deployed properly. On the other hand, those organizations that could not migrate to fast digitalization, went for government support to pay their employees. Unfortunately, the unemployment rate went high during COVID-19. During COVID-19, the cyber attackers and attacks were not stopped, no matter what their reasons were (unemployment, money, reputation, etc.). The question is whether cyber attacks increased during COVID-19 compared to the pre-pandemic period? What types of attacks are observed during COVID-19? Were new types of attacks introduced? If yes, what are those attacks? To find the answer, we used the methodology discussed in Sect. 3, created a taxonomy in Sect. 4, mentioned related work in Sect. 5, talked about setup, preconditions, and assumptions in Sect. 6, presented the results in Sect. 7, discussed analysis and evaluation in Sect. 8, referenced some mitigation techniques in Sect. 9, and concluded the paper with a summary and future work in Sect. 10.

2   Background

New Normal: There is a global transformation in working patterns after the pandemic of COVID-19. Most of the organizations having good IT infrastructure allowed their employees to work from home. And after working remotely for two years, this becomes a new normal to employees life. A number of cyber security issues has been raised because of working from home although working remotely can be convenient, promote work-life balance, and eliminate the need for a commute. Controlling security measures has become a horrible headache for an IT team as a result of employee reliance on personal networks and occasionally their own devices, new online tools and services, and more diversions. Because of the susceptibility of remote workers and the public's heightened interest in news about the coronavirus, cybercriminals perceive the epidemic as an avenue to intensify their illicit activities (e.g. malicious fake coronavirus related websites). According to statistics from the NCSC (National Cyber Security Center), there were 350 reported cases of cyber attacks (phishing, fraudulent websites, direct attacks on companies, etc.) in Switzerland in April 2020, compared to the norm of 100–150 [33]. Major contributing factors to this increase include the coronavirus pandemic and an increase in telecommuting since those who work from home lack the same amount of built-in security as those who operate in an office setting (e.g. internet security). The rise in remote working necessitates a higher attention on cybersecurity due to the increasing vulnerability to cyber risks.

Virtualization: The last ten years have seen a number of developments that have impacted industries and altered how we adjust to life, including the digitalization of services, the move to cloud infrastructure, high dependence on Internet connectivity, AI - powered analytical capabilities, deep learning, and the modern economy. Similar seismic changes in the area had been brought on by COVID-19 and the ensuing lockdowns in a matter of weeks. Long-term planning had been challenging due to the erratic and imprecise nature of regulation, policy, and guidelines. Because of the shutdown of physical offices and transit systems, many businesses and government organizations were compelled to go online. Working remotely was becoming more common than unusual. Because their previous business models relied heavily on physical encounters, food shops, gymnasium, movie theaters, and educational institutions have had to entirely reinvent their offerings. As a result, we have seen a surge in virtualization over the last two years. In many instances, up overnight, global connectivity and data mobility have overtaken physical movements of individuals and products as the foundation of globalization.

Relation of Security Relevant Terms: With rapid virtualization, the system resources need to be protected by deploying security measures against cyber attacks.

The threat actor (individual or organization) exploits system vulnerabilities for benefits/motivations. Threat actors unauthorizedly expose system entry points that may exploit system vulnerabilities [7]. And, the vulnerabilities are a prerequisite to initiating further security attacks.

System entry points are considered Attack Surface, which is very large and complex for today's business organization. The threat actor launches small to large attacks, which lead to huge risks for organizations.

3   Research Method

The question is, whether the attacks and vulnerabilities have been increased or decreased during COVID-19. The method to find out the answer is displayed in Fig. 1.

We have listed and searched 54 attack and vulnerability databases and news portals including Software Engineering Institute (SEI), Security Mailing List Archive (SECLISTS), Vulners, Defcon, Microsoft, and Network Vulnerability Database (NVD).

For each database, we listed whether it is an open source database and whether the database can be searched for a specific duration (from a specific date to a specific date). Afterwards, considering the reliability of the data in the database and open source as criteria, one vulnerability database called National Vulnerability Database (NVD) has been selected. Since NVD reflects only vulnerabilities not the attacks, therefore, we narrowed down our research to it. Then, an attack taxonomy has been created considering the definitions from the standards including NIST [21], ISO [9], and BSI [4]. Based on the taxonomy, the chosen database is searched and the obtained result is displayed and analyzed in Sect. 7 and Sect. 8 respectively.

Fig. 1. Research method

4 Taxonomy

There is no standardized cybersecurity taxonomy, however, a taxonomy has been created in Fig. 2 based on the definitions from cybersecurity standards including NIST, ISO, and BSI. The word cybersecurity is written as one word or two words where there is a space in between cyber and security [18]. ISO defined cyber security as "protection of an IT-system from the attack or damage to its hardware, software or information, as well as from disruption or misdirection of the services it provides" [18] and as "safeguarding of people, society, organizations and nations from cyber risks" [16]. Several cybersecurity topics are cyber attack potential (which increases probability of cyber attacks due to weak implementation/mechanism), cyber attack method ("Method for an attack, via cyberspace, targeting an enterprise's use of cyberspace for the purpose of disrupting, disabling, destroying, or maliciously controlling a computing environment/infrastructure; or destroying the integrity of the data or stealing controlled information" [25]), cyber security property ("property of a system or application that is crucial to achieve the security objectives defined for the system or application" [15]), cyber attack target (aim for a cyber attack), and cyber attack source (origin of a cyber attack). Various instances of cyber security potential are weak monitoring, broken access control, security misconfiguration, broken authentication, weak logging, insecure deserialization and weak permission. Different forms of cyber attack methods are exploit, privacy attack (attack on "assurance that the confidentiality of, and access to, certain information about an entity is protected" [22]), buffer overflow attack ("a method of overloading a predefined amount of memory storage in a buffer, which can potentially overwrite and corrupt memory beyond the buffer's boundaries" [23]), password attack ("a method of accessing an obstructed device by attempting multiple combinations of numeric/alphanumeric passwords" [23]), ransomware ("a type of malicious attack where attackers encrypt an organization's data and demand payment to restore access."), phishing attack (is termed as "scam by which an email user is duped into revealing personal or confidential information which the scammer can then use illicitly" [14]), spam ("unsolicited contents which are deemed to be of no relevance or long-term value" [17]), brute force attack ("in cryptography, an attack that involves trying all possible combinations to find a match" [29]), social engineering attack (is interpreted as an "act of manipulating people into performing actions or divulging confidential information" [13]), spyware ("devices or software that capture a participant's data or behaviour without obtaining consent" [10]), trojan ("apparently harmless program containing malicious logic that allows the unauthorized collection, falsification, or destruction of data" [11]) and path traversal. Integrity ("a property possessed by data items that have not been altered in an unauthorized manner since they were created, transmitted, or stored" [24]), availability ("ensuring timely and reliable access to and use of information" [27]) and confidentiality ("preserving authorized restrictions on access and disclosure, including means for protecting personal privacy and proprietary information" [27]) are three important cyber security property. Common forms of cyber attack targets are network ("a system implemented with a collection of connected components" [26]), web, database and end point. Backdoor (which is described as "an undocumented way of gaining access to computer system" [28]) and botnet (which is defined as "remote control software, specifically a collection of malicious bots, that run autonomously or automatically on compromised computers" [12]) are two common type of cyber attack sources.

5   Related Work

Numerous works on cybersecurity during COVID-19 have been done. Some of the works which are relevant to the work of this article have been mentioned here. For example, in a study [30], three categories of cybersecurity issues during COVID-19 namely phishing, malware, and DDoS and proposed some mitigations to protect from those attacks were discussed. In this article, however, we illustrated the trends of cybersecurity vulnerabilities during COVID-19 compared to pre-COVID period and we extended the mitigation which were missed in their work.

Over the course of a year, there were 377.5 million brute force attacks on the Remote Desktop Protocol (RDP), up from 93.1 million (February 2020 to February 2021) [32]. In addition, it is shown that the number of malicious files has been spread through the meeting apps (for example, Webex, Zoom, MS Teams, High-Five, Lifesize, Join.me, Slack, Flock, Gotomeeting) has been increased from 90 Thousands to 1.15 Million during the one year time period (March 2020 to February 2021). We have not done this specific type of analysis in this paper, rather, we wanted to illustrate the cybersecurity vulnerabilities during the COVID-19 period compared to pre-COVID period.

Common Vulnerabilities and Exposures (CVE) entries from the National Vulnerability Database (NVD) were retrieved and a method of classifying CVE entries into vulnerability types was presented using a naive Bayes classifier in [20]. For this classification and validation process, between 1999 and 2016, a total 77,885 CVE entries were obtained from NVD. They categorized the ten leading CWEs on the basis of CWE frequency as well as CWE-119 and CWE-79, which have the most CWEs (Common Weakness Enumeration) that have been

(a) Classification of Cyber Security

(b) Classification of Cyber Security Property and Attack Source

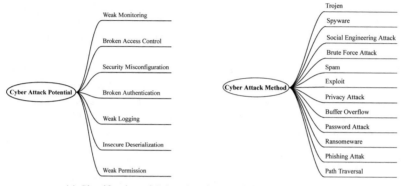

(c) Classification of Cyber Attack Potential and Attack Method

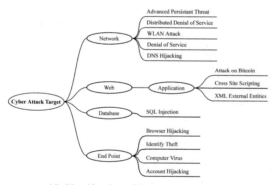

(d) Classification of Cyber Attack Target

Fig. 2. Cyber attack taxonomy

detected. Although this is one kind of taxonomy, but we have not done this type of categorizing in our paper, rather, we wanted to illustrate the cybersecurity vulnerabilities during the COVID-19 period (year 2020 to 2021) compared to pre-COVID period.

The authors of the aricle [30] emphasized the correlation between the COVID-19 pandemic and cybersecurity issues. They identified notable cyber attacks (especially phishing and ransomware) and vulnerabilities where the attackers targeted vulnerable sectors. Authors highlighted the healthcare sector as one of the victims of cyber attacks which requires strong protection. Finally, they provided various practical approaches to reducing cyber attacks.

From the standpoint of cyber-crime and a wide variety of cyber attack incidents, the COVID-19 pandemic was analyzed in [19]. The cyber-criminals and APT groups initiate cyber attacks by targeting the vulnerable system and people. As a consequence, it generates a set of special circumstances associated with cyber-crime that have an effect on society and industry.

The COVID-19 pandemic caused global digital transformation while increased threats and vulnerabilities and thus, it causes a global impact on the IT industry in terms of business and the economy [35]. Within this context, the authors of [35] examined whether the IT industry can makeover these changes or handle crises that might face in the future.

In [3], cyber attacks at the beginning of the COVID-19 pandemic in Italy was described. The pandemic moved companies to extensive use of digital services. This work found that the number of attacks has not increased but affects the small companies than earlier. Moreover, the Italian small and medium industry lacks the necessary cybersecurity culture to oblige quick reconversion during the pandemic.

The probability of companies being compromised by many folds due to the rush of working from home was remarked in [8]. The undergo analysis pinpoint weakness and barriers that deliver countermeasures against cybersecurity in organizations. This propelled safety for organizational intellectual properties and stimulates the adoption of protection systems and safety practices. Besides, [2] is expected to maintain cybersecurity standards and best practices to ensure organizational business continuity. Hence, to minimize the potential impact of cyber attacks.

In another study [31], distinct types of cybersecurity and threats during the period COVID-19 pandemic as well as recommendations and defense strategies for each type was demonstrated. Attacks have been studied in the view of the pandemic and classified into denial of service, information leakage, flow control, and injection. Moreover, researchers analyzed the efficiency of physical infrastructure as well as hacker's and criminal's behavior during the time of the pandemic.

6   Setup/Pre-condition/Assumption

The timeline of our study is given in Table 1. The reasons for selecting the time slots for our study are as follows. Employees used to go to the office for work before COVID-19 was proclaimed as a pandemic (From December 2019 to 10th March 2020). However, after the declaration, many companies allowed their employees to work from home. We selected two years (from 11th March 2020 to 10th March 2021 and from 11th March 2021 to 10th March 2022) during COVID-19 for our study to understand the difference in cybersecurity issues even during the COVID-19 period. We would like to investigate whether the IT

Table 1. Timeline of study and number of vulnerabilities

Title for the duration	Duration	Total vulnerabilities
Before COVID-19	From 11.03.2019 To 10.03.2020	13880
During COVID-19 First Year	From 11.03.2020 To 10.03.2021	13514
During COVID-19 s Year	From 11.03.2021 To 10.03.2022	16262

infrastructure was made stronger against cyber attacks with the introduction of, for example, defense mechanisms and security critical updates during the 2nd year compared to the 1st year of the pandemic.

7   Search and Found

The selected database NVD was searched by using the *Search Type* Option as *Advanced* because the *Basic* type does not allow to search using a specific time slot as it is seen in Fig. 3.

Fig. 3. National vulnerability database (basic search type)

Fig. 4. National vulnerability database (advanced search type)

The options for *Result Type* are: *Overview*, where an overview of the vulnerabilities are shown using three columns including Vulnerability (Vuln) ID, Summary (Short Description and Published Date), and CVSS Severity (Versions 2.0 and 3.X) whereas in the *Statistics* option several graphics are displayed such as *Total Matches by Year*, *Percent Matches by Year*, and *Raw Data* with three columns Year, Matches, Total, and Percentage. In the Keyword Search, the terms to be searched are inserted. In this case, we used the standard term from our taxonomy. However, we do not use the option *Exact Match* to be able to get vulnerabilities with the similar terms. The options for *Contains Hyperlinks* to search the database with the keyword which has detail information attached to it as a hyperlink from organizations such as US-CERT (Coordinators are from Software Engineering Institute from Carnegie Mellon University) Technical Alerts, US-CERT Vulnerability Notes, and OVAL (Open Vulnerability and

Assessment Language) Queries. When then *Basic Search Type* is selected, the database can be search for the all duration ("All Time") or for "Last 3 Months".

The options for the *Advanced Search Type* is shown in Fig. 4. In this type, it is possible to search a keyword in a specific time slot considering the publication(Published Data Range) and update (Last Modified Data Range) of the vulnerabilities. To note that, the date format is MM/DD/YYYY (not DD/MM/YYYY). In this search type, CVSS Metrics can be selected either for one or for both versions. This search type inherits some options from the basic type such as *Results Type, Keyword Search,* and *Contains Hyperlinks.* In addition, in this advanced search type, it is possible to search the database with CVE (Common Vulnerability Enumeration) Identifier and CWE (Common Weakness Enumeration) Category. Right now, there are 1286 CWE categories defined and two extra CWE where no info is available (NVD-CWE-noinfo) or cannot be categorized called NVD-CWE-other. Last year, we could search the NVD database for any data range, however, this is not possible now. The search data range is to limited to maximum 120 days.

8 Analysis and Evaluation

The result of cyber attack potential before and during COVID-19 (1st and 2nd year) is shown in Fig. 5 where the inner circle represents the percentage of attack potential before COVID-19, middle circle represents the same during the 1st year of COVID-19 and the outer circle represents the same during the 2nd year of COVID-19.

As it is seen from the Fig. 5, the broken authentication has been 32% before COVID-19, 42% during the 1st year of COVID-19 and 34% during the 2nd year of COVID-19. This means that, the broken authentication vulnerability has been 10% more than one year earlier. The reasons are:

Figure 6 illustrates the occurrences of various types of cyber attack target which were happened before and during (1st and 2nd year) COVID-19. We have seen a great impact in the second year of COVID-19.

Figure 7 depicts the level of occurrences of different kinds of cyber attack methods before and during (1st and 2nd year) COVID-19. It shows that cyber attack has increased during COVID-19 in comparison to before COVID-19. Privacy attack, buffer overflow attack, social engineering attack and path traversal have increased in the second year of COVID-19 in comparison to before and first year of COVID-19.

Figure 8 illustrates two types of cyber attack sources and the number of occurrences of each. It indicates that cyber attacks employing botnet increased during COVID-19 compared to before COVID-19. It also demonstrates that cyber attacks through backdoor has decreased during COVID-19 compared to before COVID-19.

It is seen from the result Table 2 that the APT attack has been increased during COVID-19 which can be proved by the [5] news titled "Microsoft says three APTs have targeted seven COVID-19 vaccine makers".

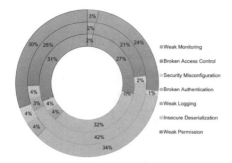

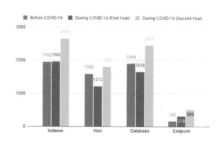

Fig. 5. Before COVID-19: inner circle, during COVID-19 (1st year): middle circle, during COVID-19 (2nd year): outer circle)

Fig. 6. Cyber attack target before COVID-19 and during COVID-19 (1st and 2nd year)

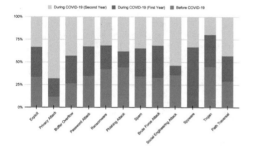

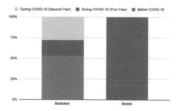

Fig. 7. Cyber attack methods before COVID-19 and during COVID-19 (1st and 2nd year)

Fig. 8. Cyber attack sources before COVID-19 and during COVID-19 (1st and 2nd year)

9 Mitigation

One of the lessons to be learned from COVID-19 is that the IT infrastructure should be ready for such pandemic at any time. Therefore, a set of security controls is recommended as follows:

Up-to-date System: All of the systems (Servers, Clients) in the IT infrastructure should be always up-to-date. Update of software components (Driver, OS, Middle ware, Application) should be done as soon as they are available. Before the update, all data and applications should be kept back up in case of update failure. Hardware update should be done when its cybersecurity features and capabilities are enhanced.

Up-to-date Network: The network should be up-to-date as well with the updated version of networking components such as switches, routers, gateways and firewalls [6,34]. In addition, these devices need to be checked and (re)configured in a regular basis with the defined cybersecurity policies and rules from the enterprises and standards/regulations. The cybersecurity policies and

Table 2. Search result of NVD database in different time periods

Search keyword	Before COVID-19	During COVID-19 (1st year)	During COVID-19 (2nd year)
(Account) hijacking	69	69	101
APT (advanced persistent threat)	200	212	236
(Attack on) bitcoin	5	15	5
Backdoor	25	9	13
Botnet	0	3	0
(Broken) access control	638	561	670
(Broken) authentication	769	1134	951
Browser hijack(ing)	7	10	11
Brute force (attack)	73	78	70
Buffer overflow	760	865	1220
Computer virus	1	1	0
CSS (cross site scripting)	41	42	66
Denial of service	1619	1571	2252
DNS (Hijacking)	125	178	150
Exploit	4361	4279	4,331
Identity (theft)	92	86	101
Information leakage	40	37	30
(Insecure) deserialization	89	98	107
Injection	1296	1233	1648
(Weak) logging	87	85	99
(Weak) monitoring	44	56	70
Password attack	368	344	345
Path traversal	162	157	241
Phishing (attack)	42	17	36
Privacy attack	3	5	17
Ransom(wear)	8	5	6
(Security) misconfiguration	15	53	31
Social engineering attack	10	3	15
Spam	20	18	20
Spyware	0	2	1
SQL injection	603	423	811
Trojan	23	18	10
(Weak) permission	738	687	847
Web (application) attack	1222	832	1605
Web-based attacks	209	231	32
WLAN attack	8	5	17
XXE (XML external entities)	108	92	97

rules need to be reviewed in a regular basis (weekly, monthly) so that new types of attacks can be detected and mitigated [1].

Establish an ISMS and Follow the Standards: An Information Security Management System (ISMS) should be established by following standards such as CISIS12® for SMEs, ISO 2700x, BSI-Standards 200-x, and Compendium and the Common Criteria from the German Federal Office of Information Security (BSI). In addition, domain-specific cybersecurity standards should be followed, such as ISO 21434 for automotive systems/components and ISO 14971 for medical systems.

Follow the Regulation: Cybersecurity regulations are available in different levels: country, continent, and world, and should be followed accordingly to avoid financial penalties. For example, the General Data Protection Regulation (GDPR) to protect private data of EU citizens is valid worldwide.

Incident Response Management (IRM): Every IT team should have a IRM subteam (CERT) to monitor and detect abnormal behavior of the IT infrastructure in 24/7 basis and implement defense mechanisms accordingly. Normally, such a CERT (Computer Emergency Response Team) team is available in every city, country, and continent level. Local CERT team should collaborate with these teams to predict the cybersecurity threats and attacks and prepare (defend) own networks accordingly.

Cybersecurity Training: Employees are one of the weak links of cybersecurity; therefore, they should be trained on a regular basis by providing cybersecurity awareness training, as well as job/role specific training. Additionally, they should be checked with periodic cybersecurity assessments about their knowledge on cybersecurity and the next training plan to be offered/recommended accordingly.

10   Conclusion and Future Work

In this work, we compared cyber attacks before and during COVID-19. We found out that, the overall cyber attacks have increased during COVID-19 for our selected database sources and timeline. Hence, the mitigation strategies proposed in this article will be helpful for a similar pandemic in the future.

In the future, researchers could concentrate on other types of cyber attacks or in other time slot. In addition, researchers could provide some recommendations/solutions how to deal with those type of attacks.

References

1. Al Asif, M.R., et al.: Stride-based cyber security threat modeling for IoT-enabled precision agriculture systems. In: 2021 3rd International Conference on STI 4.0, pp. 1–6. IEEE (2021)
2. Bejarano, M.H., et al.: Cybersecurity and business continuity in pandemic times. Ann. Roman. Soc. Cell Biol. **25**(6), 7280–7289 (2021)

3. Bozzetti, M.R., et al.: Cybersecurity impacts of the COVID-19 pandemic in Italy. In: 5th Italian Conference on Cybersecurity, ITASEC 2021, pp. 145–155 (2021)

4. BSI (2022). Standards, Training, Testing, Assessment and Certification — BSI. Accessed 15 Sep 2022

5. Cimpanu, C.: Microsoft says three APTs have targeted seven COVID-19 vaccine makers (2020). https://www.zdnet.com/article/microsoft-says-three-apts-have-targeted-seven-covid-19-vaccine-makers/. Accessed 24 Feb 2022

6. Hasan, K.F., et al.: Cognitive internet of vehicles: motivation, layered architecture and security issues. In: 2019 International Conference on STI 4.0, pp. 1–6. IEEE (2019)

7. Hasan, K.F., et al.: Security, privacy, and trust of emerging intelligent transportation: cognitive internet of vehicles. In Next-Generation Enterprise Security and Governance, pp. 193–226. CRC Press (2022)

8. Hejase, H.J., et al.: Cyber security amid COVID-19. Comput. Inf. Sci. **14**(2), 1–10 (2021)

9. ISO (2022). ISO - International Organization for Standardization. Accessed 15 Sep 2022

10. ISO 20252:2019(en): Market, opinion and social research, including insights and data analytics — Vocabulary and service requirements. Standard, International Organization for Standardization, Geneva, CH (2019)

11. ISO/IEC 2382:2015(en): Information technology — Vocabulary.Standard, International Organization for Standardization, Geneva, CH (2015)

12. ISO/IEC 27032:2012(en): Information technology — Security techniques — Guidelines for cybersecurity. Standard, International Organization for Standardization, Geneva, CH (2012)

13. ISO/IEC 27033-3:2010(en): Information technology — Security techniques — Network security — Part 3: Reference networking scenarios — Threats, design techniques and control issues. Standard, International Organization for Standardization, Geneva, CH (2010)

14. ISO/IEC 29115:2013(en): Information technology — Security techniques — Entity authentication assurance framework. Standard, International Organization for Standardization, Geneva, CH (2013)

15. ISO/IEC TS 19249:2017(en): Information technology — Security techniques — Catalogue of architectural and design principles for secure products, systems and applications. Standard, International Organization for Standardization, Geneva, CH (2017)

16. ISO/IEC TS 27100:2020(en): Information technology — Cybersecurity — Overview and concepts. Standard, International Organization for Standardization, Geneva, CH (2020)

17. ISO/TR 14873:2013(en): Information and documentation — Statistics and quality issues for web archiving. Standard, International Organization for Standardization, Geneva, CH (2013)

18. ISO/TR 22100-4:2018(en): Safety of machinery — Relationship with ISO 12100 — Part 4: Guidance to machinery manufacturers for consideration of related IT-security (cyber security) aspects. Standard, International Organization for Standardization, Geneva, CH (2018)

19. Lallie, H.S., et al.: Cyber security in the age of COVID-19: a timeline and analysis of cyber-crime and cyber-attacks during the pandemic. Comput. Secur. **105**, 102248 (2021)

20. Na, S., Kim, T., Kim, H.: A study on the classification of common vulnerabilities and exposures using Naïve Bayes. In: BWCCA 2016. LNDECT, vol. 2, pp. 657–662. Springer, Cham (2017). https://doi.org/10.1007/978-3-319-49106-6_65
21. NIST (2022). National Institute of Standards and Technology. Accessed 15 Sep 2022
22. NIST 1800–25: Data Integrity: Identifying and Protecting Assets Against Ransomware and Other Destructive Events. NIST special publication 1800–25, National Institute of Standards and Technology, US (2020)
23. NIST 800–101: Guidelines on Mobile Device Forensics. NIST special publication 800–101, National Institute of Standards and Technology, US (2014)
24. NIST 800–133. Recommendation for Cryptographic Key Generation. NIST special publication 800–133, National Institute of Standards and Technology, US (2020)
25. NIST 800–39: Information Security, Managing Information Security Risk Organization, Mission, and Information System View. NIST special publication 800–39, National Institute of Standards and Technology, US (2011)
26. NIST 800–53: Security and Privacy Controls for Information Systems and Organizations. NIST special publication 800–53, National Institute of Standards and Technology, US (2020)
27. NIST 800–59: Guideline for Identifying an Information System as a National Security System. NIST special publication 800–59, National Institute of Standards and Technology, US (2003)
28. NIST 800–82: Guide to Industrial Control Systems (ICS) Security. NIST special publication 800–82, National Institute of Standards and Technology, US (2013)
29. NISTIR 8053: De-Identification of Personal Information. NIST internal report 8053, National Institute of Standards and Technology, US (2015)
30. Pranggono, B., Arabo, A.: COVID-19 pandemic cybersecurity issues. Internet Technol. Lett. 4(2), e247 (2021)
31. Ramadan, R.A., et al.: Cybersecurity and countermeasures at the time of pandemic. J. Adv. Transp. 2021, 1–9 (2021)
32. SECURE LIST: COVID-19: Examining the threat landscape a year later (2021). https://securelist.com/covid-19-examining-the-threat-landscape-a-year-later/101154//. Accessed 24 Feb 2022
33. SWI: Jump in cyber attacks during COVID-19 confinement (2020)
34. Talukder, M.A., et al.: A dependable hybrid machine learning model for network intrusion detection. JISA 72, 103405 (2023)
35. Weil, T., Murugesan, S.: It risk and resilience-cybersecurity response to COVID-19. IT Prof. 22(3), 4–10 (2020)

Identifying Fake News in the Russian-Ukrainian Conflict Using Machine Learning

Omar Darwish[1]([✉]), Yahya Tashtoush[2], Majdi Maabreh[3], Rana Al-essa[2],
Ruba Aln'uman[2], Ammar Alqublan[2], Munther Abualkibash[1], and Mahmoud Elkhodr[4]

[1] Information Security and Applied Computing, Eastern Michigan University, Ypsilanti, USA
{odarwish,mabualki}@emich.edu
[2] Computer Science, Jordan University of Science and Technology, Irbid, Jordan
yahya-t@just.edu.jo, {raaleesa21,rsalnuman21,
aealqoblan20}@cit.just.edu.jo
[3] Information Technology, The Hashemite University, Zarqa, Jordan
majdi@hu.edu.jo
[4] Central Queensland University, Rockhampton, Australia
m.elkhodr@cqu.edu.au

Abstract. The widespread dissemination of false information made possible by social networks' universal accessibility and ease of use may be harmful to both individuals and societies. Regrettably, one of the most popular subjects on social media right now is the crisis between Russia and Ukraine. Spreading fake news about this conflict may cause serious consequences to both countries and their citizens. Therefore, we are motivated to build a fake news detection system similar to those systems already accessible in other fields like healthcare. In this study, we build a dataset of fake and real tweets about the Russian-Ukrainian conflict and evaluate the power of machine learning in this context. The pre-train BERT and five classical machine learning algorithms; namely support vector machine (SVM), Decision Tree (DT), K-nearest neighbor (KNN), logistic regression (LR), and Naïve Bayes (NB) are trained and evaluated through different scenarios of the dataset. The results show that it is possible to develop a system that can discern between real and fake news regarding the Russian-Ukrainian conflict. Support vector machine and logistic regression outperform other learning algorithms and produce comparable prediction models with approximately 76% of accuracy.

Keywords: Machine learning · Deep learning · Fake news · Bert ·
Russian-Ukrainian conflict

1 Introduction

With the rapid development of technology and the availability of online resources for news and information exchange, people now significantly rely on affordable internet-enabled devices to read and disseminate news. The simplicity of accessing such resources via websites or mobile applications makes the dissemination of news a straightforward process, whether they are correct or false. With a large amount of news being spread, interested readers will no longer be able to differentiate between what is true and what is false.

One of the hot events that people are talking about these days is the Russian-Ukrainian war, where news dominated the Internet's news feeds on multiple platforms, including social networks; Facebook, Twitter, and others. It is quite simple for individuals to acquire thoughts and attitudes about the conflict due to misleading news, especially if it is a war between two legitimate countries. This study was motivated by this issue, where the main goal is to create a form that will let interested readers check whether or not the news is accurate, and also evaluate how well machine learning algorithms handle political-related texts and news.

The main contributions of this study are the introduction of the "Russian-Ukrainian conflict" fake news dataset, which consists of both real and fake news, and the proposal of a fake news detection system that can automatically identify and detect the fake news using commonly used Machine Learning algorithms, including the Support Vector Machine (SVM), Multinomial Nave Bayes (NB), K-Nearest Neighbors (KNN), Logistic Regression (LR), and Decision Tree (DT).

2 Related Work

The issue of fake news is spreading around the globe and is seen as one of the biggest challenges to a variety of fields, including politics, education, economics, health, e-commerce, etc. [1]. Due to its success stories in academia and industry, researchers in different disciplines invest in machine learning (ML) to develop efficient solutions against fake news. During the recent pandemic, COVID-19, fake news dissemination was one of the challenges that the global agencies need to deal with besides the disease spreading and treatment. Machine learning showed promising results to fight this phenomenon in several healthcare kinds of research.

A recent study on detecting fake news about COVID-19 and its vaccine showed the power of neural network models in identifying fake news using a dataset collected from trusted resources; the World Health Organization (WHO), the International Committee of the Red Cross (ICRC), the United Nations (UN), the United Nations Children's Fund (UNICEF), and their official Twitter accounts. The fake news information was gathered from various fact-checking websites (such as Snopes, PolitiFact, and FactCheck). The system was able to detect fake news with an accuracy of 94.2% [2]. Using a hybrid model of convolutional neural network and bidirectional Gated Recurrent Units (CNN + Bi-GRU) on about 1,375,592 tweets, the results showed that COVID-19 fake news can be detected with accuracy as high as 92% [3]. A study has been carried out in 2021 to find fake news regarding COVID-19 in Arabic tweets. From January to August 2020 on Twitter, seven million Arabic messages regarding the coronavirus pandemic have been collected by using popular hashtags at the time of the epidemic. After the preprocessing of tweets; removing mentions, hyperlinks, and hashtags, removing non-Arabic and strange words, and text normalization, the Logistic Regression (LR) was trained with the n-gram-level Term Frequency Inverse Document Frequency. The model scored an F1 score of 93.3 in detecting fake Arabic tweets [4].

Twitter was chosen as the platform because its tweets have a high concentration of clear and significant terms. It can be interpreted as linguistically significant information and used to build features into trained predictive models that assist in making decisions. Several studies depend on Twitter to fuel their learning algorithms not only for health-care applications but also for other domains. In their study for identifying fake news on Twitter, Stefan et al. [5] automatically gather a sizable training dataset made up of tens of thousands of tweets. User-level features, tweet-level features, text features, subject features, and sentiment features are the five categories they consider for feature extraction. The XGBoost classifier provided the best accuracy for differentiating between false and true tweets when numerous supervised learning algorithms were employed to build the model. Convolutional neural networks and long-short-term recurrent neural networks were used to build a hybrid model for generating news messages from Twitter posts. Without prior knowledge of the domain, the approach intuitively identifies relevant features associated with fake news stories using both text and images. Using the PHEME dataset, which included 5,800 tweets centered on five rumor stories, the deep learning model achieved 82% of accuracy in detecting fake text [6]. In [7], researchers discovered that using a variety of learning algorithms, fake news sent to Twitter can be identified. To train neural networks, support vector machines, and naive Bayes models for fake message identification, a dataset of 327,784 Twitter posts was examined. Both models of neural network and support vector machine were able to achieve 99% accuracy, in contrast to Naive Bayes' high overall performance of 96% accuracy on the same dataset. Data from the 2010 Chile earthquake has been utilized in [8]. They processed the dataset with Pandas, Numpy, Scikit-Learn, and Keras libraries before applying feature extraction techniques (Count Vectors, TF-IDF, Word Embedding). To illustrate the efficiency of the classification performance on the dataset, they trained and compared five well-known Machine Learning techniques, including Support Vector Machine, Nave Bayes Method, Logistic Regression, and Recurrent Neural Network models. The SVM and Nave Bayes classifiers outperformed the other algorithms in the experiments. The Naïve Bayes classifier has also shown better performance in comparison with other machine learning algorithms applied to the Bengali dataset for fake news [9]. The efficiency of word embedding and word2vec features in Deep Neural networks has been evaluated in [10]. Feature extraction techniques, n-gram, and TF-IDF feature extraction were used to build a dataset for various deep learning and machine learning algorithms; (K nearest neighbors, Decision tree, CNN, LSTM, etc.). Up to 97% accuracy was achieved using the CNN and LSTM models in detecting false news.

In politics, the recent protests in Hong Kong sparked an outpouring of fake news articles on Twitter, which were quickly removed and collated into databases to aid research. Alexandros et al. [11] released research on a dataset published by Twitter, specifically from Twitter's Election Integrity Hub1, that dates from August and September 2019. These linguistic content datasets were previously used to discriminate between fake and real news tweets using typical machine learning methods, on a total dataset size of 13,856,454. They started by pre-processing the text of the tweets and removing hash-tags, mentions, and URLs from both Chinese and English tweets. The Chinese tweets were then translated into English using the Google Translation API. They combined deep learning algorithms with a wide range of input data, including raw text and created features. Experiments indicated that deep learning algorithms outperformed traditional approaches, where the model XLM Roberta outperformed other algorithms reaching scores as high as 99.3% F1 Score. Sachin et al. [12] also used deep learning models to detect fake news utilizing 1356 news examples from various users via Twitter and media sites like PolitiFact. The research analyzes a variety of multiple state-of-the-art approaches including convolutional neural networks (CNNs), long short-term memory (LSTMs), ensemble methods, and attention processes. The CNN + bidirectional LSTM ensembled network with attention mechanism achieved the maximum accuracy of 88.78%. A corpus of 344 articles has been built and annotated manually as accurate news or false news from well-known Pakistani news websites. The study considered the term frequency (TF) and term frequency-inverse document frequency feature extraction methods (TF-IDF). The comparison results of seven different supervised Machine Learning (ML) classification algorithms revealed that K Nearest Neighbors (KNN) had the best performing classifier, providing 70% accuracy [13].

3 Methodology

Figure 1 shows the procedure we follow to conduct the experiments. The below sections discuss the steps in detail.

3.1 Dataset

The dataset used in this study has been collected using Twitter API. We randomly annotate 500 tweets about the Russian-Ukrainian conflict by checking their authenticity on trusted sites. In our project, the number 1 stands for fact tweets, −1 for fake tweets, and 0 for neutral tweets. The available 500 labeled tweets consist of 306 fact news, 66 fake news, and 128 neutral news, Fig. 2.

Since the manual processing to classify the dataset is unfeasible, we applied (Data Augmentation) using the method of Synonym substitution on the manually labeled tweets. Synonymous replacement is a method of choosing a random word that is not a stop word and replacing this word with one of its synonyms at a random moment [4]. The reason for choosing this method over others is the nature of the data that we have; news, where it is not necessary to change it to the degree of change in the field, so it was the most appropriate method. After applying the synonym substitution method, the data becomes as follows: 306 real news, 150 false news, and 224 neutral news, Fig. 3. We used 80% of the data set for training (544 tweets) and 20% for testing (136 tweets).

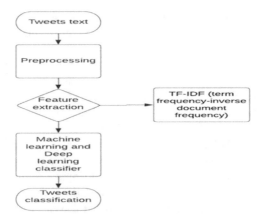

Fig. 1. The general procedure to build ML models for fake news detection used in this study.

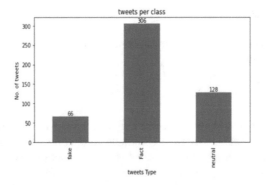

Fig. 2. The distribution of dataset classes before augmentation

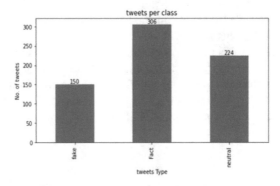

Fig. 3. The distribution of dataset classes after augmentation

3.2 Dataset Preprocessing

Preprocessing the data is very important to remove such undesired/unrelated symbols in tweets to improve the performance of the models, so we did convert text to lowercase, remove all links from tweets, remove all unimportant numbers inside the text, convert strings to a list of strings, remove frequently repeated words (stop words), remove symbols such as @, #, etc.

3.3 Learning Algorithms and Data Representation

- Deep learning

One of the most important areas in which deep learning is used is facial and speech recognition, and so on. The machine learns from a massive data set where the data go through multiple layers between the input and the output layers. Deep learning showed several successes in different research domains. It has the ability also to rank the dataset features by their importance in identifying the class [6].

- Decision Tree

Although the decision tree is distinguished from other algorithms by its speed, ease of understanding, success, and proven effectiveness even with little data, in the case of uncertain values, the decision tree is very complex. The inputs (nodes) are a set of tests, and the outputs are the possible decisions. In some cases, the decision tree is constructed according to the divide and conquer principle, the path from the starting point, the root of the tree, until the leaf node, class, or label, constructs the inference rule [14].

- Support Vector Machine (SVM)

It is a machine learning algorithm that is usually used in classification problems. This algorithm works to put each line of data in a point in the space that has a certain dimension, and its dimension depends on the features present with the theme of each feature to be a specific coordinate value and then classify by finding the highest point distinguishes the two categories well. The goal of the SVM is to separate false or fake tweets through a hyperplane [15].

- Multinomial Naïve Bayes

It is a group of algorithms based on the principle of Bayes Theorem, and they all share one principle, which is that each pair of features that are categorized are independent of each other. It is known to be used to find out the accuracy of true or false news through NB polynomial identifiers and concepts [16].

- K-Nearest Neighbors (KNN)

KNN is known for being simple and mostly unmistakable. Its idea is to take K of points from the available data set. When entering new points, it determines which category it belongs to, and based on it determines to whom it is closer [17].

- Logistic regression

It is used to address the categorization problem. It provides a binomial result since it estimates the probability based on the values of the input variables. Logistic regression is used to predict a categorical target variable because the output can be polynomial. Scaling is not required for input features. Because the degree of probability used to solve a business problem is the product of logistic regression, precise performance criteria must be set to obtain a goal classification threshold. Logistic regression is unaffected by minor data noise or polylines [18].

- BERT (Bidirectional Encoder Representations from Transformers)

BERT learns from a series of words. It consists of several switch encoders grouped, and each switch has a specific encoder, and this encoder consists of two main layers: (Base PERT) and (Big PERT). The BERT model expects a series of symbols (words) as input. In each symbolic sequence, there are two types of symbols that BERT expects as input: [CLS]: is the first symbol for each sequence that stands for the classification symbol, and [SEP]: is the token that lets BERT know which token belongs to which sequence. BERT consists of 340M parameters, trained on 3.3 billion words. Of course, the trend is always towards using larger models and more training data over time [19] In 2021 the OpenAI GPT-3 model [20] had 170 billion parameters, and Google's GShard [21] had 600 billion parameters.

- Term Frequency Inverse Document Frequency (TF-IDF)

During the pre-processing step, the texts are converted to a more manageable representation. One common approach for extracting features from the text is to use Term Frequency Inverse Document Frequency (TF-IDF). For each tweet, the presence (the frequency) of words is taken into consideration. To improve the performance of the ML models and feed them with a better data representation compared to the original text, we calculate a TF-IDF; where TF is the term frequency and the IDF is the inverse term frequency, vector for each tweet caption [22]. The formulas are given below.

$$TF(t, d) = \frac{\text{Number of times term } t \text{ appears in a document } d}{\text{Total number of terms in the document}} \tag{1}$$

$$IDF(t) = \log \frac{\text{Total number of documents}}{\text{Number of documents with the term } t} \tag{2}$$

$$TF - IDF = TF * IDF \tag{3}$$

- Term - Documents Matrix (TDM):

TDM describes how frequently certain words appear in a group of documents (tweets in this study). In a document-term matrix, columns represent terms in the collection and rows represent documents in the collection [23].

- Model evaluation

All prediction models are evaluated and assessed by various performance metrics (accuracy, precision, recall, and F1 score) [24].

4 Results and Discussion

The learning algorithms being evaluated in this study have undergone multiple training sessions, with each training session having a distinct dataset version in terms of size, balancing, preprocessing, and representation of tweets. The learning algorithms have been trained and tested using the dataset version which consists of 500 labeled tweets. They have been evaluated one more time on an augmented version of the dataset using Synonym substitution. The goal of augmentation is to balance the dataset according to the number of instances of each label; real or fake. The learning algorithms have also been evaluated on a preprocessed version and other times using the original tweets. On top of that, the representation is also another dimension taken into account while evaluating the algorithms; the learners scan the dataset as TF-IDF vectors and as a term-document matrix (TDM). The tables in this section show the overall performance of the learning algorithms using the combination of the above experiment dimensions.

Table 1 shows the performance of the learning algorithms when trained and tested on the original number of labeled tweets where each tweet is represented as a TF-IDF vector. The algorithms scan the tweets with and without preprocessing. The results in Table 1 show the poor performance of the prediction models where the overall performance is almost a chance; ~50%. Moreover, the preprocessing phase does not affect the model performance.

The comparisons between Table 3 and Table 1 revealed that the augmentation process significantly improves the training process where the learners have produced models of much better performance with or without preprocessing the tweets. The support vector machine algorithm has been able to build a model of about 76% of accuracy after preprocessing the tweets while struggling to do so when consuming the original labeled dataset, Table 1.

Table 2 Shows the overall performance of the models when the dataset is represented as a term-document matrix when the tweets were preprocessed and never preprocessed. The results do not change which simply means the representation does not affect the training efficacy.

Logistic regression ranked in second place in Table 3 where, with preprocessing phase, has been able to produce a more efficient model, in terms of the four performance metrics, compared to the one in Table 1. Even without preprocessing the tweets, the logistic regression model is comparable to the one built one preprocessed dataset.

Table 1. The overall performance of ML models without dataset augmentation and by using TF-IDF.

Without augmentation								
	Without pre-processing				With pre-processing			
	P	R	F1	Acc	P	R	F1	Acc
SVM	.47	.58	.45	.58	.35	.52	.38	.52
DT	.46	.48	.47	.48	.43	.46	.44	.46
KNN	.39	.42	.40	.42	.48	.49	.48	.49
LR	.42	.55	.44	.55	.67	.58	.50	.58
NB	.39	.56	.42	.56	.43	.54	.39	.54

Table 2. The overall performance of ML models without dataset augmentation and by using TDM.

Without augmentation								
	Without pre-processing				With pre-processing			
	P	R	F1	Acc	P	R	F1	Acc
SVM	.40	.52	.42	.52	.45	.51	.43	.51
DT	.45	.45	.44	.45	.45	.49	.46	.49
KNN	.45	.51	.45	.51	.49	.53	.47	.53
LR	.44	.53	.48	.48	.42	.47	.44	.47
NB	.51	.54	.49	.49	.52	.54	.49	.54

Table 3. The overall performance of ML models with dataset augmentation and by using TF-IDF.

	Without pre-processing				With pre-processing			
	P	R	F1	Acc	P	R	F1	Acc
SVM	.73	.72	.69	.72	.76	.76	.76	.76
DT	.63	.62	.62	.62	.69	.65	.66	.65
KNN	.67	.62	.63	.62	.69	.65	.65	.65
LR	.73	.74	.72	.74	.75	.74	.75	.74
NB	.67	.75	.74	.75	.43	.71	.71	.71

The learning algorithms successfully produced comparable models when using TDM representation of the augmented dataset, when comparing the models' performance in Table 2 and Table 4. However, the logistic regression with preprocessed tweets outperforms others, especially the support vector machine, where the overall performance of logistic regression(i.e. 76% of accuracy) is 5% higher than the support vector machine (i.e. 71% of accuracy) given a TDM representation of the dataset.

The augmentation process has a clear impact on improving the overall performance of the learning algorithms. This finding encourages the evaluation of the BERT algorithm which is a well-known google tool for NLP tasks. Table 5 shows that the BERT algorithm offers a model of 75% of accuracy on our augmented dataset after removing the URLs and stop words, while the word tokenization measurably affects its performance. The pre-trained BERT offers a model of comparable accuracy on the augmented dataset to those models built using classical machine learners, Table 6. However, given the higher precision, recall, and F1-score values of the support vector machine and logistic regression in Table 3, and logistic regression in Table 4, the BERT algorithm could not be the algorithm of choice for the task in this study.

Table 4. The overall performance of ML models with dataset augmentation and by using TDM.

	Without pre-processing				With pre-processing			
	P	R	F1	Acc	P	R	F1	Acc
SVM	.50	.56	.48	.56	.71	.71	.70	.71
DT	.58	.53	.54	.53	.72	.72	.71	.71
KNN	.71	.70	.70	.70	.68	.65	.66	.65
LR	.70	.70	.70	.70	.76	.76	.76	.76
NB	.60	.70	.70	.70	.72	.67	.67	.67

Table 5. The performance of BERT with different preprocessing methods on the augmented dataset.

	P	R	F1	Acc
Removing URLs and stop words	.65	.75	.69	.75
Word tokenization and stemming	.42	.37	.39	.37

Table 6. The overall performance of the Pre-trained BERT model After Augmentation

P	R	F1	Acc
.65	.75	.69	.75

5 Conclusion and Future Work

The findings of this study demonstrate that it is feasible to create a system that is capable of distinguishing between real and fake news about the Russian-Ukrainian conflict. Five classical machine learning algorithms; namely support vector machine (SVM), Decision Tree (DT), K-nearest neighbor (KNN), logistic regression (LR), and Naïve Bayes (NB) are trained and evaluated through different scenarios of the dataset. The dataset set has been prepared to be consumed by the learners in different versions; before and after tweets preprocessing, before and after augmentation, and in representing tweets as TF-IDF vectors and as a term-document matrix. Comparisons of overall performance among the machine learning models and also BERT algorithm on different versions of the dataset revealed that support vector machine could outperform others, including BERT, on augmented datasets represented as TF-IDF vectors, while logistic regression could be the choice of developers when the augmented dataset is preprocessed and represented as a term-document matrix (TDM). Although the BERT produced a comparable model to the above ML ones, however, the amount of data could be a major factor, that we need to improve in the future, to improve its performance in this context.

References

1. Lahby, M., Aqil, S., Yafooz, W.M.S., Abakarim, Y.: Online fake news detection using machine learning techniques: a systematic mapping study. In: Lahby, M., Pathan, A.S.K., Maleh, Y., Yafooz, W.M.S. (eds.) Combating Fake News with Computational Intelligence Techniques. SCI, vol. 1001, pp. 3–37. Springer, Cham (2022). https://doi.org/10.1007/978-3-030-90087-8_1
2. Tashtoush, Y., Alrababah, B., Darwish, O., Maabreh, M., Alsaedi, N.: A deep learning framework for detection of COVID-19 fake news on social media platforms. Data **7**(5), 65 (2022)
3. Mulahuwaish, A., Osti, M., Gyorick, K., Maabreh, M., Gupta, A., Qolomany, B.: Covid-Mis20: COVID-19 Misinformation Detection System on Twitter Tweets using Deep Learning Models. arXiv preprint arXiv:2209.05667 (2022)
4. Mahlous, A.R., Al-Laith, A.: Fake news detection in Arabic tweets during the COVID-19 pandemic. Int. J. Adv. Comput. Sci. Appl. **12**(6), 778–788 (2021). https://doi.org/10.14569/IJACSA.2021.0120691
5. Helmstetter, S., Paulheim, H.: Weakly supervised learning for fake news detection on Twitter. In: 2018 IEEE/ACM International Conference on Advances in Social Networks Analysis and Mining (ASONAM), pp. 274–277. IEEE (2018)
6. Ajao, O., Bhowmik, D., Zargari, S.: Fake news identification on twitter with hybrid CNN and RNN models. In: Proceedings of the 9th International Conference on Social Media and Society, pp. 226–230. (2018)
7. Aphiwongsophon, S., Chongstitvatana, P.: Detecting fake news with machine learning method. In: 2018 15th International Conference on Electrical Engineering/Electronics, Computer, Telecommunications and Information Technology (ECTI-CON), pp. 528–531. IEEE (2018)
8. Mahir, E.M., Akhter, S., Huq, M.R.: Detecting fake news using machine learning and deep learning algorithms. In: 2019 7th International Conference on Smart Computing & Communications (ICSCC), pp. 1–5. IEEE (2019)
9. Mugdha, S.B.S., et al.: A Gaussian naive Bayesian classifier for fake news detection in Bengali. In: Hassanien, A.E., Bhattacharyya, S., Chakrabati, S., Bhattacharya, A., Dutta, S. (eds.) Emerging Technologies in Data Mining and Information Security. AISC, vol. 1300, pp. 283–291. Springer, Singapore (2021). https://doi.org/10.1007/978-981-33-4367-2_28

10. Kaliyar, Rohit Kumar. "Fake news detection using a deep neural network." In 2018 4th International Conference on Computing Communication and Automation (ICCCA), pp. 1–7. IEEE, 2018
11. Zervopoulos, A., Alvanou, A.G., Bezas, K., Papamichail, A., Maragoudakis, M., Kermanidis, K.: Deep learning for fake news detection on twitter regarding the 2019 Hong Kong protests. Neural Comput. Appl. **34**(2), 969–982 (2021). https://doi.org/10.1007/s00521-021-06230-0
12. Kumar, S., Asthana, R., Upadhyay, S., Upreti, N., Akbar, M.: Fake news detection using deep learning models: A novel approach. Trans. Emerging Telecommun. Technol. **31**(2), e3767 (2020)
13. Kareem, I., Awan, S.M.: Pakistani media fake news classification using machine learning classifiers. In: 2019 International Conference on Innovative Computing (ICIC), pp. 1–6. IEEE (2019)
14. Quinlan, J.R.: Learning decision tree classifiers. ACM Comput. Surv. **28**(1), 71–72 (1996). https://doi.org/10.1145/234313.234346
15. Yazdi, K., et al.: Improving fake news detection using k-means and support vector machine approaches. Int. J. Electron. Commun. Eng. **14**(2), 38–42 (2020)
16. Poovaraghan, R. J., Keerti Priya, M.V., Sai Surya Vamsi, P.V., Mewara, M., Loganathan, S.: Fake news accuracy using naive bayes classifier. Int. J. Recent Technol. Eng. (IJRTE). **8**(1C2), 2277–3878 (2019)
17. Kuang, Q., Zhao, L.: A practical GPU based kNN algorithm. In: Proceedings of The 2009 International Symposium on Computer Science and Computational Technology (ISCSCI 2009), p. 151. Academy Publisher (2009)
18. Ray, S.: A quick review of machine learning algorithms. In: 2019 International conference on machine learning, big data, cloud and parallel computing (COMITCon), pp. 35–39. IEEE (2019)
19. Minaee, S., Kalchbrenner, N., Cambria, E., Nikzad, N., Chenaghlu, M., Gao, J.: Deep learning–based text classification: a comprehensive review. ACM Comput. Surv. (CSUR) **54**(3), 1–40 (2021)
20. Brown, T., et al.: Language models are few-shot learners. Adv. Neural Inf. Process. Syst. **33**, 1877–1901 (2020)
21. Lepikhin, D., et al.: Scaling giant models with conditional computation and automatic sharding. arXiv preprint arXiv:2006.16668 (2020)
22. Jing, L.-Pi., Huang, H.-K., Shi, H.-B.: Improved feature selection approach TFIDF in text mining. In: Proceedings. International Conference on Machine Learning and Cybernetics, vol. 2, pp. 944–946. IEEE (2002)
23. Barathi Ganesh, H.B., Anand Kumar, M., Soman, K.P.: Distributional semantic representation for text classification and information retrieval. In: FIRE (Working Notes), pp. 126–130 (2016)
24. Grandini, M., Bagli, E., Visani, G.: Metrics for multi-class classification: an overview. arXiv preprint arXiv:2008.05756 (2020)

Challenges of Managing an IoT-Based Biophilic Services in Green Cities

Farhad Daneshgar[1]([✉]), Rahim Foroughi[2], Nava Tavakoli-Mehr[3], and Atefa Youhangi[4]

[1] Victoria University Sydney, 160 Sussex Street, Sydney, NSW 2000, Australia
Farhad.Daneshgar@vu.edu.au
[2] University of Sunderland, Edinburgh Building, Chester Road, Sunderland Tyne and Wear, Sunderland, UK
[3] The Faculty of Environmental Design and Architecture, Iran University of Science and Technology, Tehran, Iran
nava_tavakoli@cmps2.iust.ac.ir
[4] Islamic Azad University, Payambar Azam Educational Complex, 1 Simaye Iran Street, Saadat Abad Street, Tehran, Iran
youhangifard.atefe@wtiau.ac.ir

Abstract. Biophilic services provide many benefits to the cities in terms of recreational enjoyment, health and wellbeing, and providing green assets. This study is a work-in progress analyzing the nature of biophilic services to identify high-level requirements of an IoT-based Information and Communication application for managing biophilic service-provisioning processes in green (biophilic) cities. In this study, the biophilic city is considered as a complex network of ecosystems that maintains solidarity, stability, and sustainability within the environment. This study provides high-level design directives for the development of a full-blown service-oriented application for managing biophilic service provisioning processes in biophilic cities.

1 Background and Scope of Study

It is estimated that by 2045 about 65% of the urban areas of the earth will be used for housing and urban structures (Zhang and Li 2016). At the same time, governments in general are showing strong tendency for developing more green spaces to the benefit of their nations. Benefits such as recreational enjoyment, health, and wellbeing, and various biophilic assets associated with green environments such as wood, food, and other assets. To this aim governments are trying to combine green services with urban spaces and at the same time making sure that sustainability of green sources are maintained.

The current study is an early attempt in identifying areas where the existing green service are provided to the citizens through user-centric computerized systems and applications. The study therefore explores required attributes, dimensions, and conceptualizations of green services that are required for a future development of a service-oriented computerized application for green cities.

© The Author(s), under exclusive license to Springer Nature Switzerland AG 2023
L. Barolli (Ed.): AINA 2023, LNNS 655, pp. 558–564, 2023.
https://doi.org/10.1007/978-3-031-28694-0_52

2 Green Services

Generally, research on green services fall into three groups: (i) design of green services (Nikolaidou et al. 2016), (ii) aspects and dimensions of green services (Russo et al. 2017), and (iii) benefits of green services in terms of economic, health, medical, and educational benefits (Jennings et al. 2016).

A milestone study by el-Baghdadi and Desha (2017) defines a set of decision cascade for decision-making that has three iterative components that facilitate decisions related to (i) evaluation, (ii) justification, and (iii) optimization of green services. It was further argued that such green services will enhance people's physical fitness and reduce depression.

And finally, Haase (2015) provides a holistic categorization scheme for green services. The scheme suggests the following typology of green services in four categories:

Providing green services: This is the resource-based perspective of green services and results in materialistic benefits such as food, water, medicine, wood, timber, and several other assets.

Regulating green services: This perspective incorporates materials that help to balance air conditions and may include control of air temperature and moisture, quality of water and soil, and control of storm water, flood, and diseases.

Habitat services: These green services overlap with all other categories and maintains biodiversity for organisms to live.

Cultural ecosystem services: This perspective may include non-material benefits such as psychological, cognitive, and health services. These benefits are normally obtained by citizens' contact with green spaces in their living and workspaces.

3 Operationalization of Green Services

To operationalize the service provisioning process for green services the current study proposes a Service-Oriented Architecture (SOA) as a design directive for a future application. One major requirement for applying SOA however is the existence of a standard and operational definition of green services and associated indicators. A recent approach for the operationalization of green services is the presence of a holistic urban policy that encapsulates all environmental and special urban plans with green planning practices (Scott et al., 2016). Another study by Solutions (2015) outlines a set of interrelated objectives for the conceptualization of green services:

"First objective: To enhance sustainable urbanisation through protection of essential ecosystem functions and promoting urban regeneration through the adoption of nature-based approaches.

Second objective: To restore functionality of degraded ecosystems and their services.

Third objective: To develop climate change adaptation and mitigation through redesigning human-made infrastructure and production systems as natural ecosystems, or developing nature-based "frugal technologies" for lowering energy use by integrating grey with green and blue infrastructure; and Fourth objective: To adopt appropriate and relevant risk management processes to manage crises and resilience by utilising nature-based design that combines multiple functions and benefits such as pollution reduction, carbon storage, biodiversity conservation, reducing heat stress, and enhanced water retention" (Ibid).

The above objectives are in line with the requirements and objectives of our proposed biophilic SOA application. The above objectives, among other things, imply incorporation of multiple functions and services for drainage management; habitat provision; ecological connectivity; health and well-being; recreational space; energy reduction; and climate change (Haase 2015). This in turn suggests interventions such as designing city-wide networks of biophilic SOA subsystems, multifunctional parks in urban areas with recreational facilities, cooling and flood alleviation services, and streetscapes for water retention, and integration of living with built systems including green walls and roofs to reduce heat stress (Ibid).

A more recent UN-sponsored study by Czucz et al. (2021) provided a typology of green services that can partly reduce the ambiguity surrounding the multiple current definitions of green services. This *condition typology* consists of six types sorted into three major categories 'abiotic', 'biotic', and 'landscape-level' characteristics (Ibid). The study claims that the above typology of three green services leads to a definition of a ecosystem that incorporates an 'information structure'. The current study argues that this information structure component can become a sub-part of the software service components of our proposed biophilic SOA; and this constitutes one of our future directions.

4   Green Services and Information Technology

This section is a precursor to understanding high-level requirements of the proposed biophilic SOA. It is a brief review of the past and present ICT applications in the domain of urban planning. According to Masnavi et al. (2018) one main challenge in urban planning is to plan settlements in a way that humans enjoy a quality of life guided by sustainable principles. Jennings et al. (2016) and Hunter and Luck (2015) categorize these benefits as *economic, health,* and *recreational benefits.* In other words, the three elements 'biophilic services', 'health of citizens', and 'organization of the city' are major elements of any urban plan that should be linked together (Powers et al. 2020).

Most of the current ICT-based solutions focus on solving infrastructural problems in development of urban spaces such as enhancing mobility of people and resources and eliminating negative effects of the activities of citizens (Bachanek 2018). The idea behind all these systems was to keep the cities' natural resources free from pollution and environmental degradation. These systems supported urban planning processes such as water management, air quality and various resources in urban areas, and maintaining integration addressing environmental protection and sustainable development. A brief review of the applications that have been used in urban planning is provided below.

Telecity is an early example of a computerized urban planning application that applies information technology to enhance mobility and other public services where residents can access specific IT services (Siembab 1996). Smart city is a more holistic concept that incorporates various city concepts into a single architecture (Silva et al. 2018; Alvi et al. 2016). Despite all existing developments, there seem to be inadequate conceptualization of the term smart city, and this is a barrier for fully understanding the concept.

A recent study by Kim et al. (2021) proposes a generic framework for both defining and evaluating smart cities by using three core objectives that a typical city would want for its improvement. These are "productivity, sustainability, and livability" (Ibid). The authors of the current study consider this as a major achievement in the development of user-centric urban planning computerized application; most previous applications were focusing on the off-the-shelf technologies as the first step in developing the system.

ChangeExplorer is another application that encapsulates hardware and software in a smart watch application (Wilson et al. 2019). This application manages the process of citizen feedback through a digital wearable watch that enables individuals to participate in a participatory urban planning process. Similarly, an open-source device has been designed and developed by which participants can share their thoughts. The system encourages citizens to express their thoughts and ideas by drawing and speaking words (Wilson and Tewdwr-Jones 2020).

Silva et al. (2018) proposed a Big Data Analytics experimental architecture for smart cities that enhances effectiveness of Urban Big Data (UBD) exploration in urban planning in smart cities. One recent study provides an assessment framework for the quality of municipal services that is based on SERVQUAL, AHP and Citizen's Score Card (Afroj et al. 2021).

Most of the above applications are enabled because of some technological achievements in ICT that ultimately found their way into the urban planning domain. The current study argues that the initial step in the development of the above ICT systems has been primarily rooted in the ICT innovations (technology-centric design) rather than being based on in-depth analysis of the user requirements as a starting point in system development life cycle. One distinct feature of the current study is that the requirements analysis of the users/stakeholders of the future system, that is beneficiaries of biophilic services, have guided design of our proposed biophilic SOA as a technological artefact rather than vice-versa (user-centric design). This is demonstrated in more detail in Sects. 5 and 6. A user-centric design approach is more sustainable and agile for responding to emerging opportunities and changes in the citizens' requirements. The ICT design paradigm for the proposed biophilic SOA is explained next.

5 Biophilic Service-Oriented Architecture (SOA)

As a kind of user-centric design, the Service-Oriented Architecture (SOA) is a software design paradigm for distributed systems that is primarily based on managing various service defines by its users. In an SOA design each (distributed) component provides services through a communication protocol in a network (Bogner et al. 2018). According to the SOA design paradigm, functions provided by an SOA-base application for green sources would depend on both the presence of clear definitions and defined attributes of green service, as well as the net benefit these services provide to its beneficiaries. This implies that technology providers of future biophilic SOA systems will be facing many additional challenges because of no unified operationalization scheme for green services; and this is due to the nature-based ecological condition of green services. For example, there is no agreed upon computer compiler (or translator) for mapping biophilic services into a language that is understandable by the computer system, and vice-versa.

The SOA system requirement analysis in the current study refers to the identification of, and defining, user expectations of the proposed SOA application. In an idealistic situation such requirements will have to match with the functions and services that the SOA system provides to its users. These requirements are generally divided into two groups: (i) *functional requirements* and (ii) *non-functional requirements*. These two sets of requirements are briefly discussed in the next section, but their full identification is the next future step of our current study.

6   Functional and Non-functional Requirements of Biophilic SOA

In the current study, the functional requirements refer to the behavioral aspects of the proposed biophilic SOA system that are expected to be implemented by the system for the benefit of the users of biophilic services, for the latter to accomplish their tasks or receive benefits. Non-functional requirements on the other hand are defined as how the system should perform the above functions.

As discussed before, the quality and suitability of the proposed biophilic SOA functions will depend on the cohesiveness and clarity of the existing definitions of biophilic service. Teeb (2010) provides a classification scheme for ecosystem services that provides distinct categories of biophilic services and the benefits derived from biophilic services. However, it does not explain how these services and benefits can be managed. Similarly, in another study La Notte et al. (2017) explored one major challenge in the management of biophilic service provisioning processes and that was a lack of consistency in concepts, terminology, and definitions among different ecosystem elements. Both above studies highlight serious challenges on the design and development of a biophilic SOA that expects a uniform set of service definitions for green services that can be inputted to a computer algorithm.

To partially respond to the above challenge, and to develop a uniform set of serviceable ecosystem assets, the current study identified a new conceptualization for ecosystem services with appropriate techniques for the assessment and measurement of services. It combines biomass, information, and interaction. These terms were adopted from system ecology where information is considered as a subset of interactions, biomass is biological material resulted from living/dead organisms, and interactions occur as components may affect one another (Jørgensen 2012). The current study claims that the renewed conceptualization of ecosystem services by La Notte et al. (2017) can potentially lead to a full identification of two main functional requirements of the proposed biophilic SOA, that is, (i) facilitation of the flow of information or communication, and (ii) supporting interactions through coordination of, and collaboration among, various elements of biophilic city. These are two major functions that are expected from the biophilic SOA. The future step in the current study is to develop a high-level design artefact for the implementation of the proposed biophilic SOA based on the current challenges mentioned above.

References

Afroj, S., et al.: Assessing the municipal service quality of residential neighborhoods based on SERVQUAL, AHP and Citizen's score card: a case study of Dhaka North City Corporation area, Bangladesh. J. Urban Manage. **10**(3), 179–191 (2021). https://doi.org/10.1016/j.jum. 2021.03.001

Bachanek, K.H.: Development of IT services in urban space–Smart City logistics. Europ. J. Serv. Manage. **28**, 27–33 (2018)

Bogner, J., Zimmermann, A., Wagner, S.: Analyzing the relevance of SOA patterns for microservice-based systems. In: ZEUS 2018: Workshop on Services and the Composition: proceedings of the 10th Central European Workshop on Services and their Composition: Dresden, Germany, February 8–9, 2018, (CEUR workshop proceedings; 2072), pp. 9–16. RWTH Aachen (2018)

Czúcz, B., Keith, H., Jackson, B., Nicholson, E., Maes, J.: A common typology for ecosystem characteristics and ecosystem condition variables. One Ecosyst. **6**, e58218 (2021)

elBaghdadi, O., Desha, C.: Conceptualising a biophilic services model for urban areas. Urban For. Urban Green. **27**, 399–408 (2017)

Haase, D.: Reflections about blue ecosystem services in cities. Sustainab. Water Qual. Ecol. **5**, 77–83 (2015)

Hunter, A.J., Luck, G.W.: Defining and measuring the social-ecological quality of urban greenspace: a semi-systematic review. Urban Ecosyst. **18**, 1139–1163 (2015)

Jennings, V., Larson, L., Yun, J.: Advancing sustainability through urban green space: cultural ecosystem services, equity and social determinants of health. Int. J. Environ. Res. Public Health **13**, 196 (2016)

Jørgensen, S.: Introduction to systems ecology. CRC Press. Jax, K., 2005. Function and functioning in ecology: what does it mean? Oikos **111**, 641–648 (2012)

Kim, H.M., Sabri, S., Kent, A.: Smart cities as a platform for technological and social innovation in productivity, sustainability, and livability: a conceptual framework. In: Smart Cities for Technological and Social Innovation, pp. 9–28. Elsevier (2021). https://doi.org/10.1016/B978-0-12-818886-6.00002-2

La Notte, A., et al.: Ecosystem services classification: a systems ecology perspective of the cascade framework. Ecol. Ind. **74**, 392–402 (2017)

Masnavi, M.R., Gharai, F., Hajibandeh, M.: Exploring urban resilience thinking for its application in urban planning: a review of literature. Int. J. Environ. Sci. Technol. **16**(1), 567–582 (2018). https://doi.org/10.1007/s13762-018-1860-2

Nikolaidou, S., Klöti, T., Tappert, S., Drilling, M.: Urban gardening and green space governance: towards new collaborative planning practices. Urban Plan **1**, 5 (2016)

Overby, E., Bharadwaj, A., Sambamurthy, V.: Enterprise agility and the enabling role of information technology. Eur. J. Inf. Syst. **15**(2), 120–131 (2006)

Powers, B.F., Ausseil, A.G., Perry, G.L.: Ecosystem service management and spatial prioritisation in a multifunctional landscape in the Bay of Plenty, New Zealand. Australas. J. Environ. Manage. **27**(3), 275–293 (2020)

Russo, A., Escobedo, F.J., Cirella, G.T., Zerbe, S.: Edible green infrastructure: an approach and review of provisioning ecosystem services and disservices in urban environments. Agric. Ecosyst. Environ **242**, 53–66 (2017)

Scott, M., Lennon, M., Haase, D., Kazmierczak, A., Clabby, G., Beatley, T.: Nature-based solutions for the contemporary city/Re-naturing the city/Reflections on urban landscapes, ecosystems services and nature-based solutions in cities/Multifunctional green infrastructure and climate change adaptation: brownfield greening as an adaptation strategy for vulnerable communities?/Delivering green infrastructure through planning: insights from practice in Fingal, Ireland/Planning for biophilic cities: from theory to practice. Plan. Theory Pract. **17**(2), 267–300 (2016)

Siembab, W.: Telecity development strategy for sustainable, livable communities. The blue line televillage in Compton, California. In: Proceedings from Urban Design, Telecommuting and Travel Forecasting Conference, September 8, pp. 229–238, September 1996

Silva, B.N., et al.: Urban planning and smart city decision management empowered by real-time data processing using big data analytics. Sensors **18**(9), 2994 (2018)

Solutions, E.N.B.: Final Report of the Horizon 2020 Expert Group on 'Nature-Based Solutions and ReNaturing Cities'. Directorate-General for Research and Innovation–Climate Action, Environment, Resource Efficiency and Raw Materials, 74 (2015)

Teeb 2010: The economics of ecosystems and biodiversity. Ecological and economic foundations, Routledge Abingdon. In: Kumar, P. (ed.) The Economics of Ecosystems and Biodiversity: Ecological and Economic Foundations. Routledge, UK, p.410 (2012)

Wilson, A., Tewdwr-Jones, M., Comber, R.: Urban planning, public participation and digital technology: app development as a method of generating citizen involvement in local planning processes. Environ. Plann. B Urban Anal. City Sci. **46**(2), 286–302 (2019)

Wilson, A., Tewdwr-Jones, M.: Let's draw and talk about urban change: deploying digital technology to encourage citizen participation in urban planning. Environ. Plann. B Urban Anal. City Sci. **47**(9), 1588–1604 (2020)

Zhang, L., Li, M.: Local fiscal capability and liberalization of urban Hukou. J. Contemp. China **25**(102), 893–907 (2016)

Control and Diagnosis of Brain Tumors Using Deep Neural Networks

Alireza Izadi[1], Farshid Hajati[2(✉)], Roohollah Barzamini[1], Negar Janpors[3], Babak Farjad[4], and Sahar Barzamini[5]

[1] Department of Electrical Engineering, Islamic Azad University Tehran Central Branch, Tehran, Iran
{Ali.izadi.eng,r.barzamini.eng}@iauctb.ac.ir
[2] College of Engineering and Science, Victoria University Sydney, Sydney, Australia
Farshid.hajati@vu.edu.au
[3] Department of Computer Engineering, Islamic Azad University South Tehran Branch, Tehran, Iran
janpors@alborz.kntu.ac.ir
[4] Department of Geomatics Engineering, University of Calgary, 2500 University Drive NW, Calgary, AB, Canada
bfarjad@ucalgary.ca
[5] Iran University of Medical Sciences, Tehran, Iran
sbarzamini@gmail.com

Abstract. Early recognition of various brain tumours can be useful for physicians to control and prevent the progression of the disease and can be very effective and useful in rescuing and healing patients. Computer-aided detection (CAD) plays an essential role in diagnosis and detection of numerous diseases. In this study, an artificial intelligence model, ResNet 50, is developed to diagnose three types of brain tumors (meningioma, gliomas, and pituitary tumors). The model is trained and tested based on a data set which contains 3064 MRI images. The model achieved an accuracy of 95.32% with a 95.11% precision, 95.15% recall, and 95.13% F1-score.

Keywords: Artificial neural network · Deep learning · Machine learning · Digital health · Disease control

1 Introduction

One in six deaths in the world is currently due to cancer [1]. Glioma, meningioma, and pituitary tumors are the most common types of brain tumors. The probability that an individual could develop Glioma, meningioma, and pituitary, tumors during the course of a lifetime is approximately 45%, 15%, and 15%, respectively [1]. Physicians can provide effective care to treat the tumor diseases if the type of the tumor is detected correctly and timely.

MRI is a non-invasive medical imaging technique, and is considered as one of the most accurate techniques for detecting and classifying cancer [3]. However, sometimes

© The Author(s), under exclusive license to Springer Nature Switzerland AG 2023
L. Barolli (Ed.): AINA 2023, LNNS 655, pp. 565–572, 2023.
https://doi.org/10.1007/978-3-031-28694-0_53

the type of cancer is misdiagnosed, which may result in a delay in proper medical treatment or erroneous treatment. Therefore, accurately diagnosis of the type of brain tumors is a very important matter. An integration of CAD and artificial intelligence (AI) can boost tumor detection with powered knowledge and reasoning capabilities [1, 9]. CAD systems generally consist of three components. Lesions are divided from images, the specifications of fragmented tumors are extracted based on statistic or mathematical analyses through learning procedures, a set of MRI images was labeled, then use an appropriate machine learning (ML) classification method to estimate the anomaly classes [4]. To bridge the gap between human and computer vision in pattern recognition, deep learning has been widely demonstrated, which can achieve higher classification accuracy between common ML methods [5]. Different deep learning methods have been developed in the literature to diagnose brain cancer [1].

In this study, an advanced deep learning model is developed using Residual Network (ResNet) architecture for controlling and diagnosing complex brain tumors from MRI images. The model is trained and evaluated using several performance criteria. The results of the experiment show that the model and method of our work is competitive with other approaches.

2   Model of System

Convolutional Neural Networks (CNNs) are generally utilized in image and video recognition [10–17]. Figure 1 shows the structure of CNN [7]. One of the types of CNNs architecture is Residual Network [6] which allows the CNN model to skip layers without affecting performance. A building block of this architecture with the difference between plain and Residual Network, is shown in Fig. 2. As explained in Eq. 1, m and n are, respectively, input and output vectors for the Residual mapping function, and F (m, X_i) is the Residual mapping to be learned [1].

$$n = F(m, X_i) + m \tag{1}$$

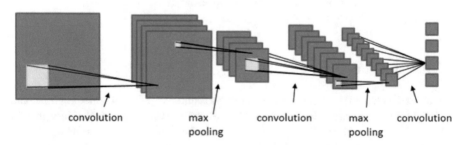

Fig. 1. Convolutional neural networks layers.

The dataset was obtained from 2 hospitals in China which includes 2D MRI images from 233 cancer patients. The dataset is available at: (https://figshare.com/articles/brain_tumor_dataset/1512427/5). This collection has been increased to 3064 contrast-enhanced

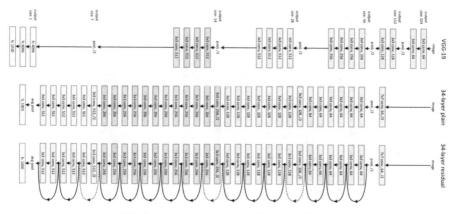

Fig. 2. A residual connection building block.

MRI images [1, 2] using relevant techniques. Figure 3 shows some samples of this collection. The contrast-enhanced MRI images include 708 slices of Meningioma, 1426 slices of Glioma, and 930 slices of Pituitary tumor. Figure 4 illustrates the number of data per segment in the brain tumor dataset.

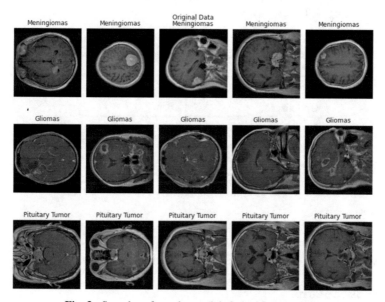

Fig. 3. Samples of our dataset labeled with their class.

The dataset was initially presented in.mat format, and provided the following image information: Label (I for meningioma, II for glioma, III for pituitary tumor), patient ID, image data, tumor border coordinates and a binary mask image with 1s marker of tumor area. The images were stored in the original size of 512 × 512 pixels while each image was located in a folder corresponding to the type of tumor. The dataset has been divided

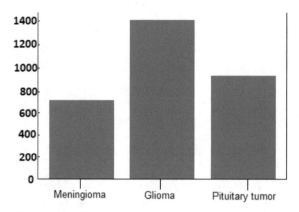

Fig. 4. The number of data per segment in the brain tumor dataset class.

into 80% training and 20% validation sets [8]. As a pre-processing step, the images were cropped to 450 × 450 pixels from center. The cropped parts did not contain remarkable information, but only the background color. In order to facilitate the processing time and achieve better results, the resized images were used as input into the ResNet50 model. The optimized Hyper Parameters used in the model are shown in Table 1. The model was trained using GPU.

Table 1. Optimized hyper parameters used for model.

Hyper parameter	Optimized value
Optimizer	Adam
Number of epochs	500
Batch size	32
Learning rate	0.00001
Learning rate decay	0

3 Results

3.1 Confusion Matrix

A confusion matrix is used to describe the performance of model. The confusion matrix represents counts from predicted and actual classes (Fig. 5) [1]. We used a normalized confusion matrix, which allows to divide the values by number of elements in each class, for a better visualization of misalignment.

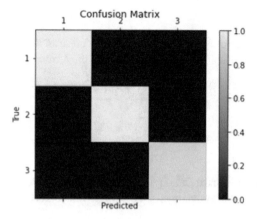

Fig. 5. Confusion Matrix results.

3.2 Accuracy and Loss

Figure 6 illustrates the training and validation results during epochs.

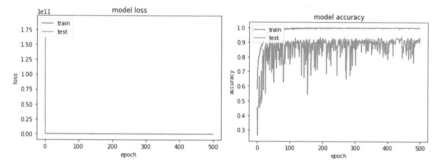

Fig. 6. Classification model results.

3.3 Accuracy

One of the most common evaluation measures for classification models is accuracy. For an imbalanced dataset, because of bias towards the class with the most sample count, we may get high accuracy. In this way, an accuracy equal to the number of frequent labels in the test set will be achieved. Nevertheless, accuracy is not a suitable measure of performance balanced accuracy that is shown in Eq. 2, can be a better measure. The model achieved an accuracy of 95.32% and 95.3% balanced accuracy of 95.3%, respectively.

$$\frac{\sum_i^l (TP_i + TN_i)(TP_i + FP_i + TN_i + FN_i)}{l} \tag{2}$$

3.4 Recall

Recall, as presented in Eq. 3, is the fraction of relevant instances that was retrieved. The developed model in this study achieved 95.15% as the Macro-average recall.

$$\frac{\sum_i^l (TP_i)(TP_i + FN_i)}{l} \tag{3}$$

3.5 Precision

Fraction of relevant instances among the retrieved instances is called precision. As illustrated in Eq. 4, a Macro-averaged precision of 95.11% is achieved for the model.

$$\frac{\sum_i^l (TP_i)(TP_i + FP_i)}{l} \tag{4}$$

3.6 F1-Score

A Macro-averaged F1-scoreof 95.13% is achieved for the model using the following Eq. 5:

$$\frac{Precision_M * Recall_M * 2}{Precision_M + Recall_M} \tag{5}$$

4 Conclusion

In this study, a deep learning model for diagnosis of brain tumors from MRI images is developed using Residual Network (ResNet). An ResNet50 architecture is implemented with accuracy of 95.3%. Table 2 shows the list of other studies which have worked on the same database, but employed different methods. While they have been able to improve the overall accuracy by 93.68%, the relative long computational time has limited their modelling applications. This study not only addressed this limitation, but also improved the accuracy of the modelling results.

Table 2. Optimized review results and records.

Author	Year	Method	Accuracy
Cheng et. al	2015	SVM SRC KNN	91.28%
J.Paul	2016	CNN	84.19%
J.Paul et. al	2017	CNN	91.43%
Parnian Afshar et. al	2018	CapsNets	86.56%
Parnian Afshar et. al	2018	CapsNets	90.89%
N.Abiwinanda et. al	2018	CNN	84.19%
A. Pashaei et al	2018	KELM + CNN	93.68%
S.A.A. Ismael	2019	ResNet50	95% (Except for increasing the number of images)

References

1. Ismael, S.A.A., Mohammed, A., Hefny, H.: An enhanced deep learning approach for brain cancer MRI images classification using residual networks. Artif. Intell. Med. (2020). https://doi.org/10.1016/j.artmed.2019.101779
2. Cheng, J., et al.: Enhanced performance of brain tumor classification via tumor region augmentation and partition. PLOS ONE **10**(10), e0140381 (2015). https://doi.org/10.1371/journal.pone.0140381
3. Kumar, S., Dabas, C., Godara, S.: Classification of brain MRI tumor images: a hybrid approach. In: Procedia Computer Science, vol. 122, pp. 510–517 (2017). https://doi.org/10.1016/j.procs.2017.11.400K. (Elissa, "Title of paper if known," unpublished)
4. Vidyarthi, A., Mittal, N.: Performance analysis of Gabor-Wavelet based features in classification of high grade malignant brain tumors. In: Proceedings of the 2015 39th National Systems Conference, NSC 2015 (2016)
5. Yuchen Qiu, H.L., Yan, S., Reddy Gundreddy, R., Wang, Y., Cheng, S., Zheng, B.: A new approach to develop computeraided diagnosis scheme of breast mass classification using deep learning technology. J. Xray Sci. Technol. **118**(24), 6072–6078 (2017)
6. He, K., Zhang, X., Sun, J.: Deep residual learning for image recognition. In: He, K., Zhang, X., Ren, S., Sun, J., CVPR, vol. 19(2), pp. 107–117 (2015)
7. Gorach, T.: Deep convolutional neural networks - a review. Int. Res. J. Eng. Technol. **56**(5), 1235–1250 (2018). http://www.dbpia.co.kr/Journal/ArticleDetail/NODE07109492
8. Rosie Dunford, E.T., Su, Q., Wintour, A.: The pareto principle. The Plymouth Student Scientist (2014)
9. Barzamini, H., Shahzad, M., Alhoori, H., Rahimi, M.: A multi-level semantic web for hard-to-specify domain concept, Pedestrian, in ML-based software. Requir. Eng. **27**, 1–22 (2021). https://doi.org/10.1007/s00766-021-00366-0
10. Izadi, A., Barzamini, R., Hajati, F., Janpors, N.: UAVs' flight control optimization for delivering motion-sensitive blood products. Neuro Quantol. **20**, 7661–7669 (2022)
11. Hajati, F., Raie, A., Gao, Y.: Pose-invariant 2.5 D face recognition using geodesic texture warping. In: 2010 11th International Conference on Control Automation Robotics & Vision, pp. 1837–1841 (2010)
12. Hajati, F., Cheraghian, A., Gheisari, S., Gao, Y., Mian, A.: Surface geodesic pattern for 3D deformable texture matching. Pattern Recognit. **62**, 21–32 (2017)

13. Ayatollahi, F., Raie, A., Hajati, F.: Expression-invariant face recognition using depth and intensity dual-tree complex wavelet transform features. J. Electron. Imaging **24**(2), 023031 (2015)
14. Pakazad, S.K., Faez, K., Hajati, F.: Face detection based on central geometrical moments of face components. In: 2006 IEEE International Conference on Systems, Man and Cybernetics, pp. 4225–4230 (2006)
15. Shojaiee, F., Hajati, F.: Local composition derivative pattern for palmprint recognition. In: 2014 22nd Iranian Conference on Electrical Engineering (ICEE), Tehran, Iran, pp. 965–970 (2014)
16. Barzamini, R., Hajati, F., Gheisari, S., Motamadine, M.B.: Short term load forecasting using multi-layer perception and fuzzy inference systems for islamic countries. J. Appl. Sci. **12**(1), 40–47 (2012)
17. Hajati, K.F., Pakazad, S.K.: An efficient method for face localization and recognition in color images. In: 2006 IEEE International Conference on Systems, Man and Cybernetics, Taipei, Taiwan, pp. 4214–4219 (2006)

Co-evolution Genetic Algorithm Approximation Technique for ROM-Less Digital Synthesizers

Soheila Gheisari[1]([✉]), Alireza Rezaee[2], and Farshid Hajati[1]

[1] College of Engineering and Science, Victoria University Sydney, Sydney, Australia
{soheila.gheisari,farshid.hajati}@vu.edu.au
[2] Mechatronic Part, Interdisciplinary Technology Group, Faculty of New Sciences
and Technologies, University of Tehran, Tehran, Iran
arrezaee@ut.ac.ir

Abstract. A new polynomial approximation technique is presented using the concept of co-evolution genetic algorithm. In the proposed technique each polynomial coefficient is considered as a set of independent populations rather being considered as one. These populations co-evolve as they try to optimize the spurious-free dynamic range (SFDR) of a direct digital frequency synthesizer (DDFS). Obtained SFDR from these optimized polynomials have outperformed that of the polynomials evaluated using deterministic approaches. Optimized polynomials for 2nd and 3rd order approximations have achieved 62 dBc and 82 dBc after hardware implementation. Also in the paper a simple genetic algorithm has been explained for optimal pipeline level insertion for a specific hardware implementation of polynomial functions in FPGA platforms.

Keywords: DDFS · SFDR · Genetic algorithm

1 Introduction

Machine learning has been widely used in different applications [4–10]. Direct digital frequency synthesis (DDFS) is a common technique for generation of single phase and quadrature sinusoid in modern digital systems. High resolution frequency selection, fast frequency switching and phase continuity are the outstanding attributes of DDFS. Stated characteristics have made DDFS the first choice in modern digital communication systems.

Basic DDFS architecture which was introduced by Tierney et al. [1] in 1971 consists of two blocks, a phase accumulator and a sine/cosine generator block. A simplified version of a basic DDFS architecture is shown in Fig. 1. The phase accumulator is an overflowing N-bit accumulator which generates a ramp output interpreted as an angle in the interval $[0, 2\pi)$. The slope of the accumulator output is determined by the value of F frequency control word ($M(n) = n \cdot F \mod 2^N$). The P most significant bits of $M(n)$ are input to the sin/cosine generator that provides an estimate of the sine and cosine of that angle. The frequency F_{out} of output sinewave signals are proportional to the frequency control word F (Fig. 1):

© The Author(s), under exclusive license to Springer Nature Switzerland AG 2023
L. Barolli (Ed.): AINA 2023, LNNS 655, pp. 573–584, 2023.
https://doi.org/10.1007/978-3-031-28694-0_54

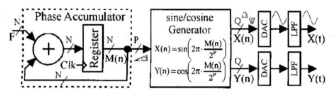

Fig. 1. DDFS with quadrature outputs.

$$F_{out} = \frac{F}{2^N} \cdot f_{clk} \qquad F \in [0, 2^N - 1] \tag{1}$$

where f_{clk} is clock frequency. It is observed that the resolution of the synthesizer can be made optionally smaller by choosing as large an N as necessary. The synthesized sinewave signals can be used directly in digital systems, or converted to analog domain via a digital to analog converter (DAC) and a low-pass construction filter (LPF).

Spectral purity of the quadrature signals generated by DDFS is characterized with a special value known as spurious free dynamic range (SFDR). SFDR is defined as the ratio (in dB) of the amplitude of the desired frequency component to that of the largest undesired frequency component.

The main part of DDFS is the sine/cosine generator block. Simple approaches were devised in [1] for implementing this block using ROMs as look up tables (LUT). Achieving acceptable SFDR demands large ROMs. These large ROMs introduce more power dissipation and slow access time which eventually degrade the timing performance of the DDFS. Due to these constraints extensive research has been going on for the reduction or omission of large ROMs. A complete list of techniques toward ROM-Less DDFS designs is available in [2]. The approach chosen in this paper for realizing the phase to sine mapping is the interpolation of the sine function using an optimized polynomial expression with constant coefficients. Optimized polynomials have been obtained using co-evolution genetic algorithm (CGA). Comparisons with polynomials obtained using deterministic approaches have been made.

2 Polynomial Approximation Technique

The sin/cosine generator block used for our DDFS design is shown in Fig. 2. The sine and cosine blocks compute the sine and cosine function in $[0, \pi/4]$. By having the sine and cosine values in their first octant, it is possible to compute these functions in $[0, 2\pi]$ interval. The sine and cosine blocks can be implemented using a single expression of degree K for each. In other words :

$$A(n) \approx a_K \cdot C(n)^K + a_{K-1} \cdot C(n)^{K-1} + \ldots + a_0$$
$$B(n) \approx b_K \cdot C(n)^K + b_{K-1} \cdot C(n)^{K-1} + \ldots + b_0 \tag{2}$$

In general since these polynomials approximate the values of sine and cosine function they ultimately decrease SFDR performance. It is obvious that the error introduced by

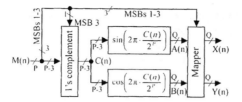

Fig. 2. Sine/cosine generator block.

these approximations can be alleviated using higher order polynomials. However, as K increase so does the hardware that realizes the polynomial functions. Thus a trade-off between hardware complexity and SFDR performance exist.

Different polynomial approximation techniques have been used for choosing the coefficients $aK,..., a0$ and $bK,..., b0$ in Eq. (2). Taylor approximation about 0 for sine and cosine functions, Tchebyshev and Legendre polynomials are among the few to name. But non of these techniques aim to maximize the SFDR of DDFS, instead these approaches mainly try to minimize the maximum error between sine and cosine function and the approximating polynomial or the mean square error caused by them. As a consequence the possibility of using an optimization technique in which ai and bi polynomial coefficients are chosen in order to maximize SFDR has been investigated in this paper.

3 Genetic Algorithm

3.1 Simple Genetic Algorithm

Genetic algorithm (GA) is a general purpose optimization technique which has become widespread in the past decades. Unlike deterministic approaches which are mainly restricted for optimizing special objective goals; genetic algorithm is capable of dealing with optimization problems with various kinds of objective goals. This special attribute has made this technique a good nominee for obtaining optimized polynomials in our investigation.

In the next paragraphs the procedure for conducting a simple genetic algorithm (SGA) for obtaining optimized set of polynomial coefficients has been explained. Obtained results were compared against deterministic approaches. Although we have carried out the following optimization technique for 2nd and 3rd degree polynomial approximations, it can also be invoked for higher orders.

First we consider the polynomial coefficients to be the variables that construct the population in SGA. So we have:

$$\begin{pmatrix} a_{K,1} & a_{K-1,1} & \cdots & a_{0,1} & b_{K,1} & b_{K-1,1} & \cdots & b_{0,1} \\ a_{K,2} & a_{K-1,2} & \cdots & a_{0,2} & b_{K,2} & b_{K-1,2} & \cdots & b_{0,2} \\ \vdots & \vdots & \cdots & \vdots & \vdots & \vdots & \cdots & \vdots \\ a_{K,m} & a_{K-1,m} & \cdots & a_{0,m} & b_{K,m} & b_{K-1,m} & \cdots & b_{0,m} \end{pmatrix} \qquad (3)$$

Variable m resembles the population size. Each individual in the population has a length of $2K$, presenting each coefficient.

In order to have an amplitude of 1 in DDFS quadrature outputs we imposed $b_0 = 1$ for all individuals. Thus the new population is:

$$\begin{pmatrix} a_{K,1} & a_{K-1,1} & \cdots & a_{0,1} & b_{K,1} & b_{K-1,1} & \cdots & b_{1,1} \\ a_{K,2} & a_{K-1,2} & \cdots & a_{0,2} & b_{K,2} & b_{K-1,2} & \cdots & b_{1,2} \\ \vdots & \vdots & \cdots & \vdots & \vdots & \vdots & \cdots & \vdots \\ a_{K,m} & a_{K-1,m} & \cdots & a_{0,m} & b_{K,m} & b_{K-1,m} & \cdots & b_{1,m} \end{pmatrix} \quad (4)$$

This makes the number of variables to be $2K$-1 for each individual in the population.

Any GA needs a fitness function for scoring and ranking of individuals in the population. Based on this information and through some of the most common operations in GA such as crossover, mutation and…a new population is generated. By using the policy of conserving the fittest, GA will gradually improve the score of the best individual, regarding the fitness function, after each generation.

The fitness function defined for our case matches the definition of SFDR. In order to find the SFDR for a specific frequency control word F, we used Fast Fourier Transform (FFT) for spectral analysis. By using FFT the amplitude of the desired frequency component along with the maximum amplitude of the undesired frequency component can be calculated. Thus the return value of the fitness function is simply the SFDR of DDFS quadrature outputs with frequency control word F.

Two distinct SGAs have been run to obtain optimized polynomial coefficients for 2nd and 3rd degree approximations. Population size m for 2nd and 3rd degree approximations were chosen to be 100 and 120 respectively. The initial population for each SGA has been created using a uniform distribution in $[-1, 1]$ interval. Best fitness values (SFDR) for each generation are depicted in Fig. 3 and Fig. 4.

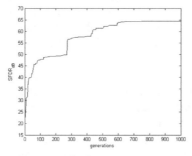

Fig. 3. Best fitness values for 2nd order approximation, $F = 2^N/128$, $N = 15$.

For comparison reasons SFDR outputs resulted from deterministic approaches have been stated in Table 1. As can be seen from Fig. 3 and Fig. 4, SFDR obtained in SGA for 2nd order approximation has outperformed that of the deterministic approaches whereas for 3rd order approximation we only observed improvement over Taylor series method.

It should be mentioned that for further verification of the results obtained from these SGAs with the cited conditions, (Due to the random behavior of GAs) they have been repeated for several times and no improvement were observed.

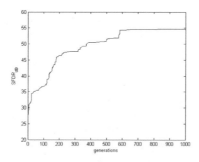

Fig. 4. Best fitness values for 3rd order approximation, $F = 2^N/1024$, $N = 19$

Table 1. Output SFDR (dB).

Technique	SFDR (dB)	F	K
Legendre	61.13	$2^N/128$	2
Tchebyshev	58.25	$2^N/128$	2
Taylor	35.78	$2^N/128$	2
Legendre	86.55	$2^N/1024$	3
Tchebyshev	86.25	$2^N/1024$	3
Taylor	51.13	$2^N/1024$	3

In the next subsection we have explained a method that not only obtains optimized polynomials for any K-th order of approximation but also outperforms SGA algorithm in terms of generations needed (speed) for obtaining the right answer.

3.2 Co-evolution Genetic Algorithm

We observed in the previous subsection that SGA was not capable of finding better SFDR than deterministic approaches for the case of 3rd order approximation. The answer to this problem might be in the increase of the population size or number of generations. These solutions come at the expense of longer simulation times which are not favored if speed is desired. Even for the case of 2nd order approximation as shown in Fig. 3, 1000 generations have been processed which is also considered to be time consuming regarding the number of FFT operations needed.

It should be mentioned that GA is trying to find optimized set of coefficients in a multidimensional search space. As the dimensions of this space increases (more coefficients) so does the complexity of searching for special values. We can overcome this complexity by restricting the search space and using more than one population for finding the best values of our interest. The premier implies proper initialization and the latter implies cooperative style genetic algorithm among distinct populations.

As for the first concept; although functions obtained from deterministic approaches were not an objective for us. But this fact didn't prohibit us from using them as proper initial guesses. As a consequence we created the initial population using a uniform distribution in $[-\varepsilon/2, \varepsilon/2]$ interval centered at the coefficient values obtained from a deterministic technique. The choice of ε depends on how close will the optimize values be to deterministic values.

The second concept can be well realized using co-evolution genetic algorithm (CGA). In this approach different populations try to get ahead (competitive CGA) of each other or cooperate (cooperative CGA) with each other to reach a certain objective. A simple cooperative CGA best suits our problem. Opposing to the case of SGA where each individual in one population consisted of $2K$-1 variables, CGA has $2K$-1 populations where every individual in each population has one variable, as shown below:

$$
\begin{pmatrix} a_{K,1} \\ a_{K,2} \\ \vdots \\ a_{K,t} \\ \vdots \\ a_{K,m} \end{pmatrix}
\begin{pmatrix} a_{K-1,1} \\ a_{K-1,2} \\ \vdots \\ a_{K-1,t} \\ \vdots \\ a_{K-1,m} \end{pmatrix}
\cdots
\begin{pmatrix} a_{0,1} \\ a_{0,2} \\ \vdots \\ a_{0,t} \\ \vdots \\ a_{0,m} \end{pmatrix}
\begin{pmatrix} b_{K,1} \\ b_{K,2} \\ \vdots \\ b_{K,t} \\ \vdots \\ b_{K,m} \end{pmatrix}
\begin{pmatrix} b_{K-1,1} \\ b_{K-1,2} \\ \vdots \\ b_{K-1,t} \\ \vdots \\ b_{K-1,m} \end{pmatrix}
\cdots
\begin{pmatrix} b_{1,1} \\ b_{1,2} \\ \vdots \\ b_{1,t} \\ \vdots \\ b_{1,m} \end{pmatrix}
\tag{5}
$$

Each population optimizes one coefficient of the polynomials. In order to optimize the coefficients, a population scoring and ranking must be done with respect to other populations. In the next paragraph procedures which we conducted for our CGA have been explained.

After making proper initialization for all populations we sort the individuals in each population from best (individual 1) to worst (individual m) by considering all populations as a whole as shown in (4). Then by choosing the first population we gave opportunity for all its individuals to match themselves with one of the top t rows consisting of individuals of other populations. Selection through ranking of 1 to t is made randomly with each of the rankings having the same probability. Scoring of each individual in the population is done using the same SFDR function defined in SGA. By using the scores obtained by the fitness function and common operations in GA a new population is created. Before proceeding to the next population we first survey that if this new population improves over the previous one. In case of improvement we replace the old population with the new one otherwise the population remains unchanged. Steps mentioned after the initialization is repeated for other populations successively till satisfactory results are obtained.

The source of such an interaction among populations has been gained by a close survey over the population at different generations in SGA. Interestingly we observed that some individuals do consist of variables close to coefficient values obtained from deterministic approaches such as Legendre polynomial approximation. But other consisting variables' values were far away from satisfactory results which made the individual drop in the ranking. Thus the chance of its participation in the regeneration process (such as crossover) was decreased. We overcome this problem in CGA by letting each variable in an individual to match it self with other variables' best categories.

Two CGAs have been run for 2nd and 3rd degree approximations. The sizes (m) of all populations were chosen to be 10 and 12 for 2nd and 3rd degree approximations respectively. The number of top ranking t was set to 3. As for the initial values, coefficients obtained from the Legendre polynomials were used. The best fitness values obtained (SFDR) for each CGA are shown in Fig. 5 and Fig. 6.

As can be seen from these figures using CGA along with proper initialization values we can immediately reach the optimum points.

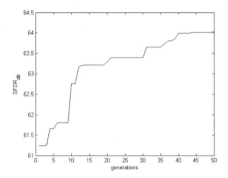

Fig. 5. Best fitness values for 2nd order approximation, $F = 2^N/128$, $N = 15$.

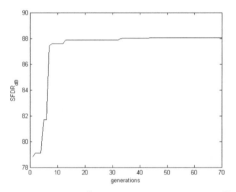

Fig. 6. Best fitness values for 3rd order approximation, $F = 2^N/1024$, $N = 19$.

Table 2. Optimized polynomials for $K = 2$ and $K = 3$.

$K = 2$	$A(n) \approx \sin(\pi/4.x) \approx -0.0018152+0.8195451.x-$ $0.1060769.x^2$ $B(n) \approx \cos(\pi/4.x) \approx 1-0.007856.x-0.2899080.x^2$	$x = 8.\dfrac{C(n)}{2^P}$ $x \in [0,1]$
$K = 3$	$A(n) \approx \sin(\pi/4.x) \approx 0.00004691+0.78642468.x -$ $0.00526649.x^2-0.07380982.x^3$ $B(n) \approx \cos(\pi/4.x) \approx 1+0.00414089.x-0.32751660.x^2$ $+0.03044189.x^3$	

Optimized coefficients outperform the best achieved SFDR obtained by deterministic approaches shown in Table 1, by 3 and 2 dB for 2nd and 3rd degree approximations respectively. Optimized coefficients are shown in Table 2.

We have optimized the SFDR for a special frequency control word F. It is also possible to optimize the value of SFDR for several F at the same time. If so, we are dealing with a multi-objective problem. But the necessity of this matter has been investigated by evaluating the SFDR for other frequency control words. Figure 7 and Fig. 8 depict the SFDR obtained for various F using optimized polynomials as the approximating functions. In addition SFDR obtained from the best and worst deterministic approaches have been plotted too. As can be observed from Fig. 7, SFDR resulted from CGA technique is not always at its optimum value for all possible F. Nevertheless it still remains higher than the SFDR obtained from deterministic approaches. On the other hand Fig. 8 implies that SFDR obtained using CGA is constant and remains higher than that of the deterministic techniques.

One last point to mention is that optimized coefficients were obtained with out considering quantization errors caused by hardware implementation. Hence attained SFDR is the upper bound of DDFS performance.

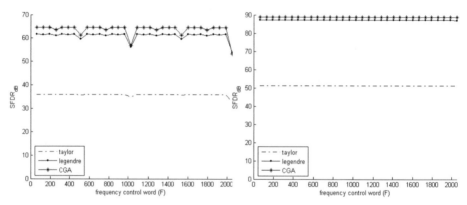

Fig. 7. SFDR vs. Frequency control word for $K = 2$.

Fig. 8. SFDR vs. Frequency control word for $K = 3$.

4 Hyper-Folding Technique

A novel method was presented in [3] for realizing a circuit that computes a K degree polynomial $P(x)$:

$$P(x) = c_K x^K + c_{K-1} x^{K-1} + \ldots + c_0 \qquad (6)$$

Authors in [3] named this technique as Hyper-Folding which was a generalization of folding technique to the computation of polynomials.

In order to properly illustrate the idea of Hyper-Folding with an example, let us consider the following polynomial:

$$P(x) = x^2 + 3 \cdot x \qquad (7)$$

where x is a 2's complement number presented with $n = 4$ bits.

$$x = -8 \cdot x_3 + 4 \cdot x_2 + 2 \cdot x_1 + x_0 \qquad (8)$$

Substituting (8) in (7) yields:

$$
\begin{aligned}
P(x) = {} & x_0.x_0 + 2x_0.x_1 + 4x_0.x_2 - 8x_0.x_3 + 2x_1.x_0 \\
& + 4x_1.x_1 + 8x_1.x_2 - 16x_1.x_3 + 4x_2.x_0 + 8x_2.x_1 \\
& + 16x_2.x_2 - 32x_2.x_3 - 8x_3.x_0 - 16x_3.x_1 - 32x_3.x_2 \\
& + 64x_3.x_3 + 3x_0 + 6x_1 + 12x_2 - 24x_3
\end{aligned}
\tag{9}
$$

By using Boolean properties of idempotent, associative and commutative we can simplify and join equal partial products, as shown below.

$$
\begin{aligned}
P(x) = {} & 7x_0 + 4x_0.x_1 + 8x_0.x_2 - 16x_0.x_3 + 10x_1 \\
& + 16x_1.x_2 - 32x_1.x_3 + 28x_2 - 64x_2.x_3 + 40x_3
\end{aligned}
\tag{10}
$$

Equations (9) and (10) show that using Hyper-folding technique a great reduction of the number of partial products that have to be added can be achieved.

The number of partial products (*NPP*) that have to be added using this technique for evaluation of a typical equation like (6) can be calculated with:

$$
NPP = C\binom{n}{K} + C\binom{n}{K-1} + \ldots + C\binom{n}{0}
$$
$$
C\binom{m}{n} = \frac{m!}{n!(m-n)!} \quad m \geq n, \quad C\binom{m}{n} = 0 \quad m < n
\tag{11}
$$

where n is the number of bits presenting x. Variable K is the degree of the polynomial. The last term in Eq. (11) resembles the constant c_0 in Eq. (6). If this constant is not present in the polynomial this last term must be avoided. As the number of bits n or the degree of polynomial K increases so does the *NPP*. Consequently as *NPP* increases so does the realizing hardware of the polynomial functions.

To realize polynomial functions with Hyper-folding technique a multi-operand adder must be used. The inputs of this adder are realized with a partial product matrix. It is obvious that the partial products can be simply implemented using and gates. Partial product matrix can be realized by replacing the partial products for each 1 of the binary presentation of their associated constants. Figure 9 is an illustrative view of Hyper-folding technique which realizes the following function.

$$
P(x) = 3x_2.x_0 + 6x_2.x_1 + 5x_1.x_0 + 2x_0
\tag{12}
$$

Addition tree just like shown in Fig. 9 is the preferred choice that can realize the multi-operand adder. In addition tree structures, the output of the adders in the initial row are input to the succeeding row of adders. The same follows for other adders in other rows until the last single adder is reached. Large NPP require more adders and hence more addition stages. Long addition stages cause performance degradation. To avoid degradation in performance, registers (pipelining) are inserted between consecutive addition rows. Performance increases by inserting more pipeline levels. However more pipelining simply implies more delay. Even insertion of pipeline levels for different stages does not take the same amount hardware (registers). Pipelining at initial stages demand more registers than at the ending stages. Thus a trade off between register usage and performance exist. In the next section we have developed a simple method based on SGA for obtaining the optimum pipeline level insertion regarding the hardware cost and system performance in FPGA platforms.

5   Pipelining

FPGAs are general purpose platforms for designing digital circuits of any kind. Unlike ASICs which have a specific performance boundary for a special circuit design. Performance of one design might vary when it is implemented in different FPGAs.

The latter reason have motivated us to generate an algorithm that optimizes the number of pipeline levels of an addition tree in a Hyper-Folding combination with respect to hardware cost (pipeline levels) and performance for FPGA platforms.

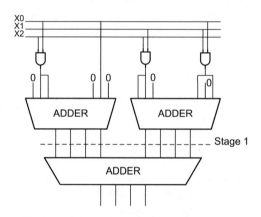

Fig. 9. Hyper-Folding realization of Eq. (12).

Let us assume that a certain Hyper-Folding design has S stages present in its multi-operand adder. We can consider the combination of pipeline levels as a binary word of length S. Each position in the binary word resembles a stage between two consecutive addition rows. A 1 in this word implies the presence of a register level between the specified rows. A 0 on the other hand simply means a wire connection between the specified addition rows. Thus for a particular Hyper-Folding architecture which has S stages available in its multi-operand adder, we have 2^S possible combinations of pipelining. For small values of S we can check all the combination in a specific FPGA and choose the best arrangement which satisfies our hardware and performance criteria. In contrast for large values (e.g. $S \geq 10$) of S this process is very lengthy and time consuming. Automation of such a process is a must for large values of S. SGA can also be invoked here again for finding the best combination regarding our criteria which we can define in our fitness function.

Each binary individual pertains to a pipeline combination. The initial value for this population is created using a random selection between 1 and 0 with each having similar likelihood. The fitness function can be defined in favor of any criteria such as finding the best performance or in general a relation that defines some kind of trade off between pipeline levels and performance.

System performance and hardware utilization for an FPGA platform can be obtained using available FPGA synthesis tools. Output reports from these tools can be utilized by the fitness function for scoring and ranking purposes.

FPGA synthesis tool entries must be given in hardware descriptive formats. Hence a special program has been written by us that generates the replica of these simple binary presentations of pipeline levels in Verilog format.

6 FPGA Implementation

The FPGA platform that was chosen for conducting the SGAs was Xilinx XC3S1000ft256-4 FPGA. The synthesis tool utilized by the fitness function was Xilinx ISE 8.1 batch mode synthesis tool software. Two SGAs have been conducted for obtaining the best performance achieved with less number of pipeline levels having higher priority, for both 2nd and 3rd order approximations. The number of stages S was 5 and 7 for 2nd and 3rd order approximations respectively. Though the answer to this optimization is known in advance (pipeline for all stages) and even the number of stages is not large enough for both cases (we considered $S \geq 10$ as large value).But it served well as a good starting point for verifications of the results obtained by this technique.

Subsequently the designer is capable of defining any fitness function accordingly. It is also worth mentioning that quantization steps were also carried out before and after the conversion of polynomials into Hyper-Folding format. Obtained SFDR after quantization have founded to be 62 and 82 dB for 2nd and 3rd degree approximations respectively. The results of SGAs confirmed the fact that by insertion of registers for all stages the best performance is achieved. Performances of the implemented designs are shown in Table 3.

Table 3. 2nd and 3rd approximation FPGA implementations.

Approx	SFDR (dB)	Pipeline levels	FPGA	MAX Clock Freq (MHz)	4in LUTs
2nd	62	5	XC3S1000–4	197.3	491
3rd	82	7	XC3S1000–4	187.6	1586

7 Conclusion

A new polynomial approximation technique that uses CGA was proposed for obtaining optimized coefficients for DDFS application. Optimized coefficients outperformed deterministic techniques by 3 and 2 dB in terms of SFDR performance for 2nd and

3rd degree approximations respectively. An optimum pipeline level insertion for Hyper-Folding realization of polynomial functions was demonstrated using SGA. The SGA procedure can be used for defining any kind of trade off between hardware cost and performance. Implemented designs in a specific FPGA platform have reached 62 and 82 dB SFDR performance for 2nd and 3rd degree approximations respectively.

References

1. Tierney, J., Rader, C.M., Gold, B.: A digital Frequency synthesizer. IEEE Trans. Audio Electroacoust, AU-19, pp 48–57 (1971)
2. Langlois, J.M.P., Al-Khalili, D.: Phase to sinusoid amplitude conversion techniques for direct digital synthesis. IEEE Proc. Circ. Dev. Syst. **151**(6) (2004)
3. De Caro, D., Napoli, E., Strollo, A.G.M.: ROM-less direct digital frequency synthesizer exploiting polynomial approximation. In: Proceedings of IEEE International Conference on Electronics, Circuits and Systems, Croatia, pp. 481–484 (2002)
4. Hajati, F, Raie, A., Gao., Y.: Pose-invariant 2.5 D face recognition using geodesic texture warping. In: 2010 11th International Conference on Control Automation Robotics & Vision, pp. 1837–1841 (2010)
5. Hajati, F., Cheraghian, A., Gheisari, S., Gao, Y., Mian, A.: Surface geodesic pattern for 3D deformable texture matching. Pattern Recogn. **62**, 21–32 (2017)
6. Ayatollahi, F., Raie, A., Hajati, F.: Expression-invariant face recognition using depth and intensity dual-tree complex wavelet transform features. J. Electron. Imaging **24**(2), 023031 (2015)
7. Pakazad, S.K., Faez, K., Hajati, F.: Face detection based on central geometrical moments of face components. In: 2006 IEEE International Conference on Systems, Man and Cybernetics, pp. 4225–4230 (2006)
8. Shojaiee, F., Hajati, F.: Local composition derivative pattern for Palmprint recognition. In: 2014 22nd Iranian Conference on Electrical Engineering (ICEE), Tehran, Iran, pp. 965–970 (2014)
9. Barzamini, R., Hajati, F., Gheisari, S., Motamadine, M.B.: Short term load forecasting using multi-layer perception and fuzzy inference systems for Islamic countries. J. Appl. Sci. **12**(1), 40–47 (2012)
10. Faez, H.K., Pakazad, S.K.: An efficient method for face localization and recognition in color images. In: 2006 IEEE International Conference on Systems, Man and Cybernetics, Taipei, Taiwan, 2006, pp. 4214–4219 (2006). https://doi.org/10.1109/ICSMC.2006.384796

Application of Generalized Deduplication Techniques in Edge Computing Environments

Ryu Watanabe[1]([⊠]), Ayumu Kubota[1], and Jun Kurihara[2]

[1] KDDI Research, Inc., 2-1-15 Ohara, Fujimino, Saitama 356-8502, Japan
{ry-watanabe,kubota}@kddi-research.jp
[2] Graduate School of Information Science, University of Hyogo,
7-1-28 Minatojima-Minamimachi, Chuo, Kobe, Hyogo 650-0047, Japan
kurihara@ieee.org

Abstract. One of the use cases of mobile networks that can be considered for use in Beyond 5G is a massive IoT environment where many IoT (Internet of Things) terminals with low power consumption and computing power are connected. In order to efficiently use network resources in this environment, it is necessary to compress and reduce the amount of data uploaded by a large number of IoT terminals. In this study, we consider data compression in a Massive IoT environment using edge servers, assuming a Multi-access Edge Computing (MEC) scenario. In particular, we consider the application of "Generalized Deduplication (GD)", a stream data compression method based on duplicate deletion, which has been attracting attention in recent years for its lightweight and efficient compression of IoT sensing data. The basic GD algorithm assumes one-to-one stream transmission and reception. In this report, we propose an extension of the GD algorithm that is suitable for one-to-multi (edge server and IoT terminals) MEC environments and has more efficient performance. Specifically, we investigate dictionary construction for the GD utilization in a one-to-multi environment and show a basic evaluation of the efficiency of the proposed algorithm.

1   Introduction

Increasing connectivity through the miniaturization of communication devices and ubiquitous networks has led to the ever-increasing use of IoT [1,4], where everything is connected to a network and communicates via the Internet. In recent years, the use of Multi-access Edge Computing (MEC), which consists of servers and peripheral terminals located at the edge area of the network, has also been progressing. Servers called "edge servers" are less powerful than centralized servers. Therefore, a resource management is required for operation [11]. As an application, a proposal and demonstration of the support for self-driving cars using sensors near the road has been reported. In addition, demonstrative

© The Author(s), under exclusive license to Springer Nature Switzerland AG 2023
L. Barolli (Ed.): AINA 2023, LNNS 655, pp. 585–596, 2023.
https://doi.org/10.1007/978-3-031-28694-0_55

experiment of AR games with multiple players has been seen[1]. However, IoT terminals that require a large amount of deployment for use with sensors and other devices often have low computing power or limited power consumption. In addition, depending on the application, a large number of various types of information, such as weather-related observation data or visual image data, are transmitted. Therefore, compressing the large amount of data uploaded from IoT terminals and reducing the communication cost are essential for efficient use of the MEC environment. In this paper, we discuss data compression in a Massive IoT environment utilizing MEC. Recently, "Generalized Deduplication (GD)", which is a stream data compression method using deduplication, has been attracting attention, because it is lightweight and suitable for IoT data compression. However, the basic GD algorithm assumes one-to-one stream transmission and reception, and is not suitable for use in such a case of the Massive IoT environment described above. Therefore, we extend the method to consider data deduplication in a one-to-multi communication environment. In particular, we provide a basic evaluation of the construction of an integrated dictionary and its efficiency, which is a characteristic of one-to-multi communication.

The rest of this paper is organized as follows. In the second section, related work and basic explanation of the GD are denoted as preliminaries. Then, in the third section, our proposal is described. Finally, the results of the paper are summarized as a conclusion in the fourth section.

The contribution of the paper is as follows: The GD is extended for one-to-multi communication environment like a MEC environment. In addition, the basic evaluation of the performance of the proposed algorithm is presented.

2   Preliminaries

2.1   Related Work

Compression of data, such as sensor data, is essential for transmitting and receiving huge amounts of data. Normal compression methods, such as those based on the LZ algorithm [14,15], compress data by applying relatively small window sizes. However, there is a growing need to compress huge data, such as sensor data, where the pattern of data occurrence is fixed throughout the data, and the use of "de-duplication" is attracting more attention than the use of ordinary compression methods [12].

Basic de-duplication reduces the overall amount of data by discarding duplicate data chunks (or files) throughout the data storage or data stream. At this time, a "dictionary" is built to recover the original data chunks, which are retained with the reduced data. In normal de-duplication, only identical data chunks can be discarded. Therefore, it cannot be simply applied to similar data chunks, such as log data, where only parts of the data are different. Generalized Deduplication (GD) [8–10] was proposed to solve this problem. The GD

---

[1] https://www.telekom.com/en/media/media-information/archive/worlds-first-mobile-edge-mixed-reality-multi-gamer-experience-564004.

can be regarded as a generalization of the data chunk de-duplication similarity algorithm proposed to solve this problem. The GD transforms individual data chunks into data tuples called (basis, deviation) using a bijection method. Duplicate deletion is then performed only on the entire set of bases, and the deviations are stored or transmitted as is. If the mapping on a dictionary is appropriately chosen according to the overall pattern of the applied data, highly similar data chunks will have the same basis, thus achieving high data reduction efficiency.

As an application of the GD, Vaucher et al. proposed a method of in-network compression of DNS data stream [7], which achieves both high compression rate and high processing speed. Hadi et al. have extended the GD and proposed a method to increase compression efficiency by using it in combination with general compression methods [5]. For data protection and privacy reasons, a method, which transmits only the basis and removes duplicate data after encryption, has been proposed [2,13]. As a sample of the GD implementation, a practical code [3] is also released. On the other hand, these methods are designed for sending and receiving one-to-one data streams or huge data on single storage, and their application for one-to-multi environment has not been considered.

2.2 Basic Concept of the Generalized Deduplication

The Generalized Deduplication (GD) is a generalization of duplicate data elimination; when utilizing the GD, a transformation function is first applied to a chunk of stream data to split it into two distinct values: a basis and the deviation from the basis. Thereafter, for each stream to be transmitted, the same portion of the previously transmitted values as the bases are eliminated[2] as duplicates. The deviation is sent so that it can be inverted for data recovery.

2.2.1 Basic Process of the Generalized Deduplication

As an example of the GD, a method based on the $(7, 4)$ Hamming code [7] is denoted.

Example 1. Assume that the size of the processing unit chunk is 7 bits. Processing the chunk to be transmitted through a $(7, 4)$ Hamming code decoder yields 4 bits of information and at most 1 bit-flip error locations. In this case, the 4 bits of information bits are the bases, while the position of the error in the bit reversal is the deviation. The location of the bit-flip error is represented by $1, \ldots, 7$, so there are 8 possible locations, including the error-free case. Therefore, the deviation is expressed in 3 bits. For example, the following eight chunks are all considered to contain at most one bit of error for the code word $0000000 \in \{0, 1\}^7$ of the $(7, 4)$ Hamming code and are mapped to the basis $0000 \in \{0, 1\}^4$.

$$0000000, 0000001, 0000010, 0000100,$$
$$0001000, 0010000, 0100000, 1000000 \in \{0, 1\}^7,$$

---

[2] In practice, the dictionary-based replacement is performed simultaneously with the elimination.

similarly,

$$1111111, 1111110, 1111101, 1111011,$$
$$1110111, 1101111, 1011111, 0111111 \in \{0,1\}^7,$$

Those chunks are mapped to a basis $1111 \in \{0,1\}^4$. Then, a "dictionary"[3] is generated based on the basis obtained by processing the chunks in order from the previous one at the time of transmission. By referring to this dictionary, the GD transmits the index on the dictionary instead of the basis for chunks themselves. The above example of a 7-bit chunk group is shown above.

A 42–bit data sequence

$$0000000\ 1111111\ 0100000\ 1111011\ 1000000\ 1011111,$$

consists of six chunks.

First, if all chunks are independently mapped to pairs of (basis, deviation),

$$0000000 \rightarrow (0000, 000),\ 1111111 \rightarrow (1111, 000),$$
$$0100000 \rightarrow (0000, 111),\ 1111011 \rightarrow (1111, 100),$$
$$1000000 \rightarrow (0000, 101),\ 1011111 \rightarrow (1111, 110),$$

and the transformed data series are obtained. Every chunk is converted into two bases $0000, 1111$. Therefore, for this data, one bit is sufficient to build a dictionary of bases and to identify each basis in the dictionary. When chunks are processed sequentially during transmission, bases that are not in the dictionary are transmitted as they are, and when they appear again, the index in the dictionary is transmitted instead of the basis. This process compresses the above 42–bit data sequence to 30–bit data for transmission as follows.[4]

$$0000|000|1111|000|0|111|1|100|0|101|1|110$$

The receiver synchronizes with the sender and processes the data in the order received, allowing the receiver to construct a dictionary in the same manner as the sender, and to recover the original chunk data from the transmitted data with the dictionary. A dictionary built based on data where a particular basis occurs frequently can represent a large number of chunks and can be compressed more efficiently.   □

Because of this mechanism, the efficiency of compression depends on how efficiently the basis can be aggregated into a dictionary and how often chunks with the same basis appear in the data. When compressing static data such as files or data on storage, dictionary construction can be performed based on the knowledge of the data structure as in general compression methods, but

---

[3] "registry" is another name of the "dictionary".

[4] In practice, a separator bit is required to clearly indicate either the basis data or the index data.

when the data is a stream such as IoT data, dynamic dictionary construction is necessary. To simplify the explanation of dictionary construction, we denote the data sent by IoT devices to an edge server as "$A, B, C \cdots$", where "A" represents the chunk data that is mapped to the basis "a". Therefore, in data processing by the GD, any sequence of data that can be represented by a basis "a" and a deviation "dev_n" is represented as "A". The index of a basis on the dictionary is denoted by "$index_*$".

2.2.2   Dictionary Creation

If the sender sends chunk data such as "A, A, B, B" (Fig. 1), the dictionary is generated as follows.

1. When sending a chunk "A", the sender converts it into a basis "a" and a deviation "dev_n", registers the basis "a" as "$index_1$" on the dictionary, and sends "$a|dev_n$" to the reciever (Fig. 2-(1)).
2. The receiver registers "a" received for the first time in the dictionary as "$index_1$" and stores the received data as "$index_1|dev_n$" (Fig. 2-(2)).
3. The sender sends the next chunk "A". The "A" is converted to basis "a" and deviation "dev_n" as in the first step.
4. The sender searches the dictionary, replaces "a" with "$index_1$" because "a" is registered as "$index_1$", and sends "$index_1|dev_n$" (Fig. 3-(1)).
5. When the receiver receives "$index_1|dev_n$", it stores it as "$index_1|dev_n$" as it is (Fig. 3-(2)).
6. The sender then sends chunk "B". At this time, "B" is converted into a basis "b" and a deviation "dev_n", the basis "b" is registered in the dictionary as "$index_2$", and "$b|dev_n$" is sent to the receiver(Fig. 4-(1)).
7. The receiver registers the first "b" received in the dictionary as "$index_2$" and stores the received data as "$index_2|dev_n$" (Fig. 4-(2)).

When the next chunk "B" is sent, the same procedure as 3–5 is used. In this way, a dictionary "$index_1$" : "a", "$index_2$" : "b" is shared between the sender and receiver. When recovering data, the original chunks can be decoded from the stored data by using the dictionary for "$index_*|dev_n$", "$*|dev_n$", and then recovered.

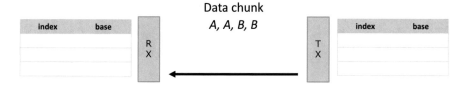

Fig. 1. Dictionary creation (initial setting).

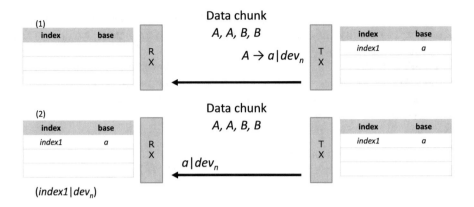

Fig. 2. Dictionary creation (index1 entry).

3   Proposed Method

First, a MEC environment, which we assume, is noted, then a dictionary creation method of the GD for the MEC scenario is described.

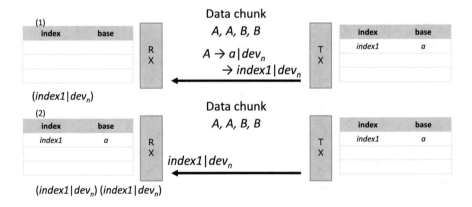

Fig. 3. Dictionary creation (index1 usage).

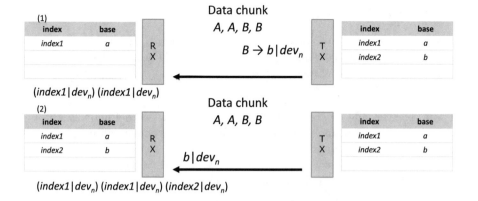

Fig. 4. Dictionary creation (index2 entry).

3.1 Multi-access Edge Computing

Figure 5 shows an overview of the envisioned use case of the MEC environment, which consists of edge servers and IoT devices. The purpose is to reduce the amount of communication between IoT devices and edge servers. As shown in the figure, the MEC environment consists of similar environments. The data from each environment aggregated at the edge server is finally aggregated to the server on the cloud. The amount of communication between the edge server and the cloud server, as well as the amount of data on the nodes, can be reduced. At this point, for high compression efficiency, it is expected that the data format at the time of transfer be identical, even though the sensors connected to the edge server may vary. If the formats are different, duplicate deletion will not work. So, the data compression will not be effective.

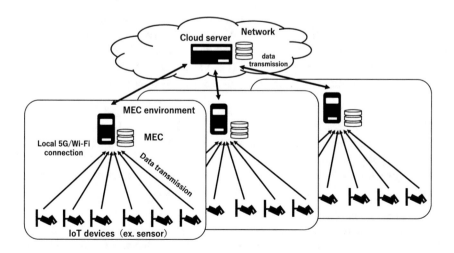

Fig. 5. Use case of MEC environment.

3.2   Proposed Method

In order to apply to data transmission in the envisioned MEC environment, we extend the dictionary construction method of the GD described in the previous section. Without making any changes to the non-powerful transmitter side, we extend the dictionary construction method on the receiver side as follows.

- The receiver side creates an integrated dictionary (we define it as a unified dictionary) in addition to the dictionaries with the coresponding senders (we define them as individual dictionaries).
- The unified dictionary preserves the relationship between the indices and the basis as in the usual method. This unified dictionary shall be the main dictionary in its MEC environment[5].
- Individual dictionaries do not retain the relationship between indices and bases, but keep the relationship between individual indices and indices in the unified dictionary[6].

With this extension, the following processes are added to the procedure described in the previous section when creating a dictionary on the receiver side.

- When a new basis is received at the receiver side, the system checks to see if there is an index for the basis on the unified dictionary.
- If there is an index in the individual dictionary corresponding to the sender, the relationship between the index in the individual dictionary and the index in the unified dictionary is noted in the individual dictionary.
- If not, the index of the basis is registered in the unified dictionary, and the relationship between the index in the unified dictionary and the index in the corresponding individual dictionary is noted in the corresponding individual dictionary.

Figures 6 and 7 explain how the dictionary is created by using the case where there are two senders. They are "α" and "β", respectively. The sender "α" sends chunk data "A, A, B, B", and the sender "β" sends chunk data "C, C". For the sake of explanation, the reception of the data sent from "β" is assumed to be after the arrival of the data sent from "α", but the actual operation is not affected by the arrival time.

1. The sender "α" converts chunk data "A, A, B, B" into bases and deviations (or indices and deviations) and sends them to the receiver. Then, "$index(u)_1 : a$", "$index(u)_2 : b$" are registered in the unified dictionary following the procedure. At the same time, "$index_1 : index(u)_1$" and "$index_2 : index(u)_2$" are registered in the individual dictionary for "α".

---

[5] In the illustration of the explanation, it is represented as "main".

[6] Individual dictionaries are dictionaries that play a supplementary role because they preserve the relationship between indices and indices, unlike the original dictionaries. Therefore, on the illustration for the explanation, it is denoted as "sub".

2. Then, when the sender "β" converts the chunk "C" to basis and deviation and sends them to the receiver, the receiver checks the unified dictionary according to the procedure, and since there is no index on it, the receiver registers "$index(u)_3 : C$" in the unified dictionary, and records "$index_1 : index(u)_3$" is registered in the individual dictionary for "β".

Thus, the unified dictionary becomes a basic dictionary, and when a new basis from each sender arrives, a new index is created in the unified dictionary, and a new index on each individual dictionary is also created at the same time.

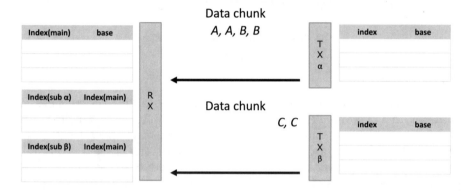

Fig. 6. Dictionary creation (initial setting).

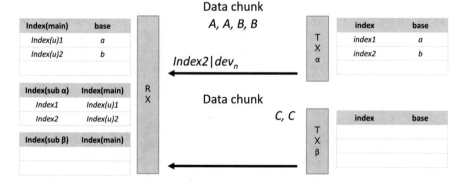

Fig. 7. Dictionary creation (α entry).

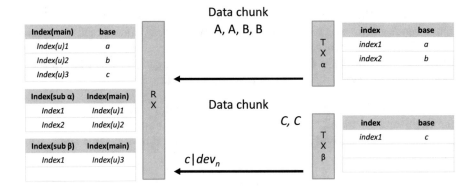

Fig. 8. Dictionary creation (β entry).

3.3   Comparison

As an evaluation of the proposed method, a comparison with the existing method used as-is in a one-to-multi environment is shown in Table 1. Although the proposed method requires a unified dictionary and processing for its creation, the size of each individual dictionary can be reduced. If the size of the basis is k bits, the size of each index is t bits ($k > t$), and the number of IoT devices is n, the size of the dictionary can be expected to be reduced by $k(n-1) - t(n+1)$ bits, at most. The proposed method does not affect the size of data transmitted or stored on the edge servers, since it only processes the dictionary. So, there is no change in the data compression size itself. However, by using proposed method, the bases can be combined into a single unified dictionary, which promotes an advantage in terms of management. On the other hand, a disadvantage is also assumed. The unified dictionary will perform index references or writes on all data receptions. The more IoT terminals there are and the more frequently they communicate, the more accesses to this unified dictionary will be made. This situation may affect the processing at the edge server. There would be a limit to the number of IoT terminals that can be connected.

3.4   Consideration

Our proposal described in this paper extends the ordinary GD by focusing on dictionary creation in the multi-source edge computing scenario. However, in actual use, security considerations such as data privacy or data protection should be taken into account. To fulfill this requirement, we can consider the following approaches: the sender may send only the basis, and the deviations may be stored separately and referred to during decompression at each terminal sensor node if original data is utilized at each one. Or else, the deviations could be encrypted in the end-to-end manner between each terminal node and the cloud server by bypassing the edge server if the original data is utilized in the cloud.

These approaches should be reasonable since the size of deviations is always fixed and relatively much smaller than basis, and hence the upload cost could be

Table 1. Evaluation.

Item	Proposal	Conventional	Note
Unified dictionary	Necessary (1)	Unnecessary	Not necessary for conventional methods, as they perform compression individually. Only the proposed method requires one
Individual dictionary	Small size	Normal size	Although individual dictionaries are required for the number of transmitting terminals regardless of which method is used, the existing method registers index and basis pairs, whereas the proposal registers index and index pairs, thus reducing the size of the dictionaries
Data size	Same		Since both methods are subject to similar compression, the data size retained on the edge server is essentially the same

negligible. This means that they can be easily aggregated at each sensor, e.g., in the example described in Sect. 2.2, the $(7, 4)$ Hamming code was used, so the deviation was 3 bits for a basis of 4 bits, but with the $(2047, 2036)$ Hamming code, the deviation can be 11 bits for a basis of 2036 bits. This makes it an efficient method for a MEC environment of IoT terminals with limited computing power. Also in these cases, we emphasize that the original data cannot be recovered only from the plaintext basis in the probabilistic manner even by the edge server as shown in [5,6]. For the application of each method, it is necessary to consider an appropriate way for each usage scenario, assuming the use cases of the MEC.

4   Conclusion

This paper describes a consideration on an application of the GD to one-to-multi communication environments. Adapting the proposed method to an environment, such as MEC, is expected to provide administrative advantages in addition to a reduction in dictionary size. In the future, we plan to further study for security features to ensure the safe use of the GD.

Acknowledgements. This work was supported in part by NICT22401 and JSPS KAKENHI Grant Number JP22K11994, JP21H03442, JP20K23329.

References

1. Atzori, L., Iera, A., Morabito, G.: The internet of things: a survey. Comput. Netw. **54**(15), 2787–2805 (2010). https://doi.org/10.1016/j.comnet.2010.05.010

2. Bellare, M., Keelveedhi, S., Ristenpart, T.: Message-locked encryption and secure deduplication. In: Johansson, T., Nguyen, P.Q. (eds.) EUROCRYPT 2013. LNCS, vol. 7881, pp. 296–312. Springer, Heidelberg (2013). https://doi.org/10.1007/978-3-642-38348-9_18

3. Kurihara, J.: rust-gd: a rust implementation of generalized deduplication (2022). https://github.com/junkurihara/rust-gd

4. Minerva, R., Biru, A., Rotondi, D.: Towards a definition of the Internet of Things (IoT). IEEE Internet Initiative 1(1), 1–86 (2015)

5. Sehat, H., Kloborg, A.L., Mørup, C., Pagnin, E., Lucani, D.E.: Bonsai: a generalized look at dual deduplication. arXiv:2202.13925v2 (2022)

6. Sehat, H., Pagnin, E., Lucani, D.E.: Yggdrasil: privacy-aware dual deduplication in multi client settings. In: the Proceedings of IEEE International Conference on Communications, pp. 1–6 (2021)

7. Vaucher, S., Yazdani, N., Felber, P., Lucani, D.E., Schiavoni, V.: Zipline: In-network compression at line speed (2021). arXiv:2101.05323

8. Vestergaard, R., Lucani, D.E., Zhang, Q.: Generalized deduplication: lossless compression for large amounts of small IOT data. In: European Wireless 2019; 25th European Wireless Conference, pp. 1–5 (2019)

9. Vestergaard, R., Zhang, Q., Lucani, D.E.: Generalized deduplication: bounds, convergence, and asymptotic properties. In: 2019 IEEE Global Communications Conference (GLOBECOM), pp. 1–6 (2019). https://doi.org/10.1109/GLOBECOM38437.2019.9014012

10. Vestergaard, R., Zhang, Q., Lucani, D.E.: Lossless compression of time series data with generalized deduplication. In: Proceedings of IEEE GLOBECOM 2019, pp. 1–6 (2019)

11. Watanabe, R., Kubota, A., Kurihara, J.: Resource authorization patterns on edge computing. In: IN2020-68, vol. 120, pp. 85–90. IEICE (2021)

12. Xia, W., et al.: A comprehensive study of the past, present, and future of data deduplication. Proc. IEEE 104(9), 1681–1710 (2016)

13. Zhou, Y., Yu, Z., Gu, L., Feng, D.: An efficient encrypted deduplication scheme with security-enhanced proof of ownership in edge computing. BenchCouncil transactions on benchmarks, Stand. Eval. 2(2), 100,062 (2022). https://doi.org/10.1016/j.tbench.2022.100062

14. Ziv, J., Lempel, A.: A universal algorithm for sequential data compression. IEEE Trans. Inf. Theory 23(3), 337–343 (1977)

15. Ziv, J., Lempel, A.: Compression of individual sequences via variable-rate coding. IEEE Trans. Inf. Theory 24(5), 530–536 (1978)

On the Realization of Cloud-RAN
on Mobile Edge Computing

Andres F. Ocampo[1(✉)] and Haakon Bryhni[2]

[1] SimulaMet - Simula Metropolitan Center for Digital Engineering Oslomet - Oslo
Metropolitan University, Oslo, Norway
`andres@simula.no`
[2] SimulaMet - Simula Metropolitan Center for Digital Engineering, Oslo, Norway
`haakonbryhni@simula.no`

Abstract. The cellular network architecture is evolving to support a
wide variety of applications with different traffic characteristics expected
for 5G and beyond. Providing shared computing and network resources,
Cloud based Radio Access Network (Cloud-RAN) in conjunction with
Mobile Edge Computing (MEC) are considered key enablers to building
5G networks in a cost-efficient way. Understanding the limits and constraints of deploying the Cloud-RAN on MEC servers, allows the system
to be engineered meeting latency and capacity requirements. By conducting a literature review, this paper discusses sharing computing and
networking resources in MEC servers, which run software implementation of the Base Band Unit (vBBU) along with collocated applications.

Keywords: Cloud-RAN · Mobile edge computing · virtualization ·
Resource management

1 Introduction

Unlike its predecessors, the fifth generation of mobile networks 5G aims to
cater for a wide spectrum of services with diverse requirements. The IMT-2020
requirements by ITU envisions three broad use cases. Enhanced mobile broadband (eMBB), requiring peak data rates up to 20 Gbps downlink (DL) and 10
Gbps uplink (UL); ultra-reliable low latency communications (URLLC), providing high availability and reliability with a maximum latency of 1 millisecond;
and massive machine type communication (MMTC), supporting connectivity
and mobility in highly dense scenarios like the Internet of Things (IoT). It is
expected that services under these use cases increase the demand for wireless
traffic by 1000 times today's demand [1].

The design and architecture of the radio access network (RAN) is key to
realizing the 5G vision. For instance, dense deployments of small cells seems to be
the most likely network scenario for 5G [2]. One reason is that small cells increase
substantially the transmission capacity and, therefore, the peak data rate per
cell. Deploying small cells close to the users is also ideal for low energy devices

L. Barolli (Ed.): AINA 2023, LNNS 655, pp. 597–608, 2023.
https://doi.org/10.1007/978-3-031-28694-0_56

as in the IoT use case. Nevertheless, such scenarios of dense small cells make the traditional distributed RAN (DRAN) architecture not economically feasible. The number of deployed BSs in dense scenarios of small cells would increase dramatically both capital expenditure (CAPEX) and operational expenditure (OPEX).

To cope with such unprecedented wireless traffic demand with low latency requirements in 5G and beyond, both the Cloud-RAN architecture and Mobile (also known as multi access) Edge computing (MEC) provide an important paradigm shift on how the RAN is designed. Leveraging both software defined wireless networking and virtualization technology, the Cloud-RAN deploys several virtual Base Band Unit (vBBU) on the same MEC server, in the so-called vBBU pool [3], along with collocated applications. Providing cloud computing capabilities at the very edge of the mobile network, MEC significantly reduces latency of collocated mobile services while easing both processing and traffic pressure over the mobile system [4]. Furthermore, by sharing network and processing resources, the Cloud RAN architecture and MEC brings optimized operation and maintenance benefits to mobile network operators (MNO) and service providers.

Despite its appeal as a key enabler for 5G and beyond generations of mobile systems, running the Cloud-RAN on MEC imposes stringent latency constraints and limitations that need to be understood in order to determine the scalability of instantiating vBBUs on top of MEC servers. Moreover, addressing these limitations is key to design robust RAN that meets future wireless traffic demands with latency requirements. This paper presents a literature review on the realization of the Cloud-RAN using MEC servers. The goal is to further the understanding and implications of sharing processing and network resources when deploying the Cloud-RAN on top of MEC servers.

In summary, the main contributions of this paper are as follows.

- Presenting a general overview of the Cloud-RAN architecture using MEC.
- Discussing running RT applications using virtualization technology on MEC.
- Presenting a literature review of MEC deploying the Cloud-RAN.

2   The Cloud-RAN Architecture

The Cloud-RAN architecture provides a new paradigm to the RAN design. As shown in Fig. 1, the RAN consists of: user equipment(s) (UE); the air interface; antennas; the base station, which is decoupled into the Remote Radio Unit (RRU), which converts radio waves into digital waveform, and the Baseband Unit (BBU), which performs signal and network protocols processing; and a network link connecting the RRU and the BBU known as Fronthaul. The Fronthaul is typically deployed using fiber optic cable through one of the following protocols: CPRI [5], eCPRI [6], OBSAI [7]. A transport network known as Backhaul connects the RAN to the Core Network (CN), which provides access to mobile services. In Cloud-RAN, the vBBU is implemented as a software-defined wireless networking application. Leveraging virtualization technologies, several vBBUs could be deployed on top of a centralized MEC server sharing processing and network resources [8].

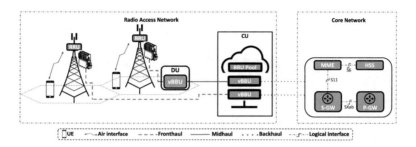

Fig. 1. Cloud Radio Access Network architecture.

Nevertheless, moving the BBU from the cell tower poses stringent latency constraints and capacity requirements on the Fronthaul. Although a fiber or WDM solution would meet such requirements, deploying dedicated fibers per cell would make the Cloud-RAN not a cost-efficient architecture. To tackle these challenges, two promising solutions are being considered and standardized.

Functional Split

Processing part of the BBU protocol stack locally close to the antennas instead of transporting in-phase and quadrature (IQ) signal samples between the RRU and BBU, decreases the requirements of bandwidth and latency in the Fronthaul. Different functional splits of the BBU protocol stack have been proposed by the industry and standardization bodies. For instance, the 3GPP proposes a functional split reference model based on the LTE protocol stack [9], as shown in Fig. 2. Moreover, the IEEE 1914 working group has defined two logical split points placement [10]: the distributed unit (DU), which is located near the cell tower; and the centralized unit (CU), which is located at the MNO's MEC. Introducing both split points redefines the mobile transport network segments and their requirements in terms of latency and capacity, as depicted in Fig. 1: the Fronthaul is the segment between the RRU and the DU, which transports time-domain IQ signal samples; the Midhaul is the segment connecting the DU and the CU, which data rate and requirements depend upon the chosen functional split; and the Backhaul, which connects the Cloud-RAN with the CN. These transport segments are referred to as mobile Crosshaul (XHaul).

As illustrated in Fig. 2, the dotted red line highlights the split option as defined by 3GPP. Functions on the left side of a given option are instantiated at the CU, whereas right side functions are left for the DU. The more functions are instantiated at the DU, the less stringent network delay requirements over the Midhaul; the more functions are instantiated at the CU, the higher the requirements of capacity and latency on the Midhaul.

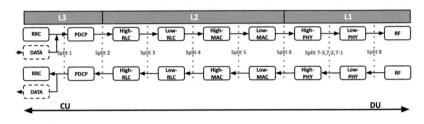

Fig. 2. 3GPP functional split of the LTE-BBU functionality [9].

The survey paper by Larsen et al. [11] summarizes the works implementing and analyzing the different functional splits on the realization of Centralized/Cloud RAN. Recently, a debate has arisen over which functional split is optimal. However, defining the optimal split depends on different technical and business parameters that include network topology, number of users, type of service, among others. As stated in the blog entry from RCR Wireless [12], because of the intrinsic difficulty of meeting the Fronthaul requirements, the 3GPP is considering splits 2 and 7.x for deployment, in addition to the traditional monolithic BBU.

Switched Ethernet Mobile X-Haul Network

Because deploying dedicated optical fiber links in the Fronthaul does not scale for 5G dense-small antenna cells, switched Ethernet is being standardized as a candidate solution for the mobile XHaul in 5G. The main reason is that switched Ethernet statistical multiplexing enables a shared transport network, aggregating traffic flows from multiple BBUs into the same multihop network, instead of deploying a dedicated fiber solution per BBU [13].

The IEEE 1914 Working Group-Next Generation Fronthaul Interface (NGFI) through the project 1914.1 [14], is standardizing the interfaces for the Fronthaul segments based on Ethernet. For instance, the project 1914.3 [15] defines the signal encapsulation into Ethernet frames according to the functional split. Another project for Ethernet encapsulation of radio signals, is the eCPRI from the CPRI consortium [16]. This project is intended for the encapsulation of radio IQ samples from-to the RRU.

Because switched Ethernet uses queuing to aggregate traffic flows, switched Ethernet induces stochastic delay issues that might affect mobile services. To mitigate this problem, switched Ethernet requires mechanisms that allow bounded delay and delay variation for mobile traffic. The IEEE P802.1CM standards group is standardizing mechanisms to support Time-Sensitive Networking (TSN) for the Fronthaul [17]. Based on low priority traffic preemption, IEEE P802.1CM TSN standardized the mechanism IEEE 802.1Qbu [18] that enables high priority frames to preempt low priority frames. To guarantee bounded delay when aggregating traffic flows, the IEEE P802.1CM TSN standardized a time-aware scheduling mechanism IEEE 802.1Qbv [19], which manages link access to a set of queues. The idea behind this mechanism is to guarantee bounded delay to time sensitive traffic by round robin enqueued frames based on contending time windows. While frames in high priority queues are assigned a high priority window, low priority queues are assigned best effort windows. In the context of vRAN, the work in [20] evaluates the IEEE P802.1CM TSN mechanisms in the Fronthaul, while transporting mobile traffic in a controlled experimental setup using TSN switches. Similarly, the work in [21] uses TSN switched Ethernet as mobile Fronthaul transporting data plane and control plane traffic for a network slicing based mobile network.

3   Running the Cloud-RAN on MEC

MEC provides a unique ecosystem for MNOs, services providers and applications developers. By deploying cloud computing capabilities at the very edge of the mobile network, MEC significantly reduces latency of mobile services while easing both processing and traffic pressure over the mobile system [4]. Because mobile services often impose stringent low latency requirements, MEC servers run multiple applications with diverse execution time requirements using virtualization technology [22]. For instance, running the vBBU which performs a combination of both time sensitive processing (e.g., L1 and lower L2 functions in Fig. 2) and general purpose processing (e.g., upper L2 and L3 functions in Fig. 2), imposes different execution time requirements.

To cope with the execution time requirements of time sensitive services, MEC servers instantiate those services as RT applications. In computing, an RT application is defined by the upper-bound execution time constraint (i.e., deadline) in which the process should run [23]. The way the system behaves when a deadline is missed, defines the classification of the RT application. For example, when missing a deadline does not imply critical consequences but the service degradation, the application is classified as soft-RT, e.g., video streaming. Conversely, if missing a deadline implies critical consequences, the application is classified as hard-RT, e.g., autonomous navigation, the RAN. To meet those requirements, both the MEC server OS and the virtualization environment hosting RT applications must provide RT guarantees, i.e., preemption of non-RT processes on behalf of RT processes, and prioritized resource allocation for RT applications.

Linux RT-Kernel as MEC's OS Running the vBBU and RT Applications

To run RT applications on top of a MEC system, the host OS must provide RT guarantees, i.e., preemption and a scheduling policy that focuses on meeting timing constraints of individual processes rather than maximizing the average amount of scheduled processes. Nevertheless, the incurred cost of development, maintenance, and licensing of a RT OS, has motivated the adoption of a general purpose OS like Linux to run RT systems [24]. Linux open source licensing and Kernel modularity, ease the development, customization and maintenance of any feature into the source code, thus reducing costs [25]. Moreover, several mechanisms have been proposed in recent years to provide RT support in the Linux Kernel (e.g., RTLinux [26], Low-Latency patch [27], PREEMPT_RT [28] patch), opening up the possibility of its use for RT systems [23], particularly, for RT signal processing of vBBU functions in [29]. For instance, the Linux RTAI (Real Time Application Interface) [30], has been used in mobile system testbeds in [29,31,32]. On the other hand, the PREEMPT_RT patch has been used in few works in the context of Cloud-RAN [33,34]. Included in the mainline code from the Ubuntu distribution, the Low-Latency Kernel patch [35] has been widely adopted by researchers using the OpenAirInterface (OAI) code [36–39]. The main reason is that OAI's developers have optimized their code to provide full compatibility with the low-latency Kernel. In this paper, because the testbed used for experimentation is based on OAI, the Low-Latency Kernel patch has been adopted as MEC's OS Kernel.

Virtualization Technology with RT Support for Running the vBBU and RT Applications

Virtualization technology (e.g., hypervisor, containers) enables multiple applications running as isolated processes on the same computing infrastructure. Running RT applications using virtualization has been addressed in related research. For instance, the terms RT VMs and RT containers refer to VMs or containers providing RT support, respectively [40,41]. To provide RT guarantees in hypervisor based virtualization, while the hypervisor defines the RT scheduling mechanism that allocates CPU time to virtual machines (VM), the guest OS deploys an RT Kernel that preempts non-RT tasks on behalf of RT ones [42] (e.g., Linux with RT support).

In containerized virtualization, on the other hand, containers do not deploy such guest OS. Instead, containers rely on Kernel's features Cgroups and namespaces to isolate processes [43]. Hence, to provide RT guarantees, containers rely on the host OS which adopts a RT-Kernel supporting preemption and RT Scheduling [44].

When running RT processes, containers provide better RT performance than VMs [41]. The reason has to do with the overhead generated by Hypervisor and the guest OS. Conversely, containerized virtualization is considered a lightweight virtualization as containers are isolated using Kernel features namespaces and

Cgroups [45]. In the context of Cloud-RAN, for instance, the works in [46,47] evaluated different virtualization environments and compared them with the bare-metal deployment: hypervisor, Docker containers, and Linux containers (LXC). Evidence from those works show that containers achieve lower processing time than hypervisor VMs. More specifically, LXC achieves similar processing time as the bare-metal deployment.

4   Sharing Computing Resources with the Cloud-RAN

This section discusses the deployment of vBBUs on top of MEC servers using virtualization technology while sharing computing resources.

4.1   Sharing Computing Resources with RT Applications in MEC Servers Managed by the Linux RT-Kernel

While running RT processes on a system managed by the Linux RT-Kernel, RT processes may be (are) sharing computing resources either with collocated user-space processes or with Kernel threads. As a result, collocated workloads potentially induce processing interference, either from sharing physical resources [48] or from Kernel space processing [49] that could impact the performance of RT applications [50]. Sharing physical resources (e.g., CPU, I/O, memory) among applications with diverse execution time requirements (e.g., mixed time-critically services) collocated on general purpose processors (GPP) have been widely investigated in the literature [51]. Few works, though, have investigated the processing interference caused by Kernel space processing. For instance, Reghenzani et al. in [52] characterized the processing interference caused by different Kernel subsystems under different workloads of mixed time critical services. Because Kernel threads can not be preempted, Kernel threads could potentially impact the performance of RT applications if they happen to be scheduled on the same CPU. Despite the research efforts to understand the impact of resource sharing among mixed time-sensitive applications on GPP, there remains the need of investigating the impact that sharing resources with collocated workloads causes on time-sensitive applications deployed using virtualization technology, particularly in containerized virtualization. Moreover, there remains the need for resource management solutions that enables resource sharing among collated applications while providing RT guarantees to time-sensitive applications in MEC [53].

Single Server Runtime Resource Management for Heterogeneous Time-Sensitive Applications. To provide processing quality of service to time-sensitive applications while sharing computing resources with collocated workloads on GPP, runtime resource management provides an alternative solution for resource contention among collocated applications [54]. Typically deployed either at the Kernel level (e.g., task scheduling solutions [55]), or at the user-space (i.e., running as a daemon process [56]), runtime/dynamic resource management allow efficient resource utilization while providing differentiated processing guarantees to heterogeneous applications [57]. For instance, the work in [58] proposes adaptable runtime mapping of resources for RT applications running in embedded platforms along with collocated workloads. These ideas were extended in [59] to enable energy efficient vBBU processing, through a design model approach used in embedded systems to provide hybrid and flexible resource mapping.

Cloud-Based/Multi-server Resource Management for Heterogeneous Time-Sensitive Applications. In the context of MEC and multi-Cloud Computing in general, resource management caters to a wide spectrum of problems [60], e.g., resource provisioning, resource allocation, resource mapping, service migration, among others. Typically implemented on resource orchestration platforms, solutions to these problems aim at providing dynamic multi-server/multi-Cloud resource management that optimize the utilization of physical resources among applications instantiated using virtualization technology [61]. Resource management solutions in MEC/Multi-Cloud computing considering time-critical services commonly relies on dynamic/static resource provisioning of RT VMs/containers [62]. For instance, the work in [63] proposes a dynamic resource provisioning mechanism for VMs running time-critical services, which prioritizes VMs according to their application deadlines. Nevertheless, the inherent migration time cost of dynamically provisioning applications among servers could impact the RT performance of time-sensitive services [64].

To avoid RT performance degradation of services due to resource provisioning migration among MEC servers, we consider that hard RT applications (e.g., vBBU) are deployed on a single MEC server within a MEC system. This approach is based on the evidence from embedded computing platforms running multiple applications with diverse execution time requirements [59]. Nevertheless, RT applications may (are) sharing computing resources with mixed time-critical and general purpose applications that are subject to be migrated and provisioned on different servers. Such scenarios open the door for integrating runtime resource management mechanisms for applications running on virtualization technology on a single server, with Cloud resource provisioning and allocation mechanisms. While the former enables efficient resource sharing while providing QoS guarantees to RT applications within MEC servers (e.g., vBBU), the latter enables efficient service provisioning according to dynamic user demands. The work in [65] integrates these two paradigms to enable vBBU processing on a MEC server along with general purpose applications. An external mobile network controller computes and predicts worst case execution time (WCET) of vBBUs deployed on GPP, according to current performance and user traffic demands. Based on WCET predictions, the external controller dynamically defines the amount of CPUs allocated to the vBBU, while enabling collocated applications.

5   Conclusions

Throughout this paper, we have discussed the feasibility of running vBBUs on MEC servers in the Cloud-RAN architecture. Through a literature review of both the Cloud-RAN architecture and MEC, this work sheds light on sharing computing and network resources to the vBBU on MEC servers.

This paper highlights the need to develop run-time algorithms aiming to dynamically allocate physical resources (e.g., memory, disk) while prioritizing RT services in MEC servers. Moving forward this research could integrate run-time resource management mechanisms with multi Cloud resource management. While the former enables sharing computing resources with RT applications in MEC server, the latter enables flexible and efficient resource allocation for user's applications. Those applications, though, can afford server migration and provisioning times. Finally, to conclude on the benefits of adopting resource management mechanisms in MEC, it is interesting to include user quality of experience for 5G use cases, including ultra low latency applications.

References

1. Bhushan, N., et al.: Network densification: the dominant theme for wireless evolution into 5g. IEEE Commun. Mag. **52**(2), 82–89 (2014)
2. Ge, X., Song, T., Mao, G., Wang, C.-X., Han, T.: 5g ultra-dense cellular networks. IEEE Wirel. Commun. **23**(1), 72–79 (2016)
3. Checko, A., et al.: Cloud ran for mobile networks-a technology overview. IEEE Commun. Surv. Tutor. **17**(1), 405–426 (2015)
4. Hu, Y.C., Patel, M., Sabella, D., Sprecher, N., Young, N.: Mobile edge computing-a key technology towards 5g. ETSI White Paper **11**(11), 1–16 (2015)
5. CPRI. Common public radio interface: ECPRI interface specification. CPRI Specification V7.0 (2015)
6. CPRI: Common public radio interface: ECPRi interface specification. eCPRI Specification V2.0 (2019)
7. OBSAI: Open base station architecture initiative. BTS System Reference Document version, 2 (2006)
8. Checko, A., et al.: Cloud ran for mobile networks-a technology overview. IEEE Commun Surv. Tutor. **17**(1), 405–426 (2015)
9. 3GPP TR 38.801. Study on new radio access technology: Radio access architecture and interfaces (2017)
10. IEEE. IEEE STD 1914.1-2019: Standard for packet-based fronthaul transport network. IEEE Standards (2019)
11. Larsen, L.M.P., Checko, A., Christiansen, H.K.: A survey of the functional splits proposed for 5g mobile crosshaul networks. IEEE Commun. Surv. Tutor. **21**(1), 146–172 (2019)
12. RCRWireless. Exploring functional splits in 5g ran: Tradeoffs and use cases. Accessed Dec 2021
13. Assimakopulos, P., Birring, G.S., Kenan Al-Hares, M., Gomes, N.J. Ethernet-based fronthauling for cloud-radio access networks. In: 2017 19th International Conference on Transparent Optical Networks (ICTON), pp. 1–4 (2017)

14. IEEE. IEEE standard for packet-based fronthaul transport networks. IEEE Std. 1914.1-2019, pp. 1–94 (2020)
15. IEEE. IEEE standard for radio over ethernet encapsulations and mappings. IEEE Std. 1914.3-2018, pp. 1–77 (2018)
16. Gomes, N.J., Chanclou, P., Turnbull, P., Magee, A., Jungnickel. V.: Fronthaul evolution: from CPRI to ethernet. Opt. Fiber Technol. **26**, 50–58 (2015)
17. Finn, N.: Introduction to time-sensitive networking. IEEE Commun. Stand. Mag. **2**(2), 22–28 (2018)
18. IEEE-P802.1CM. IEEE 802.1qbu-2016 - IEEE standard for local and metropolitan area networks - bridges and bridged networks - amendment 26: Frame preemption. IEEE Std. 802.1Q-2014 (2016)
19. IEEE-P802.1CM. IEEE 802.1qbv-2015 - IEEE standard for local and metropolitan area networks - bridges and bridged networks - amendment 25: Enhancements for scheduled traffic. IEEE Std. 802.1Q-2014 (2015)
20. Bhattacharjee, S., Schmidt, R., Katsalis, K., Chang, C.-Y., Bauschert, T., Nikaein, N.: Time-sensitive networking for 5g fronthaul networks. In: ICC 2020–2020 IEEE International Conference on Communications (ICC), pp. 1–7 (2020)
21. Bhattacharjee, S., et al.: Network slicing for TSN-based transport networks. IEEE Access **9**, 62788–62809 (2021)
22. Tomaszewski, L., Kukliński, S., Kołakowski, R.: A new approach to 5G and MEC integration. In: Maglogiannis, I., Iliadis, L., Pimenidis, E. (eds.) AIAI 2020. IAICT, vol. 585, pp. 15–24. Springer, Cham (2020). https://doi.org/10.1007/978-3-030-49190-1_2
23. Reghenzani, F., Massari, G., Fornaciari, W.: The real-time linux kernel: A survey on preempt_rt. ACM Comput. Surv. **52**(1), 36 (2019)
24. Mosnier, A.: Embedded/real-time linux survey (2005)
25. Timmerman, M.: Real-time capabilities in the standard linux kernel: How to enable and use them? Int. J. Recent Innov. Trends Comput. Commun. **3**(1), 131–135 (2015)
26. Yodaiken, V., et al.: The rtlinux manifesto. In: Proceedings of the 5th Linux Expo (1999)
27. Molnar, I.: Linux low latency patch. Accessed Dec 2021
28. The Linux Foundation. Preempt_rt patch. https://wiki.linuxfoundation.org/realtime/preempt_rt_versions
29. Nikaein, N., et al.: Openairinterface: an open LTE network in a PC. In: Proceedings of the 20th Annual International Conference on Mobile Computing and Networking, pp. 305–308 (2014)
30. Giacobbi, G.: The GNU Netcat project. Accessed Nov 2021
31. Kaltenberger, F., Wagner, S.: Experimental analysis of network-aided interference-aware receiver for LTE MU-MIMO. In: 2014 IEEE 8th Sensor Array and Multichannel Signal Processing Workshop (SAM), pp. 325–328, June 2014
32. Alyafawi, I., Schiller, E., Braun, T., Dimitrova, D., Gomes, A., Nikaein, N.: Critical issues of centralized and cloudified LTE-FDD radio access networks. In: 2015 IEEE International Conference on Communications (ICC), pp. 5523–5528. IEEE (2015)
33. Bhaumik, S., et al.: Cloudiq: a framework for processing base stations in a data center. In: Proceedings of the 18th Annual International Conference on Mobile Computing and Networking, pp. 125–136. ACM (2012)
34. Fajjari, I., Aitsaadi, N., Amanou, S.: Optimized resource allocation and RRH attachment in experimental SDN based cloud-ran. In: 2019 16th IEEE Annual Consumer Communications& Networking Conference (CCNC), pp. 1–6. IEEE (2019)

35. Molnar, I.: Linux low latency patch. Accessed Dec 2021
36. Huang, S.-C., Luo, Y.-C., Chen, B.L., Chung, Y.-C., Chou, J.: Application-aware traffic redirection: a mobile edge computing implementation toward future 5g networks. In: 2017 IEEE 7th International Symposium on Cloud and Service Computing (SC2), pp. 17–23 (2017)
37. Younis, A., Tran, T.X., Pompili, D.: Bandwidth and energy-aware resource allocation for cloud radio access networks. IEEE Trans. Wireless Commun. **17**(10), 6487–6500 (2018)
38. Nikaein, N.: Processing radio access network functions in the cloud: Critical issues and modeling. In: Proceedings of the 6th International Workshop on Mobile Cloud Computing and Services, MCS 2015, pp. 36–43, New York, NY, USA, Association for Computing Machinery (2015)
39. Foukas, X., Nikaein, N., Kassem, M.M., Marina, M.K., Kontovasilis, K.: Flexran: a flexible and programmable platform for software-defined radio access networks. In: Proceedings of the 12th International on Conference on Emerging Networking EXperiments and Technologies, CoNEXT 2016, pp. 427–441, New York, NY, USA Association for Computing Machinery (2016)
40. Kim, H., Rajkumar, R.: Predictable shared cache management for multi-core real-time virtualization. ACM Trans. Embed. Comput. Syst. **17** (1) (2017)
41. Reghenzani, F., Massari, G., Fornaciari, W.: The real-time linux kernel: a survey on preempt_rt. ACM Comput. Surv. **52**(1), 36 (2019)
42. Xi, S., et al.: Real-time multi-core virtual machine scheduling in xen. In: 2014 International Conference on Embedded Software (EMSOFT), pp. 1–10 (2014)
43. Pahl, C.: Containerization and the PAAS cloud. IEEE Cloud Comput. **2**(3), 24–31 (2015)
44. Struhár, V., Behnam, M., Ashjaei, M., Papadopoulos, A.V.: Real-time containers: a survey. In: Cervin, A., Yang, Y. (eds.) 2nd Workshop on Fog Computing and the IoT (Fog-IoT 2020), volume 80 of OpenAccess Series in Informatics (OASIcs), pp. 7:1–7:9, Dagstuhl, Germany. Schloss Dagstuhl-Leibniz-Zentrum fuer Informatik (2020)
45. Li, Z., Kihl, M., Lu, Q., Andersson, J.A.: Performance overhead comparison between hypervisor and container based virtualization. In: 2017 IEEE 31st International Conference on Advanced Information Networking and Applications (AINA), pp 955–962 (2017)
46. Nikaein, N., Schiller, E., Favraud, R., Knopp, R., Alyafawi, I., Braun, T.: Towards a cloud-native radio access network. In: Mavromoustakis, C.X., Mastorakis, G., Dobre, C. (eds.) Advances in Mobile Cloud Computing and Big Data in the 5G Era. SBD, vol. 22, pp. 171–202. Springer, Cham (2017). https://doi.org/10.1007/978-3-319-45145-9_8
47. Mao, C.N., et al.: Minimizing latency of real-time container cloud for software radio access networks. In: 2015 IEEE 7th International Conference on Cloud Computing Technology and Science (CloudCom), pp. 611–616 (2015)
48. Cavicchioli, R., Capodieci, N., Bertogna, N.: Memory interference characterization between CPU cores and integrated GPUs in mixed-criticality platforms. In: 2017 22nd IEEE International Conference on Emerging Technologies and Factory Automation (ETFA), pp. 1–10 (2017)
49. De, P., Mann, V., Mittaly, U.: Handling OS jitter on multicore multithreaded systems. In: 2009 IEEE International Symposium on Parallel & Distributed Processing, pp. 1–12 (2009)

50. Barletta, M.C., De Simone, L., Corte, R.D.: Achieving isolation in mixed-criticality industrial edge systems with real-time containers. In: 34th Euromicro Conference on Real-Time Systems (ECRTS 2022). Schloss Dagstuhl-Leibniz-Zentrum für Informatik (2022)

51. Burns, A., Davis, R.I.: Mixed Criticality Systems-A Review (February 2022). (2022)

52. Reghenzani, F., Massari, G., Fornaciari. W.: Mixed time-criticality process interferences characterization on a multicore linux system. In: 2017 Euromicro Conference on Digital System Design (DSD), pp. 427–434 (2017)

53. Shekhar, S., Gokhale, A.: Dynamic resource management across cloud-edge resources for performance-sensitive applications. In: 2017 17th IEEE/ACM International Symposium on Cluster, Cloud and Grid Computing (CCGRID), pp. 707–710 (2017)

54. Singh, A.K., Shafique, M., Kumar, A., Henkel, J.: Mapping on multi/many-core systems: survey of current and emerging trends. In: 2013 50th ACM/EDAC/IEEE Design Automation Conference (DAC), pp. 1–10 (2013)

55. Fried, J., Ruan, Z., Ousterhout, A., Belay, A.: Caladan: Mitigating interference at microsecond timescales. In: Proceedings of the 14th USENIX Conference on Operating Systems Design and Implementation, OSDI 2020, USA. USENIX Association (2020)

56. Fornaciari, W., Pozzi, G., Reghenzani, F., Marchese, A., Belluschi, M.: Runtime resource management for embedded and HPC systems. In: PARMA-DITAM 2016, pp. 31–36, New York, NY, USA. Association for Computing Machinery (2016)

57. Niknafs, M., Ukhov, I., Eles, P., Peng, Z.: Runtime resource management with workload prediction. In: Proceedings of the 56th Annual Design Automation Conference 2019, DAC 2019, New York, NY, USA. Association for Computing Machinery (2019)

58. Khasanov, R., Castrillon, J.: Energy-efficient runtime resource management for adaptable multi-application mapping. In: 2020 Design, Automation & Test in Europe Conference & Exhibition (DATE), pp. 909–914 (2020)

59. Khasanov, R., Robledo, J., Menard, C., Goens, A., Castrillon, J.: Domain-specific hybrid mapping for energy-efficient baseband processing in wireless networks. ACM Trans. Embed. Comput. Syst. **20**(5s), (2021)

60. Manvi , S.S., Shyam, G.K.: Resource management for infrastructure as a service (IAAS) in cloud computing: a survey. J. Netw. Comput. Appl. **41**, 424–440 (2014)

61. Alves, M.P., Delicato, F.C., Santos, I.L., Pires, P.F.: Lw-coedge: a lightweight virtualization model and collaboration process for edge computing. World Wide Web **23**(2), 1127–1175 (2020)

62. Azarmipour, M., Elfaham, H., Grothoff, J., von Trotha, C., Gries, G., Epple, U.: Dynamic resource management for virtualization in industrial automation. In: IECON 2018–44th Annual Conference of the IEEE Industrial Electronics Society, pp. 2878–2883 (2018)

63. Begam, R., Wang, W., Zhu, D.: Timer-cloud: Time-sensitive VM provisioning in resource-constrained clouds. IEEE Trans. Cloud Comput. **8**(1), 297–311 (2020)

64. Doan, T.V., et al.: Containers vs virtual machines: choosing the right virtualization technology for mobile edge cloud. In: 2019 IEEE 2nd 5G World Forum (5GWF), pp. 46–52 (2019)

65. Foukas, X., Radunovic, B.: Concordia: teaching the 5g VRAN to share compute. In: Proceedings of the 2021 ACM SIGCOMM 2021 Conference, SIGCOMM '21, pp. 580–596, New York, NY, USA. Association for Computing Machinery (2021)

TEATOM: A True Zero Touch Intent Based Multi-cloud Framework

B. Ramesh Ramanathan[1] and P. Preethika[2(✉)]

[1] Tata Elxsi, Open Source MANO (OSM) Technical Steering Member (TSC),
ETSI SOL006 Rapporteur, Bangalore, India
ramerama@tataelxsi.co.in
[2] Tata Elxsi, Bangalore, India
preethika.p@tataelxsi.co.in

Abstract. The modern day network is a software centric, virtualized network that can be built on COTS hardware or on any public clouds. Despite the openness such ecosystems bring in, there are practical challenges in achieving this network and get the benefit out of it. A large reason for this is the large number of components, complex configurations and an entire new set of skill required. An elegant solution to this problem is end to end automation of rollout and operations of such networks. This paper delves into TEATOM as a framework that can enable such an automation journey that can lead to zero touch operations and maintenance.

1 Introduction and Motivation

Traditional Enterprise IT and Telecom Operators have long suffered from proprietary vendor lock in solutions that are difficult to scale, change and evolve. This escalated the total cost of ownership (TCO) as well as made it difficult for them to launch new services and business models. This problem statement gave birth to the era of virtualization and network programmability. Such software centric networks helped IT and network systems to be built on COTS hardware with best of breed vendors for the various solutions and services. The key technologies that drove this revolution were NFV for virtualized applications/network elements, SDN for network programmability and virtualization for new age infrastructure management. These software centric systems were initially deployed in private clouds on premise of the enterprise or the telecom operator. The next wave of softwarization bought in public clouds, hybrid clouds and multi clouds. These various options of deployment further gave flexibility and choice for end to end rollouts. These notions of evolutions created a true converge of IT and Network.

However on the flip side of things, the number of moving parts and options has dramatically increased. Now enterprises and telecom operators are staring multiple modes of deployment, multiple vendors to manage, continuous software upgrades from vendors, heterogeneous environments to manage and a matrix of provisioning, configuration and assurance that needs to be carried out. The problem is further compounded by lack of skills to manage such complex systems, security concerns due to the intense software centric nature and overall difficultly in managing these new age networks. Hence automation of these networks is the key to truly leverage the power of the new age

L. Barolli (Ed.): AINA 2023, LNNS 655, pp. 609–618, 2023.
https://doi.org/10.1007/978-3-031-28694-0_57

technologies and systems. However, the new age software centric networks can be split into infrastructure management, service/app management and overall operational work-flow management. This drives a need for automation to span across all these 3 layers to truly achieve zero touch automation. There are standardization efforts ongoing by organizations like ETSI [1, 2], however there is a distinct lack of reference architecture and solution to illustrate this concept.

In this paper we introduce TEATOM – A true intent based multi cloud framework which only requires the intent and automates the same end to end starting from vendor and license management to deployment and operations. The same is illustrated in Fig. 1 below.

With TEATOM Enterprises and telecom operators can now truly leverage the bene-fit of software centric systems in terms of new business models (launching new ser-vices/apps) and deployment models (private, public, hybrid and multi cloud). With TEATOM it is possible to deploy apps/services in one cloud, stretch it across multi clouds, fold and migrate across the clouds as well.

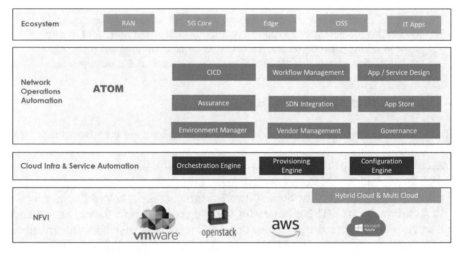

Fig. 1. Layered stack view of TEATOM

2   TEATOM Architecture and Building Blocks

TEATOM drives true end to end automation and helps to ease management across three layers namely service (application deployment and management across multi cloud), Operations (certification and rollouts) and NFVI layer (infrastructure creation as per requirements on demand).

2.1 Cloud Infra and Service Automation

Services and application need the base infrastructure to be set up, e.g. setting up a Kubernetes cluster to deploy a cloud native app. It is important to have this infrastructure to be setup in a cloud agnostic way for multi cloud portability. This notion of portability enables customers with choice of cloud and also from an app perspective high availability and deployment in the right proximity to meet the SLAs required.

TEATOM framework provides such on demand creation of infrastructure as per business needs by having a strong notion of GitOps integration. TEATOM uses GIT, Terraform and Ansible for provisioning and configuration of the environment.

Once the environment is ready, for deploying the app multiple options are provided. For Telco apps an Orchestrator can be used to model the same as per NFV standards and for IT apps a mix of Helm, Terraform and Ansible.

2.2 Network Operations Automation

TEATOM Operations Automation helps to ease management, starting from vendor onboarding with their license and performing continuous certification of software released by the vendor against various cloud platforms. The certified applications are stored in app store which is used further in deployments. This helps to ease roll out and backup of application.

2.3 Ecosystem

With the power of TEATOM, Enterprises can manage their IT workloads across clouds. Telecom operators can manage apps across IT and network and across domains like Core, Edge and ORAN.

2.4 TEATOM – Process View

This section provides a process viewpoint on how the layers described in previous section interact with one another. The workflows in TEATOM can be split into two main categories – Continuous Delivery and Continuous Deployment (Fig. 2).

The **Continuous delivery** phase begins with onboarding the vendor and their software, licenses and environment requirements. License flow in TEATOM is a key component since composite services/apps built from multiple vendors will have several licenses that need to be managed and checked together. Once the vendor is successfully onboarded, their software and associated versions are either uploaded or taken from the vendor's CICD or other delivery mechanisms in a periodic or event driven manner. The software and the environments required is deployed in the test/preproduction setup and tested for the expected certification criteria (functional, non-functional, security). Every version of the software goes through this process; this notion of continuous certification is key part of TEATOM architecture to maintain the required quality. Issues found in testing are reported using an incident management tool in an automatic manner. Certified versions are given to Governance, upon approval are pushed to an app store. The app store is a combination of GIT to store the Terraform state information, Harbor/glance/cloud

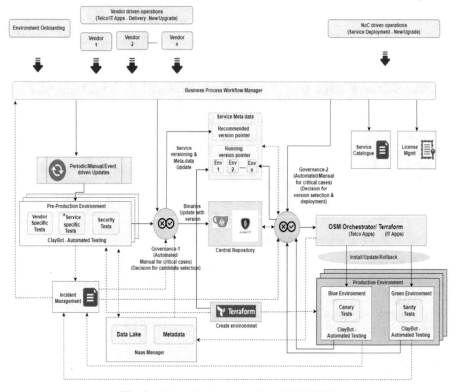

Fig. 2. End to End process view of TEATOM

tool for storing the image and MongoDB for storing the Meta data like Helm chart, version number and service descriptors.

The next stage of TEATOM workflow is **Continuous Deployment**. Certified apps from app store are available for Governance decisions on when and where to deploy. Deployment can be via an Orchestrator like OSM for Telco apps or Terraform for IT apps. Data from certification and production is feed to a data lake. This data is used for both historical correlation and analysis, as well as to use the data to build digital twins for ease of overall lifecycle management.

3   TEATOM and the Enterprise and Telco Cloud Use Case

TEATOM provides end to end workflow automation in a highly customizable and use case friendly form factor. This section will delve into a sample workflow to illustrate TEATOM in real world usage. The following figure gives a high level view of such a workflow (Fig. 3).

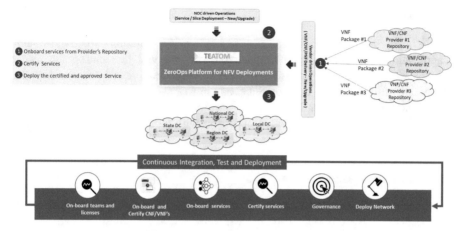

Fig. 3. Workflow view of TEATOM

3.1 Continuous Certification and Delivery Workflow

Following figure illustrates the users, role players and interactions to certify apps and build an app store (Fig. 4).

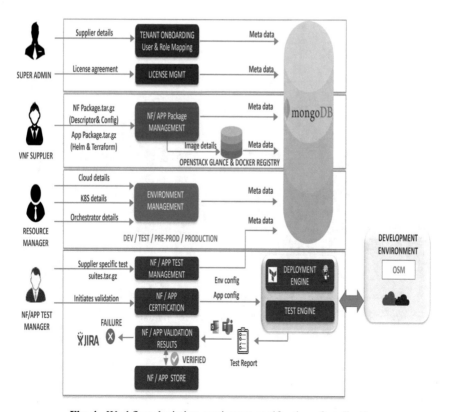

Fig. 4. Workflow depicting continuous certification of application

3.2   Vendor Onboarding

Vendor onboarding is the initial stage in the implemented solution, where the vendors of application or network function is added to the system with their appropriate license, this ensures the validity of license is met for all application tested and certified. The onboarding of the vendor is done by Super admin.

The next step is onboard all necessary entities such as descriptors with config templates in case of telco application and helm with terraform in case of Enterprise application. This is done by the vendor.

3.3   Environment Management

Environment Manager helps Resource Managers/Cloud Ops Team to onboard various clouds and their respective Orchestrators available for usage. Orchestrators are added to help to deploy application in case of telecom based application.

The cloud (public/private) onboarded to system should also have the respective Terraform scripts provided with tags for specific deployment stages. This makes sure when Operations team initiates any stage like validation and certification or deployment the right infrastructure is created. Existing infrastructure like Kubernetes clusters can also be discovered and onboarded if present and tagged to the right stage.

A test manager takes up the testing of the apps and certified apps land up in the app store (Fig. 5).

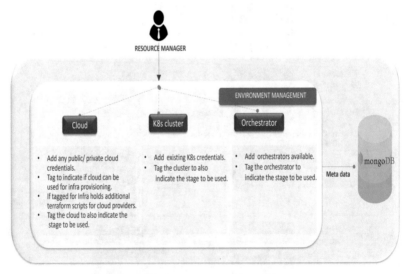

Fig. 5. Environment Management implemented

3.4   Continuous Deployment Workflow

Apps present in app store can be used standalone or super apps/services/slices can be built using them. In the latter scenario, the newly created app/service is certified and taken to production. This is illustrated in Fig. 6 below.

Service Test Manager picks up the service, the test suites, cloud infra configurations via Terraform templates and proceeds with service certification. Jenkins is used as pipeline engine which helps to deploy the infrastructure in respective cloud with help of Terraform templates and further uses the infrastructure created to deploy application.

Test engine performs the required test against the application, upon successful validation; the application is ready for deployment governance by a HOD or Ops Manager.

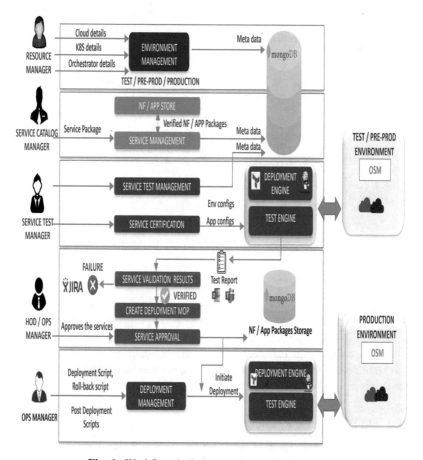

Fig. 6.   Workflow depicting continuous deployments

Approved applications can be deployed by the Operations Manager. Following diagram depicts a typical deployment workflow pattern (Fig. 7).

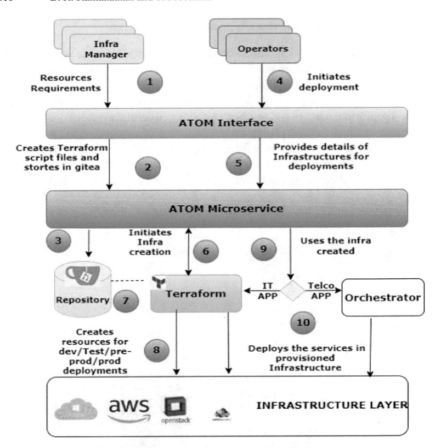

Fig. 7. On demand environment and app deployment

3.5 Multi Cloud Deployments

Apps and services can be deployed to an environment as described in above section. TEATOM supports a notion of app/service update which includes update of the app as well as the environment requirements. The update workflow is very similar to a fresh install since the only change is instead of creating meta data in TEATOM, existing meta data will be updated. With this simple approach use cases like migration an app from a cloud to another and deploying apps across clouds become easy to achieve.

3.6 Test Bed Details and Results

TEATOM was implemented using pure stateless microservices architecture with choreography by Apache Kafka. With TEATOM a variety of apps (CNFs/VNFs) in a variety of scenarios (private, hybrid and multi cloud) was deployed and tested.

The mean time to deliver new apps/services shrank from weeks to minutes. This was due to the end to end automation capability of TEATOM across infrastructure, services and operations. On a similar note, updating applications was a single click operation that can otherwise mean days of planning. With TEATOM the requirement to man a network for operations dropped by 90%. TEATOM ability to collect data (Prometheus and Grafana) in Data Lake helped to send out alarms and metrics to an external system like NMS for end to end AI/ML based assurance policy decisions. This helped establish closed loop assurance and help reduce MTTR by 65%. The zero touch nature of TEATOM helped reduce rollout time of a data center by 70%. By harnessing the power of open platform, end to end automation and the workflow cutting across silos in the organization, an overall reduction in TCO by 60% was observed.

4 Conclusion and Future Work

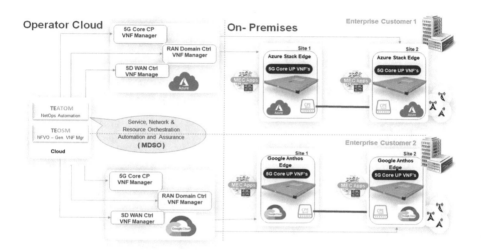

Fig. 8. TEATOM and multi-clouds

TEATOM end to end workflow automation capability across infrastructure, services/apps and Operations help to deliver a true zero touch experience. A good addition to integrate to complete the zero touch journeys is SDN integration. SDN integration is viewed as very important in some scenarios, e.g. a private cloud implementation. As part of the environment required, networking requirements like VLAN and subnets can be added and delivered by TEATOM talking to a SDN controller.

TEATOM has been deployed to orchestrate variety of cloud scenarios, including the multi cloud scenario illustrated in Fig. 8 [3, 4]. Next step is to expand such deployments to edge scenarios for app placement according to requirements like latency and bandwidth.

Final work is to collect data from the network and use it to build a digital twin of the same. This will help to evolve TEATOM to a sandbox – not just certify services/apps but the entire end to end journey before any deployment.

References

1. ETSI IFA 011 and IFA 014 v4.3.1 (2022)
2. ETSI SOL 006 v4.3.1 (2022)
3. AWS: Architecture and design guide (2021)
4. Azure: Architecture and design guide (2021)

Utility Function Creator for Cloud Application Optimization

Marta Różańska[1(✉)], Kyriakos Kritikos[2], Jan Marchel[3], Damian Folga[3], and Geir Horn[1]

[1] Department of Informatics, University of Oslo,
P.O. Box 1080, Blindern, 0316 Oslo, Norway
`martaroz@ifi.uio.no`, `Geir.Horn@mn.uio.no`
[2] Aegean University, Karlovassi, Samos, Greece
`kritikos@ics.forth.gr`
[3] 7Bulls.com, Al. Szucha 8, 00-582 Warsaw, Poland
`{jmarchel,dfolga}@7bulls.com`

Abstract. Cloud computing promises unlimited elasticity and allows Cloud applications to scale or change the configuration in response to demand. To automate the application management one must capture the goals and preferences of the application owner, and the most flexible way to represent these is as a utility function. However, it is often difficult for the application owner to define a such mathematical function. Therefore, we propose the Utility Function Creator, a software component that can create the utility function for Cloud application optimization based on a set of predefined utility function policies as well as by template utility function shapes.

1 Introduction

Modern applications are often distributed and deployed in the Cloud so they can be scaled dynamically according to the changing demand. Hummaida *et al.* [1] and Kephart and Das [8] all claimed that the utility based optimization allows the most flexible representation of the application owner's adaptation preferences. Given a representative utility function, it is possible to use concepts from autonomic computing [7] to manage Cloud applications as implemented in a framework like the Multi-cloud Execution ware for Large scale Optimised Data Intensive Computing(MELODIC)[1] [4]. More specifically, a set of some high-level goals, such as keep the cost as low as possible or optimize the response time to the application users, can be specified. It should be modelled as a part of the application model.

However, it is often difficult for the DevOps engineer to provide a utility function formula by hand [6]. What is more, these days the need for DevOps engineers is bigger than the number of qualified people such that low-code and high-level approaches must be developed. Sousa *et al.* showed that it is possible for DevOps

---

[1] https://melodic.cloud/.

L. Barolli (Ed.): AINA 2023, LNNS 655, pp. 619–630, 2023.
https://doi.org/10.1007/978-3-031-28694-0_58

engineers to define their goals by fixing parameters of a given template utility function [9]. That is the reason for the investigation of more user-friendly utility function modelling, in terms of less effort to generate the utility function and its implementation as the Utility Function Creator service. As the modelling language Cloud Application Modelling and Execution Language (CAMEL) was chosen to be used, because it is one of the most advanced Cloud application modelling languages that allows for complex utility modeling [10].

The first task was to identify and model the most common and needed high-level policies for the users such as "optimize response time" or "minimize cost" with the help of the use case partners of the project Modelling and Orchestrating heterogeneous Resources and Polymorphic applications for Holistic Execution and adaptation of Models In the Cloud (MORPHEMIC)[2]. These policies are often formulated with limited knowledge of how it translates to actual application configurations and how it relates to monitoring measurements taken from the deployed application. The results showed that even different software applications may have similar high-level optimization policies that can be eventually modelled as similar utility functions.

The paper is structured as follows: the description of the utility based Cloud application optimization is provided in Sect. 2 followed by the description of the utility function modelling in CAMEL in Sect. 3. Then, the predefined utility functions and utility function templates are presented in Sect. 4. The Utility Function Creator architecture is discussed in Sect. 5, as well as utility function creation process. Finally, the approach is assessed in Sect. 6.

2   Cloud Application Optimization

Cloud applications can be considered as a set of communicating *components*, $\mathbb{C}$, where the term component should be understood a part of the application, which can be a software component, a container, a serverless function, platform provided software like databases or data lakes, or third party external web servers. A persisted application must be able to respond to variation in workload and other changes in its environment happening during its extended execution time. This requires that there is some variability allowed for the configuration of the components. In general, a component, $C_i \in \mathbb{C}$, has a set of quality attributes, $\mathbb{A}_i$, that allows the behaviour of the component to be adapted. Examples of an attribute, $a_{i,j} \in \mathbb{A}_i$, may be the number of cores given to the component, the encoding format of a video stream that can be changed to save power or bandwidth etc. Each attribute takes its *value* from its domain, $a_{i,j} \in \mathbb{V}_{i,j}$, which can be a continuous range or a discrete set of options. The domain is fixed since it represents the various configuration options for that particular attribute. The Cartesian product of the attribute domains represents the variability space of the component, i.e. the number of different variants that can be deployed, $\mathbb{V}_i = \mathbb{V}_{i,1} \times \mathbb{V}_{i,2} \times \cdots \times \mathbb{V}_{i,|\mathbb{A}_i|}$. Similarly, the variability space for the whole application will be the Cartesian product of the variability spaces for all its components, $\mathbb{V} = \mathbb{V}_1 \times \mathbb{V}_2 \times \cdots \times \mathbb{V}_{|\mathbb{C}|}$.

---

[2] https://morphemic.cloud.

The values assigned to all application attributes at a particular time is called the application's *configuration* and can be ordered as a vector $c(t) \in \mathbb{V}$.

The application must be monitored to detect when there is a need to reconfigure the application. The measurement events occur at discrete time points, t_k, and one cannot assume that the time points are equidistant, *e.g.*, consider for instance the monitoring event representing the arrival or departure of a user of the application. The most recent measurements, the metrics, available at the time point t_k can either depend on the application's configuration or be independent of the configuration. The number of active users is an example of the latter type, and the response time recorded for a user's request is an example of the former. The vector of the dependent metrics will be denoted $\boldsymbol{\theta}_D\left(t_k \mid c(t_k)\right)$ where the vertical line indicates that the values are conditioned on the given configuration, and $\boldsymbol{\theta}_I(t_k)$ will be used to indicate the vector of the independent metric values. The overall metric vector is just the concatenation of the two types of metrics, $\boldsymbol{\theta}(t_k \mid c(t_k)) = [\boldsymbol{\theta}_I(t_k)^T, \boldsymbol{\theta}_D\left(t_k \mid c(t_k)\right)^T]^T$. The metric values may not be directly useful for the application management if they change too quickly as reconfigurations take time, and so they should better be filtered by various functional combinations as performance indices denoted $\boldsymbol{\psi}(t_k \mid c(t_k))$. For instance, the response time measure for an individual user request may mostly indicate the complexity of that request whereas the windowed average of the most recent response times for all users says something about the application performance and the user experience. These functional combinations of the metric values together with the metric vector itself jointly form the *application's execution context*.

The application is deployed for a reason, and the application's owner has goals and preferences for the management of the application. These must be captured for autonomic and optimised operations, and the rational choice would be to optimise the owner's *utility* [5]. Kephart and Das have convincingly argued that the best way to capture utility is through a utility *function* mapping from the application's configuration to a goodness value in the unit interval, $U : \mathbb{V} \mapsto [0, 1]$, see [8]. Hence, finding the optimal application configuration, $c^*(t_k)$, implies solving the following mathematical programme for the application's execution context.

$$c^*(t_k) = \underset{c \in \mathbb{V}}{\operatorname{argmax}} \, U\left(c \mid \boldsymbol{\theta}(t_k \mid c), \boldsymbol{\psi}(t_k \mid c)\right) \qquad (1)$$

3   Utility Function Modelling in CAMEL

CAMEL is a Domain Specific Language (DSL) covering multiple domains, developed for the modelling of multi-cloud applications. This language captures, in a rich and integrated manner, all relevant domains to the (multi-) Cloud application life-cycle, including the deployment, requirement, and metric domains. The deployment domain covers the abstract and concrete deployment architecture of an application, allowing the specification of the former by the application owner and the automatic production of the latter by the application management platform. This follows the type-instance pattern which fits well the models@run.time paradigm [3].

The requirement domain covers all sorts of application requirements, including resource, provider, location, horizontal scalability, and optimisation requirements. The Quality of Service (QoS) requirements are specified as constraints on metrics at the application, component, or resource level where the violation of such constraints leads to the automatic application reconfiguration by the application management platform. The metric domain covers all the necessary details for expressing how metrics and functions are computed, how often, and according to which measurements. In the sequel, we analyse in more detail how CAMEL expresses metrics and utility functions.

CAMEL is able to specify all necessary details for the computation of any kind of metric, see Fig. 1. Such details include the quantity measured based and its measurement unit, *e.g.*, seconds, and value type, *e.g.*, $[0.0, \infty)$. The CAMEL metrics are classified raw or composite. A raw metric, *e.g.*, raw response time, which we will simply call *metric* in this paper, is measured directly through the use of a sensor. On the other hand, a composite metric is called a *performance indicator* and calculated through the use of mathematical formulas, *e.g.*, the average, over other metrics, *e.g.*, raw response time. Such mathematical formulas are expressed using the syntax of the MathParser[3].

A metric context supplies three important pieces of information: (a) the measurement schedule, *i.e.*, how often to compute the metric, (b) the measurement window, *i.e.*, how many measurements to consider of this metric in order to calculate the performance indicator value, and (c) which object is being actually measured, *i.e.*, an application component

In this respect, we can associate the same raw metric with different contexts. For instance, we can express, on one hand, that we compute the average response time for the whole application every minute with a window of the last 10 raw application response time measurements. While, on the other hand, we can express that we compute the average response time for one application component every 30 s with a window of the last 5 component response time measurements.

CAMEL enables both high-level optimisation requirements and low-level ones. The high-level optimisation requirements are expressed in the form of optimisation goals to maximise or minimise a specific performance indicator identified as the utility function. This mathematical formula is encapsulated by the concept of a *Variable*. A variable has two meanings in CAMEL. On one hand, it can express a variable in a mathematical optimisation problem. For instance, a variable can express the number of cores for an application component, which will take a specific value eventually deciding on the actual Virtual Machine (VM) offering that will be selected to host that component. In this sense, it is a variable as its value varies to maximize the utility function. On the other hand, it can be a mathematical formula which can include both metrics and other variables

---

[3] www.mathparser.org.

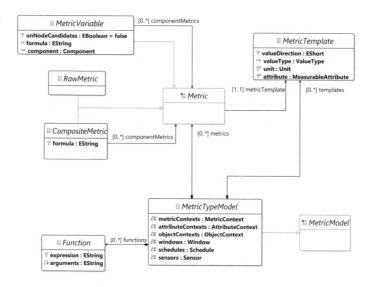

Fig. 1. The class diagram for the metric domain in CAMEL

as its operands along with functions playing the role of the operators. Thus, in this particular case, a variable can be either a utility function, a partial utility function, or a part from which a utility function is constructed. This enables creating template utility functions that can be re-used for new applications, or as components of more composite utility functions for such applications. Suppose that one wants to express the following utility function:

$$U\left(\boldsymbol{c}\,\middle|\,[\alpha,\beta,B,P]\right) = exp\left(\frac{\alpha}{(B-P)^{\beta}} - \frac{\alpha}{\left(B - c_1 \cdot \mathrm{Price}(c_2)\right)^{\beta}}\right)$$

where $\phi = [\alpha, \beta, B, P]^T$ are constants with B being the user budget, and P is the price of the least expensive node candidate for a single-component application. The other symbols map to metric variables of the application configuration: c_1 is the number of instances of that component for the current solution examined and $\mathrm{Price}(c_2)$ is the minimum price among the VMs with c_2 cores.

4   Predefined Utility Functions

In this section, we present the directly modelled utility functions for the utility dimensions corresponding to the identified high-level policies: expected response time, finish simulation on time, locality, Central Processing Unit (CPU) usage, and cost. These directly modelled utility functions from high-level policies can be re-used as utility function for a Cloud application. What is more, we present a set of templates also available in the Utility Function Creator. The idea is that the overall utility function can be constructed as an affine combination of these dimensional utility functions where each dimension is weighted according to the importance of that dimension such that the sum of the weights is unity.

It is important to notice that all functions are modelled for one or two component applications and in case of more components, more decision variables representing parameters of all components should be used. One should also note that each utility function for each dimension is modelled independently, and the notation and symbols used will be local to that utility dimension. Thus, when re-using these dimensional utility functions, proper variable substitutions must be applied when representing the overall utility function of the managed application.

Solving the optimisation problem of Eq. (1) at the current time point t_k for the given the measurements, $\boldsymbol{\theta}(t_k)$, and the derived performance indices, $\boldsymbol{\psi}(t_k)$, yields the optimal configuration for the application's current execution context. However, there will be a delay in deploying the found solution, as it will only be active at a time $t > t_k$. Ideally, it will be active before the next measurement event, or one must assume that the next measurement events do not change the application execution too much to invalidate the solution under deployment. The remedy is to predict the execution context [13], but this is beyond the scope of the present paper focusing on the modelling of the utility function.

4.1   Expected Response Time

It is appropriate to define the expected response time $\bar{T}$ and the maximum acceptable response time T^+, and the default timeout time, T_D, after which the server returns the timeout and the request is not served at all. What is more, there is a need to measure the current average response time, $\theta_1(t_k)$. The decision variables can be changed to be more complete, but for the simplification of the example, it is assumed that there are two main decision variables: c_1, which is the number of instances, and c_2, which is the number of cores available at each instance. We propose to represent this utility function as a probability density function with bounded support, such as the Beta distribution, $\beta(\tau \mid a, b)$ for $\tau \in [0, 1]$, and take its two constant parameters a and b with respect to T_D and $\bar{T}$. The current configuration is defined as $c_1^*(t_k)$, $c_2^*(t_k)$. The performance utility can therefore be expressed as follows:

$$\nu_{\text{time}}\left(\boldsymbol{c}(t) \mid \boldsymbol{\theta}(t_k)\right) = \beta \left(\frac{\theta_1(t_k) \cdot c_1^*(t_k) \cdot c_2^*(t_k)}{T^+ \cdot c_1(t) \cdot c_2(t)} \,\middle|\, \phi_1, \phi_2 \right) \Big/ \beta^+(\phi_1, \phi_2)$$

Note that the above utility function is normalized by $\beta^+(a, b) = \text{argmax}_\tau \, \beta(\tau \mid a, b)$ to ensure that the utility is a value within the unit interval. This function was used for optimization of a typical e-commerce application [4,12].

4.2   Finish Simulation on Time

This utility function can be used, for example, for data farming applications running simulations or other data parallel applications. By the start of the computation, the number of jobs, N, will be given. However, the time it takes to execute one job will depend on the data being processed and the complexity the calculations. Thus, at the event time point t_k then k jobs have completed, and the number of jobs remaining, $\theta_1(t_k) = N - k$, will be a change in the application's execution context. It is also possible to measure the upper $1 - \alpha/2$, $0 < \alpha < 1$ quantile of the empirical job duration distribution, $\theta_2(t_k)$. Thus, the predicted time needed for the parallel computations of the remaining jobs is $(\theta_1(t_k) \cdot \theta_2(t_k))/(c_1 \cdot c_2)$, where the nominator is an upper bound for the time needed to complete the remaining jobs under the assumption that each calculation takes as long as the upper quantile currently observed. The denominator represents the number of available cores for the job as the number of worker machines c_1 times the number of cores per worker c_2. To know the predicted overall completion time for all calculations one must add to $(\theta_1(t_k) \cdot \theta_2(t_k))/(c_1 \cdot c_2)$ the total time one has spent for the jobs until now, $\theta_3(t_k)$. The utility of the deployment should be close to unity for a given set of workers, c_1, if this estimated overall completion time is less than the deadline. The farther beyond the deadline the predicted completion is, the lower the utility should be. Thus, a natural utility function form would be a sigmoid function $1/(1 + e^{-ax})$ with a negative argument x scaled so that the utility will be $1/2$ when the deadline is exactly met. The performance utility for the timeliness of the application is then

$$\nu_{\text{deadline}}\Big(\mathbf{c}(t) \,|\, \boldsymbol{\theta}(t_k), \boldsymbol{\phi}\Big) =$$

$$\left[1 + \exp\left(-\phi_2\left(\phi_1 - \frac{\theta_1(t_k) \cdot \theta_2(t_k)}{c_1(t) \cdot c_2(t)} + \theta_3(t_k)\right)\right)\right]^{-1}$$

where $\boldsymbol{\phi}$ is a vector of fixed parameters; here, the target soft deadline $\phi_1 = T^+$ and scaling parameter $\phi_2 > 0$. It is worth mentioning that this type of the function was used to optimize the Genome Big Data application [13].

4.3 Locality Utility

One may want to include the distance between two deployed components into the utility function which may also give an estimation about the possible access latency seen from the user perspective [11]. The high-level goal is to keep the components as close as possible. The decision variables are c_1 being the latitude and c_2 being the longitude of the first component, and c_3 representing latitude and c_4 the longitude of the second component, given in degrees The geodesic distance between two deployed components can be calculated using spherical law of cosines and normalized to be within zero and one. R is the earth's radius The locality utility is then:

$$\nu_{\text{locality}}(\mathbf{c}(t) \,|\, \boldsymbol{\phi}) = \Big(1 + \arccos\Big(\sin(c_1(t)) \cdot \sin(c_3(t))$$

$$+ \cos(c_1(t)) \cdot \cos(c_2(t)) \cdot \cos(c_4(t) - c_2(t))\Big) \cdot R\Big)^{-1}$$

4.4   CPU Usage

The utility function that considers the percentage CPU usage can be a very similar function to the previously described response time utility. It may have a form of the Beta distribution function, because Beta distribution is defined on an interval from 0 to 1. The desired percentage should be specified by the user. The decision variables are c_1, which is the number of component instances, and c_2, which is the number of cores available in one component instance. The measured CPU usage on the one instance, $\theta_1(t_k)$, is better considered as the performance index, ψ_1, calculated as the average CPU usage for all instances. The utility function can be defined as:

$$\nu_{\text{CPU}}\left(c(t) \,|\, \theta(t_k), \psi(t_k), \phi\right) =$$
$$\beta\left(\left.\frac{\psi_1(t_k) \cdot c_1^*(t_k) \cdot c_2^*(t_k)}{c_1(t) \cdot c_2(t)} \,\right|\, \phi_1, \phi_2\right) \Big/ \beta^+(\phi_1, \phi_2)$$

4.5   Core Cost Utility

A negative exponential is a good candidate for the cost utility dimension as the utility increases when the cost decreases. It needs to be scaled so that the exponent is zero when the cost is at the minimum, and then the exponent should be more and more negative as the deployment cost increases. We use $\text{Price}(c)$ to denote the on-demand price of the machine with c cores. The same function was provided as the example in Sect. 3. Let ϕ_3 be the available budget and let ϕ_4 be the price of the least expensive VM possible. Furthermore, it is reasonable to assume that the price of a machine is decided by the number of cores, c_2, it offers as for many cases it is the most important factor. A reasonable cost utility function is therefore

$$\nu_{\text{core}}\left(c(t) \,|\, \theta(t_k), \phi\right) = \exp\left(\frac{\phi_5}{(\phi_3 - \phi_4)^{\phi_6}} - \frac{\phi_5}{\phi_3 - c_1(t) \cdot \text{Price}(c_2(t))^{\phi_6}}\right)$$

where $\phi_5 > 0$ is a scale parameter and $\phi_6 > 0$ is a shape parameter. Note that the denominator of each term of the exponent represents the remaining budget raised to the power of ϕ_6. In the first term the remaining budget is defined by deploying a single virtual machine with the least cost, whereas in the second term the remaining budget is given by the number of component instances multiplied by the price of each instance, $i.e.$, the total cost of the deployment configuration.

4.6   Cost Per User

This cost utility function takes into account the current workload and the number of potentially served users. The utility is best captured using the cost per served user as a parameter, because the same price for the VM is considered to be a lower cost if more users can be served using it. This function was used in typical e-commerce application optimization [12]. Let P^+ and P^- be the maximum and the minimum prices for available VMs, c_1^+ and c_1^- be the maximum and the minimum number of application servers, and $\text{Price}(\boldsymbol{c})$ be the price of deployment represented by $\boldsymbol{c}$. The cost utility can be expressed as:

$$\nu_{\text{user}}\left(\boldsymbol{c}(t) \mid \boldsymbol{\theta}(t_k), \boldsymbol{\psi}(t_k), \boldsymbol{\phi}\right)$$
$$= \frac{\theta_1(l_k) \cdot P^| \cdot c_1^+ - \text{Price}\left(\boldsymbol{c}(t)\right) / \psi_1(t_k)}{\theta_1(t_k) \cdot P^+ \cdot c_1^+ - P^- \cdot c_1^-}$$

The main parameter is the price of configuration divided by the ratio of correctly served requests $\psi_1(t_k)$. The maximum value is calculated as the most expensive deployment, which is the biggest possible number of the most costly VMs divided by the worst possible ratio of served requests, which is only one from all requests $\theta_1(t_k)$ received in one minute. The minimum price for the deployment is calculated as the minimum number of the least expensive machines divided by the best possible ratio of served requests that is one.

4.7   Utility Function Templates

Additionally to the directly modelled utility functions, the Utility Function Creator allows for creation of the function by a template over a performance indicator ψ. It supports a set of useful templates which provide the most popular shapes of the functions [12]. For all templates, there is a need to provide parameters ϕ_1, ϕ_2 for which the function value should be 0.5 or close to 1. The S-shaped sigmoid function (2) can accurately represent the goal to have the argument value as high as possible while a reversed S-shaped sigmoid function (3) can represent the desire to keep the performance indicator value as low as possible. The Bell-shaped (4) function can represent the goal to keep the argument value as close as possible to the desired value. Finally, the U-shaped function template (5) can represent a similar goal, but to avoid some particular value.

$$g_1(\psi \mid \phi_1, \phi_2) = 1 - \left[1 + e^{((\psi - \phi_2)\ln[(1-\epsilon)/\epsilon])/(\phi_1 - \phi_2)}\right]^{-1} \tag{2}$$

$$g_2(\psi \mid \phi_1, \phi_2) = \left[1 + e^{((\psi - \phi_2)\ln[\epsilon/(1-\epsilon)])/(\phi_1 - \phi_2)}\right]^{-1} \tag{3}$$

$$g_3(\psi \mid \phi_1, \phi_2) = e^{\ln \epsilon (\psi - \phi_1)^2/(\phi_2 - \phi_1)^2} \tag{4}$$

$$g_4(\psi \mid \phi_1, \phi_2) = 1 - e^{-\ln \epsilon (\psi - \phi_1)^2/(\phi_2 - \phi_1)^2} \tag{5}$$

5   Utility Function Creator

This section presents the Utility Function Creator[4], which is an open source service with a Graphical User Interface (GUI) created to help the application owners developing a utility function. It is integrated with MORPHEMIC Cloud management platform. The internal architecture for this module is presented in Fig. 2.

---

[4] https://gitlab.ow2.org/melodic/uf-creator.

It contains, among others, a GUI component responsible for the interactions with the user. It also stores the utility function into the MORPHEMIC's Connected Data Objects (CDO) database with the application's CAMEL application model. The Utility Function Creator consists of the following sub-components:

The CAMEL Model Service is responsible for retrieving the application's CAMEL model and storing the created utility function formula

The Metric CAMEL analyser is responsible for analysing the CAMEL model to retrieve the defined metrics

The Weights Calculator is responsible for calculating the weights of each utility dimension. It may use a simple weighted sum method, but it can also be extended to use various methods to retrieve and calculate weights

The Formula Creator is responsible for creating the utility function formula for the given utility metrics, templates, parameters, and weights

The Function Optimizer is responsible for optimizing the parameters of the utility function. It aims at improving the initial utility function created by the application owner by making it more accurate.

Before creating the utility function, the application owner must define a CAMEL Model for the application using the CAMEL Designer [14] or any textual editor. Then, the model is stored in the CDO database. The process of utility function creation starts with choosing the CAMEL model. The flow depends if the function is created from predefined functions or from the available templates. The sequence of utility function creation from the set of predefined functions starts with presenting the set of predefined functions (see Sect. 4). Then, the application owner chooses functions, and variables and metrics that should be used as arguments to the utility function. For the utility function creation with templates the flow is different and the application owner chooses

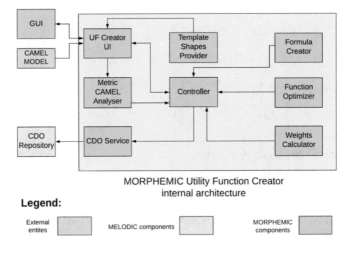

Fig. 2. Internal architecture of Utility Function Creator.

the performance indicators defined in CAMEL model that should be considered. Then, for all indicators chosen, the application owner chooses the most suitable template shape (see Sect. 4.7) and needed parameters. Finally, the user defines the weights that indicates the importance of each utility dimension. The overall utility function formula is presented in GUI and saved in the CAMEL Model.

6 Conclusion and Future Work

This paper has discussed the utility function modelling process supported by the new open source Utility Function Creator service. It allows for creation of utility function using predefined utility function, or by template utility function shape. We provided the definitions of supported predefined functions as well as available templates. What is more, we have described the design of the service. According to our knowledge, this is the first service for construction of utility function for Cloud application optimization. Furthermore, it is integrated with the MORPHEMIC platform and it is now being evaluated with MORPHEMIC use case providers.

Acknowledgements. This work has received funding from the European Union's Horizon 2020 research and innovation programme under grant agreement No 871643 MORPHEMIC *Modelling and Orchestrating heterogeneous Resources and Polymorphic applications for Holistic Execution and adaptation of Models In the Cloud.*

References

1. Hummaida, A.R., Paton, N.W., Sakellariou, R.: Adaptation in cloud resource configuration: a survey. J. Cloud Comput. **5**(1), 1–16 (2016). https://doi.org/10.1186/s13677-016-0057-9
2. Rossini, A., et al.: The cloud application modelling and execution language (camel), p. 39 (2017). https://doi.org/10.18725/OPARU-4339
3. Blair, G., Bencomo, N., France, R.B.: Models@ run.time. Computer **42**(10), 22–27 (2009). ISSN 1558-0814, https://doi.org/10.1109/MC.2009.326
4. Horn, G., Skrzypek, P.: MELODIC: utility based cross cloud deployment optimisation. In: Proceedings of the 32nd International Conference on Advanced Information Networking and Applications Workshops (WAINA), pp. 360–367. IEEE Computer Society (2018). https://doi.org/10.1109/WAINA.2018.00112
5. Gilboa, I.: Rational Choice. MIT Press, Cambridge (2012). ISBN 978-0-262-01400-7
6. Floch, J., et al.: Playing music – building context-aware and self-adaptive mobile applications. Softw. Pract. Experience **43**(3), 359–388 (2012). ISSN 1097-024X, https://doi.org/10.1002/spe.2116
7. Kephart, J.O., Chess, D.M.: The vision of autonomic computing. Computer **36**(1), 41–50 (2003). ISSN 0018-9162, https://doi.org/10.1109/MC.2003.1160055
8. Kephart, J.O., Das, R.: Achieving self-management via utility functions. Internet Comput. **36**(1), 41–50 (2003)
9. Sousa, J.P., Balan, R.K., Poladian, V., Garlan, D., Satyanarayanan, M.: User guidance of resource-adaptive systems. In: Proceedings of the Third International Conference on Software and Data Technologies (ICSOFT 2008), vol. 3, pp. 36–44. SciTePress - Science and and Technology Publications (2008). ISBN 978-989-8111-52-4, https://doi.org/10.5220/0001881500360044

10. Kritikos, K., Skrzypek, P.: Are cloud modelling languages ready for multi-cloud? In: Proceedings of the 12th IEEE/ACM International Conference on Utility and Cloud Computing Companion, UCC 2019 Companion, pp. 51–58. Association for Computing Machinery (2019). ISBN 978-1-4503-7044-8, https://doi.org/10.1145/3368235.3368840

11. Li, J., Zhang, T., Jin, J., Yang, Y., Yuan, D., Gao, L.: Latency estimation for fog-based internet of things. In: 2017 27th International Telecommunication Networks and Applications Conference (ITNAC), pp. 1–6 (2017). https://doi.org/10.1109/ATNAC.2017.8215403, ISSN: 2474-154X

12. Różańska, M., Horn, G.: Marginal metric utility for autonomic cloud application management. In: Proceedings of the 14th IEEE/ACM International Conference on Utility and Cloud Computing Companion (UCC 2021), pp. 21:1–21:8. Association for Computing Machinery (ACM) (2021). ISBN 978-1-4503-9163-4, https://doi.org/10.1145/3492323.3495587

13. Różańska, M., Horn, G.: Proactive autonomic cloud application management. In: Proceedings of the 15th IEEE/ACM International Conference on Utility and Cloud Computing (UCC 2022). IEEE/ACM (2022)

14. Moussaoui, A., Bagnato, A., Brosse, E., Krasnodebska, J.: The MORPHEMIC project and its unified user interface (2022)

A Review of Monitoring Probes for Cloud Computing Continuum

Yiannis Verginadis[1,2(✉)]

[1] Department of Business Administration, Athens University of Economics and Business, Patission 76, 10434 Athens, Greece
jverg@aueb.gr
[2] Institute of Communications and Computer Systems, Iroon Polytechniou 9, 15780 Zografou, Greece

Abstract. Nowadays, we consider optimizing data-intensive applications imperative for the digital enterprise to exploit the vast amounts of available data and maximize its business value. This fact necessitated the broad adoption of multi-cloud and fog deployment models towards enhanced use of distributed hosting resources that may reach the edge of the network. However, this poses significant research challenges concerning how one can automatically discover the best initial deployment of such an application and then continuously adapt it according to the defined Service Level Objectives (SLOs), even in extreme scenarios of workload fluctuations. Among the key tools for managing such multicloud applications are advanced distributed monitoring mechanisms. In this work, we consider some of their fundamental components, which refer to the means for efficiently measuring and propagating information on the application components and their hosting. Specifically, we analyze the 20 most well-known monitoring tools and compare them against several criteria. This comparison allows us to discuss their fit for the distributed complex event processing frameworks of the future that can efficiently monitor applications and trigger reconfigurations across the Cloud Computing Continuum.

1 Introduction

It is well-known that monitoring is an imperative process to ensure the efficient performance of a computing system. It can be distinguished in resource and application monitoring. The first refers to the cumbersome task of collecting data from physical or virtualized environments (e.g., CPU utilization, network jitter, packet losses, etc.). Such data can assist in maintaining the resources' availability at high efficiency and also detecting abnormal situations (e.g., SLO violations) that require immediate mitigation actions (i.e., reconfiguration). We refer to software that tracks and publishes such data as monitoring probes that involve at least two basic dimensions when it comes to resource monitoring: computation and network. Computation monitoring aspects refer to monitoring data that allow the inference on the context of bare metal or virtualized resources (e.g., CPU Speed, CPU utilization, RAM utilization, disk throughput, etc.). Network monitoring aspects are related to the tracking of network-layer metrics such as one-way and two-way delays, network jitter, throughput, packet loss, available bandwidth, capacity, and traffic volume [1].

On the other hand, application monitoring refers to equivalently important means of measuring the health status and performance aspects of the involved application components. This monitoring kind is imperative since it can provide valuable monitoring information to multicloud management platforms [2, 3] in order to improve the efficiency and timing of computing jobs. One of the main goals of application monitoring is to extract application patterns [1]. Such patterns can be used to provide feedback and aggregate important knowledge for the optimization of application components. The objective is to attest that the application service level objectives (SLOs) are met, and in order to do so, a complete awareness of the applications' running context is required. This is not an easy task, especially for multicloud applications and Big Data frameworks [4], as it requires a cross-layer monitoring scheme in order to determine the main factors that impact the quality and performance of different application types [5]. In general, cross-layer monitoring implies the use of different types of monitoring probes depending on the cloud technology stack: i) IaaS level – monitor how resources are utilized involving information such as CPU and Memory usage; ii) PaaS and SaaS level – provide information on the system context with respect to running services, uptime/downtime, availability, etc. For example, in the case of a Hadoop deployment, there are metrics such as MapReduce processing time, job turnaround, shuffle operations, etc. [4]. Therefore, the monitoring probes that should be used along with their configuration rely mainly on the application type used (e.g., the data transfer rate for video streaming applications, process and network latencies for batch processing applications).

This cross-layer monitoring is even more challenging when discussing big data frameworks deployed on multicloud-based and fog environments. This is evident since individual application components can be dispersed across different layers of the cloud stack on various VMs or other hosting resources [4]. For this purpose, the monitored probes should be automatically set up and connected for transmitting across all the cloud resources used by the applications in order to provide an overview of the current context of applications at a certain time.

In this paper, we focus on a technical comparative analysis of the most prominent monitoring probes that could be integrated in multicloud management platforms (MCMPs) [6] for tracking all the monitoring parameters (both for applications and hosting resources) that are required for guaranteeing the SLOs provided by the DevOps. The objective is to pinpoint the most appropriate monitoring probes that can enhance a federated event management system with configurable volumes of raw monitoring data. Such systems, such as Melodic Event Management System (EMS) [7, 8], are critical for aggregating and efficiently processing monitoring data from dispersed resources in a distributed manner that avoids bottlenecks. Nevertheless, they are usually based on a limited list of resource monitoring probes and a number of application-specific ones that are statically installed. An extension is required so that any available monitoring metrics can be measured and aggregated in a unified way with any application-specific metrics that may be defined by DevOps. We note that this comparative technical review of monitoring probes can be valuable for deciding on the monitoring sources to be used in any modern event management system that aims to aggregate and process monitoring data in the cloud computing continuum. Therefore, this work does not focus on analysing monitoring processing technologies or efforts that introduce advanced adaptive monitoring techniques (like federated event processing [7, 8] or adaptive sampling and filtering [9–12]). The rest of the sections are structured as follows: in Sect. 2, we provide a technical overview of prominent monitoring tools, and we present a comprehensive comparison based on a number of important criteria for the cloud computing continuum monitoring. Based on this analysis, in Sect. 3, we discuss the most promising approach for coping with the significant monitoring challenges in this domain, and in Sect. 4, we conclude with future work aspects.

2 Monitoring Probes for Cloud Computing Continuum

2.1 Overview of Prominent Monitoring Tools

According to several classifications [13], a monitoring tool is mainly responsible for the following actions: (i) collection, (ii) propagation, (iii) processing, (iv) storage, and (v) visualization of monitoring data. A data collector or else a monitoring probe refers to either a daemon operating on the monitored host or an agent scraping data from monitoring APIs exposed by the resources being monitored (e.g., Java Management Extensions (JMX)) [11]. We note that although our objective is not to focus on the processing capabilities of the monitoring technologies that we reviewed, we have included their processing capabilities aspect for completion reasons. Below we briefly introduce, in alphabetical order, each one of the most prominent monitoring tools that we examined.

Amazon CloudWatch[1] is a vendor-specific web service for AWS services (e.g., EC2) monitoring in real-time. Specifically, it provides data on how resources are utilized such as CPU, disk and network as well as monitoring information on the status of Amazon RDS, DynamoDB, Amazon EBS volumes, Amazon S3, AWS Lambda, etc. For providing memory and disk usage or average load metrics, one needs to run supplementary software on the instance side, i.e., CloudWatch Agents that can be deployed on Windows and Linux operating systems. Moreover, Amazon CloudWatch supports a number of custom metrics that can be programmatically loaded via a dedicated API to be monitored in a similar way to all the rest internal metrics (e.g., CPU usage). The internal metrics can be published with a rate of up to 1 min, while the custom metrics with up to 1 s granularity. It also supports the setting up of basic alarms on internal and custom metrics that enable the dynamic addition or removal of EC2 instances through the auto-scaling feature. As a commercial service, there is charging according to the observed number of monitoring instances.

Apache Chukwa[2] is a monitoring system focused on the Hadoop Distributed File System (HDFS) and the Map/Reduce framework. This system is open-source and comprises a flexible and scalable toolkit for displaying and analysing monitoring data. In addition, Apache Chukwa is able to analyse the collected data through a dedicated toolkit [4]. This means that it is designed to analyse and display the outcomes of different runs of

---

[1] https://aws.amazon.com/cloudwatch/.

[2] http://chukwa.apache.org.

the monitored software. It uses agents to acquire HDFS-related data which are persisted into files until a 64MB chunk size is reached or a given time interval has passed. Then the DemuxManager and the PostProcessManager filter and analyse the data collected.

Azure Cloud Monitoring[3] is a Microsoft service to monitor applications, analyse log files, and identify security threats. This vendor-specific service provides a detailed view of the utilization, performance, and health of applications, infrastructure, and workloads. It delivers services for collecting, analysing, and acting on telemetry data from cloud and on-premises environments while it offers advanced analysis of monitoring data for correlating infrastructure issues. Furthermore, this service supports the addition of any custom metric deemed necessary for the user and presents alerts on them. All data collected by this service belongs to metrics or logs. Metrics are numerical near real-time values that describe several aspects of a system at a given time in a lightweight way. Logs contain different data organized into records with distinct sets of properties for each type.

Cacti[4] is a tool that monitors and graphs network performance data, following the data push model. It is open-source and uses the round robin database tool (RRDtool) while it polls monitoring data at set intervals. Cacti primarily uses Simple Network Management Protocol (SNMP) to collect data from network switches and routers. It can be configured with either a PHP script or a C-based poller. It supports custom data-gathering scripts and allows for data collection on non-standard timespans. Cacti is useful for tracking metrics like CPU load and network bandwidth usage.

Collectd[5] is an UNIX daemon that collects, relays and stores data concerning performance and network status. It is an open-source tool offered directly to DevOps or through other monitoring platforms that mainly use it for collecting metrics. It uses JMX to extract valuable metrics related to the JVM, which also enables the monitoring of Big Data frameworks (since most of them are Java-based complex frameworks). Furthermore, it follows a modular design which means that data acquisition and storage are handled by appropriate plugins as shared objects, while the daemon implements the capabilities of filtering and relaying the data. Therefore, especially due to minimum resources requirements it is useful for embedded devices.

Datadog[6] is a well-known commercial monitoring solution that provides full-stack monitoring for cloud applications as a SaaS-based data analytics platform, which covers server machines, databases, and services. It offers a dashboard, alerting, and visualizations of metrics and covers all major public and private cloud technology providers, Google Cloud Platform, AWS, Microsoft Azure, Red Hat OpenShift, VMware, and OpenStack (more than 350 integrations offered out-of-the-box). Through a Go-based agent, it can handle monitoring of both the infrastructure and the application. Its backend is based on several open and closed source technologies, including Apache Cassandra, Kafka and PostgreSQL, among others.

---

[3] https://docs.microsoft.com/en-us/azure/azure-monitor/overview.

[4] https://www.cacti.net/.

[5] https://collectd.org/.

[6] https://www.datadoghq.com/.

Dynatrace[7] is a commercial software that uses AI to optimise and monitor application performance. It provides a comprehensive overview of the entire solution stack, as it can monitor microservice-based applications, container management platforms, and any other hosts running in multicloud or hybrid cloud environments. It features OneAgent for the automated data collection, SmartScape for visualising topology mapping, PurePath for code-level distributed tracing, and an AI engine for automatic root-cause fault tree analysis. It can be deployed as a SaaS or managed service.

Google Cloud Monitoring[8] is a vendor-specific service that provides visibility into the performance, availability, and health of applications and infrastructure in the Google Cloud environment (Cloud, Anthos, Kubernetes, Istio, etc.). It offers an automatic out-of-the-box metric collection for Google Cloud services which are visualized in dashboards, while it supports monitoring of hybrid and multicloud environments through the installation of an open-source agent[9]. This agent is a Collectd[10]-based daemon that gathers system and application metrics. Furthermore, this service supports custom metrics to monitor application and business-level information while it can automatically infer.

Hadoop Performance Monitoring [14] is a big data framework-focused tool capable of finding issues in Hadoop installations while providing visualizations of the available tunable parameters that may increase the performance of Hadoop. It is considered to be a lightweight monitoring tool for Hadoop servers. An advantage of this tool is that it is built into the Hadoop ecosystem and is easy to use [4]. Although it is built-in, it presents performance issues, especially with respect to garbage collection [4]. It provides YARN REST APIs and is capable of visualising node-level log hierarchy (i.e., stderr, stdout, and syslog).

Icinga[11] is a popular open-source monitoring tool that appeared as a fork of the Nagios monitoring tool. It is written in C++ and PHP for cross-platform monitoring with additional database connectors and features in comparison to Nagios. It requires a master component to be hosted only on Linux in order to aggregate, process, and visualize monitoring data. Its REST API allows administrators to integrate extensions while it permits the funnelling of monitoring data from other domain-specific monitoring applications. It offers an alerting service that, based on certain preconfigured thresholds, it notifies the appropriate contact persons for performing mitigation actions. Moreover, it offers analytics to discover relations and patterns through visualizations. Icinga can monitor network services and aspects of the host resources while it is open to extensions that might aggregate application-level monitoring data. Nevertheless, Icinga seems to have a high learning curve, while dynamically adding new hosts and services is not straightforward. Last, it offers dedicated modules to monitor private, public and hybrid clouds.

LogicMonitor[12] is a commercial cloud-based monitoring platform that is able to cope with hybrid infrastructures. It is an extensible platform that is able to monitor

---

[7] https://www.dynatrace.com/.

[8] https://cloud.google.com/monitoring/.

[9] https://cloud.google.com/monitoring/agent.

[10] https://collectd.org/.

[11] https://icinga.com/.

[12] https://www.logicmonitor.com/.

several aspects of the IT stack and applies machine learning to further process and exploit monitoring data to optimize the IT environment. It provides many integrations to monitor cloud resources from different vendors (e.g., AWS, Azure, Google Cloud, etc.). Specifically, it supports the installation of collectors on Virtual Machines or Servers in any cloud instance to be able to use out-of-the-box infrastructure performance metrics. Moreover, it provides metrics for container and microservices monitoring as well as network, database, and application-level performance aspects.

Munin[13] is an open-source network monitoring and infrastructure monitoring software application. It is written in Perl and provides graphs that are accessible over a web interface. It offers plug-and-play capabilities through 500 monitoring plugins. It is lightweight enough to monitor even edge devices. Munin follows a master/node architecture in which the master is connected to all the nodes to pull data at regular intervals. It then stores the data in RRD files and updates the graphs accordingly. Moreover, TLS-based communication can be enabled for secure master node exchange of monitoring data. The Munin master and nodes are running on Perl and can be installed on Windows or Unix-based platforms.

Nagios[14] is an open-source application able to monitor systems, networks, and infrastructure. It comprises of a free basic edition called Nagios Core and an enterprise edition called Nagios XI, which includes, among others, a built-in web configuration GUI, an integrated DB, a backend API, SNMP Trap Support, and Database support. In the comparison Table 1, we report on the Nagios XI. It follows a centralized approach in which a centralized server collects data, but it is possible to create a static processing hierarchy that cope with the disadvantages of centralized approaches. Nagios provides a way of monitoring both cloud-based resources as well as in-house infrastructures through SNMP monitoring. It can be used as an event scheduler, event processor, or alert manager, and it is deployed as a daemon written in C. Therefore, it is capable of running natively on Unix-based systems. It has been successfully used in mission-critical applications that require per-second monitoring, but it requires complex configurations by experienced Linux engineers.

Netdata[15] is a popular open-source tool that is able to operate across bare metal systems, virtual machines, applications, and IoT devices to aggregate real-time metrics (e.g., CPU/RAM usage, disk or bandwidth usage, etc.). It offers fine-grained monitoring capabilities (about hosts, networks, containers, databases, etc.) to enable DevOps to quickly identify and troubleshoot issues and make data-driven decisions to maintain the quality of applications. Netdata is built as a lightweight daemon able to gather and display real-time information. Its subcomponents are written in C, Python, and JavaScript, and it is optimized to use minimal resources (i.e., about 2% on a single-core system). The Netdata daemon (i.e., Netdata agent) can run on any GNU kernel to monitor entire systems or application components. Moreover, it is capable of running on PCs, servers, and embedded Linux devices. The agents, although they are autonomous, their metrics can be aggregated by the Netdata Cloud, which is a cloud-based centralized management system. In addition, it can be installed without interrupting any of the running applications

---

[13] http://munin-monitoring.org/.

[14] https://www.nagios.org/.

[15] https://www.netdata.cloud/.

while it abides by the specified memory requirements, using only idle CPU cycles. It contains a number of default plugins to gather key system metrics, but it is highly extensible to any other metric through a dedicated plugin API. The metrics collected are unlimited and can reach per-second granularity. One of its most popular plugins involves the integration to Vnstat[16] (a popular open-source console-based network traffic monitor).

New Relic[17] is a proprietary SaaS solution for application and infrastructure monitoring over traditional, hybrid, or cloud environments. It is a monitoring tool implemented as serverless application that can support milliseconds speed with respect to the monitoring data. New Relic manages complex and constantly changing cloud applications and infrastructure while it presents one network monitoring dashboard that follows a centralised approach. Nevertheless, the insights-based queries that it supports are quite limited for enterprise-scale purposes.

Openstack Telemetry Service[18] is an open-source tool which aggregates and persists monitoring data from private clouds (i.e., Openstack). Telemetry comprises three sub-projects: i) Aodh[19] which exploits monitoring metrics to issue predefined alarms; ii) Ceilometer[20] which is the main data collection service, and iii) Panko[21] which is an event metadata indexing service acting as a REST API for Ceilometer. All these systems are mainly used for billing purposes, but they are valuable for the following up on the health of the deployed infrastructural nodes. Therefore, they enable the delivery of the so called counters and events (relevant to the QoS of the hosting and networking environment) in traceable and auditable way. All of them are provided in extensible YAML format to support new data collection agents who are independent of the monitoring system. Such monitoring data are published to various target services, including data stores and message queues.

Opsview[22] offers commercial products that specialise in IT infrastructure and cloud monitoring software for on-premises, cloud, and hybrid IT environments. It offers the Opsview Cloud and Opsview Monitor products as scalable monitoring platforms that can scale to tens of thousands of hosts. The Opsview Monitor exploits the open-source Ansible automation framework for deploying and managing the lifecycle of the monitoring agents. Furthermore, it offers auto-discovery features to automatically monitor new workloads as they are deployed on bare metal, virtual machines, or in the cloud. Last, it provides a wide and extensible range of monitoring metrics that target applications in public, private, and hybrid clouds, containers, networks, and databases. The Opsview monitor collector uses a TLS-encrypted message bus (i.e., RabbitMQ) to push periodically monitoring data to the Opsview monitor orchestrator.

---

[16] https://humdi.net/vnstat/.
[17] https://newrelic.com/.
[18] https://github.com/openstack/telemetry-specs.
[19] https://github.com/openstack/aodh.
[20] https://github.com/openstack/ceilometer.
[21] https://github.com/openstack/panko.
[22] https://www.opsview.com/.

Prometheus[23] is a widely used open-source system for monitoring and alerting, which is often used with Grafana[24] for monitoring data visualisation. It has an active developer and user community, and it offers several useful features. It produces time series data (i.e., key-value pairs) per monitoring metric and per autonomous node and supports a flexible query language (i.e., PromQL). The measurements are collected through the HTTP protocol, following a pull model, while it can push time series through an intermediary gateway to allow ephemeral and batch jobs to expose their metrics. The monitoring targets are usually statically configured, but there is also the option of service discovery. It is integrated with special-purpose exporters for services like HAProxy, StatsD, and Graphite. Last, it allows the definition of alerting rules which generate alerts that are sent to a centralised Alertmanager.

SequenceIQ[25] is an open-source tool, for monitoring Hadoop clusters, which has been merged since 2019 with the Cloudera Manager[26]. The SequenceIQ architecture is based on Docker containers and the ELK stack, that is, Elasticsearch[27], Logstash[28] and Kibana[29]. The use of client and server containers provides separation of concerns with respect to Hadoop deployment and monitoring. The server container implements the aggregators of monitoring data, and in particular, Kibana is used for visualization and Elasticsearch for consolidation of the monitoring metrics. Through Elasticsearch, the horizontal scaling and clustering of multiple monitoring components are supported [4]. The client container comprises Logstash and collectd modules.

Shinken[30] is an open-source solution for network monitoring that was developed as a branch of Nagios 4. It is written in Python and primarily monitors network services, but also tracks basic hosting resource metrics like CPU load and disk usage. It allows for the creation of a host hierarchy to differentiate between down and unreachable hosts, while it supports custom service checks. Shinken requires all hosts running a specific daemon with the same versions of python packages and distribution package cores. It can calculate KPIs based on state and performance data and exports data to visualization modules like PNP4Nagios[31] and Graphite[32]. It also supports secure remote monitoring through SSH or SSL-encrypted tunnels.

Zabbix[33] is a well-known open source monitoring tool that tracks network devices, servers, VMs and cloud services. It uses SNMP and supports metrics like bandwidth, CPU and memory usage and other aspects related to hosting devices health. Zabbix can monitor thousands of servers, VMs, networks, or IoT devices using agentless monitoring techniques. It also provides root cause analysis and uses XML-based templates for

---

[23] https://prometheus.io/.

[24] https://grafana.com/.

[25] https://github.com/sequenceiq.

[26] shorturl.at/enBP3.

[27] https://www.elastic.co.

[28] http://logstash.net.

[29] https://www.elastic.com/producs/kibana.

[30] http://www.shinken-monitoring.org/

[31] https://docs.pnp4nagios.org/

[32] http://graphiteapp.org/.

[33] https://www.zabbix.com/.

monitoring configuration. Zabbix supports notification methods like XMPP, and it can run local scripts in response to alerts.

Table 1. Monitoring tools comparative analysis

Monitoring Tool	Monitoring Level	Processing	Output Types	Distribution	Number of metrics	Configu ration	Custom Metrics	Foot print	Provider Depen dent
Amazon CloudWatch	host/database/ap plication(- functions)	basic/alerts	push-based/ persisted (DB)	centralized	medium	medium	yes	medium	yes
Apache Chukwa	big data framework	basic	pull-based/ persisted (file)	centralized	low	low	yes	medium	no
Azure Cloud Monitoring	host/application/ database	basic/alerts	pull-based/ persisted (file/DB)	centralized	high	medium	yes	medium	yes
Cacti	network/partially host	alerts	pull-based/persisted (DB)	statically distributed	low	medium	yes	low/ medium	no
Collectd	host/network/ext ensible to big data frameworks	no	push-based/ persisted (file)	centralized	medium	low	yes	low	no
Datadog	host/network/app lication/ database	advanced/root cause analysis	push-based/persisted (DB)/time-series	centralized	high	high	yes	medium	no
Dynatrace	host/application/ network/ database	alerts/root cause analysis/ forecasting	pull-based/ persisted(file/DB)/ time-series	distributed	high	high	yes	medium	no
Google Cloud Monitoring	host/application/ network/ database	basic/alerts	push-based/ persisted (file/DB)	centralized	high	medium	yes	medium	yes
Hadoop Perf. Monitoring	big data framework	basic	pull-based/ persisted (DB)	statically distributed	low	low	yes	medium	no
Icinga	host/network	alerts	push-based/ persisted (DB)	dynamically distributed	high	medium	yes	medium	no
Logic Monitor	host/application/ network/ database	alerts/ forecasting	pull-based/ persisted (DB)	centralized	high	high	yes	medium	no
Munin	host/network	alerts	pull-based/ persisted (file)	centralized	high	medium	yes	low	no
Nagios	host/network	basic/alerts	pull-based/ persisted (DB)	centralized/ statically distributed	high	medium	no	medium	no
Netdata	host/network/dat abase/ extensible to application, big data frameworks	basic/define thresholds/ alerts	pull-based/ persisted(file/DB)/ time-series	centralized	high	high	yes	low	no
New Relic	host/network/app lication	advanced/root cause analysis	pull-based/ persisted(DB)/ time-series	centralized	high	high	yes	medium	no
Openstack Telemetry	host/network	alerts	push-based/ persisted (DB)	centralized	medium	low	yes	medium	yes
Opsview	host/application/ network/ database	alerts	push-based/persisted (file/DB)/time-series	statically distributed	high	high	yes	medium	no
Prometheus	host/network/app lication	alerts	pull-based/push-base/persisted (DB)/time-series	centralized/ statically distributed	high	high	yes	low	no
SequenceIQ	big data framework	basic	push-based/ persisted (DB)	dynamically distributed	medium	low	yes	medium	no
Shinken	network/partially host	basic	pull-based/ persisted (file/DB)	statically distributed	medium	medium	yes	medium	no
Zabbix	host/network	basic/alerts	pull-based/ persisted (DB)	statically distributed	medium	medium	no	low	no

2.2 Monitoring Tools Comparison

We provide a brief overview of 20 well-known monitoring tools and compare them by listing the following information in Table 1:

Monitoring Level – at which level the monitoring is focused on: host, application, network, database, or big data framework. These levels were recognised as the superset of available monitoring types that we found in this analysis.

Processing – which refers to the kind of processing on collected monitoring data supported: basic (i.e., statistical aggregation capabilities such as maximum, minimum, average), define thresholds, cep (complex event processing), root cause analysis, define actions (e.g., alerts) or forecasting.

Output Types – which refers to how data can be collected (either pushed by a monitoring agent to a monitoring server or pulled from monitoring agents by a specific monitoring server or monitoring stack) and whether or not they are persisted in files, databases, and or as time-series.

Distribution – refers to the monitoring topology used for aggregating monitoring data, i.e., centralized when a unique monitor server is used, statically distributed where a number of statically defined peers are considered for processing monitoring data coming from neighbour monitoring nodes, and dynamically distributed when these peers can change at run-time according to the reconfigurations of the distributed application or the changes in the monitoring topology.

Number of metrics supported – mentioned as low, medium, or high depending on how monitoring metrics are provided out-of-the-box.

Configuration – which indicates if it is easily allowed to change the measurement rate of certain monitoring parameters and/or the output periodicity in which monitoring values are transmitted, persisted, and processed. The alternative indications are: Low (static output/measurement rate), Medium (either output or measurement rate are configured dynamically), High (both can be configured at run-time).

Custom metrics supported – a Boolean indication on whether or not custom monitoring probes can be accommodated (especially useful for appending application-related metrics).

Footprint – which refers to how lightweight the monitoring agents are (i.e., Low/Medium/High – where low indicates the possible installation on resource-constrained devices, e.g., Raspberry Pi).

Hosting Provider Dependent – which refers to whether or not the monitoring agents introduced can be used on only one specific cloud or not.

3   Discussion

Based on the above-mentioned comparative analysis of prominent monitoring tools, we examined the main aspects of several commercial and open-source projects that are widely used. We found out that the majority of these tools follow a centralised approach while only two of them (i.e., Icinga and SequenceIQ) provide the appropriate means to support a real dynamically distributed monitoring topology, a fact that denotes the importance of federated monitoring processing frameworks (e.g., EMS approach [7, 8]). We consider the support for dynamic and distributed monitoring topology management as imperative for the dispersed and vigorous cloud computing continuum environments. Also, the majority of these tools focus mainly on the aggregation of monitoring data to visualise them and let the DevOps decide on mitigation actions based on the current status of the application. Only a few of them support advanced processing capabilities and root cause analysis (i.e., Datadog, Dynatrace, New Relic), while just one of them supports proactive mitigation actions suggestions (Dynatrace). Last, the majority focus on hosting resources and network indicating the diversity required to cope with different application monitoring types.

As mentioned above, we conducted this comparative analysis in order to recognise the most promising software tool with respect to its monitoring probes in order to use it as the fundamental building block of a federated event management system that can cope with the challenges of the cloud computing continuum. Based on our analysis, we consider Netdata as the most appropriate solution for our purposes. It is a popular open-source monitoring solution that introduces lightweight monitoring agents that are flexible enough to be integrated with an event-driven architecture (EDA) that will be able to follow and process monitoring information in a highly distributed and resilient manner. Specifically, Netdata agents require just 2% of a single core system to collect a vast amount of different monitoring parameters, which are highly and dynamically configurable in terms of measurement rate and periodicity of collection. It involves a very low footprint and cloud/resource provider independence, along with a straightforward way to introduce custom application-level probes. Last, it consists of a wide and very active community of GitHub contributors, which results in constant improvements and the addition of new monitoring capabilities.

We note that the outcome of this analysis was successfully validated with real experimentation and testing of the Netdata capabilities over heterogeneous resources such as public and private cloud resources, resource-constrained devices (i.e., Raspberry Pi), and installation on accelerators such as FPGAs. This work was concluded with successful integration of Netadata as multiple, extensible and dynamically configurable monitoring probes to EMS [7, 8].

4   Conclusions

In this technical analysis, we focused on the cloud computing continuum and specifically on significant challenges with respect to continuously monitoring the defined SLOs, and in general, the health status of cloud applications and their hosting infrastructure. We introduced an extensive technical review of well-known monitoring tools that are able to propagate and process measurements. Then, we discussed which of these tools is appropriate for building a flexible and distributed monitoring framework for coping with the monitoring challenges in this field. The next steps of this work involve the introduction of a novel monitoring framework that will be able to aggregate data from dedicated monitoring probes and enforce complex event processing on their monitoring streams. The target will be to do so across cloud computing continuum resources in an efficient and resilient manner that will avoid any bottlenecks that become apparent in centralized approaches.

Acknowledgments. The research leading to these results has received funding from the EU's Horizon 2020 research and innovation programme under grant agreement No. 871643 MORPHEMIC project.

References

1. Aceto, G., Botta, A., de Donato, W., Pescape, A.: Cloud monitoring: a survey. Comput. Netw. **57**(9), 2093–2115 (2013). https://doi.org/10.1016/j.comnet.2013.04.001
2. Horn, G., Skrzypek, P., Prusinski, M., Materka, K., Stefanidis, V., Verginadis, Y.: MELODIC: selection and integration of open source to build an autonomic cross-cloud deployment platform. In: International Conference on TOOLS 50+1: Technology of Object-Oriented Languages and System, 14–19 October, Innopolis, Russia (2019)
3. Verginadis, Y., et al.: PrEstoCloud - a novel framework able to dynamically manage data-intensive multi-cloud, fog, and edge function-as-a-service applications. IGI Inf. Resour. Manage. J. (IRMJ) **34**(1), Article 4, 66–85 (2021)
4. Drăgan, I., Iuhasz, G., Petcu, D.: A scalable platform for monitoring data intensive applications. J. Grid Computing **17**(3), 503–528 (2019). https://doi.org/10.1007/s10723-019-094 83-1
5. Bautista Villalpando, L.E., April, A., Abran, A.: Performance analysis model for big data applications in cloud computing. J. Cloud Comput. **3**(1), 1–20 (2014). https://doi.org/10. 1186/s13677-014-0019-z
6. Verginadis, Y., Kritikos, K., Patiniotakis, I.: Data and cloud polymorphic application modelling in multi-clouds and fog environments. In: La Rosa, M., Sadiq, S., Teniente, E. (eds.) CAiSE 2021. LNCS, vol. 12751, pp. 449–464. Springer, Cham (2021). https://doi.org/10. 1007/978-3-030-79382-1_27
7. Baur, D., Griesinger, F., Verginadis, Y., Stefanidis, V., Patiniotakis, I.: A model driven engineering approach for flexible and distributed monitoring of cross-cloud applications. In: 2018 IEEE/ACM 11th International Conference on Utility and Cloud Computing (UCC), pp. 31–40 (2018). https://doi.org/10.1109/UCC.2018.00012
8. Stefanidis, V., Verginadis, Y., Patiniotakis, I., Mentzas, G.: Distributed complex event processing in multiclouds. In: Kritikos, K., Plebani, P., de Paoli, F. (eds.) ESOCC 2018. LNCS, vol. 11116, pp. 105–119. Springer, Cham (2018). https://doi.org/10.1007/978-3-319-99819-0_8
9. Trihinas, D., Pallis, G., Dikaiakos, M.D.: Low-cost adaptive monitoring techniques for the Internet of Things. IEEE Trans. Serv. Comput. **14**(2), 487–501 (2021). https://doi.org/10. 1109/TSC.2018.2808956
10. Trihinas, D., Pallis, G., Dikaiakos, M.D.: Monitoring elastically adaptive multi-cloud services. IEEE Trans. Cloud Comput. **6**(3), 800–814 (2018). https://doi.org/10.1109/TCC.2015.251 1760

11. Demirbaga, U., et al.: AutoDiagn: an automated real-time diagnosis framework for big data systems. IEEE Trans. Comput. **71**, 1035–1048 (2021). https://doi.org/10.1109/TC.2021.307 0639

12. Do, N.H., Van Do, T., Farkas, L., Rotter, C.: Provisioning input and output data rates in data processing frameworks. J. Grid Comput. **18**(3), 491–506 (2020). https://doi.org/10.1007/s10 723-020-09508-0

13. Tamburri, D.A., Miglierina, M., Di Nitto, E.: Cloud applications monitoring: an industrial study. Inf. Softw. Technol. **127**, 106376 (2020). https://doi.org/10.1016/j.infsof.2020.106376

14. Venner, J., Wadkar, S., Siddalingaiah, M.: Pro Apache Hadoop, 2nd edn. Apress, New York (2014)

Multi Languages Pattern Matching-Based Scraping of News and Articles Websites

Hamza Salem[(✉)] and Manuel Mazzara

Software and Service Engineering Lab, Innopolis University,
Innopolis, Republic of Tatarstan, Russia
h.salem@innopolis.university, m.mazzara@innopolis.ru

Abstract. Web scraping refers to the extraction of data from a specific website. Every website will include web pages and each page has a source code containing HTML tags that show a representation of the data. The problem with any scraping method for a web page is how the page is structured, each page has a different structure. That's why the process of data extraction requires more knowledge about web page structure. To solve such a problem users should know the content and the structure of any page. In this paper, we propose a multi-language pattern mining technique to scrap news and articles websites by recognising title, body and thumbnail image based on a content structure pattern. This approach is an improvement for our previous work that has been done before. By using the same method we have added thumbnail Image as new parameters to be extracted. This approach can be applied to several data-sets, in our case we have prepared 550 web pages as a dataset to test it in both languages Arabic and English.

1 Introduction

Web scraping is the process of using code or bots to extract content and data from a website. It is different from screen scraping, in which only content of the viewed part in pixels displayed on screen, web scraping extracts underlying HTML code and content at the same time [1]. Web scraping is used in several digital businesses that rely on data mining for example, search engine bots, price comparison sites and Market research companies who are using data for machine learning models training [2,3]. The process of web scraping for a specific page will start with a URL to scrape data from. Then the scraper will extract all the data on the page and then the scraper script can be customized to select the specific data based on the requirement. Data can be stored into database, spreadsheet or JSON objects handled by developers. Scrapers can be software or script written by developers. Web scraping software can be an application or web service provided online using a website, some of these software are paid based on the usage and some of these already have plugin on browsers like WebScraper.io that provide UI for user to choose data to be scraped [4]. On the other hand, writing code by developers using scripting languages like Pyhton [5], in this case

L. Barolli (Ed.): AINA 2023, LNNS 655, pp. 644–648, 2023.
https://doi.org/10.1007/978-3-031-28694-0_60

users should identify data needed to be scraped and developers write code for that, and the cost is very high based on the requirements.

In this paper, we will introduce an improvement for a technique that was already developed in our previous work [6] to scrap news and articles websites using pattern mining. We have extended the script to accept all languages and to extract thumbnail images and test it in 550 web pages. By using this technique, we have improved the scraper to have more accurate data from the page itself without human interaction and knowing the content or the structure.

2   Literature Review

Different approaches have been proposed to extract news and articles from the web using several methods such as structured based scraper or Machine learning too. For example, in [7] the authors created a script to construct pattern of Regex and implementing the patterns as a set of rule to extract article title, article author, and publication date, however, the main limitation for such as method is the dataset that used in the experiment; it was three websites only. On the other hand, Authors [8] presented a method to use supervised machine learning algorithms like Naive Bayes and Logistic Regression to retrieve news site content, such as method is effective when you have good dataset for all languages and also labeled data to recognize the main element in each page. Finally, we have presented in previous paper pattern mining techniques to scrap news and blog websites by recognizing title and body based on a content structure pattern and in this paper we extend the work to include thumbnail image and provide a large dataset supporting both languages Arabic and English [6].

3   Methods

The Arabic and English languages were used in this study report. Both the language dataset acquired using web scraping script applied on Google news. The following steps presented the methodology that have been applied in the script:

1. Data collection using web scraping Script on Google News
2. Extract full HTML code in all web pages and download it
3. Identify News Title, News description and Thumbnail Image based on the pattern
4. Compare the Thumbnail Image with Real Thumbnail Image in Each page.

From our previous work on [6] the biggest text size on the page is a title and the biggest (div) character length is the body. Using the same pattern, the biggest image size (width, height) will be the thumbnail image for each article or news. As simple as this method appears, it was the most effective method to recognize the news thumbnail image. The uniqueness of such a method is the adaptability for other categories like news, articles and blogs, most of these websites follow the same structure for Title, description and thumbnail and because of that it is easy to identify more information about the scraped data.

This paper focuses on how to recognize thumbnail images in news and articles based on pattern mining in the structure and the style of the shown content. We have used 550 web pages to test the pattern and we have improved the previous script to get a thumbnail image. In the next sections, we will show the new part in the script that retrieves the thumbnail and show how it will work and also, we will present the result based on the new dataset.

4   Implementation

There are several programming languages to scrape content from the internet like JavaScript, Java and Python. In our previous work [6], We have used Beautiful Soup [10] scrape content HTML and using the same pattern. First of all, we have scraped Title and description using the same rules that have been shown before. As seen in Fig. 1, the script will identify all images in the web page and fetch source link for each page and by simple observation, we will conclude that the biggest image in width and height is the thumbnail for the news in that page. By using the same rule we have identified all thumbnails for almost all 550 web pages that have been extracted.

Fig. 1. Fetch all images in web page

```
soup = BeautifulSoup(html)
for tag in soup.find_all('img'):
    print(tag.attrs.get('height', None),
    tag.attrs.get('width', None))
```

Code Snippet 1: Biggest width, height Images

As shown in Code Snippet 1, the script fetched all width and height for all images and using simple comparison in the code we fetched the biggest image in that page. By using such a method we have generated the full preview for any news including Title, Description and thumbnail image to be used as metadata for any other system automatically.

5   Results and Discussion

We had analyzed 550 web pages from different websites scraped using google news aggregator in Arabic and English as source. The results were 100% accurate in all pages, and these results can be generalized because most of the websites are using the visual standard for all news websites. Table 1 shows the results for 550 web pages in English and Arabic and shows the count and the accuracy for getting the right Thumbnails.

Table 1. Results of thumbnail images

Website language	Web pages count	Script accuracy
Arabic	320	100%
English	230	100%

This result has further verified the assumption in the previous work to use pattern mining techniques in web scraping [6]. By using the same technique the same script can be generalized to other languages and other parameters such as date or author name. By defining the base for each parameter the dataset can be used in other tests. The same result can be replicated in other domains such as e-commerce or ads listing websites. This paper is an improvement for pattern mining in news and blog websites and adds thumbnail image as new parameter that can be extracted without knowing anything about the HTML structure of the scraped website.

6   Conclusions

In this paper, we have presented an improvement for an existing model that has been used to scrape news and blog websites using pattern mining. In this paper, we have used pattern mining scrapers to recognize thumbnail Images in news and articles websites without knowing the structure of the page. By using the same pattern on 550 pages, the script shows 100% accuracy to recognize thumbnail images in all dataset that have been used both in Arabic and English language. It was a correct assumption that has been made as the biggest image in any news web page is the thumbnail of that news and the same for articles. However, Our work clearly have to be tested on big dataset .and more languages to verify the results and have more chance to be extended that's why we have provide our dataset on public repository to be used by researchers [10].

References

1. Zhao, B.: Web scraping. In: Encyclopedia of Big Data, pp. 1–3 (2017)
2. Slamet, C., et al.: Web scraping and Naïve Bayes classification for job search engine. In: IOP Conference Series: Materials Science and Engineering, vol. 288, no. 1. IOP Publishing (2018)
3. Julian, L.R., Natalia, F.: The use of web scraping in computer parts and assembly price comparison. In: 2015 3rd International Conference on New Media (CON-MEDIA). IEEE (2015)
4. Sirisuriya, D.S.: A comparative study on web scraping (2015)
5. Nair, V.G.: Getting Started with Beautiful Soup. Packt Publishing Ltd., Birmingham (2014)
6. Salem, H., Mazzara, M.: Pattern matching-based scraping of news websites. J. Phys. Conf. Ser. **1694**(1), 012011 (2020)
7. Maududie, A., Retnani, W.E.Y., Rohim, M.A.: An approach of web scraping on news website based on regular expression. In: 2018 2nd East Indonesia Conference on Computer and Information Technology (EIConCIT). IEEE (2018)
8. Prehanto, D.R., et al.: Implementation of web scraping on news sites using the supervised learning method. Ilkogretim Online 20.3 (2021)
9. Richardson, L.: Beautiful soup documentation. Dosegljivo (2007). https://www.crummy.com/software/BeautifulSoup/bs4/doc/. Dostopano 7 July 2018
10. Enghamzasalem. Enghamzasalem/Websegmentation. GitHub (n.d.). https://github.com/enghamzasalem/websegmentation/. Retrieved 14 Nov 2022

Decoding COVID-19 Vaccine Hesitancy Using Multiple Regression Analysis with Socioeconomic Values

Wei Lu[1(✉)], Ling Xue[2], and Bria Shorten[1]

[1] Department of Computer Science, Keene State College, The University System of New Hampshire, USNH, Keene, NH, USA
wlu@usnh.edu

[2] Department of Health Management and Policy, University of New Hampshire, USNH, Durham, NH, USA

Abstract. With the growth and development of COVID-19 and its variants, reaching a level of herd immunity is critically important for national security in public health. To deal with COVID-19, the United States has implemented phased plans to distribute COVID-19 vaccines. As of November 2022, over 80% of Americans had received their first shot to guard against COVID-19, and 68.6% were considered fully vaccinated, according to the dataset provided by CDC. However, a significant number of American people still hesitate to receive a shot of the COVID-19 vaccine. This paper aims to demystify COVID-19 vaccine hesitancy by analyzing various socioeconomic characteristics among individuals and communities, including unemployment rate, age groups, median household income, and education level. A multiple regression modeling and data visualization analysis show patterns with an increasing trend of vaccine hesitancy associated with a lower median household income, a younger age group, and a lower education level, which would help policymakers to make policies accordingly to target vaccine support information and remove this hurdle to end the COVID-19 pandemic effectively.

1 Introduction

Since COVID-19 has spread into a global pandemic, many countries have developed vaccines to deal with COVID-19. The COVID-19 vaccines have been widely recognized for their effectiveness in reducing the severity and death caused by this virus. The United States has implemented phased distribution plans. As of November 2022, over 80% of Americans had received their first shot to guard against COVID-19, and 68.6% were considered fully vaccinated, according to the most recent CDC data [1]. The number of COVID-19 vaccine doses administered in the United States as of November 9, 2022, is summarized in [2], from which we can see 404,714,328 Pfizer doses, 253,590,880 Moderna doses, and 18,942,185 J&J/Janssen doses have been distributed. Most American people trusted and chose Pfizer when they did COVID-19 vaccination; there are, however, a significant amount of American people who still hesitate to receive any shot of the COVID-19 vaccine, mainly because (1) they worry about possible side effects and

© The Author(s), under exclusive license to Springer Nature Switzerland AG 2023
L. Barolli (Ed.): AINA 2023, LNNS 655, pp. 649–659, 2023.
https://doi.org/10.1007/978-3-031-28694-0_61

would like to wait to see if it is safe; and (2) they don't trust the government and don't believe they need it because they think COVID is not a significant threat.

To address the issue of vaccine hesitancy, in this paper, we investigate four social-economic questions related to the COVID-19 vaccine rate using data visualization, namely, (1) the relationship between the percentage of senior people in a region and the COVID-19 booster vaccination rate in that region? (2) what is the relationship between the unemployment rate and the COVID-19 booster vaccination rate by the state? (3) what is the connection between median household income and the COVID-19 booster vaccination rate? and (4) are college-educated people more willing to get vaccinated? The visualization results using the data collected from the Centers for Disease Control and Prevention (CDC) and the Economic Research Service of the US Department of Agriculture (USDA) show that the poorer and more vulnerable social groups in terms of income and education level have a disproportionately lower number of vaccine rate, validating that an increase in income and education level contribute unequally to the rise of COVID-19 vaccination rate. These visualization results are also proved in a multiple linear regression model, including the vaccination rate, the household median income, the education level, and age groups by county, where unemployment rate is removed from the model due to anomalies generated during data collection [3–6].

As a result, we conclude that in addition to the top six reasons illustrated in [7], there is also an association between COVID-19 vaccination rate and senior people, median household income, and educated level, which would help our policymakers to make policies accordingly, such as increasing income, stimulating the economy, raising people's education, to target vaccine support information and remove this hurdle to end the COVID-19 pandemic effectively.

2   Related Works

Studies on the socio-economic aspects and their influence on the COVID-19 vaccination rate during the pandemic are new. In [8], Willis et al. explored what proportion of Arkansas residents are hesitant to get the COVID-19 vaccine and how it changed across sociodemographic groups by analyzing a survey result collected from 1,205 responders. The study showed that younger respondents with lower incomes had a higher prevalence and odds of vaccine hesitancy. Conversely, those older, White, in higher income brackets, or had a four-year degree were less hesitant to get the vaccine. Similarly, Yasmin et al. in [9] compared COVID-19 vaccine uptake across the United States while investigating predictors of vaccine hesitancy and acceptance of different groups. They found that several factors led to low or high vaccine acceptance, including (1) males and individuals with a college degree or higher education was more accepting, (2) non-Hispanic African Americans had the lowest vaccine acceptance rate, while Whites and Asians had the most positive attitude toward the vaccine, (3) People over the age of 45 were more accepting compared to younger people, and (4) Lower-income individuals were less likely to get the vaccine.

In [10], Mewhirter et al. looked to employ a machine learning model to analyze online survey data and predict vaccine hesitancy, where predictors of hesitancy included support of former President Donald Trump, especially among White Christian Evangelicals. In contrast, predictors of acceptance include comfort in a healthcare setting, a greater risk of catching COVID-19, and the likelihood of passing on the illness to vulnerable others. In [11], Morales et al. looked to see if gender intersects with socioeconomic status to co-produce inequalities in people's intent to take vaccines. The study used four groups of explanatory measures: demographics, socioeconomic status, household structure, and health-related variables, and found that women, in general, were more hesitant to get the vaccine, particularly women in households below the poverty threshold, working women, and women with children. For both men and women, lack of education (no college degree) and type of residency (mobile home, RC, van, boat) increase vaccine hesitancy.

In [12], Dror et al. anonymously surveyed 1941 healthcare workers and members of the general Israeli population about their acceptance of a potential COVID-19 vaccine. They found that people were more accepting of a vaccine if they believed they were at high risk for severe COVID-19 infection or were already vaccinated against seasonal influenza were positive predicting factors. Males were more likely to accept the vaccine. All adults with children were less accepting. Doctors were more accepting than nurses. The most significant concern among all respondents in this study was the vaccine's safety, given the speed of its development.

3   Datasets on Vaccination, Education, Income, and Age Group

This paper addresses the COVID-19 vaccination rate and its associations with the unemployment rate, age groups, median household income, and education level. In particular, the dataset includes 3 CSV files, namely (1) *us-COVID-19-vaccinations.csv* is the county-level data for the COVID-19 vaccination rate, including the vaccination percentage of COVID-19 booster doses, the total number of population by county in the 2019 census, and the number of the population whose age is older than 65 years old [13]; (2) *education.csv* is data about educational attainment for adults age 25 and older for the U.S., States, and counties from 1970 to 2019; and (3) *unemployment.csv* is the data about the unemployment and median household income for the United States, States, and counties from 2000 to 2020, where the median income data collected in the past 12 months of 2019 is based on the inflation-adjusted dollars.

The size of the dataset of COVID-19 vaccinations in the US by county is large, including 1.58 million rows and 66 columns. However, what we need to use in our study are seven columns described in Table 1. The *us-COVID-19-vaccinations.csv* has the age group information to help analyze the relation between vaccination rates and age groups. The *unemployment.csv* summarizes unemployment and median household income data for counties that are used to analyze the relationship between vaccination and income level, and the *education.csv* collects educational attainment for the counties that are used to analyze the relationship between vaccination and education level [14].

Table 1. Feature Description in COVID-19 Vaccination Dataset

Feature	Description
Date	Date data are reported on CDC COVID Data Tracker
FIPS	Federal Information Processing Standard State Code
Recip_County	County of residence
Recip_State	Recipient State
Booster_Doses_Vax_Pct	Percent of people who completed a primary series and have received a booster (or additional) dose
Census2019	2019 Census Population
Census2019_65PlusPop	2019 Census Population for $\geq$65 years of age

4   Data Visualization

We use Tableau Prep Builder to connect to our three datasets. There are no potential problems for each dataset file, i.e., there are not too many null values, and data distributions look good. We used the FIPS code to join the tables during cleaning. After dataset wrangling, we have a combined full dataset that can be found in [15].

Figure 1 visualizes the data on the percentage of senior people in a region and the vaccination rate in that region, which shows a trend that the more senior people, the higher the vaccination rate in that region. As illustrated in Fig. 1, Chattahoochee County, GA, has the lowest number of senior people at 4.86%, its COVID-19 vaccination rate is also the lowest rate, which is only 1.9%, and Sumter County, FL, has the largest number of senior people of 58.17%, its COVID-19 vaccination rate is 56.10%, which is on the top 20% compared all the other 3218 regions/counties. From the most concentrated counties in the middle part of the chart, we can observe a trend of positive correlation between the percentage of senior people in a county and the COVID-19 vaccination rate in this county.

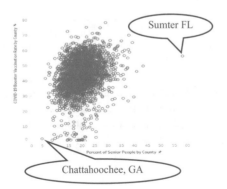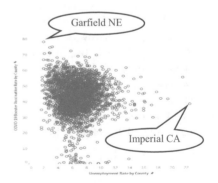

Fig. 1. Percent of senior people in a county vs. COVID-19 vaccination rate in this county

Fig. 2. Relationship between the unemployment rate and COVID-19 vaccination rate by county

Figure 2 visualizes the data on the percentage of unemployed people in a region and the COVID-19 vaccination rate in that region, which shows a slight trend that the more unemployed people, the lower the vaccination rate. As illustrated in Fig. 2, Garfield County, Nebraska, has the lowest unemployment rate of 2.0%; its COVID-19 vaccination rate, however, achieves the highest rate of 78.2%. Imperial County, CA, has the largest employment rate of 22.50%; its COVID-19 vaccination rate is 38.5%, which is about half of the highest vaccination rate of 78.2% achieved by Garfield County. From the edge of the curve in the chart, we can roughly observe a negative correlation trend between the county's unemployment rate and the COVID-19 vaccination rate in this county.

Figure 3 illustrates the relationship between the median family household income and the COVID-19 vaccination rate by state. The State of Rhode Island has the highest vaccination rate of 56.56%, while its median family household income is also ranked top six after the District of Columbia, New Jersey, Massachusetts, Connecticut, and Maryland; on the other hand, Mississippi has the lowest number median household income of $41,703 and has a low number of vaccination rate of 36.81% which is on the bottom of six states and is merely better than Virginia, North Carolina, Alabama, Georgia, and Texas. This illustrates that the higher the income, the higher the vaccination rate, and there is a trend of positive correlation between the median family household income rate and the vaccination rate.

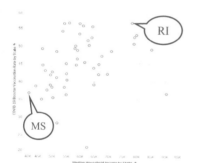
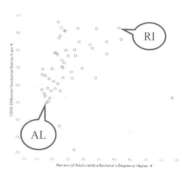

Fig. 3. Median household income vs. COVID-19 vaccination rate by state

Fig. 4. Percentage of people who hold a bachelor's or higher vs. vaccination rate by state

Similarly, in Fig. 4, we visualize the relationship between the percentage of people who hold a bachelor's degree or higher and the COVID-19 vaccination rate by state. Again Rhode Island is ranked top 3 in terms of the percentage of people who hold a bachelor's degree or higher (i.e., 41.10%), and it also achieves a second place in vaccination rate (i.e., 56.56%) over all the other states. Alabama, on the other hand, has one of the lowest vaccination rates (i.e., 35.04%), which is merely better than Virginia and North Carolina, and its percentage of people who hold a bachelor's degree or higher is 18.05% which is located on the bottom lists, better than Alabama, Louisiana, Mississippi, Tennessee, Kentucky, and West Virginia. High-level education is tied to a higher income in America, so it's no surprise that the higher the education level, the higher the salary, and the higher the vaccination rate.

There are some outliers, though, in these visualizations, such as the region of D.C. has the highest percentage of people who hold a bachelor's degree or higher (i.e., 58.5%); its vaccination rate, however, is 36.6% which is just slightly higher than the rate in Alabama. Moreover, the unemployment rate of 21.5% in Skagway, AK, is the second highest, but its vaccination rate is 55.4% in the top 20% compared to all other 3,180 counties. Network anomalies could simply cause this due to the pollution of data collection [16, 17]. As a result, to address these issues and translate the association among COVID-19 vaccination rate, median household income, education level, age group, and unemployment rate, we conduct a complete multiple regression modeling in the following section.

5   Multiple Regression Modeling

The multiple regression model is a probabilistic model that includes more than one independent variable. It is an extension of a first-order straight-line-based linear model aiming to make accurate predictions by incorporating more potentially important independent variables [18,19]. The general form of the multiple regression model is illustrated in the following:

$$y = \beta_0 + \beta_1 \times x_1 + \beta_2 \times x_2 + \ldots + \beta_k \times x_k + \varepsilon$$

where y is the dependent variable, $x_1, x_2, \ldots x_k$ are the independent variables, and β_k determines the contribution of the independent variable x_k.

The visualization results show that the covid-19 vaccination rate most likely depends on the age group, the median household family income, the education level, and the unemployment rate. By running the regression model, we find that the factor of the unemployment rate has a large p-value value of 0.0578, which is larger than the significant threshold p-value of 0.05, meaning that it is not substantial in predicting the vaccination rate. Therefore, we need to remove it from our regression model to ensure that all the parameters are significant.

As a result, in our analysis, we have:

$x_1 =$ *the median household income in a county*

$x_2 =$ *percentage of adults with a bachelor's degree or higher in a county*

$x_3 =$ *percentage of senior people with age older than 65 in a county*

$y =$ *the COVID vaccination rate by county*

Therefore, we can hypothesize the regression model in the following:

$$y = \beta_0 + \beta_1 \times x_1 + \beta_2 \times x_2 + \beta_3 \times x_3 + \varepsilon$$

where y, x_1, x_2, x_3 are illustrated above, and $\beta_1, \beta_2, \beta_3$ determines the contribution of the independent variable x_1, x_2, x_3 respectively.

$$E(y) = \beta_0 + \beta_1 \times x_1 + \beta_2 \times x_2 + \beta_3 \times x_3$$

Table 2. Parameter estimates

Term	Estimate	Std. Error	t Ratio	Prob > |t|	Lower 95%	Upper 95%
Intercept	16.75	1.147	14.61	< 0.0001	14.5	18.99
Income	0.145	0.017	20.95	< 0.0001	0.705	0.85
Education	0.139	0.025	5.47	< 0.0001	0.0892	0.1888
Age	0.777	0.037	20.95	< 0.0001	0.7045	0.85

and it is the deterministic portion of the regression model, while ε is a random error component. The modeling with least-squares gives the parameter estimates as llustrated in Table 2 where can find that:

$$\beta_0 = 16.75 \quad \beta_1 = 0.145 \quad \beta_2 = 0.139 \quad \beta_3 = 0.777$$

Thus, our least squares prediction is:

$$y = 16.75 + 0.145 \times x_1 + 0.139 \times x_2 + 0.777 \times x_3$$

Figure 5 is a graph regarding the probability distribution of the random error component ε, in which we can see that the residuals do not follow the solid red line exactly, but they are within the Lilliefors confidence bounds and do not contain any pattern [20]. So, the assumption of normality is verified. Table 3 is the analysis of variance showing that s^2, estimator of variance σ^2 of random error term ε, is 91.0.

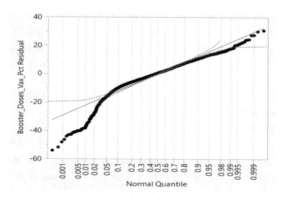

Fig. 5. Residual normal quantile plot

Table 3. Analysis of Variance

Source	DF	Sum of Squares	Mean Square	F Ratio
Model	3	56096.21	18698.7	205.5882
Error	3122	283953.29	91.0	**Prob > F**
C. Total	3125	340049.49		< 0.0001

Given the multiple regression model:

$$Y = \beta_0 + \beta_1 \times x_1 + \beta_2 \times x_2 + \beta_3 \times x_3$$

We evaluate the utility of the model by conducting the following hypothesis testing:

$$H_0 : \beta_1 = \beta_2 = \beta_3 = 0$$

H_a : at least one of the coefficients is nonzero

As illustrated in Table 3, the p-value of the hypothesis test is smaller than 0.0001. So, at $\alpha = 0.05$, the p-value is smaller than α. As a result, in this case, we have sufficient evidence to reject the null hypothesis, therefore we can conclude that there is strong evidence that one of the coefficients is nonzero and the overall model is useful.

The value of s (i.e. $s = sqrt(s^2) = 9.54$) represents the root mean square error (RMSE) that is the square root of the mean square error (MSE) and is an estimate of the standard deviation of the random error because the true error is generally not known. The s measures the spread of the distribution of y-values about the least-square and these errors are assumed to be normally distributed. As a result, we should usually find that most (about 95%) of the observations lie within $2 \times s$ of the least-square, i.e. $y \pm 19.08$, and most of the y-values fall within 19.08 of their respective predicted values using the least-square regression.

The coefficient $\beta_1 = 0.145$ estimates mean value of the vaccination rate in a county (i.e. $E(y)$) to decrease 0.145% for each one thousand-dollar family household income increase (x_1) when the percentage of adults with a bachelor's degree or higher (x_2) and percentage of senior people with age older than 65 (x_3) are held constant. At $\alpha = 0.05$, the p-value is less than 0.0001 which is less than α. . As a result, in this case we have sufficient evidence to reject the null hypothesis, therefore we can conclude that the mean value of the vaccination rate in a county $(E(y))$ will increase as family household income (x_1) increases while all the other variables are held constant.

The coefficient $\beta_2 = 0.139$ estimates the mean value of the vaccination rate in a county (i.e. $E(y)$) to increase 0.139% for each one percent increase of people who hold with a bachelor's degree or higher (x_2) when Income (x_1) and Age group (x_3) are held constant. At $\alpha = 0.05$, the p-value is less than 0.0001 which is less than α. . As a result, in this case we have sufficient evidence to reject the null hypothesis, therefore we can conclude that the mean value of the vaccination rate in a county $(E(y))$ will increase as Education level (x_2) increases while all the other variables are held constant.

The coefficient $\beta_3 = 0.777$ estimates mean value of the vaccination rate in a county (i.e. $E(y)$) to increase 0.777% for each one percent increase of people who are older than 65 (x_3) when Income (x_1) and Education (x_2) are held constant. At $\alpha = 0.05$, the p-value is les than 0.0001 which is less than α.. As a result, in this case we have sufficient evidence to reject the null hypothesis, therefore we can conclude that the mean value of the vaccination rate in a county ($E(y)$) will increase as the senior people's rate (x_3) increases while all the other variables are held constant.

The intercept value of β_0 is a marker and in most cases, unless all the $x's$ are zero then it is mainly used to set up the value of y points. A 95% confidence interval for β_2, is in the range of 0.0892 to 0.1888, that is to say there is a 95% chance that our estimation for β_2 is going to occur in that range and we have the mean value of the vaccination rate to increase in anywhere from 0.09% to 0.19% for every one percentage increase of people who hold with bachelor's degree or higher (x_2) when Income (x_1) and Age_Group (x_3) are held constant.

6   Conclusions and Future Work

The poorer and more vulnerable social groups have a disproportionately higher number of deaths, implying that a decrease of income and education level inequality contributes to the increase of COVID-19 vaccination hesitancy. According to a recent study on COVID-19 Vaccine Hesitancy Worldwide in [21], the United States has a COVID-19 vaccine acceptance rate of 56.9%, one of the lowest rates compared to other 32 countries, and is merely better than Kuwait (23.6%), Jordan 28.4%), Italy (53.7%), Russia (54.9%), Poland (56.3%). In this paper we validate this inequality phenomena using data visualization and multiple regression modeling. One of the major contributions of this work is that we create a COVID-19 vaccination dataset including the population by county in the 2019 census, the number of the population whose age is older than 65 years old, the educational attainment data for adults age 25 and older for counties in 2019, and the 2020 median household income for counties. This dataset we build is publicly available so people in the healthcare professional domain can use it as a reference to help policy makers make decisions wisely. The other contributions include findings that: (1) the higher the median household income in a region, the lower the vaccination hesitancy rate in that region; (2) the higher the percentage of senior people in a region, the higher the vaccination acceptance rate in that region, and (3) the higher the percentage of college-educated people in a region, the lower the vaccination hesitancy rate in that region.

In the future we will integrate a much larger variety of features such as immigration status, family size, and rural vs urban areas and then apply multiple regression and other machine learning technologies, such as E-means clustering algorithm [22–24], to build a predictive model to predict vaccination hesitancy rate for certain group of patients so healthcare officials and professionals can get notified proactively when making corresponding policies.

Acknowledgments. This research is supported by New Hampshire - INBRE through an Institutional Development Award (IDeA), P20GM103506, from the National Institute of General Medical Sciences of the NIH.

References

1. COVID-19 vaccination rates. https://www.usnews.com/news/best-states/articles/these-states-have-the-best-covid-19-vaccination-rates, Accessed 25 Nov 2022
2. Number of COVID-19 vaccine doses administered in the United States as of November 9, 2022, by vaccine manufacturer. https://www.statista.com/statistics/1198516/covid-19-vaccinations-administered-us-by-company/. Accessed 25 Nov 2022
3. Lu, W., Tavallaee, M., Ghorbani, A.A.: detecting network anomalies using different wavelet basis functions. In: the 6th Annual Communication Networks and Services Research Conference (CNSR 2008), pp. 149–156 (2008). https://doi.org/10.1109/CNSR.2008.75
4. Lu,W.: A Novel framework for network intrusion detection using learning techniques. In: PACRIM. 2005 IEEE Pacific Rim Conference on Communications, Computers and signal Processing, 2005, pp. 458–461 (2005). https://doi.org/10.1109/PACRIM.2005.1517325
5. Lu, W., Traore, I.: A new unsupervised anomaly detection framework for detecting network attacks in real-time. In: Desmedt, Y.G., Wang, H., Mu, Y., Li, Y. (eds.) Cryptology and Network Security. CANS 2005. LNCS, vol. 3810, pp. 96–109. Springer, Berlin (2005). https://doi.org/10.1007/11599371_9
6. Lu, W., Traore, I.: An unsupervised approach for detecting DdoS attacks based on traffic based metrics. In: Proceedings of IEEE Pacific Rim Conference on Communications, Computers and Signal Processing (PACRIM 2005), pp. 462–465, Victoria, B.C., August 2005
7. Americans don't getting vaccinated. https://www.vox.com/2021/6/2/22463223/covid-19-vaccine-hesitancy-reasons-why, Accessed 25 Nov 2022
8. Willis, D.E., et al.: COVID-19 vaccine hesitancy: race/ethnicity, trust, and fear. Clin. Transl. Sci. **14**(6), 2200–2207 (2021). https://doi.org/10.1111/cts.13077. Epub 2021 Jul 2. PMID: 34213073; PMCID: PMC8444681
9. Yasmin, F., et al.: COVID-19 vaccine hesitancy in the United States: a systematic review. Front Public Health **23**(9), 770985 (2021). https://doi.org/10.3389/fpubh.2021.770985.PMID:348 88288;PMCID:PMC8650625
10. Mewhirter, J., Sagir, M., Sanders, R.: Towards a predictive model of COVID-19 vaccine hesitancy among American adults. Vaccine **40**(12), 1783–1789. https://doi.org/10.1016/j.vaccine.2022.02.011. Epub 2022 Feb 7. PMID: 35164989; PMCID: PMC8832389
11. Morales, D.X., Beltran, T.F., Morales, S.A.: Gender, socioeconomic status, and COVID-19 vaccine hesitancy in the US: an intersectionality approach. Soc. Health Illn. **44**(6), 953–971. https://doi.org/10.1111/1467-9566.13474. Epub 2022 May 2. PMID: 35500003; PMCID: PMC9348198
12. Dror, A.A., et al.: Vaccine hesitancy: the next challenge in the fight against COVID-19. Eur. J. Epidemiol. **35**, 775–779 (2020)
13. COVID-19 vaccinations in the US County. https://data.cdc.gov/Vaccinations/COVID-19-Vaccinations-in-the-United-States-County/8xkx-amqh. Accessed 25 Nov 2022
14. USDA county-level datasets. https://www.ers.usda.gov/data-products/county-level-data-sets/county-level-data-sets-download-data/. Accessed 25 Nov 2022
15. Vaccination dataset. https://unh.box.com/s/qdlnhdpqsqk3h8lam6x0236g2gzp789q. Accessed 25 Nov 2022
16. Lu, W., Tong, H.: Detecting network anomalies using CUSUM and EM clustering. In: Cai, Z., Li, Z., Kang, Z., Liu, Y. (eds.) ISICA 2009. LNCS, vol. 5821, pp. 297–308. Springer, Heidelberg (2009). https://doi.org/10.1007/978-3-642-04843-2_32
17. Lu, W., Traore, I.: Unsupervised anomaly detection using an evolutionary extension of K-means algorithm. Int. J. Inf. Comput. Secur. **2**(2), 107–139 (2008)
18. Alexopoulos, E.C.: Introduction to multivariate regression analysis. Hippokratia **14**(Suppl. 1), 23–28. PMID: 21487487; PMCID: PMC3049417

19. DeBono, A.: Research Methods and Statistics for the Social Sciences: A Brief Introduction. Cognella Academic Publishing (1 June 2020), ISBN:13-978-1516537389
20. Multiple regression. https://www.jmp.com/en_nl/learning-library/topics/correlation-and-regression/multiple-linear-regression.html. Accessed 25 Nov 2022
21. Sallam, M.: COVID-19 vaccine hesitancy worldwide: a concise systematic review of vaccine acceptance rates. Vaccines (Basel) **9**(2), 160 (2021). https://doi.org/10.3390/vaccines9020160.PMID:33669441;PMCID:PMC7920465
22. Lu, W., Tong, H., Traore, I.: E-means: an evolutionary clustering algorithm. In: Kang, L., Cai, Z., Yan, X., Liu, Y. (eds.) ISICA 2008. LNCS, vol. 5370, pp. 537–545. Springer, Heidelberg (2008). https://doi.org/10.1007/978-3-540-92137-0_59
23. Lu, W., Traore, I.: Determining the optimal number of clusters using a new evolutionary algorithm. In: Proceedings of IEEE International Conference on Tools with Artificial Intelligence (ICTAI 2005), pp. 712–713, Hongkong, November 2005
24. Lu, W., Traore, I.: A new evolutionary algorithm for determining the optimal number of clusters. In: Proceedings of IEEE International Conference on Computational Intelligence for Modeling, Control and Automation (CIMCA 2005), vol. 1, pp. 648–653 (2005)

Video Indexing for Live Nature Camera
on Digital Earth

Hiroki Mimura[1], Masaya Tahara[1], Kosuke Takano[1(✉)], Nobuya Watanabe[2],
and Kin Fun Li[3]

[1] Department of Information and Computer Sciences, Kanagawa Institute of Technology,
1030 Shimo-Ogino, Atsugi 243-0292, Kanagawa, Japan
{s1921032,s1921123}@cco.kanagawa-it.ac.jp,
takano@ic.kanagawa-it.ac.jp
[2] International GIS Center, Chubu University, 1200 Matsumoto-Cho, Kasugai 487-8501, Aichi,
Japan
nov@isc.chubu.ac.jp
[3] Department of Electrical and Computer Engineering, Faculty of Engineering,
University of Victoria, 3800 Finnerty Road, Victoria, BC V8P 5C2, Canada
kinli@uvic.ca

Abstract. This paper presents a method of indexing video stored from live nature cameras on digital Earth and its application. Since two feature vectors from scene images are not necessary identical in the metrical space even if they are similar, it is hard to get benefit of indexing feature vectors. That is, the search system consequently has to fully access to almost whole feature vectors in the storage. The feature of the proposed system is that for indexing feature vectors, similarity-based indexing by clustering feature vectors is applied, where the feature vectore can be indexed with a centorid vector of the corresponding cluster. Consequently, the computing cost for accsessing to the objective images can be majorly reduced without the almost full access to whole feature vectors. For making our concept clear, we show an application scenario using video data from live streaming cameras that are opened to the public and some aerial or satellite images.

1 Introduction

A digital earth [1, 2] is a concept to project every observable thing on the earth into a cyber information space in several kinds of digital formats such as images, audio, and videos. In the digital earth, building an adequate cyber information space that can be a metaphor of the real world, without lacking necessary information and capturing vast quantities of geospatial data is needed [2]. However, a massive amount of digital information on the digital earth platform would make it difficult for us to find and access the necessary information according to the given purposes.

In this study, in order to solve such problems, we design and develop a prototype of a fast scene search system for massive video data streamed from environmental monitoring cameras on the digital earth platform. Our system stores video data into the corresponding strage for each camera. The video data has time window and is separated by a certain time.

© The Author(s), under exclusive license to Springer Nature Switzerland AG 2023
L. Barolli (Ed.): AINA 2023, LNNS 655, pp. 660–667, 2023.
https://doi.org/10.1007/978-3-031-28694-0_62

For the scence search, a feature vector is extracted from each image frame in the video data, so that the simirarity between images can be calculated with the smilarity metrics such as dot product, consine, and so on. However, since two feature vectors are not necessesary identical in the metrical space even if they are similar, it is hard to get benefit of indexing feature vectors, that means, the search system consequently has to fully access to almost whole feature vectors in the storage. The feature of the proposed system is that for indexing feature vector, similarity-based indexing by feature vector clustering is applied, where the feature vectore can be indexed with a centorid vector of the corresponding cluster. The clusters of feature vectors are constructed by a time widow in each storage, so that the acual calucualation of the similarity can be focused on the limited set of feature vectors from the whole strages. Consequently, the computing cost for accsessing to the objective images can be majorly reduced.

For making our concept clear, we show an application scenario using video data from live streaming cameras that are opened to the public and some aerial or satellite images.

2 Rerated Work

In this section, we review researches focused on the similarity computation for multi-dimenal vectors, which are especially extracted from a neural network.

The importance of embedding-search is increasing for nearest kneeboard search using embedding matrix, which is obtained through neural network, mathematical operations such as SVD, matrix factorization approach, and so on. In [3], Shrivastava and Li propose a hashing algorithm for searching with inner product as the underlying similarity measure. In [4], a score-aware quantization loss function is proposed to compress feature vectors for the fast approximate distance calculation.

Metrics learning [5, 6] is an machine learning approach for constructing distance metrics based on the similarity between data. Since the capability of neural network to extract a feature vector resenting semantics of data, researches regarding metrics leaning with neural network are also active. In [8], a distance metric leaning between a sub-image of a photograph and an iconic product image of that object is proposed with Siamese network [7] applying convolutional networks for caluculating visual similarity of product design. In [9], Hoffer and Ailon propose a triplet network model for obtaining useful representations by distance comparisons.

Meanwhile, clustering methods for feature vectors extracted from neural network have been researched to achieve unsupervised learning. In [10], deep embedded clustering (DEC) is proposed. DEC simultaneously learns feature representations and cluster assignments using deep neural networks. In [11], Caron et al. propose a clustering method called DeepCluster for jointly learning the parameters of a neural network and the cluster assignments of the resulting features. In [12], Nailussa'ada et al. propose a method for indexing and finding deep neural networks for image recognition, where a meta neural network indexes sub neural networks that can answer the input image query.

For those conventional researches, the basic idea of our approach is to index sets of feature vectors based on their similarities, where the similar feature vectors have the same index vector, so that the fast search for the objective images stored in several databases can be realized.

3   Proposed Method

3.1   Overview

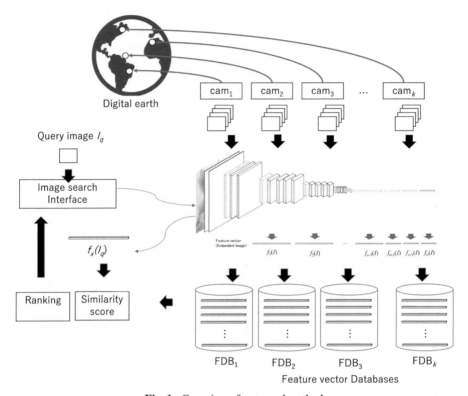

Fig. 1. Overview of proposed method

Figure 1 shows an overview of proposed method. In Fig. 1, video cameras for monitoring the environment of some areas are mapped on the digital earth. Images in video data recorded in each campera are transformed to feature vectors by a pre-trained neural network. The feature vectors are stored in feature databases FDB_1, FDB_2,..., FDB_k for the image retrieval process. Furthermore, according to a time window T with Δt width of the video data, and feature vectors are clustered into a set of feature vectors by a certain time.

Since two feature vectors are not necessesary identical in the metrical space even if they are similar, it is hard to get benefit of indexing feature vectors, that means, the search system consequently has to fully access to almost whole feature vectors in the storage. Therefore, we apply a similarity-based indexing by feature vector clustering, where the feature vectore can be indexed with a centorid vector of the corresponding cluster. The clusters of feature vectors are constructed by a time widow in each storage, so that the acual calucualation of the similarity can be focused on the limited set of feature vectors from the whole strages.

In the image retrieval process, an input query image I_q by a user is transformed to a feature vector with the same neural network used to extract feature vectors in the feature vector databases. The similarity scores between I_q and I in FDB are calculated based on each feature vector, and then the ranking result is shown to the user.

3.2 Feature Vector Extraction

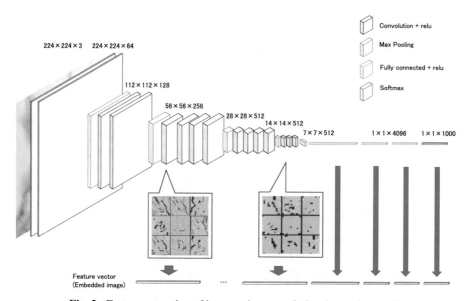

Fig. 2. Feature extraction of image using convolutional neural network

For the feature extraction from an image, a convolutional neural network is applied as the feature extractor. As shown in Fig. 2, for an input image I, a feature vector f_x in a layer l_x is extracted by specifying a set of channels C_x in the layer l_x.

$$f_x(I) = NN\big(l_x, C_y, I\big) \tag{1}$$

$$C_x = \{c_a, c_b, ..., c_z\}, \tag{2}$$

3.3 Image Similarity

Suppose that we have a set of cameras *Camera* that are mapping on the digital earth. Each camera cam_i stores video data v_i, which includes k frame of images I.

$$Camera = \{cam_1, cam_2, ..., cam_i\} \tag{3}$$

$$V_k = \{I_1, I_2, ..., I_k\} \tag{4}$$

Using (1), each image I_k can be transformed to a feature vector as follows:

$$f_x(V_k) = \{f_x(I_1), f_x(I_2), \ldots, f_x(I_k)\} \tag{5}$$

For an input query image I_q, the image search is executed by calculating the similarity between feature vectors extracted from the query image I_q and the image frame I_p. For the similarity metric, we can apply vector operations such as a dot product, cosine, and so on.

$$score(I_q, I_p) = similarity(f_x(I_q), f_x(I_p)) \tag{6}$$

Thorough the neural network, metadata $M(I)$ for an image I can be represented with a pair of the feature vector $f_x(I)$ and term labels $T(I) = \{t_1, t_2, \ldots, t_k\}$.

$$M(I) = \{f_x(I), T(I)\} \tag{7}$$

Then, the equal (6) is alternatively represented as follows. Basically, the term similarity between two sets of terms is calculated based on a vector space model.

$$score(I_q, I_p) = similarity(f_x(I_q), f_x(I_p)) + term\_similarity(T(I_q), T(I_p)) \tag{8}$$

3.4 Similarity-Based Indexing

For indexing a feature vector, we classify a set of feature vectors into some clusters based on the similarity. By clustering feature vectors, each feature vectore can be indexed with a centorid vector of the corresponding cluster as shown in Fig. 3.

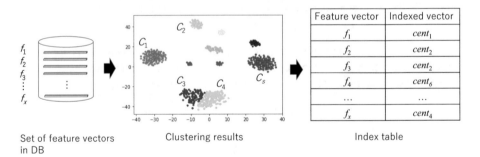

Feature vector	Indexed vector
f_1	$cent_1$
f_2	$cent_2$
f_3	$cent_2$
f_4	$cent_6$
...	...
f_x	$cent_4$

Set of feature vectors in DB   Clustering results   Index table

Fig. 3. Similarity-based indexing

Suppose that a feature vecor $f_x(I)$ in a set of feature vectors is classified into a cluster C_s. For the clustring, we can apply clustring algrorims from standard k-means to neural-based metrics learning methods [8, 9] and clustring algorithms [10, 11]. The domain of feature vectorus for the clustering can be within each database, or within each time-window in a video data.

$$f_x(I)C_s \tag{9}$$

In this case, when a centroid of the cluster C_s is represented as $cent_s$, the index tuple of the feature vecor $f_x(I)$ in an index table is obtained as a pair of $f_x(I)$ and $cent_s$.

$$index\_tuble((f_x)(I)) = (f_x(I), \ cent_s) \tag{10}$$

Using the index tuple, the system can fastly find the set of similar feature vectors that can be candidates of the answer for the user query image, since focusing on the limited set of feature vectors from the whole strages makes the computation time for accsessing to the objective images faster.

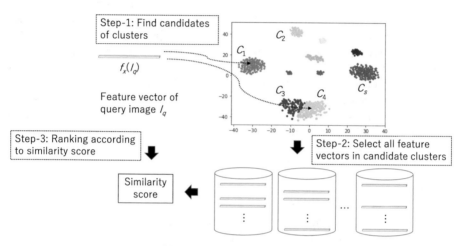

Fig. 4. Image search process

Figure 4 shows the image search process with the similarity-based indexing. In the image search process, first, the sysetem search for the index table to find candidates of clusters C_s suitable to the answer for the user query image.

$$score(I_q, cent_s) = \ similarity(f_x(I_q), cent_s) \tag{11}$$

Based on the score calculated by (11), the candidates of clusters C_s is ranked and extracted when the score is greater than a threshold δ. After that, for all feature vectors in the extracted clusters $C_1, C_2,..., C_s$, the similarity is calcurated with (6) and the ranking result is shown to the request user.

4   Application Scenarios

4.1   Device Setup

For implementing live nature cameras, low-power edge devices should be selected. For example, a spresense [13] produced by Sony with a porabe solar charger can be applied for the long-term unmanned observation of nature in many spots. Meanwhile, aerial video from drones should be considered for the wide range of observation. Many drone maker releases their drone products with a high quality camera [14, 15].

4.2 Scene Search of Nature Video on Digital Earth

The importance of earth observation is worldwidely increasing due to the effect of global climate exchange and so on. For example, destruction of coral reefs, progress of desertification, change of crop yield, sea level rise due to ice melting in the pole area are caused by the global warming. Meanwhile, it is said that the destruction of forest accelarizes the global warming.

In order to protect the global environment, it would be useful to offer archives of video data of monitoring natural environment in as many spots as possible and visually analyze what is happening in each spot. To provide such a facility, construction of monitoring environment on the digital earth with live nature cameras and drones described as Sect. 4.1 is going to be planned. By incorporating the proposed method described in Sect. 3 into the scene search system, the fast scene search over massive amount of video archive can be realized.

In the application scenarios that we designs, in addition to the basic scene search, joining the similar scenes over worldwide spots is one of the promising functions. By joining the similar scenes between remote spots, we can find the phenomenon progessing simultaneously in the world and swiftly share the facts with scientist and goveners for leading the solution collaboratively.

5   Conclusion

In this study, we have proposed a method of indexing video stored from live nature cameras on digital Earth. The feature of the proposed method is that feature vector clustering is applied for the similarity-based indexing, where the feature vectore can be indexed with a centorid vector of the corresponding cluster. The proposed method allows the fast scence search by calculating the simiraty between feature vectors within candidete clusters suitable to a query image.

We will implement the prototype of the proposed system according to the application scenarios as showin in Sect. 4, and evaluate the feasibility and effectiveness in our future work.

References

1. Guo, H., Goodchild, M.F., Annoni, A.: Manual of Digital Earth, Springer, Singapore (2020). https://doi.org/10.1007/978-981-32-9915-3
2. Fukui, H., Man, D.C., Phan, A.: Digital earth: a platform for the SDGs and green transformation at the global and local level, employing essential SDGs variables. Big Earth Data 5(4), 476–496 (2021)
3. Shrivastava, A., Li, P.: Asymmetric Lsh (ALSH) for sublinear time maximum inner product search (MIPS), Adv. Neural Inf. Process. Syst. 2321–2329 (2014)
4. Guo, R., et al.: Accelerating large-scale inference with anisotropic vector quantization. In: International Conference on Machine Learning (2020). https://arxiv.org/abs/1908.10396
5. Chechik, G., Sharma, V., Shalit, U., Bengio, S.: Large scale online learning of image similarity through ranking. J. Mach. Learn. Res. 11, 1109–1135 (2010)
6. Kulis, B.: Metric learning: a survey, foundations and trends in machine learning (2013)

7. Hadsell, R., Chopra, S., LeCun, Y.: Dimensionality reduction by learning an invariant mapping. In: IEEE Computer Society Conference on Computer Vision and Pattern Recognition (CVPR'06) (2006)

8. Bell, S., Bala, K.: Learning visual similarity for product design with convolutional neural networks, ACM Trans. Graph. **34**(4), 98, 1-10 (2015)

9. Feragen, A., Pelillo, M., Loog, M. (eds.): SIMBAD 2015. LNCS, vol. 9370. Springer, Cham (2015). https://doi.org/10.1007/978-3-319-24261-3

10. Xie, J., Girshick, R., Farhadi, A.: Unsupervised deep embedding for clustering analysis. In: International Conference on Machine Learning (2016)

11. Caron, M., Bojanowski, P., Joulin, A., Douze, M.: Deep clustering for unsupervised learning of visual features. In: Ferrari, V., Hebert, M., Sminchisescu, C., Weiss, Y. (eds.) Computer Vision – ECCV 2018. ECCV 2018. LNCS, vol. 11218. Springer, Cham (2018). https://doi.org/10.1007/978-3-030-01264-9_9

12. Ada, N., Bintang, F.R., Harsono, T., Barakbah, A.R., Takano, T.: Indexing and finding deep neural networks for image recognition. Inf. Model. Knowl. Bases **XXXI**, 458–469 (2019)

13. Sony, S. (2023). https://developer.sony.com/develop/spresense/

14. DJI, Phantom series (2023): https://www.dji.com/jp/products/phantom

15. Parrot, A.D. (2023). https://www.parrot.com/us/drones

Sports Data Mining for Cricket Match Prediction

Antony Anuraj, Gurtej S. Boparai, Carson K. Leung$^{(\boxtimes)}$, Evan W. R. Madill, Darshan A. Pandhi, Ayush Dilipkumar Patel, and Ronak K. Vyas

University of Manitoba, Winnipeg, MB, Canada
Carson.Leung@UManitoba.ca

Abstract. Millions of people around the globe are fond of cricket. With such a big fan base, teams are sure to be competitive. Teams usually try to get their players to play at their full potential. Like many other sports, cricket also has a large amount of data available per match. Data collected from these matches can be used to provide insights on why a specific match resulted in a certain outcome. Data related to team performance and individual player statistics can be retrieved from each match. Analysis of these data helps the team improve their player performance. However, sometimes factors like venue, toss winning, and ranking can also affect the result of a match. In this paper, we design and implement sports data mining approaches to predict match results. In particular, we focus on cricket. Evaluation on real-life T20 International World Cup data—which examine and analyze factors like venue, toss winning, team ranking, and matches won against a specific opponent to predict the winner of a given match—demonstrates the practicality and effectiveness of our sport data mining and prediction approaches.

Keywords: Advanced information networking and applications · Big data · Cricket · Data mining · Data science · Sports data mining

1 Introduction

Nowadays, big data [1,2] are everywhere. Examples include disease data [3–6], employment statistics [7], environmental data [8], music [9], news [10], social networks [11–13], sports statistics [14], and transportation data [15,16]. In general, *data mining* [17] is a process of extraction of useful information and patterns from these big data. In other words, it is also known as knowledge extraction or data and pattern analysis. Data mining is also helpful when it comes to identifying trends and possible future insights about a specific topic based on the data provided. We currently live in a time where data is one of the key factors that could influence the direction of a field of study, business, sports, and other sectors of life [18–20]. Understanding data and analyzing them could provide an added advantage that could help people excel in that area.

L. Barolli (Ed.): AINA 2023, LNNS 655, pp. 668–680, 2023.
https://doi.org/10.1007/978-3-031-28694-0_63

Sport is an area where each event contains many statistics that represent the performance of players, participants and teams. As an area of data mining, *sports data mining* [21] can be seen as influential to sports-based activities, especially those that have a competitive nature present in them. Sports data mining can provide a more in-depth analysis that could not have been seen in the past. It is important to use sports mining because, in the past, coaches, managers and analysts used to make decisions and predictions based on their observation while watching the player/teams play. These decisions may not be supported by any evidence but just based on their subjective insights. Currently, data mining could help predict the outcomes of matches, provide in-depth analysis on a player based on past performances also coaches could use information collected from training sessions to identify areas that each player would need to improve to perform to their full potential. Sports data mining has been applied to various sports activities, including:

- individual sports like cycling [22] and swimming [23];
- team sports like basketball [14], hockey [24], soccer (aka football) [25], and cricket [26–28].

As the popularity of cricket has grown rapidly, we conduct sports data mining on cricket. Many existing works [29,30] related to cricket data mining focused on mining event sequences from videos of cricket matches. In contrast, we focus on cricket match prediction. For works related to cricket match prediction, many of them [27,31] used features like player statistics. Here, we explore other features like ranking, head-to-head matches, venue, and toss outcomes. Our *key contributions* of the current paper include our design and development of an algorithm that uses feature (e.g., team ranking, head-to-head matches, venue, toss outcomes) for predicting cricket match results. Evaluation results on real-life International Cricket Council (ICC)'s T20 World Cup tournament data show the practicality of our algorithm.

The remainder of the current paper is organized as follows. The next section presents the background and related works. Section 3 describes our sports data mining algorithm for predicting cricket match results. Section 4 shows evaluation results on applying our algorithm to real-life data; Sect. 5 draws conclusions.

2   Background and Related Works

Cricket [32] is a popular team-based bat-and-ball sports played around the world. Various cricket events take place throughout the year. These include tournaments among multiple teams. A series usually involves two teams playing multiple matches against each other, and the winner of the series being the team with the most number of winning matches. In general, cricket matches are usually made up of two teams competing against each other. Each team consists of 11 players on a field. A rectangular *pitch*—of length 22 yards (approximately, 20 meters) and width 10 feet (approximately, 3 meters)—is located at the center of the oval field, and a *wicket* comprising two bails balanced on three stumps is located at each of the two ends. The *batting team* scores "runs" by striking the ball bowled at one of the two wickets with the bat and then running between the wickets. As the opponent of the batting team, the *bowling/fielding team* tries to:

- prevent the ball from leaving the field and getting to either wicket, and
- dismiss each batter (i.e., "out").

Specifically, the batter is dismissed if any of the following three actions takes place:

1. the batter bowls a ball (i.e., when the ball hits the stumps and dislodges the bails),
2. the bowling team catches the ball after it is hit by the bat but before it hits the ground, or
3. the bowling team hits a wicket with the ball before a batter can cross the crease in front of the wicket.

When 10 batters have been dismissed, an *inning* (i.e., a division of a cricket match) ends and the teams swap roles.

As a global governing body for international cricket representing 108 member states, the ICC governs and administrates many cricket events/games. According to the ICC[1], three common formats of cricket games played at the international level include:

- Test matches, which has been the traditional form of the cricket game (since 1877) and is now settled in a five-day format comprising two innings each.
- One-Day Internationals (ODIs), which have been a pacier form of the cricket game (since 1971) and have gained popularity since the 1980s. ODIs are one-inning matches of 50 "overs" (i.e., each "over" consists of six legal deliveries bowled—by mostly a single bowler—from one end of a cricket pitch to the player batting at the other end) per side.
- *Twenty20 (T20) Internationals*, which have been shortest and fastest form of the cricket game (since 2005) with 20 "overs" per side. T20 matches are usually competed within three hours, with lots of bowling (i.e., propelling the ball toward a wicket defended by a batter), batting (i.e., defending the wicket by hitting the ball with a bat to score runs), and fielding (i.e., collecting/recovering the ball after it is struck by the batter).

In this paper, we focus on the 20-over game of the T20 Internationals. In which, the first six "overs" are the *power play*, in which two fielders are allowed outside the circle and the remaining fielders are inside the circle during these six "overs". Hence, the batters have more advantage to score more runs during the power play. Once the sixth "over" is completed, all fielders can spread out in the circle. A goal of the bowling team would be to get the batting team out and prevent them from setting a large score. A bowler can take a batter out (i.e., dismissal of the batter) by taking any of the three aforementioned actions. The team that bowled first would have to try making more runs than their opponent makes within the allocated 20 "overs" (or fewer if possible) to win the match.

---

[1] https://www.icc-cricket.com/about/cricket/game-formats/the-three-formats.

Note that there exists a vast amount of historical and past cricket game statistics. One could mine cricket data to have a better performance of the cricket players (e.g., select players who may have the best outcome of the results). With the rising importance of data in the world and advancements in technology, cricket data can be promising. People may be curious to know if their favorite teams/players are going to perform according to their expectations. With huge leagues like Indian Premier League (IPL) having a massive fan base, fans are driven towards online fantasy platforms providing prizes for people who win while playing on this online platform. The benefits of mining cricket-based data managers, coaches, and fans could keep track of player performances and be able to identify the trends in player performance to predict the performance in the upcoming matches. Moreover, cricket could use statistics from past player performance to get insights on whether a player is likely to perform in a specific venue and/or against a particular team. Coaches could also use statistics (e.g., marginal wins) to model an ordinary team performance with (or without) a specific player. This, in turn, could help coaches take appropriate actions to the playing squad for a match [21].

As mentioned in Sect. 1, many existing works [29,30] related to cricket data mining focused on mining event sequences from videos of cricket matches. In contrast, we focus on cricket match prediction. For works related to cricket match prediction, many of them [27,31] used features like player statistics. In contrast, we explore other features like team ranking, head-to-head matches, venue, and toss outcomes. In general, the result of a match can be affected by a lot of factors such as:

- the venue at which the match was played,
- the team that won the toss,
- the strength of the team in comparison to its opponent (i.e., team ranking), and
- other factors.

Factors like toss are important because it allows the toss-winning team to elect to bat or to field (whichever is more advantageous to the team based on conditions of the field and pitch).

3   Our Sports Data Mining Method

Here, we describe our cricket match prediction algorithm. It considers factors like (a) team rankings, (b) head-to-head matches, (c) venue evaluation, and (d) toss evaluation. More specifically, our algorithm incorporates head-to-head matches. In situations where two teams have never played a match (i.e., no data are available for head-to-head matches between the two teams), it incorporates team rankings. Moreover, it also incorporates information like where the game is played (i.e., venue evaluation) and/or which team won the toss (i.e., toss evaluation), whenever these data are available.

3.1   Team Rankings

Team ranking is a necessary aspect in most of the sports, especially cricket, because it gives a glimpse of how a particular team has performed over a time period. The ICC keeps a track of the team rankings to determine where these teams are in terms of performance. As the T20 World Cup matches are held every two years, these matches alone may not be sufficient to rank teams. Hence, we take into consideration every series of T20 matches that are played between any two teams in between the two years to calculate their rankings. Rankings are updated in two ways:

1. Updated after every match played between any two teams: Their rankings are updated within 24 or 48 h accordingly on whether the team won or lost.
2. Updated annually: Assess the three-year team performances at the end of every year. In other words, at the end of a year, the previous three years are considered including the year that has just ended. This eliminates the oldest year and includes the newest year that just ended. This way all the team performances are looked at over the period of three years and their rankings are recalculated.

Since these rankings determine which team has been at their peak or slacked off, they act as a useful factor for predicting a match outcome. We then collect and mine T20 rankings data to compute rankings between any two teams. Consequently, the prediction percentage of win for both teams can be calculated by:

$$\%\text{win for Team A} = \frac{\text{Team A rating}}{\text{Team A rating} + \text{Team B rating}} \times 100\% \qquad (1)$$

$$\%\text{win for Team B} = \frac{\text{Team B rating}}{\text{Team A rating} + \text{Team B rating}} \times 100\% \qquad (2)$$

These equations provide prediction percentages of win for both teams by relying on their rankings as of the day they would be playing against another team.

3.2   Head-to-Head Matches

Data are gathered every match played between two specific teams. These include head-to-head matches played in the past, which help assess team performance over the past years and predict a future match between two certain teams. We first retrieve a list of all match IDs that represent matches played by two specific teams, and then analyze the number of wins and losses against each other. This information helps determine which team has had a greater number of wins than others, and thus increasing its chances of winning in a future match. The predicted percentage of win can be calculated by:

$$\%\text{win for Team A} = \frac{\text{Total \#matches won by Team A}}{\text{Total \#matches between Teams A\&B}} \times 100\% \qquad (3)$$

$$\%\text{win for Team B} = \frac{\text{Total \#matches won by Team B}}{\text{Total \#matches between Teams A\&B}} \times 100\% \qquad (4)$$

These equations provide prediction percentages of win for both teams, which reveal chances of either team's winning when playing against each other based on their previous encounters.

3.3   Venue Evaluation

The venue of the sports is another important factor in predicting a cricket match. In a T20 World Cup event, all the matches are usually played in one chosen country. A team that plays in its own country's ground (i.e., home team) often has advantages of knowing the pitch and/or ground well. Team players have trained on the venue long enough to know how well the pitch might be suitable for bowling and/or batting. For instance, home team's bowlers are able to adapt on how to bowl and work their way out to get the batters in the opponent team (i.e., away/visiting team) out. Similar comments apply to the home team's batters who know what shots to play and what not to hit, and perform as many runs as they can.

Here, we collect and mine venue data by first collecting a list of match IDs and their corresponding venues. Afterwards, we compute the total number of matches played at a certain venue (e.g., a certain cricket stadium). The prediction percentage of a team t winning at a certain venue v can then be computed by:

$$\%\text{win for Team } t \text{ at venue } v = \frac{\text{Total \#matches won by Team } t \text{ at venue } v}{\text{Total \#matches played by Team } t \text{ at venue } v} \times 100\% \quad (5)$$

This equation provides a prediction percentage of win for Team t at a particular venue v.

3.4   Toss Evaluation

In cricket, toss evaluation involves a coin toss where the team that wins the toss gets to choose whether to bat or bowl first. This is one of the main factors chosen for predicting a match because winning a toss is not just about choosing what to do first, it can be a team's strategy to win a match because:

- The pitch a team play on can be analyzed and determined if its condition is favorable for batting or bowling. Depending on this information, the toss-winning captain of the team can choose wisely whether to bowl or bat first.
- Evaluating and knowing the opposite team's strength whether they are good in either bowling or batting, the team captain would then have to choose accordingly to their advantage.

Here, we collect toss data from team experience in winning or losing. We then compute a list of match IDs. We also collect the corresponding toss results of the matches that pertains to the teams. The number of times a team has won a toss is accounted for. Finally, we also take into account for the number of times a team has won after winning a toss. Consequently, the prediction percentage of a team t winning the match after winning the toss can be calculated by:

$$\%\text{win for Team} t = \frac{\text{Total \#matches won by Team } t}{\text{Total \#tosses won by Team } t} \times 100\% \qquad (6)$$

This equation provides a prediction percentage of Team t winning in match when it wins the toss, based on its previous toss results.

4   Evaluation

To evaluate our sports data mining algorithm in predicting cricket match results, we applied our algorithm to real-life crickets data[2]. These structured data for cricket include ball-by-ball data international and T20 League cricket matches. In particular, we focused on T20 World Cup data from 2005 to 2021. Data reveal that there have been 87 teams with a total of 1,134 matches played in the T20 World Cup tournaments held during a period of 17 February 2005 to 16 October 2021.

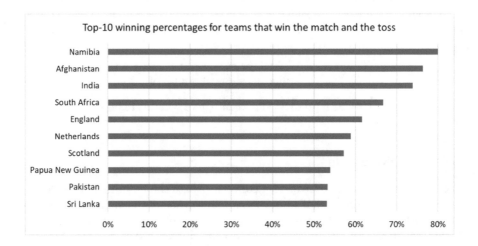

Fig. 1. Top-10 winning percentages for teams that win the match and the toss

4.1   Sports Data Mining on Historical Cricket Data

Analyses on historical data (17 February 2005–16 October 2021) reveal that toss outcomes can be a factor contributing to the winning of a cricket match. When applying Eq. (6), the percentage for a team t winning matches when it wins the toss is around 57.59%. This inclines to support that the lucky team captain who wins the toss may have a slight advantage on deciding whether his team would bat or bowl first (considering information like weather, pitch, strengths of

---

[2] https://cricsheet.org/downloads/.

his own team, strengths and/or weaknesses of his opponent team, their whims). As a concrete example, in a first round warm-up match held 12 October 2021, Oman—who won the toss and elected to bat—won the match against Namibia. Similarly, in another first round warm-up match held the same day, Scotland— who won the toss and elected to bat—also won the match against Netherlands.

Figure 1 shows the top-10 winning percentages for teams that win the match and the toss. For instance, the percentage for Team Namibia winning matches when it won the toss was 80%. The percentage for Team Afghanistan winning matches when it won the toss was 76%. These show the slight advantage for winning the toss (when compared to losing the toss).

Similarly, the same historical data also reveal that, when applying Eq. (5), the average winning percentages for:

- the home team is around 64%, and
- the away/visiting team is around 46%.

This inclines to support that the home team has a slight advantage over the away (aka visiting) team. This is because the home team is more familiar with the venue and playing conditions, receives more support from home crowd support, acquires more extensive training and experience on the said pitches. As a concrete example, in a semi-final match between Sri Lanka (home team) and Pakistan (away/visiting team) held 04 October 2012 in Sri Lanka, the host country/home team won the match and advanced to the final.

Figure 2 shows the top-10 winning percentages for home teams. For instance, the percentage for Team Namibia winning home games was 100%, and the percentage for Team Namibia winning away games dropped to 80%. Similarly, the percentage for Team Papua New Guinea winning home games was 75%, and the percentage for Team Papua New Guinea winning away games dropped to 50%. These show the slight advantage for the home team over the away (aka visiting) team.

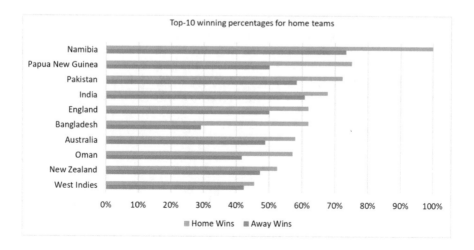

Fig. 2. Top-10 winning percentages for home teams

The same historical data also reveal that, among the four identified factors contributing to the winning of a cricket match, head-to-head matches shone. When applying Eqs. (3) and (4), prediction based on head-to-head matches were accurate 7.6 out of 10 times. Hence, we consider head-to-head as the base predictor. Incorporating other factors like venue and toss can further improved this base result. As a concrete example in the Sixth Twenty20World Cup 2016, knowing that West Indies won a head-to-head match against England in Group 1 of Super 10, West Indies defeated England again in the final match. Moreover, in both matches, West Indies won the toss and elected to field.

4.2   Prediction on Future Cricket Match Results

Based on analyses on the historical data from 17 February 2005 to 16 October 2021 (i.e., before the start day of the T20 International World Cup 2021 held 17 October–14 November 2021 in United Arab Emirates and Oman), our algorithm predicts the future cricket match results.

For T20 International World Cup 2021, there are three stages to win for the world cup. In first stage (i.e., first round) where there is a total number of 8 teams being split into Groups A and B (with 4 teams each). Within each group, a team competed with every team, for a total 6 matches per group. Top-2 winners from each group (i.e., 4 qualified teams) advanced to to the second stage (i.e., Super 12) to join 8 high-ranked teams. Again, these 12 teams were split into Groups 1 and 2 (with 6 teams each). Within each group, a team competed with every team, for a total of 15 matches per group. Top-2 winners from each group (i.e., 4 teams) advanced to to the third stage (i.e., knockout stage), which consists of two semi-final matches and the final match.

Figure 3 shows our predicted cricket match results when compared with the actual results. Note that our algorithm made the prediction based on historical data prior to the start of the game. The figure shows some prediction results, in which our predicted winners are highlighted. Teams highlighted in green are correctly predicted winners of the matches, whereas teams highlighted in red are those losers that were incorrectly predicted as winners. For instance, the figure shows that our algorithm accurately predicted 14 of the 15 winners in Group 2 of Super 20, as well as the winner in the final match. Statistically, our algorithm managed to correctly predict 78% of results of the 45 matches in T20 International World Cup 2021, including the final winner—Australia.

	win		loss	(cf. predicted winner)
	First stage (First round)			
Group A	Sri Lanka	v.	Namibia	Sri Lanka
	Sri Lanka	v.	Ireland	Sri Lanka
	Namibia	v.	Ireland	Ireland
	Sri Lanka	v.	Netherlands	Sri Lanka
	...	v.	...	...
Group B	...	v.	...	...
	Scotland	v.	Papua New Guinea	Scotland
	Bangladesh	v.	Oman	Bangladesh
	Bangladesh	v.	Papua New Guinea	Bangladesh
	Scotland	v.	Oman	Scotland
	Second stage (Super 12)			
Group 1	Australia	v.	South Africa	Australia
	England	v.	West Indies	West Indies
	...	v.	...	...
Group 2	...	v.	...	...
	Pakistan	v.	India	India
	Afghanistan	v.	Scotland	Afghanistan
	Pakistan	v.	New Zealand	Pakistan
	Namibia	v.	Scotland	Namibia
	Pakistan	v.	Afghanistan	Pakistan
	Afghanistan	v.	Namibia	Afghanistan
	New Zealand	v.	India	New Zealand
	Pakistan	v.	Namibia	Pakistan
	New Zealand	v.	Scotland	New Zealand
	India	v.	Afghanistan	India
	New Zealand	v.	Namibia	New Zealand
	India	v.	Scotland	India
	New Zealand	v.	Afghanistan	New Zealand
	Pakistan	v.	Scotland	Pakistan
	India	v.	Namibia	India
	Third stage (Knockout stage)			
Semi-finals	...	v.	...	...
Final	Australia	v.	New Zealand	Australia

Fig. 3. Predicted vs. actual winning teams

5 Conclusions

In this paper, we conducted sports data mining. In particular, we focused on cricket. Specifically, we designed and implemented sports data mining algorithm to predict cricket match results. The algorithm takes into account features like team rankings, matches won against a specific opponent (i.e., head-to-head matches), venue, and toss winning to predict the winner of a given match. We evaluated our algorithm with real-life T20 International World Cup data. Specifically, historical data prior to Seventh ICC Men's T20 World Cup tournament 2021 (i.e., before 17 October 2021) reveal the importance of the aforementioned features. These historical data were then used for predicting cricket match results of the 2021 tournament. Comparisons on the predictions against the actual cricket match results demonstrate the practicality of our sports data mining algorithm.

As ongoing and future work, we would incorporate additional features such as weather conditions, most runs scored on the power play, pitch conditions, cricket crowd in the venue, highest runs, most number of catches, most number of wickets taken, most sixes and fours (i.e., six and four runs scored by the batting team), batting strike rate and average, bowling economy rate, etc. Moreover, we would also transfer our algorithm to predicting match results for (a) related formats of crickets like test matches and One Day Internationals (ODIs), as well as (b) related events like ICC Champions Trophy, Indian Premier League (IPL) and Pakistan Super League (PSL).

Acknowledgements. This work is partially supported by NSERC (Canada) and University of Manitoba.

References

1. Dhaouadi, A., Bousselmi, K., Monnet, S., Gammoudi, M.M., Hammoudi, S.: A multi-layer modeling for the generation of new architectures for big data warehousing. In: Barolli, L., Hussain, F., Enokido, T. (eds.) AINA 2022. LNNS, vol. 450, pp. 204–218. Springer, Cham (2022). https://doi.org/10.1007/978-3-030-99587-4_18
2. Di Martino, B., D'Angelo, S., Esposito, A., Lupi, P.: Anomalous witnesses and registrations detection in the Italian justice system based on big data and machine learning techniques. In: Barolli, L., Hussain, F., Enokido, T. (eds.) AINA 2022. LNNS, vol. 451, pp. 183–192. Springer, Cham (2022). https://doi.org/10.1007/978-3-030-99619-2_18
3. Fung, D.L.X., et al.: Self-supervised deep learning model for COVID-19 lung CT image segmentation highlighting putative causal relationship among age, underlying disease and COVID-19. J. Transl. Med. **19**(1), 1–18 (2021)
4. Leung, C.K., et al.: Explainable data analytics for disease and healthcare informatics. In: IDEAS 2021, pp. 12:1–12:12 (2021)
5. Liu, Q., et al.: A two-dimensional sparse matrix profile DenseNet for COVID-19 diagnosis using chest CT images. IEEE Access **8**, 213718–213728 (2020)
6. Souza, J., Leung, C.K., Cuzzocrea, A.: An innovative big data predictive analytics framework over hybrid big data sources with an application for disease analytics. In: Barolli, L., Amato, F., Moscato, F., Enokido, T., Takizawa, M. (eds.) AINA 2020. AISC, vol. 1151, pp. 669–680. Springer, Cham (2020). https://doi.org/10.1007/978-3-030-44041-1_59
7. Leung, C.K., et al.: Explainable artificial intelligence for data science on customer churn. In: IEEE DSAA 2021, pp. 235–244 (2021)
8. Anderson-Grégoire, I.M., et al.: A big data science solution for analytics on moving objects. In: Barolli, L., Woungang, I., Enokido, T. (eds.) AINA 2021. LNNS, vol. 226, pp. 133–145. Springer, Cham (2021). https://doi.org/10.1007/978-3-030-75075-6_11
9. Barkwell, K.E., et al.: Big data visualisation and visual analytics for music data mining. In: IV 2018, pp. 235–240 (2018)

10. Cabusas, R.M., Epp, B.N., Gouge, J.M., Kaufmann, T.N., Leung, C.K., Tully, J.R.A.: Mining for fake news. In: Barolli, L., Hussain, F., Enokido, T. (eds.) AINA 2022. LNNS, vol. 450, pp. 154–166. Springer, Cham (2022). https://doi.org/10.1007/978-3-030-99587-4_14

11. Cameron, J.J., et al.: Finding strong groups of friends among friends in social networks. In: IEEE DASC 2011, pp. 824–831 (2011)

12. Leung, C.K., Jiang, F., Poon, T.W., Crevier, P.É.: Big data analytics of social network data: who cares most about you on Facebook? In: Moshirpour, M., Far, B., Alhajj, R. (eds.) Highlighting the Importance of Big Data Management and Analysis for Various Applications. SBD, vol. 27, pp. 1–15. Springer, Cham (2018). https://doi.org/10.1007/978-3-319-60255-4_1

13. Leung, C.K., et al.: Personalized DeepInf: enhanced social influence prediction with deep learning and transfer learning. In: IEEE BigData 2019, pp. 2871–2880 (2019)

14. Isichei, B.C., et al.: Sports data management, mining, and visualization. In: Barolli, L., Hussain, F., Enokido, T. (eds.) AINA 2022. LNNS, vol. 450, pp. 141–153. Springer, Cham (2022). https://doi.org/10.1007/978-3-030-99587-4_13

15. Balbin, P.P.F., et al.: Predictive analytics on open big data for supporting smart transportation services. Procedia Comput. Sci. **176**, 3009–3018 (2020)

16. Leung, C.K., Braun, P., Hoi, C.S.H., Souza, J., Cuzzocrea, A.: Urban analytics of big transportation data for supporting smart cities. In: Ordonez, C., Song, I.-Y., Anderst-Kotsis, G., Tjoa, A.M., Khalil, I. (eds.) DaWaK 2019. LNCS, vol. 11708, pp. 24–33. Springer, Cham (2019). https://doi.org/10.1007/978-3-030-27520-4_3

17. Han, J., et al.: Data Mining: Concepts and Techniques, 4th edn. Morgan Kaufmann, San Francisco (2022)

18. Leung, C.K., et al.: Distributed uncertain data mining for frequent patterns satisfying anti-monotonic constraints. In: IEEE AINA Workshops 2014, pp. 1–6 (2014)

19. Leung, C.K.-S., Hayduk, Y.: Mining frequent patterns from uncertain data with MapReduce for big data analytics. In: Meng, W., Feng, L., Bressan, S., Winiwarter, W., Song, W. (eds.) DASFAA 2013. LNCS, vol. 7825, pp. 440–455. Springer, Heidelberg (2013). https://doi.org/10.1007/978-3-642-37487-6_33

20. Rahman, M.M., et al.: Mining weighted frequent sequences in uncertain databases. Inf. Sci. **479**, 76–100 (2019)

21. Schumaker, R.P.: Sports Data Mining. Springer, New York (2010). https://doi.org/10.1007/978-1-4419-6730-5

22. Steyaert, M., et al.: Sensor-based performance monitoring in track cycling. In: Brefeld, U., Davis, J., Van Haaren, J., Zimmermann, A. (eds.) Machine Learning and Data Mining for Sports Analytics, MLSA 2021. Communications in Computer and Information Science, vol. 1571, pp. 167–177. Springer, Cham (2021). https://doi.org/10.1007/978-3-031-02044-5_14

23. Jacquelin, N., et al.: Detecting swimmers in unconstrained videos with few training data. In: Brefeld, U., Davis, J., Van Haaren, J., Zimmermann, A. (eds.) Machine Learning and Data Mining for Sports Analytics, MLSA 2021. CCIS, vol. 1571, pp. 145–154. Springer, Cham (2021). https://doi.org/10.1007/978-3-031-02044-5_12

24. Moura, H.D., et al.: Low cost player tracking in field hockey. In: Brefeld, U., Davis, J., Van Haaren, J., Zimmermann, A. (eds.) Machine Learning and Data Mining for Sports Analytics, MLSA 2021. CCIS, vol. 1571, pp. 103–115. Springer, Cham (2021). https://doi.org/10.1007/978-3-031-02044-5_9

25. Leung, C.K., Joseph, K.W.: Sports data mining: predicting results for the college football games. Procedia Comput. Sci. **35**, 710–719 (2014)

26. Behera, S.R., Saradhi, V.V.: Learning strength and weakness rules of cricket players using association rule mining. In: Brefeld, U., Davis, J., Van Haaren, J., Zimmermann, A. (eds.) Machine Learning and Data Mining for Sports Analytics, MLSA 2021. CCIS, vol. 1571, pp. 79–92. Springer, Cham (2021). https://doi.org/10.1007/978-3-031-02044-5_7
27. Tirtho, D., et al.: Cricketer's tournament-wise performance prediction and squad selection using machine learning and multi-objective optimization. Appl. Soft Comput. **129**, 109526:1–109526:14 (2022)
28. Vetukuri, V.S., et al.: A multi-aspect analysis and prediction scheme for cricket matches in standard T-20 format. Int. J. Knowl.-Based Intell. Eng. Syst. **23**(3), 149–154 (2019)
29. Gupta, A., Muthiah, S.B.: Learning cricket strokes from spatial and motion visual word sequences. Multimedia Tools Appl. **82**(1), 1237–1259 (2023)
30. Raval, K.R., Goyani, M.M.: A survey on event detection based video summarization for cricket. Multimedia Tools Appl. **81**(20), 29253–29281 (2022)
31. Vetukuri, V.S., et al.: Generic model for automated player selection for cricket teams using recurrent neural networks. Evol. Intel. **14**(2), 971–978 (2021)
32. Longmore, A., et al.: Cricket. In: Encyclopedia Britannica (2021). https://www.britannica.com/sports/cricket-sport

Author Index

A

Abahussein, Suleiman 48
Abdelkader, Manel 60
Abidemi, Sarumi Usman 321
Abualkibash, Munther 546
Agrawal, Ankit 206
Aguili, Taoufik 458
Akanni, Olukayode 321
Akhalaia, Giorgi 480
Alalmaie, Abeer Z. 181
Alayan, Mohrah Saad 181
Al-essa, Rana 546
Aliyu, Aliyu Lawal 435
Aln'uman, Ruba 546
Alqublan, Ammar 546
Al-Turjman, Fadi 292
Al-turjman, Fadi 321
Amato, Alba 251, 261
Ameen, Zubaida Said 321
Angelucci, Simone 271
Anuraj, Antony 668
Aoueileyine, Mohamed Ould-Elhassen 60
Aral, Atakan 136
Asif, Md. Rashid Al 532
Aslam, Hamna 169
Aversa, Rocco 241
Azuma, Masaya 406

B

Balzano, Walter 396
Barolli, Leonard 28, 376, 406, 427
Barzamini, Roohollah 565
Barzamini, Sahar 565
Benkner, Siegfried 136
Bensaid, Rahil 415
Bezerra, Marcus M. 193
Bhatia, Ashutosh 206
Bhatia, Ritika 206
Bochicchio, Mario A. 136
Bocu, Dorin 1
Bocu, Razvan 1, 480

C

Bonchis, Cosmin 157
Boparai, Gurtej S. 668
Bouchoucha, Yosra 458
Boujemaa, Hatem 415
Branco, Dario 127, 241
Bryhni, Haakon 597

Cacciagrano, Diletta 311, 321
Cheng, Zishuo 48
Cirillo, Giuseppe 261
Corradini, Flavio 311
Cosconati, Sandro 127
Culmone, Rosario 292

D

D'Angelo, Salvatore 127, 146
Dahaoui, Ibrahim 351
Daneshgar, Farhad 558
Darwish, Omar 546
Davidov, Maxim 12
de Alencar, Allender V. 193
de Matos, Everton 518
Di Giacomo, Emilio 90
Di Martino, Beniamino 70, 80, 90, 127, 136, 146
Didimo, Walter 90
Diockou, Jim 435
Duma, Alecsandru 100

E

Elkhodr, Mahmoud 546
Esposito, Antonio 80, 90, 136, 146

F

Farjad, Babak 565
Folga, Damian 619
Foroughi, Rahim 558
Franchi, Fabio 271

© The Editor(s) (if applicable) and The Author(s), under exclusive license
to Springer Nature Switzerland AG 2023
L. Barolli (Ed.): AINA 2023, LNNS 655, pp. 681–683, 2023.
https://doi.org/10.1007/978-3-031-28694-0

G

Galib, Syed Md.   446
Gheisari, Soheila   573
Gorgônio, Kyller C.   490
Graziosi, Fabio   271
Grüner, Andreas   471

H

Haider, Furqan   169
Hajati, Farshid   565, 573
Hasan, Khondokar Fida   532
Hattab, Siham   331
He, Xiangjian   181
Hessel, Fabiano   357, 518
Hirata, Aoto   427
Horn, Geir   146, 619

I

Iavich, Maksim   480
Ikeda, Makoto   406
Islam, Md. Zahirul   446
Iuhasz, Gabriel   100
Izadi, Alireza   565

J

Janpors, Negar   565
Jouini, Oumayma   340

K

Kadric, Adnan   471
Karmous, Neder   60
Khabou, Nesrine   36
Khondoker, Rahamatullah   446, 532
Kranzlmueller, Dieter   127
Krichene, Mohamed   36
Kritikos, Kyriakos   619
Kubota, Ayumu   585
Kudrenok, Ilya   12
Kurihara, Jun   585

L

Lanuto, Antonio   396
Leung, Carson K.   668
Li, Kin Fun   660
Liotta, Giuseppe   90
Longo, Antonella   115

Low, Warren   70
Lu, Wei   649

M

Maabreh, Majdi   546
Madill, Evan W. R.   668
Marchel, Jan   619
Marouf, Rabab   169
Martin, Marlon Rodrigues   357
Matsui, Tomoaki   376
Mazzara, Manuel   12, 169, 331, 644
Mazzeo, Oronzo   115
Mbarek, Bacem   504
Meddeb, Aref   504
Meinel, Christoph   471
Mim, Sharmin Akter   532
Mimura, Hiroki   660
Mnaouer, Adel Ben   415
Montecchiani, Fabrizio   90
Mosbah, Mohamed   351
Mostarda, Leonardo   279, 301
Mubarak, Auwalu Saleh   321
Mühle, Alexander   471

N

Nagai, Yuki   28, 376, 427
Nagiyev, Andrey   136
Nanda, Priyadarsi   181
Nath, Amar   218
Nie, Hongrui   384
Niihara, Masahiro   28
Niyogi, Rajdeep   218

O

Ocampo, Andres F.   597
Oda, Tetsuya   28, 376, 427
Omri, Dorsaf   458

P

Pandhi, Darshan A.   668
Panica, Silviu   100
Patel, Ayush Dilipkumar   668
Perkusich, Angelo   193, 490
Pezzullo, Gennaro Junior   80
Piangerelli, Marco   311
Pinna, Andrea   279

Pitner, Tomáš 504
Preethika, P. 609
Prosciutto, Erasmo 396

R
Rahim, Md. Abdur 446
Rahman, Roksana 532
Ramanathan, B. Ramesh 609
Rezaee, Alireza 573
Rinaldi, Claudia 271
Rochian, Vlad 157
Rodriguez, Ismael Bouassida 36
Różańska, Marta 619
Rümmler, Nils 471

S
Saha, Souradip 70
Salahuddin, Md. 446
Salama, Ramiz 292
Salem, Hamza 331, 644
Santos, Danilo F. S. 193
Sashank, Yadagiri Shiva Sai 206
Scala, Emanuele 301
Scotto di Covella, Biagio 396
Sestili, Davide 279
Sethom, Kaouthar 340
Shibata, Yoshitaka 370
Shorten, Bria 649
Singh, Tajinder 218
Sobrinho, Álvaro 490
Soula, Meriem 504
Stranieri, Silvia 396

T
Tabuchi, Kei 376
Tahara, Masaya 660
Takano, Kosuke 660
Tashtoush, Yahya 546
Tavakoli-Mehr, Nava 558
Tepeneu, Ionut 157

Tiburski, Ramão 518
Tiwari, Kamlesh 206
Tonelli, Roberto 279
Toyoshima, Kyohei 376, 427
Toyosima, Kyouhei 28

U
Uchimura, Shota 406
Ueda, Akiko 370
Ueda, Masahiro 370

V
Valadares, Dalton C. G. 193, 490
Venticinque, Salvatore 241
Verginadis, Yiannis 631
Viegas, Eduardo 518
Vyas, Ronak K. 668

W
Watanabe, Nobuya 660
Watanabe, Ryu 585
Will, Newton C. 490

X
Xue, Ling 649

Y
Yamashita, Yuma 28
Ye, Dayong 48
Youhangi, Atefa 558
Youssef, Neji 60
Yukawa, Chihiro 28, 376, 427

Z
Zappatore, Marco 115
Zemmari, Akka 351
Zhou, Wanlei 48
Zhu, Tianqing 48
Zhu, Zitong 228

Printed in the United States
by Baker & Taylor Publisher Services